Cloud, Grid and High Performance Computing:

Emerging Applications

Emmanuel Udoh
Indiana Institute of Technology, USA

Senior Editorial Director: Kristin Klinger
Editorial Director: Lindsay Johnston
Director of Book Publications: Julia Mosemann
Acquisitions Editor: Erika Carter
Development Editor: Hannah Abelbeck
Production Editor: Sean Woznicki
Typesetters: Michael Brehm, Keith Glazewski, Milan Vracarich, Jr.
Print Coordinator: Jamie Snavely
Cover Design: Nick Newcomer

Published in the United States of America by
in (an imprint of IGI Global)
701 E. Chocolate Avenue
Hershey PA 17033
Tel: 717-533-8845
Fax: 717-533-8661
E-mail: cust@igi-global.com
Web site: http://www.igi-global.com

Library of Congress Cataloging-in-Publication Data
Cloud, grid and high performance computing: emerging applications / Emmanuel Udoh, editor.
p. cm.
Includes bibliographical references and index.
Summary: "This book offers new and established perspectives on architectures, services and the resulting impact of emerging computing technologies, including investigation of practical and theoretical issues in the related fields of grid, cloud, and high performance computing"--Provided by publisher.
ISBN 978-1-60960-603-9 (hardcover) -- ISBN 978-1-60960-604-6 (ebook) 1. Cloud computing. 2. Computational grids (Computer systems) 3. Software architecture. 4. Computer software--Development. I. Udoh, Emmanuel, 1960-
QA76.585.C586 2011
004.67'8--dc22

2011013282

British Cataloguing in Publication Data
A Cataloguing in Publication record for this book is available from the British Library.

Table of Contents

Preface.......... xvi

Section 1
Introduction

Chapter 1
Supercomputers in Grids.......... 1
Michael M. Resch, University of Stuttgart, Germany
Edgar Gabriel, University of Houston, USA

Chapter 2
Porting HPC Applications to Grids and Clouds.......... 10
Wolfgang Gentzsch, Independent HPC, Grid, and Cloud Consultant, Germany

Chapter 3
Grid-Enabling Applications with JGRIM.......... 39
Cristian Mateos, ISISTAN - UNCPBA, Argentina
Alejandro Zunino, ISISTAN - UNCPBA, Argentina
Marcelo Campo, ISISTAN - UNCPBA, Argentina

Section 2
Scheduling

Chapter 4
Moldable Job Allocation for Handling Resource Fragmentation in Computational Grid.......... 58
Kuo-Chan Huang, National Taichung University of Education, Taiwan
Po-Chi Shih, National Tsing Hua University, Taiwan
Yeh-Ching Chung, National Tsing Hua University, Taiwan

Detailed Table of Contents

Preface......xvi

Section 1
Introduction

Chapter 1
Supercomputers in Grids......1
Michael M. Resch, University of Stuttgart, Germany
Edgar Gabriel, University of Houston, USA

This article describes the state of the art in using supercomputers in Grids. It focuses on various approaches in Grid computing that either aim to replace supercomputing or integrate supercomputers in existing Grid environments. We further point out the limitations to Grid approaches when it comes to supercomputing. We also point out the potential of supercomputers in Grids for economic usage. For this, we describe a public-private partnership in which this approach has been employed for more than 10 years. By giving such an overview we aim at better understanding the role of supercomputers and Grids and their interaction.

Chapter 2
Porting HPC Applications to Grids and Clouds......10
Wolfgang Gentzsch, Independent HPC, Grid, and Cloud Consultant, Germany

A Grid enables remote, secure access to a set of distributed, networked computing and data resources. Clouds are a natural complement to Grids towards the provisioning of IT as a service. To “Grid-enable” applications, users have to cope with: complexity of Grid infrastructure; heterogeneous compute and data nodes; wide spectrum of Grid middleware tools and services; the e-science application architectures, algorithms and programs. For clouds, on the other hand, users don’t have many possibilities to adjust their application to an underlying cloud architecture, because of its transparency to the user. Therefore, the aim of this chapter is to guide users through the important stages of implementing HPC applications on Grid and cloud infrastructures, together with a discussion of important challenges and their potential solutions. As a case study for Grids, we present the Distributed European Infrastructure for Supercomputing Applications (DEISA) and describe the DEISA Extreme Computing Initiative (DECI)

for porting and running scientific grand challenge applications on the DEISA Grid. For clouds, we present several case studies of HPC applications running on Amazon's Elastic Compute Cloud EC2 and its recent Cluster Compute Instances for HPC. This chapter concludes with the author's top ten rules of building sustainable Grid and cloud e-infrastructures.

The development of massively distributed applications with enormous demands for computing power, memory, storage and bandwidth is now possible with the Grid. Despite these advances, building Grid applications is still very difficult. We present JGRIM, an approach to easily gridify Java applications by separating functional and Grid concerns in the application code, and report evaluations of its benefits with respect to related approaches. The results indicate that JGRIM simplifies the process of porting applications to the Grid, and the Grid code obtained from this process performs in a very competitive way compared to the code resulting from using similar tools.

In a computational Grid environment, a common practice is to try to allocate an entire parallel job onto a single participating site. Sometimes a parallel job, upon its submission, cannot fit in any single site due to the occupation of some resources by running jobs. How the job scheduler handles such situations is an important issue which has the potential to further improve the utilization of Grid resources, as well as the performance of parallel jobs. This paper adopts moldable job allocation policies to deal with such situations in a heterogeneous computational Grid environment. The proposed policies are evaluated through a series of simulations using real workload traces. The moldable job allocation policies are also compared to the multi-site co-allocation policy, which is another approach usually used to deal with the resource fragmentation issue. The results indicate that the proposed moldable job allocation policies can further improve the system performance of a heterogeneous computational Grid significantly.

The execution of data intensive Grid applications raises several questions regarding job scheduling, data migration, and replication. This paper presents new scheduling algorithms using more sophisticated job behaviour descriptions that allow estimating job completion times more precisely thus improving scheduling decisions. Three approaches of providing input to the decision procedure are discussed: a) single job description, b) multiple job descriptions, and c) multiple job descriptions with mutation. The proposed Grid middleware components (1) monitor the execution of jobs and gather resource access information, (2) analyse the compiled information and generate a description of the behaviour of the job, (3) refine the already existing job description, and (4) use the refined behaviour description to schedule the submitted jobs.

Grid supports heterogeneities of resources in terms of security and computational power. Applications with stringent security requirement introduce challenging concerns when executed on the grid resources. Though grid scheduler considers the computational heterogeneity while making scheduling decisions, little is done to address their security heterogeneity. This work proposes a security aware computational grid scheduling model, which schedules the tasks taking into account both kinds of heterogeneities. The approach is known as Security Prioritized MinMin (SPMinMin). Comparing it with one of the widely used grid scheduling algorithm MinMin (secured) shows that SPMinMin performs better and sometimes behaves similar to MinMin under all possible situations in terms of makespan and system utilization.

Grid is a parallel and distributed computing network system comprising of heterogeneous computing resources spread over multiple administrative domains that offers high throughput computing. Since the Grid operates at a large scale, there is always a possibility of failure ranging from hardware to software. The penalty paid of these failures may be on a very large scale. System needs to be tolerant to various possible failures which, in spite of many precautions, are bound to happen. Replication is a strategy often used to introduce fault tolerance in the system to ensure successful execution of the job, even when some of the computational resources fail. Though replication incurs a heavy cost, a selective degree of

of these tasks. This performance-oriented task dispatch policy is evaluated with two real world trace data sets on a simulator. Evaluation results demonstrate the effectiveness of this policy.

In order to create a successful grid infrastructure, sites and resource providers must be able to publish information about their underlying resources and services. This information enables users and virtual organizations to make intelligent decisions about resource selection and scheduling, and facilitates accounting and troubleshooting services within the grid. However, such an outbound stream may include data deemed sensitive by a resource-providing site, exposing potential security vulnerabilities or private user information. This study analyzes the various vectors of information being published from sites to grid infrastructures. In particular, it examines the data being published and collected in the Open Science Grid, including resource selection, monitoring, accounting, troubleshooting, logging and site verification data. We analyze the risks and potential threat models posed by the publication and collection of such data. We also offer some recommendations and best practices for sites and grid infrastructures to manage and protect sensitive data.

One of the most successful working examples of virtual organizations, computational Grids need authentication mechanisms that inter-operate across domain boundaries. Public Key Infrastructures (PKIs) provide sufficient flexibility to allow resource managers to securely grant access to their systems in such distributed environments. However, as PKIs grow and services are added to enhance both security and usability, users and applications must struggle to discover available resources-particularly when the Certification Authority (CA) is alien to the relying party. This chapter presents a successful story about how to overcome these limitations by deploying the PKI Resource Query Protocol (PRQP) into the grid security architecture. We also discuss the future of Grid authentication by introducing the Public Key System (PKS) and its key features to support federated identities.

Mobile Grid includes the characteristics of the Grid systems together with the peculiarities of Mobile Computing, with the additional feature of supporting mobile users and resources in a seamless, transparent, secure, and efficient way. Security of these systems, due to their distributed and open nature, is considered a topic of great interest. We are elaborating a process of development to build secure mobile Grid systems considering security on all life cycles. In this chapter, we present the practical results applying our development process to a real case, specifically we apply the part of security requirements analysis to obtain and identify security requirements of a specific application following a set of tasks defined for helping us in the definition, identification, and specification of the security requirements on our case study. The process will help us to build a secure Grid application in a systematic and iterative way.

Chapter 13
Trusted Data Management for Grid-Based Medical Applications 208
Guido J. van 't Noordende, University of Amsterdam, The Netherlands
Silvia D. Olabarriaga, Academic Medical Center - Amsterdam, The Netherlands
Matthijs R. Koot, University of Amsterdam, The Netherlands
Cees T.A.M. de Laat, University of Amsterdam, The Netherlands

Existing Grid technology has been foremost designed with performance and scalability in mind. When using Grid infrastructure for medical applications, privacy and security considerations become paramount. Privacy aspects require a re-thinking of the design and implementation of common Grid middleware components. This chapter describes a novel security framework for handling privacy sensitive information on the Grid, and describes the privacy and security considerations which impacted its design.

Section 4
Applications

Chapter 14
Large-Scale Co-Phylogenetic Analysis on the Grid 222
Heinz Stockinger, Swiss Institute of Bioinformatics, Switzerland
Alexander F. Auch, University of Tübingen, Germany
Markus Göker, University of Tübingen, Germany
Jan Meier-Kolthoff, University of Tübingen, Germany
Alexandros Stamatakis, Ludwig-Maximilians-University Munich, Germany

Phylogenetic data analysis represents an extremely compute-intensive area of Bioinformatics and thus requires high-performance technologies. Another compute- and memory-intensive problem is that of host-parasite co-phylogenetic analysis: given two phylogenetic trees, one for the hosts (e.g., mammals) and one for their respective parasites (e.g., lice) the question arises whether host and parasite trees are more similar to each other than expected by chance alone. CopyCat is an easy-to-use tool that allows biologists to conduct such co-phylogenetic studies within an elaborate statistical framework based on the highly optimized sequential and parallel AxParafit program. We have developed enhanced versions of these tools that efficiently exploit a Grid environment and therefore facilitate large-scale data analyses. Furthermore, we developed a freely accessible client tool that provides co-phylogenetic analysis

capabilities. Since the computational bulk of the problem is embarrassingly parallel, it fits well to a computational Grid and reduces the response time of large scale analyses.

Wide-area distributed systems offer new opportunities for executing large-scale scientific applications. On these systems, communication mechanisms have to deal with dynamic resource availability and the potential for resource and network failures. Connectivity losses can affect the execution of workflow applications, which require reliable data transport between components. We present the design and implementation of π-channels, an asynchronous and fault-tolerant pipe mechanism suitable for coupling workflow components. Fault-tolerant communication is made possible by persistence, through adaptive caching of pipe segments while providing direct data streaming. We present the distributed algorithm for implementing: (a) caching of pipe data segments; (b) asynchronous read operation; and (c) communication state transfer to handle dynamic process joins and leaves.

A Wireless Grid is an augmentation of a wired grid that facilitates the exchange of information and the interaction between heterogeneous wireless devices. While similar to the wired grid in terms of its distributed nature, the requirement for standards and protocols, and the need for adequate Quality of Service; a Wireless Grid has to deal with the added complexities of the limited power of the mobile devices, the limited bandwidth, and the increased dynamic nature of the interactions involved. This complexity becomes important in designing the services for mobile computing. A grid topology and naming service is proposed which can allow self-configuration and self-administration of various possible wireless grid layouts.

Remote memory sharing systems aim at the goal of improving overall performance using distributed computing nodes with surplus memory capacity. To exploit the memory resources connected by the high-speed network, the user nodes, which are short of memory, can obtain extra space provision. The performance of remote memory sharing is constrained with the expensive network communication cost. In order to hide the latency of remote memory access and improve the performance, we proposed the push-based prefetching to enable the memory providers to push the potential useful pages to the user nodes. For each provider, it employs sequential pattern mining techniques, which adapts to the characteristics of memory page access sequences, on locating useful memory pages for prefetching. We have verified the effectiveness of the proposed method through trace-driven simulations.

You-Fu Yu, National Taichung University, Taiwan, ROC
Po-Jung Huang, National Taichung University, Taiwan, ROC
Kuan-Chou Lai, National Taichung University, Taiwan, ROC

P2P Grids could solve large-scale scientific problems by using geographically distributed heterogeneous resources. However, a number of major technical obstacles must be overcome before this potential can be realized. One critical problem to improve the effective utilization of P2P Grids is the efficient load balancing. This chapter addresses the above-mentioned problem by using a distributed load balancing policy. In this chapter, we propose a P2P communication mechanism, which is built to deliver varied information across heterogeneous Grid systems. Basing on this P2P communication mechanism, we develop a load balancing policy for improving the utilization of distributed computing resources. We also develop a P2P resource monitoring system to capture the dynamic resource information for the decision making of load balancing. Moreover, experimental results show that the proposed load balancing policy indeed improves the utilization and achieves effective load balancing.

Tao Gu, University of Southern Denmark, Denmark
Daqing Zhang, Institut Telecom SudParis, France
Hung Keng Pung, National University of Singapore, Singapore

This article presents an ontology-based peer-to-peer network that facilitates efficient search for data in wide-area networks. Data with the same semantics are grouped together into one-dimensional semantic ring space in the upper-tier network. This is achieved by applying an ontology-based semantic clustering technique and dedicating part of node identifiers to correspond to their data semantics. In the lower-tier network, peers in each semantic cluster are organized as Chord identifier space. Thus, all the nodes in the same semantic cluster know which node is responsible for storing context data triples they are looking for, and context queries can be efficiently routed to those nodes. Through the simulation studies, the authors demonstrate the effectiveness of our proposed scheme.

Djamel Tandjaoui, Center of Research on Scientific and Technical Information, Algeria
Messaoud Doudou, University of Science and Technology Houari Boumediène, Algeria
Imed Romdhani, Napier University School of Computing, UK

In this article, the authors propose a new hybrid MAC protocol named H-MAC for wireless mesh networks. This protocol combines CSMA and TDMA schemes according to the contention level. In

Preface

Cloud computing has emerged as the natural successor of the different strands of distributed systems - concurrent, parallel, distributed, and Grid computing. Like a killer application, cloud computing is causing governments and the enterprise world to embrace distributed systems with renewed interest. In evolutionary terms, clouds herald the third wave of Information Technology, in which virtualized resources (platform, infrastructure, software) are provided as a service over the Internet. This economic front of cloud computing, whereby users are charged based on their usage of computational resources and storage, is driving its current adoption and the creation of opportunities for new service providers. As can be gleaned from press releases, the US government has registered strong interest in the overall development of cloud technology for the betterment of the economy.

The transformation enabled by cloud computing follows the utility pricing model (subscription/metered approach) in which services are commoditized as practiced in electricity; water, telephony and gas industries. This approach follows a global vision in which users plug their computing devices into the Internet and tap into as much processing power as needed. Essentially, a customer (individual or organization) gets computing power and storage, not from his/her computer, but over the Internet on demand.

Cloud technology comes in different flavors: public, private, and hybrid clouds. Public clouds are provided remotely to users from third-party controlled data centers, as opposed to private clouds that are more of virtualization and service-oriented architecture hosted in the traditional settings by corporations. It is obvious that the economies of scale of large data centers (vendors like Google) offer public clouds an economic edge over private clouds. However, security issues are a major source of concerns about public clouds, as organizations will not distribute resources randomly on the Internet, especially their prized databases, without a measure of certainty or safety assurance. In this vein, private clouds will persist until public clouds mature and garner corporate trust.

The embrace of cloud computing is impacting the adoption of Grid technology. The perceived usefulness of Grid computing is not in question, but other factors weigh heavily against its adoption such as complexity and maintenance as well as the competition from clouds. However, the Grid might not be totally relegated to the background as it could complement research in the development of cloud middleware (Udoh, 2010). In that sense, this book considers and foresees other distributed systems not necessarily standing alone as entities as before, but largely subordinate and providing research stuff to support and complement the increasingly appealing cloud technology.

The new advances in cloud computing will greatly impact IT services, resulting in improved computational and storage resources as well as service delivery. To keep educators, students, researchers, and professionals abreast of advances in the cloud, Grid, and high performance computing, this book series *Cloud, Grid, and High Performance Computing: Emerging Applications* will provide coverage of topical issues in the discipline. It will shed light on concepts, protocols, applications, methods, and tools in this emerging and disruptive technology. The book series is organized in four distinct sections, covering wide-ranging topics: (1) Introduction (2) Scheduling (3) Security and (4) Applications.

Section 1, **Introduction**, provides an overview of supercomputing and the porting of applications to Grid and cloud environments. Cloud, Grid and high performance computing are firmly dependent on the information and communication infrastructure. The different types of cloud computing - software-as-a-service (SaaS), platform-as-a-service (PaaS), infrastructure-as-a-service (IaaS), and the data centers exploit commodity servers and supercomputers to serve the current needs of on-demand computing. The chapter *Supercomputers in Grids* by Michael M. Resch and Edgar Gabriel, focuses on the integration and limitations of supercomputers in Grid and distributed environments. It emphasizes the understanding and interaction of supercomputers as well as its economic potential as demonstrated in a public-private partnership project. As a matter of fact, with the emergence of cloud computing, the need for supercomputers in data centers cannot be overstated. In a similar vein, *Porting HPC Applications to Grids and Clouds* by Wolfgang Gentzsch guides users through the important stages of porting applications to Grids and clouds as well as the challenges and solutions. Porting and running scientific grand challenge applications on the DEISA Grid demonstrated this approach. This chapter equally gave an overview of future prospects of building sustainable Grid and cloud applications. In another chapter, *Grid-Enabling Applications with JGRIM*, researchers Cristian Mateos, Alejandro Zunino, and Marcelo Campo recognize the difficulties in building Grid applications. To simplify the development of Grid applications, the researchers developed JGRIM, which easily Gridifies Java applications by separating functional and Grid concerns in the application code. JGRIM simplifies the process of porting applications to the Grid, and is competitive with similar tools in the market.

Section 2, **Scheduling**, is a central component in the implementation of Grid and cloud technology. Efficient scheduling is a complex and an attractive research area, as priorities and load balancing have to be managed. Sometimes, fitting jobs to a single site may not be feasible in Grid and cloud environments, requiring the scheduler to improve allocation of parallel jobs for efficiency. In *Moldable Job Allocation for Handling Resource Fragmentation in Computational Grid*, Huang, Shih, and Chung exploited the moldable property of parallel jobs in formulating adaptive processor allocation policies for job scheduling in Grid environment. In a series of simulations, the authors demonstrated how the proposed policies significantly improved scheduling performance in heterogeneous computational Grid. In another chapter, *Speculative Scheduling of Parameter Sweep Applications Using Job Behavior Descriptions,* Ulbert, Lőrincz, Kozsik, and Horváth demonstrated how to estimate job completion times that could ease decisions in job scheduling, data migration, and replication. The authors discussed three approaches of using complex job descriptions for single and multiple jobs. The new scheduling algorithms are more precise in estimating job completion times.

Furthermore, some applications with stringent security requirements pose major challenges in computational Grid and cloud environments. To address security requirements, in *A Security Prioritized Computational Grid Scheduling Model: An Analysis*, Rekha Kashyap and Deo Prakash Vidyarthi proposed a security aware computational scheduling model that modified an existing Grid scheduling algorithm. The proposed Security Prioritized MinMin showed an improved performance in terms of makespan and system utilization. Taking a completely different bearing in scheduling, Zahid Raza and Deo Prakash Vidyarthi in the chapter *A Replica Based Co-Scheduler (RBS) for Fault Tolerant Computational Grid,* developed a biological approach that incorporates genetic algorithm (GA). This natural selection and evolution method optimizes scheduling in computational Grid by minimizing turnaround time. The developed model, which compared favorably to existing models, was used to simulate and evaluate clusters to obtain the one with minimum turnaround time for job scheduling. As the cloud environments expand to the corporate world, improvements in GA methods could find use in some search problems.

Section 3, **Security**, is one of the major hurdles cloud technology must overcome before any widespread adoption by organizations. Cloud vendors must meet the transparency test and risk assessment in information security and recovery. Falling short of these requirements might leave cloud computing frozen in private clouds. Preserving user privacy and managing customer information, especially personally identifiable information, are central issues in the management of IT services. Wolfgang Hommel, in the chapter *A Policy-Based Security Framework for Privacy-Enhancing Data Access and Usage Control*, discusses how recent advances in privacy enhancing technologies and federated identity management can be incorporated in Grid environments. The chapter demonstrates how existing policy-based privacy management architectures could be extended to provide Grid-specific functionality and integrated into existing infrastructures (demonstrated in an XACML-based privacy management system).

In *Adaptive Control of Redundant Task Execution for Dependable Volunteer Computing*, Wang, Murata, Takizawa, and Kobayashi examined the security features that could enable Grid systems to exploit the massive computing power of volunteer computing systems. The authors proposed the use of cell processor as a platform that could use hardware security features. To test the performance of such a processor, a secure, parallelized, K-Means clustering algorithm for a cell was evaluated on a secure system simulator. The findings point to possible optimization for secure data mining in the Grid environments.

To further provide security in Grid and cloud environments, Shreyas Cholia and R. Jefferson Porter discussed how to close the loopholes in the provisioning of resources and services in *Publication and Protection of Sensitive Site Information in a Grid Infrastructure*. The authors analyzed the various vectors of information being published from sites to Grid infrastructures, especially in the Open Science Grid, including resource selection, monitoring, accounting, troubleshooting, logging, and site verification data. Best practices and recommendations were offered to protect sensitive data that could be published in Grid infrastructures.

Authentication mechanisms are common security features in cloud and Grid environments, where programs inter-operate across domain boundaries. Public key infrastructures (PKIs) provide means to securely grant access to systems in distributed environments, but as PKIs grow, systems become overtaxed to discover available resources especially when certification authority is foreign to the prevailing environment. Massimiliano Pala, Shreyas Cholia, Scott A. Rea, and Sean W. Smith proposed, in *Federated PKI Authentication in Computing Grids: Past, Present, and Future* a new authentication model that incorporates PKI resource query protocol into the Grid security infrastructure that will as well find utility in the cloud environments. Mobile Grid systems and its security are a major source of concern, due to its distributed and open nature. Rosado, Fernández-Medina, López, and Piattini present a case study of the application of a secured methodology to a real mobile system in *Identifying Secure Mobile Grid Use Cases*.

Furthermore, Noordende, Olabarriaga, Koot, and de Laat developed a trusted data storage infrastructure for Grid-based medical applications. In *Trusted Data Management for Grid-Based Medical Applications*, while taking cognizance of privacy and security aspects, they redesigned the implementation of common Grid middleware components, which could impact the implementation of cloud applications as well.

Section 4, **Applications**, are increasingly deployed in the Grid and cloud environments. The architecture of Grid and cloud applications is different from the conventional application models and, thus requires a fundamental shift in implementation approaches. Cloud applications are even more unique as they eliminate installation, maintenance, deployment, management, and support. These cloud applications are considered Software as a Service (SaaS) applications. Grid applications are forerunners to clouds and are still common in scientific computing. A biological application was introduced by Heinz Stockinger and co-workers in a chapter titled *Large-Scale Co-Phylogenetic Analysis on the Grid*. Phylogenetic data analysis is known to be compute-intensive and suitable for high performance computing. The authors improved upon an existing sequential and parallel AxParafit program, by producing an efficient tool that facilitates large-scale data analysis. A free client tool is available for co-phylogenetic analysis.

In chapter *Persistence and Communication State Transfer in an Asynchronous Pipe Mechanism* by Philip Chan and David Abramson, the researchers described distributed algorithm for implementing dynamic resource availability in an asynchronous pipe mechanism that couples workflow components. Here, fault-tolerant communication was made possible by persistence through adaptive caching of pipe segments while providing direct data streaming. Ashish Agarwal and Amar Gupta in another chapter, *Self-Configuration and Administration of Wireless Grids*, described the peculiarities of wireless Grids such as the complexities of the limited power of the mobile devices, the limited bandwidth, standards and protocols, quality of service, and the increasingly dynamic nature of the interactions involved. To meet these peculiarities, the researcher proposed a Grid topology and naming service that self-configures and self-administers various possible wireless Grid layouts. In computational Grid and cloud resource provisioning, memory usage may sometimes be overtaxed. Although RAM Grid can be constrained sometimes, it provides remote memory for the user nodes that are short of memory. Researchers Rui Chu, Nong Xiao, and Xicheng Lu, in the chapter *Push-Based Prefetching in Remote Memory Sharing System*, propose the push-based prefetching to enable the memory providers to push the potential useful pages to the user nodes. With the help of sequential pattern mining techniques, it is expected that useful memory pages for prefetching can be located. The authors verified the effectiveness of the proposed method through trace-driven simulations.

In chapters *Distributed Dynamic Load Balancing in P2P Grid Systems* by Yu, Huang, and Lai and *An Ontology-Based P2P Network for Semantic Search* by Gu, Zhang, and Pung, the researchers explored the potentials and obstacles confronting P2P Grids. Lai, Wu, and Lin described the effective utilization of P2P Grids in efficient scheduling of jobs by examining a P2P communication model. The model aided job migration technology across heterogeneous systems and improved the usage of distributed computing resources. On the other hand, Gu, Zhang, and Pung dwelt on facilitating efficient search for data in distributed systems using an ontology-based peer-to-peer network. Here, the researchers grouped together data with the same semantics into one-dimensional semantic ring space in the upper-tier network. In the lower-tier network, peers in each semantic cluster were organized as chord identifier space. The authors demonstrated the effectiveness of the proposed scheme through simulation experiment.

In this final section, there are other chapters that capture the research trends in the realm of high performance computing. In a high performance computing undertaking, researchers Djamel Tandjaoui, Messaoud Doudou, and Imed Romdhani proposed a new hybrid MAC protocol, named H-MAC, for wireless mesh networks. The protocol exploits channel diversity and a medium access control method in ensuring the quality of service requirement. Using ns-2 simulator, the researchers implemented and compared H-MAC with other MAC protocol used in Wireless Network and found that H-MAC performs better compared to Z-MAC, IEEE 802.11 and LCM-MAC.

IP telephony has emerged as the most widely used peer-to-peer-based application. Although success has been recorded in decentralized communication, providing a scalable peer-to-peer-based distributed directory for searching user entries still poses a major challenge. In a chapter titled *A Decentralized*

Directory Service for Peer-to-Peer-Based Telephony, researchers - Fabian Stäber, Gerald Kunzmann, and Jörg P. Müller, proposed the Extended Prefix Hash Tree algorithm that can be used to implement an indexing infrastructure supporting range queries on top of DHTs.

In conclusion, cloud technology is the latest iteration of information and communications technology driving global business competitiveness and economic growth. Although relegated to the background, research in Grid technology fuels and complements activities in cloud computing, especially in the middleware technology. In that vein, this book series is a contribution to the growth of cloud technology and global economy, and indeed the information age.

Emmanuel Udoh
Indiana Institute of Technology, USA

Section 1
Introduction

Chapter 1
Supercomputers in Grids

Michael M. Resch
University of Stuttgart, Germany

Edgar Gabriel
University of Houston, USA

ABSTRACT

This article describes the state of the art in using supercomputers in Grids. It focuses on various approaches in Grid computing that either aim to replace supercomputing or integrate supercomputers in existing Grid environments. We further point out the limitations to Grid approaches when it comes to supercomputing. We also point out the potential of supercomputers in Grids for economic usage. For this, we describe a public-private partnership in which this approach has been employed for more than 10 years. By giving such an overview we aim at better understanding the role of supercomputers and Grids and their interaction.

INTRODUCTION

Supercomputers have become widely used in academic research (Nagel, Kröner and Resch, 2007) and industrial development over the past years. Architectures of these systems have varied over time. For a long time special purpose systems have dominated the market. This has changed recently. Supercomputing today is dominated by standard components.

A quick look at the list of fastest computers worldwide (TOP500, 2008) shows that clusters built from such standard components have become the architecture of choice. This is highlighted by the fact that the fraction of clusters in the list has increased from about 2% in 2000 to about 73% in 2006. The key driving factor is the availability of competitive processor technology in the mass market on the one hand and a growing awareness of this potential in the user community on the other hand.

These trends have allowed using the same technology from the level of desktop systems to departmental systems and up to high end supercomputers. Simulation has hence been brought

DOI: 10.4018/978-1-60960-603-9.ch001

deep into the development process of academia and industrial companies.

The introduction of standard hardware components was accompanied by a similar trend in software. With Linux there is a standard operating system available today. It is also able to span the wide range from desktop systems to supercomputers. Although we still see different architectural approaches using standard hardware components, and although Linux has to be adapted to these various architectural variations, supercomputing today is dominated by an unprecedented standardization process.

Standardization of supercomputer components is mainly a side effect of an accelerated standardization process in information technology. As a consequence of this standardization process we have seen a closer integration of IT components over the last years at every level. In supercomputing, the Grid concept (Foster and Kesselman, 1998) best reflects this trend. First experiments coupling supercomputers were introduced by Smarr and Catlett (1992) fairly early – at that time still being called metacomputing. DeFanti et al. (1996) showed further impressive metacomputing results in the I-WAY project. Excellent results were achieved by experiments of the Japan Atomic Energy Agency (Imamura et al., 2000). Resch et al. (1999) carried out the first transatlantic metacomputing experiments. After initial efforts to standardize the Grid concept, it was finally formalized by Foster et al. (2001).

The promise of the Grid was twofold. Grids allow the coupling of computational and other IT resources to make any resource and any level of performance available to any user worldwide at anytime. On the other hand, the Grid allows easy access and use of supercomputers and thus reduces the costs for supercomputing simulations.

DEFINITIONS

When we talk about supercomputing we typically consider it as defined by the TOP500 list (TOP500, 2008). This list, however, mainly summarizes the fastest systems in terms of some predefined benchmarks. A clear definition of supercomputers is not given. For this article we define the purpose of supercomputing as follows:

- We want to use the fastest system available to get insight that we could not get with slower systems. The emphasis is on getting insight rather than on achieving a certain level of speed.

Any system (hardware and software combined) that helps to achieve this goal and fulfils the criteria given is considered to be a supercomputer. The definition itself implies that supercomputing and simulations are a third pillar of scientific research and development, complementing empirical and theoretical approaches.

Often, simulation complements experiments. To a growing extent, however, supercomputing has reached a point where it can provide insight that cannot even be achieved using experimental facilities. Some of the fields where this happens are climate research, particle physics or astrophysics. Supercomputing in these fields becomes a key technology if not the only possible one to achieve further breakthroughs.

There is also no official scientific definition for the Grid as the focus of the concept has changed over the years. Initially, supercomputing was the main target of the concept. Foster & Kesselman (1998) write:

A computational grid is a hardware and software infrastructure that provides dependable, consistent, pervasive, and inexpensive access to high-end computational capabilities.

This definition is very close to the concept of metacomputing coupling supercomputers to increase the level of performance. The Grid was intended to replace the local supercomputer. Soon, however it became clear that the Grid concept could and should be extended and Foster, Kesselman & Tuecke (2001) describe the Grid as

... flexible, secure, coordinated resource sharing among dynamic collections of individuals, institutions, and resources.

This is a much wider definition of the concept which goes way beyond the narrow problem of supercomputing. For the purpose of this article we use this second definition. We keep in mind though that the Grid started out as a concept to complement the existing supercomputing architectures.

GRIDS AND SUPERCOMPUTERS

Today the main building blocks to create a real scientific Grid are mainly in place. High speed wide area networks provide the necessary communication performance. Security procedures have been established which meet the limited requirements of scientists. Data management issues have been addressed to handle the large amount of data created e.g. in the high energy physics community (LHC, 2008). As of today, virtually every industrially developed nation has created its own national Grid infrastructure with trans-national Grids rapidly evolving (DEISA, 2008; PRAGMA-Grid 2008).

From the point of view of supercomputing, the question arises which role Grids can play in high performance computing simulation. Some aspects are briefly discussed in the following.

Grids Do Support Supercomputing

The idea of the Grid is mainly an idea of coordination and consolidation. These aspects have been widely ignored by the supercomputing community for a long time. A supercomputer was – and still is today – a one of a kind system. It is only available to a small number of users. Its mode of operation can be compared to the exclusive usage of an experimental facility. Typically, a supercomputer has no free resources. The user typically has to wait to use a supercomputer system – not the other way round.

Access to a supercomputer is hence not seen to be a standard service and no specific measures are taken to provide supercomputing at a comparable level of service as is done for other IT-services.

The Grid has, however, changed our view of supercomputers. From stand-alone systems, they have turned into "large nodes" of a mesh of resources. Although they are still unique in their potential to solve large problems the Grid has integrated them now into an ecosystem in which they play an important role. Being part of such a larger IT-landscape supercomputers have started to benefit substantially from lower level systems technology. This is in a sense a change of paradigm since so far supercomputers have typically been ahead of smaller systems in terms of complexity and level of technology. The flow of innovation – that traditionally was directed from supercomputers towards PCs – has at least partially been reversed.

The current situation can be described as follows: Supercomputers have been integrated into an ecosystem of IT-services. The quality of service for users has been improved. Aspects like security, accounting and data management have been brought in by the Grid community and the supercomputing community has picked them up. The notable exceptions are dedicated large scale system in classified installations. It remains to be seen whether these can remain in splendid isolation without losing contact with the technological drivers of the main stream IT-technology development.

Grids Cannot Replace Supercomputers

Sometimes the Grid is considered to be a replacement for supercomputers. The reasoning behind this idea is that the Grid provides such a massive amount of CPU cycles that any problem can easily be solved "on the Grid". The basic concept for such reasoning is the premise that a given problem can be described in terms of required CPU cycles needed. On the other hand, any given Grid configuration can be described in terms of CPU cycles provided. If one can match compute demand and compute supply, the problem is assumed to be solved.

This is, however, a deeply flawed view of supercomputing. The purpose of a supercomputer is to provide the necessary speed of calculation to solve a complex problem in an acceptable time. Only when being able to focus a huge resource on a single problem can we achieve this goal. So, two aspects are important here.

The size of a problem: We know of a number of problems that we call large which can actually be split into several small problems. For such embarrassingly parallel problems the Grid typically is a very good solution. A number of approaches have been developed among which Berkeley Open Infrastructure for Network Computing (BOINC 2008) and the World Community Grid (2008) are the most interesting ones. Both provide access to distributed resources for problems that can be split into very small junks of work. These small problems are sent out to a mass of computers (virtually every PC can be used). Doing this, the systems are able to tap into the Petaflops of performance available across the globe in an accumulation of small computers. However, there are other *large scale problems* that cannot be split into independent smaller parts. These truly *large scale problems* (high resolution CFD, high resolution complex scenario crash,) by nature cannot be made embarrassingly parallel and any distributed Grid solution has so far failed on them.

The time to solution: Most of the *large scale problems* mentioned above actually can run on smaller systems. However, on such smaller systems their solution may take weeks or even months. For any practical purpose such simulations would make little sense. The Grid is hence unable to provide scientists with a tool for these simulation experiments if it aims to replace supercomputers by a large amount of distributed systems.

THE ROLE OF SUPERCOMPUTERS IN GRIDS

The Grid has often been compared to the power grid (Chetty and Buyya, 2002). It actually is useful to look at the power grid as an analogy for any Grid to be set up. Power Grids are characterized by:

- A core of view production facilities providing a differing level of performance much higher than the need of any single user. Small facilities complement the overall power Grid.
- A very large number of users that typically require a very small level of performance compared to the production capacity of the providers.
- A standardized way of bringing suppliers and users together.
- A loosely coordinated operation of suppliers across large geographic areas.
- Breakdowns of the overall system if coordination is too loose or if single points of failure are hit.
- Special arrangements for users requiring a very high level of performance on a permanent basis. These are typically large scale production facilities like aluminum production.

When comparing the power grid to the compute Grid we notice a number of differences that have to be considered.

- Electrical power production can be changed at request (depending on the level of usage) with a maximum level of power defined. Depending on the type of power plant the performance may be increased to maximum or decreased to zero within minutes to days. Compute power, on the other hand, is always produced regardless of its usage. We speak of idle processors.
- Resources for electrical power production can be stored and used later. Even electricity that is produced can be stored for later usage by transferring it to hydro power plants' storage systems or using hydrogen storage devices. Compute power can never be stored.
- The lifetime of an electrical power plant is measured in tens of years. Powering up and powering down such plants can economically make sense. The lifetime of a supercomputer is more like three to five years. In order to make sense economically a supercomputer has to run 7x24 for this short period of life. Given the increase in speed of standard computer components this situation will not change over the next years.

When we analyze the analogy between the compute Grid and the power Grid carefully we find:

- A number of concepts that make sense in a large scale power Grid do not work in compute Grids.
- The economy of supercomputing differs substantially from the economy of the power Grid.
- Supercomputers are similar to large scale suppliers in the power grid as they provide a high level of performance.
- Supercomputer users are like special purpose users in the power grid that need a permanent supply of a high level of performance.

From this, we can conclude that supercomputers have to be part of a cyber-infrastructure. They have to be seen as large scale instruments that are available to a small number of users with large scale problems. In that sense supercomputers are special nodes in any compute Grid.

In the following we describe a prototype Grid that was developed over long time. It is characterized by:

- Integration of a small set of supercomputers and high-end compute-servers
- Dual use by academia and industry
- A commercial approach to supercomputing

A PUBLIC-PRIVATE SUPERCOMPUTING-GRID PARTNERSHIP

The University of Stuttgart is a technically oriented university with one of the leading mechanical engineering departments in Germany. The university has created a strong long term relationship with various companies in the region of Stuttgart. The most important ones are Daimler, Porsche and Bosch. The computing center of the university has hence been working closely with these companies since the early days of high performance computing in Stuttgart.

The computing center had been running HPC systems for some 15 years when in the late 1980s it decided to collaborate directly with Porsche in HPC operations. The collaboration resulted in shared investment in vector supercomputers for several years. Furthermore, the collaboration helped to improve the understanding of both sides and helped to position high performance computing as a key technology in academia and industry. The experiment was successful and was continued for about 10 years.

First attempts of the computing center to attract also usage from Daimler initially failed. This changed when in 1995 both the CEO of Daimler and the prime minister of the state of Baden-Württemberg gave their support for a collaboration of Daimler and the computing center at the University of Stuttgart in the field of high performance computing. The cooperation was realized as a public-private partnership. In 1995, hww was established with hww being an acronym for Höchstleistungsrechner für Wissenschaft und Wirtschaft (HPC for academia and industry)

The initial share holders of hww were:

- Daimler Benz had concentrated all its IT activities in a subsidiary called debis. So debis became the official share holder of hww holding 40% of the company.
- Porsche took a minority share of 10% of the company mainly making sure to continue the partnership with the University of Stuttgart and its computing center.
- The University of Stuttgart took a share of 25% and was represented by the High Performance Computing Center Stuttgart (HLRS).
- The State of Baden-Württemberg took a share of 25% being represented by the Ministry of Finance and the Ministry of Science.

The purpose of hww was not only to bring together academia and industry in using high performance computers, but to harvest some of the benefits of such collaboration. The key advantages were expected to be:

- Leverage of market power: Combining the purchasing power of industry and academia should help to achieve better price/performance for all partners both for purchase price and maintenance costs.
- Sharing of operational costs: Creating a group of operational experts should help to bring down the staff cost for running systems. This should be mainly achieved by combining the expertise of a small group of people and by being able to handle vacation time and sick leave much easier than before.
- Optimize system usage: Industrial usage typically comes in bursts when certain stages in the product development cycle require a lot of simulations. Industry then has a need for immediate availability of resources. In academia most simulations are part of long term research and systems are typically filled continuously. The intent was to find a model to intertwine the two modes for the benefit of both sides.

Prerequisites and Problems

A number of issues had to be resolved in order to make hww operational. The most pressing ones were: Security related issues: This included the whole complex of trust and reliability from the point of view of industrial users. While for academic users data protection and availability of resources are of less concern, it is vital for industry that its most sensitive data are protected and no information leaks to other users. Such information may even include things as the number and size of jobs run by a competitor. Furthermore, permanent availability of resources is a must in order to meet internal and external deadlines. While academic users might accept a failure of resources once in a while, industry requires reliable systems.

Data and communication: This includes the question of connectivity and handling input and output data. Typically network connectivity between academia and industry is poor. Most research networks are not open for industry. Most industries are worried about using public networks for security reasons. Accounting mechanisms for research networks are often missing. So, even to connect to a public institution may be difficult for industry. The amount of data to be transferred is another big issue as the size of output data can get prohibitively high. Both issues were addressed by increasing speed of networks and were helped by a tendency of German and local research networks opening up to commercial users.

Economic issues: One of the key problems was the establishment of costs for the usage of various resources. Until then no sound pricing mechanism for the usage of HPC system had been established either at the academic or industrial partners. Therefore, the partners had to agree on a mechanism to find prices for all resources that are relevant for the usage of computers.

Legal and tax issues: The collaboration of academia and industry was a challenge for lawyers on both sides. The legal issues had to be resolved and the handling of taxes had to be established in order to make the company operational.

After sorting out all these issues, the company was brought to life and its modes of operation had to be established.

Mode of Operation

In order to help achieve its goals, a lean organization for hww was chosen. The company itself does not have any staff. It is run by two part time directors. Hww was responsible for operation of systems, security, and accounting of system usage. In order to do this, work was outsourced to the partners of hww.

A pricing mechanism has been established that guarantees that any service of hww is sold to share holders of hww at cost price minimizing the overhead costs to the absolutely necessary. Costs and prices are negotiated for a one year period based on the requirements and available services of all partners. This requires an annual planning process for all services and resources offered by the partners through hww. The partners specifically have to balance supply and demand every year and have to adapt their acquisition strategy to the needs of hww.

Hww is controlled by an advisory board that meets regularly (typically 3 times a year). The board approves the budget of hww and discusses future service requirements of the overall company. The partners of hww have agreed that industrial services are provided by industry only while academic services are provided by academic partners only.

The Public-Private Grid

Over the life time of hww, a Grid infrastructure was set up that today consist of the following key components:

- A national German supercomputer facility, a number of large clusters and a number of shared memory systems.
- File system providing short and long term data storage facilities.
- Network connectivity for the main partners at the highest speed available.
- A software and security concept that meets the requirements of industrial users without restraining access for academic users.

The cyber-infrastructure created through the cooperation in hww is currently used by scientists from all over Germany and Europe and engineers in several large but also small and medium sized enterprises. Furthermore, the concept has been integrated into the German national D-Grid project and the state-wide Baden-Württemberg Grid. It thus provides a key backbone facility for simulation in academia and industry.

DISCUSSION OF RESULTS

We now have a 13 years experience with the hww concept. The company has undergone some changes over the years. The main changes are:

- Change of partners: When Daimler sold debis, the shares of an automotive company were handed over to an IT company. The new partner T-Systems further diversified its activities creating a subsidiary (called T-Systems SfR) together with the German Aerospace Center. T-Systems SfR took 10% of the 40% share of T-Systems. On the public side, two other universities were included with the four public partners holding 12.5% each.
- Change of operational model: Initially systems were operated by hww which outsourced task to T-Systems and HLRS at the beginning. Gradually, a new model was used. Systems are operated by the owners of the systems following the rules and regulations of hww. The public-private partnership gradually moves from being an operating company towards being a provider of a platform for the exchange of services and resources for academia and industry.

These organizational changes had an impact on the operation of hww. Having replaced an end user (Daimler) by a re-seller hww focused more on the re-selling of CPU cycles. This was emphasized by public centers operating systems themselves and only providing hww with CPU time. The increase in number of partners, on the other hand, made it more difficult to find consensus.

Overall, however, the results of 13 years of hww are positive. With respect to the expected benefits and advantages both of hww and its Grid like model the followings are noticeable:

The cost issue: Costs for HPC can potentially be reduced for academia if industry pays for usage of systems. Overall, hww was positive for its partners in this respect over the last 13 years. Additional funding was brought in through selling CPU time but also because hardware vendors had an interest to have their systems used by industry through hww. At the same time, however, industry takes away CPU cycles from academia increasing the competition for scarce resources. The other financial argument is a synergistic effect that actually allowed achieving lower prices whenever academia and industry merged their market power through hww to buy larger systems together.

Improved resource usage: The improved usage of resources during vacation time quickly is optimistic at best as companies – at least in Europe - tend to schedule their vacation time in accordance with public education vacations. As a result, industrial users are on vacation when scientists are on vacation. Hence, a better resource usage by anti-cyclic industrial usage turns out to be not achievable. Some argue that by reducing prices during vacation time for industry one might encourage more industrial usage when resources are available. However, here one has to compare costs: the costs for CPU time are in the range of thousands of Euro that could potentially be saved. On the other side, companies would have to adapt their working schedules to the vacation time of researchers and would have to make sure that their staff – very often with small children - would have to stay at home. Evidence shows that this is not happening

The analysis shows that financially the dual use of high performance computers in a Grid can be interesting. Furthermore, a closer collaboration between industry and research in high performance computing has helped to increase the awareness for the problems on both sides. Researchers understand what the real issues in simulation in industry are. Industrial designers understand how they can make good use of academic resources even though they have to pay for them.

CONCLUSION

Supercomputers can work as big nodes in Grid environments. Their users benefit from the software developed in general purpose Grids. Industry and academia can successfully share such Grids.

REFERENCES

BOINC - Berkeley Open Infrastructure for Network Computing. (2008). http://boinc.berkeley.edu/ (1.5.2008)

Chetty, M., & Buyya, R. (2002). Weaving Computational Grids: How Analogous Are They with Electrical Grids? [CiSE]. *Computing in Science & Engineering*, *4*(4), 61–71. doi:10.1109/MCISE.2002.1014981

DeFanti, T., Foster, I., Papka, M. E., Stevens, R., & Kuhfuss, T. (1996). Overview of the I-WAY: Wide Area Visual Supercomputing . *International Journal of Super-computing Applications*, *10*, 123–131. doi:10.1177/109434209601000201

DEISA project. (2008). http://www.deisa.org/ (1.5.2008)

Foster, I., & Kesselman, C. (1998). *The Grid – Blueprint for a New Computing Infrastructure*. Morgan Kaufmann.

Foster, I., Kesselman, C., & Tuecke, S. (2001). The Anatomy of the Grid: Enabling Scalable Virtual Organizations. *The International Journal of Supercomputer Applications*, *15*(3), 200–222. doi:10.1177/109434200101500302

Imamura, T., Tsujita, Y., Koide, H., & Takemiya, H. (2000). An Architecture of Stampi: MPI Library on a Cluster of Parallel Computers . In Dongarra, J., Kacsuk, P., & Podhorszki, N. (Eds.), *Recent Advances in Parallel Virtual Machine and Message Passing Interface* (pp. 200–207). Springer. doi:10.1007/3-540-45255-9_29

LHC – Large Hadron Collider Project. (2008). http://lhc.web.cern.ch/lhc/

Nagel, W. E., Kröner, D. B., & Resch, M. M. (2007). *High Performance Computing in Science and Engineering 07*. Berlin, Heidelberg, New York: Springer.

PRAGMA-Grid. (2008). http://www.pragma-grid.net/ (1.5.2008)

Resch, M., Rantzau, D., & Stoy, R. (1999). Metacomputing Experience in a Transatlantic Wide Area Application Test bed. *Future Generation Computer Systems*, *5*(15), 807–816. doi:10.1016/S0167-739X(99)00028-X

Smarr, L., & Catlett, C. E. (1992). Metacomputing. *Communications of the ACM*, *35*(6), 44–52. doi:10.1145/129888.129890

TOP500 List. (2008). http://www.top500.org/ (1.5.2008).

World Community Grid. (2008). http://www.worldcommunitygrid.org/ (1.5.2008).

This work was previously published in International Journal of Grid and High Performance Computing (IJGHPC), Volume 1, Issue 1, edited by Emmanuel Udoh & Ching-Hsien Hsu, pp. 1-9, copyright 2009 by IGI Publishing (an imprint of IGI Global).

Chapter 2
Porting HPC Applications to Grids and Clouds

Wolfgang Gentzsch
Independent HPC, Grid, and Cloud Consultant, Germany

ABSTRACT

A Grid enables remote, secure access to a set of distributed, networked computing and data resources. Clouds are a natural complement to Grids towards the provisioning of IT as a service. To "Grid-enable" applications, users have to cope with: complexity of Grid infrastructure; heterogeneous compute and data nodes; wide spectrum of Grid middleware tools and services; the e-science application architectures, algorithms and programs. For clouds, on the other hand, users don't have many possibilities to adjust their application to an underlying cloud architecture, because of its transparency to the user. Therefore, the aim of this chapter is to guide users through the important stages of implementing HPC applications on Grid and cloud infrastructures, together with a discussion of important challenges and their potential solutions. As a case study for Grids, we present the Distributed European Infrastructure for Supercomputing Applications (DEISA) and describe the DEISA Extreme Computing Initiative (DECI) for porting and running scientific grand challenge applications on the DEISA Grid. For clouds, we present several case studies of HPC applications running on Amazon's Elastic Compute Cloud EC2 and its recent Cluster Compute Instances for HPC. This chapter concludes with the author's top ten rules of building sustainable Grid and cloud e-infrastructures.

DOI: 10.4018/978-1-60960-603-9.ch002

INTRODUCTION

Over the last 40 years, the history of computing is deeply marked of the affliction of the application developers who continuously are porting and optimizing their applications codes to the latest and greatest computing architectures and environments. After the von-Neumann mainframe came the vector computer, then the shared-memory parallel computer, the distributed-memory parallel computer, the very-long-instruction word computer, the workstation cluster, the meta-computer, and the Grid (never fear, it continues, with SOA, Cloud, Virtualization, Many-core, and so on). There is no easy solution to this, and the real solution would be a separation of concerns between discipline-specific content and domain-independent software and hardware infrastructure. However, this often comes along with a loss of performance stemming from the overhead of the infrastructure layers. Recently, users and developers face another wave of complex computing infrastructures: the Grid.

Let's start with answering the question: What is a Grid? Back in 1998, Ian Foster and Carl Kesselman (1998) attempted the following definition: "A computational Grid is a hardware and software infrastructure that provides dependable, consistent, pervasive, and inexpensive access to high-end computational capabilities." In a subsequent article (Foster, 2002), "The Anatomy of the Grid," Ian Foster, Carl Kesselman, and Steve Tuecke changed this definition to include social and policy issues, stating that Grid computing is concerned with "coordinated resource sharing and problem solving in dynamic, multi-institutional virtual organizations." The key concept is the ability to negotiate resource-sharing arrangements among a set of participating parties (providers and consumers) and then to use the resulting resource pool for some purpose. This definition seemed very ambitious, and as history has proven, many of the Grid projects with a focus on these ambitious objectives did not lead to a sustainable Grid production environment. The simpler the Grid infrastructure, and the easier to use, and the sharper its focus, the bigger is its chance for success. And it is for a good reason (which we will explain in the following) that currently Clouds are becoming more and more popular (Amazon, 2007 and 2010).

Over the last ten years, hundreds of applications in science, industry and enterprises have been ported to Grid infrastructures, mostly prototypes in the early definition of Foster & Kesselman (1998). Each application is unique in that it solves a specific problem, based on modeling, for example, a specific phenomenon in nature (physics, chemistry, biology, etc.), presented as a mathematical formula together with appropriate initial and boundary conditions, represented by its discrete analogue using sophisticated numerical methods, translated into a programming language computers can understand, adjusted to the underlying computer architecture, embedded in a workflow, and accessible remotely by the user through a secure, transparent and application-specific portal. In just these very few words, this summarizes the wide spectrum and complexity we face in problem solving on Grid infrastructures.

The user (and especially the developer) faces several layers of complexity when porting applications to a computing environment, especially to a compute or data Grid of distributed networked nodes ranging from desktops to supercomputers. These nodes, usually, consist of several to many loosely or tightly coupled processors and, more and more, these processors contain few to many cores. To run efficiently on such systems, applications have to be adjusted to the different layers, taking into account different levels of granularity, from fine-grain structures deploying multi-core architectures at processor level to the coarse granularity found in application workflows representing for example multi-physics applications. Not enough, the user has to take into account the specific requirements of the grid, coming from the different components of the Grid services architecture, such as security, resource management, information services, and data management.

Obviously, in this article, it seems impossible to present and discuss the complete spectrum of applications and their adaptation and implementation on grids. Therefore, we restrict ourselves in the following to briefly describe the different application classes, present a checklist (or classification) with respect to grouping applications according to their appropriate grid-enabling strategy. Also, for lack of space, here, we are not able to include a discussion of mental, social, or legal aspects which sometimes might be the knock-out criteria for running applications on a grid. Other show-stoppers such as sensitive data, security concerns, licensing issues, and intellectual property, were discussed in some detail in Gentzsch (2007a).

In the following, we will consider the main three areas of impact on porting applications to grids: infrastructure issues, data management issues, and application architecture issues. These issues can have an impact on effort and success of porting, on the resulting performance of the Grid application, and on the user-friendly access to the resources, the Grid services, the application, the data, and the final processing results, among others.

APPLICATIONS AND THE GRID INFRASTRUCTURE

As mentioned before, the successful porting of an application to a Grid environment highly depends on the underlying distributed resource infrastructure. The main services components offered by a Grid infrastructure are security, resource management, information services, and data management. Bart Jacob et al. suggest that each of these components can affect the application architecture, its design, deployment, and performance. Therefore, the user has to go through the process of matching the application (structure and requirements) with those components of the Grid infrastructure, as described here, closely following the description in Jacob at al. (2003).

Applications and Security

The security functions within the Grid architecture are responsible for the authentication and authorization of the user, and for the secure communication between the Grid resources. Fortunately, these functions are an inherent part of most Grid infrastructures and don't usually affect the applications themselves, supposed the user (and thus the user's application) is authorized to use the required resources. Also, security from an application point of view might be taken into account in the case that sensitive data is passed to a resource to be processed by a job and is written to the local disk in a non-encrypted format, and other users or applications might have access to that data.

Applications and Resource Management

The resource management component provides the facilities to allocate a job to a particular resource, provides a means to track the status of the job while it is running and its completion information, and provides the capability to cancel a job or otherwise manage it. In conjunction with Monitoring and Discovery Service (described below) the application must ensure that the appropriate target resource(s) are used. This requires that the application accurately specifies the required environment (operating system, processor, speed, memory, and so on). The more the application developer can do to eliminate specific dependencies, the better the chance that an available resource can be found and that the job will complete. If an application includes multiple jobs, the user must understand (and maybe reduce) their interdependencies. Otherwise, logic has to be built to handle items such as inter-process communication, sharing of data, and concurrent job submissions. Finally, the job management provides mechanisms to query the status of the job as well as perform operations such as canceling the job. The application may need to utilize these capabilities to provide feedback to the user or to clean up or free up resources when required. For instance, if one job within an application fails, other jobs that may be dependent on it may need to be cancelled before needlessly consuming resources that could be used by other jobs.

Applications and Resource Information Services

An important part of the process of grid-enabling an application is to identify the appropriate (if not optimal) resources needed to run the application, i.e. to submit the respective job to. The service which maintains and provides the knowledge about the Grid resources is the Grid Information Service (GIS), also known as the Monitoring and Discovery Service (e.g. MDS in Globus (Jacob, 2003). MDS provides access to static and dynamic information of resources. Basically, it contains the following components:

- Grid Resource Information Service (GRIS), the repository of local resource information derived from information providers.
- Grid Index Information Service (GIIS), the repository that contains indexes of resource information registered by the GRIS and other GIISs.
- Information providers, translate the properties and status of local resources to the format defined in the schema and configuration files.
- MDS client which initially performs a search for information about resources in the Grid environment.

Resource information is obtained by the information provider and it is passed to GRIS. GRIS registers its local information with the GIIS, which can optionally also register with another GIIS, and so on. MDS clients can query the resource information directly from GRIS (for local resources) and/or a GIIS (for grid-wide resources).

It is important to fully understand the requirements for a specific job so that the MDS query can be correctly formatted to return resources that are appropriate. The user has to ensure that the proper information is in MDS. There is a large amount of data about the resources within the Grid that is available by default within the MDS. However, if the application requires special resources or information that is not there by default, the user may need to write her own information providers and add the appropriate fields to the schema. This may allow the application or broker to query for the existence of the particular resource/requirement.

Applications and Data Management

Data management is concerned with collectively maximizing the use of the limited storage space, networking bandwidth, and computing resources. Within the application, data requirements have been built in which determine, how data will be move around the infrastructure or otherwise accessed in a secure and efficient manner. Standardizing on a set of Grid protocols will allow to communicate between any data source that is available within the software design. Especially data intensive applications often have a federated database to create a virtual data store or other options including Storage Area Networks, network file systems, and dedicated storage servers. Middleware like the Globus Toolkit provide GridFTP and Global Access to Secondary Storage data transfer utilities in the Grid environment. The GridFTP facility (extending the FTP File Transfer Protocol) provides secure and reliable data transfer between Grid hosts.

Developers and users face a few important data management issues that need to be considered in application design and implementation. For large datasets, for example, it is not practical and may be impossible to move the data to the system where the job will actually run. Using data replication or otherwise copying a subset of the entire dataset to the target system may provide a solution. If the Grid resources are geographically distributed with limited network connection speeds, design considerations around slow or limited data access must be taken into account. Security, reliability, and performance become an issue when moving data across the Internet. When the data access may be slow or prevented one has to build the required logic to handle this situation. To assure that the data is available at the appropriate location by the time the job requires it, the user should schedule the data transfer in advance. One should also be aware of the number and size of any concurrent transfers to or from any one resource at the same time.

Beside the above described main requirements for applications for running efficiently on a Grid infrastructure, there are a few more issues which are discussed in Jacob (2003), such as scheduling, load balancing, Grid broker, inter-process communication, and portals for easy access, and non-functional requirements such as performance, reliability, topology aspects, and consideration of mixed platform environments.

The Simple API for Grid Applications (SAGA)

Among the many efforts in the Grid community to develop tools and standards which simplify the porting of applications to Grids by enabling the application to make easy use of the Grid middleware services as described above, one of the more predominant ones is SAGA, a high-level Application Programmers Interface (API), or programming abstraction, defined by the Open Grid Forum (OGF, 2008), an international committee that coordinates standardization of Grid middleware and architectures. SAGA intends to simplify the development of grid-enabled applications, even for scientists without any background in computer science or Grid computing. Historically, SAGA was influenced by the work on the GAT Grid Application Toolkit, a C-based API developed in the EU-funded project GridLab (GAT, 2005). The purpose of SAGA is two-fold:

1. Provide a simple API that can be used with much less effort compared to the interfaces of existing Grid middleware.
2. Provide a standardized, portable, common interface for the various Grid middleware systems.

According to Goodale (2008) SAGA facilitates rapid prototyping of new Grid applications by allowing developers a means to concisely state very complex goals using a minimum amount of code.

SAGA provides a simple, POSIX-style API to the most common Grid functions at a sufficiently high-level of abstraction so as to be able to be independent of the diverse and dynamic Grid environments. The SAGA specification defines interfaces for the most common grid-programming functions grouped as a set of functional packages. Version 1.0 (Goodale, 2008) defines the following packages:

- File package - provides methods for accessing local and remote file systems, browsing directories, moving, copying, and deleting files, setting access permissions, as well as zero-copy reading and writing
- Replica package - provides methods for replica management such as browsing logical file systems, moving, copying, deleting logical entries, adding and removing physical files from a logical file entry, and search logical files based on attribute sets.
- Job package - provides methods for describing, submitting, monitoring, and controlling local and remote jobs. Many parts of this package were derived from the largely adopted DRMAA Distributed

Resource Management Application API specification, an OGF standard.

- Stream package - provides methods for authenticated local and remote socket connections with hooks to support authorization and encryption schemes.
- RPC package - is an implementation of the OGF GridRPC API definition and provides methods for unified remote procedure calls.

The two critical aspects of SAGA are its simplicity of use and the fact that it is well on the road to becoming a community standard. It is important to note, that these two properties are provide the added value of using SAGA for Grid application development. Simplicity arises from being able to limit the scope to only the most common and important grid-functionality required by applications. There a major advantages arising from its simplicity and imminent standardization. Standardization represents the fact that the interface is derived from a wide-range of applications using a collaborative approach and the output of which is endorsed by the broader community.

More information about the SAGA C++ Reference Implementation (developed at the Center for Computation and Technology at the Louisiana State University) and various aspects of Grid enabling toolkits is available on the SAGA implementation home page (SAGA, 2006). It also provides additional information with regard to different aspects of Grid enabling toolkits.

GRID APPLICATIONS AND DATA

Any e-science application at its core has to deal with data, from input data (e.g. in the form of output data from sensors, or as initial or boundary data), to processing data and storing of intermediate results, to producing final results (e.g. data used for visualization). Data has a strong influence on many aspects of the design and deployment of an application and determines whether a Grid application can be successfully ported to the grid. Therefore, in the following, we present a brief overview of the main data management related aspects, tasks and issues which might affect the process of grid-enabling an application, such as data types and size, shared data access, temporary data spaces, network bandwidth, time-sensitive data, location of data, data volume and scalability, encrypted data, shared file systems, databases, replication, and caching. For a more in-depth discussion of data management related tasks, issues, and techniques, we refer to Bart Jacob's tutorial on application enabling with Globus (Jacob, 2003).

Shared Data Access

Sharing data access can occur with concurrent jobs and other processes within the network.

Access to data input and the data output of the jobs can be of various kinds. During the planning and design of the Grid application, potential restrictions on the access of databases, files, or other data stores for either read or write have to be considered. The installed policies need to be observed and sufficient access rights have to be granted to the jobs. Concerning the availability of data in shared resources, it must be assured that at run-time of the individual jobs the required data sources are available in the appropriate form and at the expected service level. Potential data access conflicts need to be identified up front and planned for. Individual jobs should not try to update the same record at the same time, nor dead lock each other. Care has to be taken for situations of concurrent access and resolution policies imposed.

The use of federated databases may be useful in data Grids where jobs must handle large amounts of data in various different data stores, you. They offer a single interface to the application and are capable of accessing data in large heterogeneous environments. Federated database systems contain information about location (node, database, table, record) and access methods (SQL, VSAM, privately defined methods) of connected data sources. Therefore, a simplified interface to the user (a Grid job or other client) requires that the essential information for a request should not include the data source, but rather use a discovery service to determine the relevant data source and access method.

Data Topology

Issues about the size of the data, network bandwidth, and time sensitivity of data determine the location of data for a Grid application. The total amount of data within the Grid application may exceed the amount of data input and output of the Grid application, as there can be a series of sub-jobs that produce data for other sub-jobs. For permanent storage the Grid user needs to be able to locate where the required storage space is available in the grid. Other temporary data sets that may need to be copied from or to the client also need to be considered.

The amount of data that has to be transported over the network is restricted by available bandwidth. Less bandwidth requires careful planning of the data traffic among the distributed components of a Grid application at runtime. Compression and decompression techniques are useful to reduce the data amount to be transported over the network. But in turn, it raises the issue of consistent techniques on all involved nodes. This may exclude the utilization of scavenging for a grid, if there are no agreed standards universally available.

Another issue in this context is time-sensitive data. Some data may have a certain lifetime, meaning its values are only valid during a defined time period. The jobs in a Grid application have to reflect this in order to operate with valid data when executing. Especially when using data caching or other replication techniques, it has to be assured that the data used by the jobs is up-to-date, at any given point in time. The order of data processing by the individual jobs, especially the production of input data for subsequent jobs, has to be carefully observed.

Depending on the job, the authors Jacob at al. (2003) recommend to consider the following data-related questions which refer to input as well as output data of the jobs within the Grid application:

- Is it reasonable that each job or set of jobs accesses the data via the network?
- Does it make sense to transport a job or set of jobs to the data location?
- Is there any data access server (for example, implemented as a federated database) that allows access by a job locally or remotely via the network?
- Are there time constraints for data transport over the network, for example, to avoid busy hours and transport the data to the jobs in a batch job during off-peak hours?
- Is there a caching system available on the network to be exploited for serving the same data to several consuming jobs?
- Is the data only available in a unique location for access, or are there replicas that are closer to the executable within the grid?

Data Volume

The ability for a Grid job to access the data it needs will affect the performance of the application. When the data involved is either a large amount of data or a subset of a very large data set, then moving the data set to the execution node is not always feasible. Some of the considerations as to what is feasible include the volume of the data to be handled, the bandwidth of the network, and logical interdependences on the data between multiple jobs.

Data volume issues: In a Grid application, transparent access to its input and output data is required. In most cases the relevant data is permanently located on remote locations and the jobs are likely to process local copies. This access to the data results in a network cost and it must be carefully quantified. Data volume and network bandwidth play an important role in determining the scalability of a Grid application.

Data splitting and separation: Data topology considerations may require the splitting, extraction, or replication of data from data sources involved. There are two general approaches that are suitable for higher scalability in a Grid application: Independent tasks per job and a static input file for all jobs. In the case of independent tasks, the application can be split into several jobs that are able to work independently on a disjoint subset of the input data. Each job produces its own output data and the gathering of all of the results of the jobs provides the output result by itself. The scalability of such a solution depends on the time required to transfer input data, and on the processing time to prepare input data and generate the final data result. In this case the input data may be transported to the individual nodes on which its corresponding job is to be run. Preloading of the data might be possible depending on other criteria like timeliness of data or amount of the separated data subsets in relation to the network bandwidth. In the case of static input files, each job repeatedly works on the same static input data, but with different parameters, over a long period of time. The job can work on the same static input data several times but with different parameters, for which it generates differing results. A major improvement for the performance of the Grid application may be derived by transferring the input data ahead of time as close as possible to the compute nodes.

Other cases of data separation: More unfavorable cases may appear when jobs have dependencies on each other. The application flow may be carefully checked in order to determine the level of parallelism to be reached. The number of jobs that can be run simultaneously without dependences is important in this context. For independent jobs, there needs to be synchronization mechanisms in place to handle the concurrent access to the data.

Synchronizing access to one output file: Here all jobs work with common input data and generate their output to be stored in a common data store. The output data generation implies that software is needed to provide synchronization between the jobs. Another way to process this case is to let each job generate individual output files, and then to run a post-processing program to merge all these output files into the final result. A similar case is that each job has its individual input data set, which it can consume. All jobs then produce output data to be stored in a common data set. Like described above, the synchronization of the output for the final result can be done through software designed for the task.

Hence, thorough evaluation of the input and output data for jobs in the Grid application is needed to properly handle it. Also, one should weigh the available data tools, such as federated databases, a data joiner, and related products and technologies, in case the Grid application is highly data oriented or the data shows a complex structure.

PORTING AND PROGRAMMING GRID APPLICATIONS

Besides taking into account the underlying Grid resources and the application's data handling, as discussed in the previous two paragraphs, another challenge is the porting of the application program itself. In this context, developers and users are facing mainly two different approaches when implementing their application on a grid. Either they port an existing application code on a set of distributed Grid resources. Often, in the past, the application previously has been developed and optimized with a specific computer architecture in mind, for example, mainframes or servers, single- or multiple-CPU vector computers, shared- or distributed-memory parallel computers, or loosely coupled distributed systems like workstation clusters, for example. Or developers start from scratch and design and develop a new application program with the Grid in mind, often such that the application architecture respectively its inherent numerical algorithms are optimally mapped onto the best-suited (set of) resources in a grid.

In both scenarios, the effort of implementing an application can be huge. Therefore, it is important to perform a careful analysis beforehand on: the user requirements for running the application on a Grid (e.g. cost, time); on application type (e.g. compute or data intensive); application architecture and algorithms (e.g. explicit, or implicit) and application components and how they interact (e.g. loosely or tightly coupled, or workflows); what is the best way to map the application onto a grid; and which is the best suited Grid architecture to run the application in an optimally performing way. Therefore, in the following, we summarize the most popular strategies for porting an existing application to a grid, and for designing and developing a new Grid application.

Many scientific papers and books deal with the issues of designing, programming, and porting Grid applications, and it is difficult to recommend the best suited among them. Here, we mainly follow the books from Ian Foster and Carl Kesselman (1999 & 2004), the IBM Redbook (Jacob, 2003), the SURA Grid Technology Cookbook (SURA, 2007), several research papers on programming models and environments, e.g. Soh (2006), Badia (2003), Karonis (2002), Seymour (2002), Buyya (2000), Venugopal (2004), Luther (2005), Altintas (2004), and Frey (2005), and our own experience at Sun Microsystems and MCNC (Gentzsch, 2004), RENCI (Gentzsch, 2007), D-Grid (Gentzsch, 2008, and Neuroth, 2007), and currently in DEISA-2 (Lederer, 2008).

Grid Programming Models and Environments

Our own experience in porting applications to distributed resource environments is very similar to the one from Soh et al. (2006) who present a useful discussion on Grid programming models and environments which we briefly summarize in the following. In their paper, they start with differentiating application porting into resource composition and program composition. Resource composition, i.e. matching the application to the Grid resources needed, has already been discussed in paragraphs 2 and 3 above.

Concerning program composition, there is a wide spectrum of strategies of distributing an application onto the available Grid resources. This spectrum ranges from the ideal situation of simply distributing a list of, say, n parameters together with n identical copies of that application program onto the grid, to the other end of the spectrum where one has to compose or parallelize the program into chunks or components that can be distributed to the Grid resources for execution. In the latter case, Soh (2006) differentiates between implicit parallelism, where programs are automatically parallelized by the environment, and explicit parallelism which requires the programmer to be responsible for most of the parallelization effort such as task decomposition, mapping tasks to processors and inter-task communication. However, implicit approaches often lead to non-scalable parallel performance, while explicit approaches often are complex and work- and time-consuming. In the following we summarize and update the approaches and methods discussed in detail in Soh (2006):

Superscalar (or STARSs), sequential applications composed of tasks are automatically converted into parallel applications where the tasks are executed in different parallel resources. The parallelization takes into account the existing data dependences

between the tasks, building a dependence graph. The runtime takes care of the task scheduling and data handling between the different resources, and takes into account the locality of the data between other aspects. There are several implementations available, like GRID Superscalar (GRIDSs) for computational Grids (Badia, 2003), which is also used in production at the MareNostrum supercomputer at the BSC in Barcelona; or Cell Superscalar (CellSs) for the Cell processor

(Perez, 2007) and SMP Superscalar (SMPSs) for homogeneous multicores or shared memory machines.

Explicit Communication, such as Message Passing and Remote Procedure Call (RPC). A messages passing example is MPICH-G2 (Karonis, 2002), a grid-enabled implementation of the Message Passing Interface (MPI) which defines standard functions for communication between processes and groups of processes, extended by the Globus Toolkit. An RPC example is GridRPC, an API for Grids (Seymour, 2002), which offers a convenient, high-level abstraction whereby many interactions with a Grid environment can be hidden.

Bag of Tasks, which can be easily distributed on Grid resources. An example is the Nimrod-G Broker (Buyya, 2000) which is a grid-aware version of Nimrod, a specialized parametric modeling system. Nimrod uses a simple declarative parametric modeling language and automates the task of formulating, running, monitoring, and aggregating results. Another example is the Gridbus Broker (Venugopal, 2004) that permits users access to heterogeneous Grid resources transparently.

Distributed Objects, as in ProActive (2005), a Java based library that provides an API for the creation, execution and management of distributed active objects. Proactive is composed of only standard Java classes and requires no changes to the Java Virtual Machine (JVM) allowing Grid applications to be developed using standard Java code.

Distributed Threads, for example Alchemi (Luther, 2005), a Microsoft .NET Grid computing framework, consisting of service-oriented middleware and an application program interface (API). Alchemi features a simple and familiar multithreaded programming model.

Grid Workflows. Many Workflow Environments have been developed in recent years for grids, such as Triana, Taverna, Simdat, P-Grade, and Kepler. Kepler, for example, is a scientific workflow management system along with a set of Application Program Interfaces (APIs) for heterogeneous hierarchical modeling (Altintas, 2004). Kepler provides a modular, activity oriented programming environment, with an intuitive GUI to build complex scientific workflows.

Grid Services. An example is the Open Grid Services Architecture (OGSA), (Frey, 2005), which is an ongoing project that aims to enable interoperability between heterogeneous resources by aligning Grid technologies with established Web service technology. The concept of a Grid service is introduced as a Web service that provides a set of well defined interfaces that follow specific conventions. These Grid services can be composed into more sophisticated services to meet the needs of users.

GRID-ENABLING APPLICATION PROGRAMS AND NUMERICAL ALGORITHMS

In many cases, restructuring (grid-enabling, decomposing, parallelizing) the core algorithm(s) within a single application program doesn't make sense, especially in the case of a more powerful higher-level grid-enabling strategy. For example, in the case of parameter jobs (see below), many identical copies of the application program together with different data-sets can easily be distributed onto many Grid nodes, or where the application program components can be mapped onto a workflow, or where applications (granularity, run time, special dimension, etc.) simply are too small to efficiently run on a grid, and the Grid latencies and management overhead become too dominant. In other cases, however, where e.g. just one very long run has to be performed, grid-enabling the application program itself can lead to dramatic performance improvements and, thus, time savings. In an effort to better guide the reader through this complex field, in the following, we will briefly present a few popular application codes and their algorithmic structure and provide recommendations for some meaningful grid-enabling strategies.

General Approach. First, we have to make sure that we gain an important benefit form running our application on a grid. And we should start asking a few more general questions, top-down. Has this code been developed in-house, or is it a third-party code, developed elsewhere? Will I submit many jobs (as e.g. in a parameter study), or is the overall application structure a workflow, or is it a single monolithic application code? In case of the latter, are the core algorithms within the application program of explicit or of implicit nature? In many cases, grid-enabling those kinds of applications can be based on experience made in the past with parallelizing them for the moderately or massively parallel systems, see e.g. Fox et al. (1994) and Dongarra et al. (2003).

In-house Codes. In case of an application code developed in-house, the source code of this application is often still available, and ideally the code developers are still around. Then, we have the possibility to analyze the structure of the code, its components (subroutines), dependencies, data handling, core algorithms, etc. With older codes, sometimes, this analysis has already been done before, especially for the vector and parallel computer architectures of the 1980ies and 1990ies. Indeed, some of this knowledge can be re-used now for the grid-enabling process, and often only minor adjustments are needed to port such a code to the grid.

Third-Party Codes licensed from so-called Independent Software Vendors (ISVs) cannot be grid-enabled without the support from these ISVs. Therefore, in this case, we recommend to contact the ISV. In case the ISV receives similar requests from other customers as well, there might be a real chance that the ISV will either provide a grid-enabled code or completely change its sales strategy and sell its software as a service, or develops its own application portal to provide access to the application and the computing resources. But, obviously, this requires patience and is thus not a solution if you are under a time constraint.

Parameter Jobs. In science and engineering, often, the application has to run many times: same code, different data. Only a few parameters have to be modified for each individual job, and at the end of the many job runs, the results are analyzed with statistical or stochastic methods, to find a certain optimum. For example, during the design of a new car model, many crash simulations have to be performed, with the aim to find the best-suited material and geometry for a specific part of the wire-frame model of the car.

Application Workflows. It is very common in so-called Problem Solving Environments that the application program consists of a set of components or modules which interact with each other. This can be modeled in Grid workflow environments which support the design and the execution of the workflow representing the application program. Usually, these Grid workflow environments contain a middleware layer which maps the application modules onto the different resources in the grid. Many Workflow Environments have been developed in recent years for grids, such as Triana (2003), Taverna (2008), Simdat (2008), P-Grade (2003), and Kepler (Altintas, 2004). One application which is well suited for such a workflow is climate simulation. Today's climate codes consist of modules for simulating the weather on the continent with mesoscale meteorology models, and include other modules for taking into account the influence from ocean and ocean currents, snow and ice, sea ice, wind, clouds and precipitation, solar and terrestrial radiation, absorption, emission, and reflection, land surface processes, volcanic gases and particles, and human influences. Interactions happen between all these components, e.g. air-ocean, air-ice, ice-ocean, ocean-land, etc. resulting in a quite complex workflow which can be mapped onto the underlying Grid infrastructure.

Highly Parallel Applications. Amdahl's Law states that the scalar portion of a parallel program becomes a dominant factor as processor number increases, leading to a loss in application scalability with growing number of processors. Gustafson (1988) proved that this holds only for fixed problem size, and that in practice, with increasing number of processors, the user increases problem size as well, always trying to solve the largest possible problem on any given number of CPUs. Gustafson demonstrated this on a 1028-processor parallel system, for several applications. For example, he was able to achieve a speed-up factor of over 1000 for a Computational Fluid Dynamics application with 1028 parallel processes on the 1028-processor system. Porting these highly parallel applications to a grid, however, has shown that many of them degrade in performance simply because overhead of communication for message-passing operations (e.g. send and receive) drops from a few microseconds on a tightly-coupled parallel system to a few milliseconds on a (loosely-coupled) workstation cluster or grid. In this case, therefore, we recommend to implement a coarse-grain Domain Decomposition approach, i.e. to dynamically partition the overall computational domain into sub-domains (each consisting of as many parallel processes, volumes, finite elements, as possible), such that each sub-domain completely fits onto the available processors of the corresponding parallel system in the grid. Thus, only moderate performance degradation from the reduced number of inter-system communication can be expected. A prerequisite for this to work successfully is that the subset of selected parallel systems is of homogeneous nature, i.e. architecture and operating system of these parallel systems should be identical. One Grid infrastructure which offers this feature is the Distributed European Infrastructure for Supercomputing Applications (DEISA, 2010), which (among others) provides a homogeneous cluster of parallel AIX machines distributed over several of the 11 European supercomputing centers which are part of DEISA (see also Section 5 in this Chapter).

Moderately Parallel Applications. These applications, which have been parallelized in the past, often using Message Passing MPI library functions for the inter-process communication on workstation clusters or on small parallel systems, are well-suited for parallel systems with perhaps a few dozen to a few hundreds of processors, but they won't scale easily to a large number of parallel processes (and processors). Reasons are a significant scalar portion of the code which can't run in parallel and/or the relatively high ratio of inter-process communication to computation, resulting in relatively high idle times of the CPUs waiting fore the data. Many commercial codes fall in this category, for example finite-element codes such as Abaqus, Nastran, or Pamcrash. Here we recommend to check if the main goal is to analyze many similar scenarios with one and the same code but on different data sets, and run as many codes in parallel as possible, on as many moderately parallel sub-systems as possible (this could be virtualized sub-systems on one large supercomputer, for example).

Explicit versus Implicit Algorithms. Discrete Analogues of systems of partial differential equations, stemming from numerical methods such as finite difference, finite volume, or finite element discretizations, often result in large sets of explicit or implicit algebraic equations for the unknown discrete variables (e.g. velocity vectors, pressure, temperature). The explicit methods are usually slower (in convergence to the exact solution vector of the algebraic system) than the implicit ones but they are also inherently parallel, because there is no dependence of the solution variables among each other, and therefore there are no recursive algorithms. In case of the more accurate implicit methods, however, solution variables are highly inter-dependent leading to recursive sparse-matrix systems of algebraic equations which cannot easily split (parallelized) into smaller systems. Again, here, we recommend to introduce a Domain Decomposition approach as described in the above section on Highly Parallel

Algorithms, and solve an implicit sparse-matrix system within each domain, and bundle sets of 'neighboring' domains into super-sets to submit to the (homogeneous) grid.

Domain Decomposition. This has been discussed in the paragraphs on Highly Parallel Applications and on Explicit versus Implicit Algorithms.

Job Mix. Last but not lease, one of the most trivial but most widely used scenarios often found in university and research computer centers is the general job mix, stemming from hundreds or thousands of daily users, with hundreds or even thousands of different applications, with varying requirements for computer architecture, data handling, memory and disc space, timing, priority, etc. This scenario is ideal for a Grid which is managed by an intelligent Distributed Resource Manager (DRM), for example GridWay (2008) for a global grid, Sun Grid Engine Enterprise Edition (Chaubal, 2003) for an enterprise grid, or the open source Grid Engine (2001) for a departmental Grid or a simple cluster. These DRMs are able to equally balance the overall job load across the distributed resource environment and submit the jobs always to the best suited and least loaded resources. This can result in overall resource utilization of 90% and higher.

Applications and Grid Portals

Grid portals are an important part of the process of grid-enabling, composing, manipulating, running, and monitoring applications. After all the lower layers of the grid-enabling process have been performed (described in the previous paragraphs), often, the user is still exposed to the many details of the Grid services and even has to take care of configuring, composing, provisioning, etc. the application and the services "by hand". This however can be drastically simplified and mostly hidden from the user through a Grid portal, which is a Web-based portal able to expose Grid services and resources through a browser to allow users remote, ubiquitous, transparent and secure access to Grid services (computers, storage, data, applications, etc). The main goal of a Grid portal is to hide the details and complexity of the underlying Grid infrastructure from the user in order to improve usability and utilization of the grid, greatly simplifying the use of grid-enabled applications through a user-friendly interface.

Grid portals have become popular in research and the industry communities. Using Grid portals, computational and data-intensive applications such as genomics, financial modeling, crash test analysis, oil and gas exploration, and many more, can be provided over the Web as traditional services. Examples of existing scientific application portals are the GEONGrid (2008) and CHRONOS (2004) portals that provide a platform for the Earth Science community to study and understand the complex dynamics of Earth systems; the NEESGrid project (2008) focuses on earthquake engineering research; the BIRN portal (2008) targets biomedical informatics researchers; and the MyGrid portal (2008) provides access to bioinformatics tools running on a back-end Grid infrastructure. As it turns out, scientific portals are usually being developed inside specific research projects. As a result they are specialized for specific applications and services satisfying project requirements for that particular research application area.

In order to rapidly build customized Grid portals in a flexible and modular way, several more generic toolkits and frameworks have been developed. These frameworks are designed to meet the diverse needs and usage models arising from both research and industry. One of these frameworks is EnginFrame, which simplifies development of highly functional Grid portals exposing computing services that run on a broad range of different computational Grid systems. EnginFrame (Beltrame, 2006) has been adopted by many industrial companies, and by organizations in research and education.

Example: The EnginFrame Portal Environment

EnginFrame (2008) is a Web-based portal technology that enables the access and the exploitation of grid-enabled applications and infrastructures. It allows organizations to provide application-oriented computing and data services to both users (via Web browsers) and in-house or ISV applications (via SOAP/WSDL based Web services), thus hiding the complexity of the underlying Grid infrastructure. Within a company or department, an enterprise portal aggregates and consolidates the services and exposes them to the users, through the Web. EnginFrame can be integrated as Web application in a J2EE standard application server or as a portlet in a JSR168 compliant portlet container.

As a Grid portal framework, EnginFrame offers a wide range of functionalities to IT developers facing the task to provide application-oriented services to the end users. EnginFrame's plug-in mechanism allows to easily and dynamically extend its set of functionalities and services. A plug-in is a self-contained software bundle that encapsulates XML Extensible Markup Language service descriptions, custom layout or XSL Extensible Stylesheet Language and the scripts or executables involved with the services actions. A flexible authentication delegation offers a wide set of pre-configured authentication mechanisms: OS/NIS/PAM, LDAP, Microsoft Active Directory, MyProxy, Globus, etc. It can also be extended throughout the plug-in mechanism.

Besides authentication, EnginFrame provides an authorization framework that allows to define groups of users and Access Control Lists (ACLs), and to bind ACLs to resources, services, service parameters and service results. The Web interface of the services provided by the portal can be authorized and thus tailored to the specific users' roles and access rights.

EnginFrame supports a wide variety of compute Grid middleware like LSF, PBS, Sun Grid Engine, Globus, gLite and others. An XML virtualization layer invokes specific middleware commands and translates results, jobs and Grid resource descriptions into a portable XML format called GridML that abstracts from the actual underlying Grid technology. For the GridML, as for the service description XML, the framework provides pre-built XSLs to translate GridML into HTML. EnginFrame data management allows for browsing and handling data on the client side or remotely archived in the Grid and then to host a service working environment in file system areas called spoolers.

The EnginFrame architecture is structured into three tiers, Client, Resource, Server. The Client Tier normally consists of the user's Web browser and provides an easy-to-use interface based on established Web standards like XHTML and JavaScript, and it is independent from the specific software and hardware environment used by the end user. When needed, the client tier also provides integration with desktop virtualization technologies like Citrix Metaframe (ICA), VNC, X, and Nomachine NX. The Resource Tier consists of one or more Agents deployed on the back-end Grid infrastructure whose role is to control and provide distributed access to the actual computing resources. The Server Tier consists of a server component that provides resource brokering to manage resource activities in the back-end.

The EnginFrame server authenticates and authorizes incoming requests from the Web, and asks an Agent to execute the required actions. Agents can perform different kind of actions that range from the execution of a simple command on the underlying Operating System, to the submission of a job to the grid. The results of the executed action are gathered by the Agent and sent back to the Server which applies post processing transformations, filters the output according to ACLs and transforms the results into a suitable format according to the nature of the client: HTML for Web browsers and XML in a SOAP message for Web services client applications.

GRID CASE STUDY: HPC ON THE DEISA E-INFRASTRUCTURE

As one example, in the following, we will briefly discuss the DEISA Distributed European Infrastructure for Supercomputing Applications. A more detailed description can be found in (Gentzsch, 2010, 2011). DEISA is different from many other Grid initiatives which aim at building a general purpose Grid infrastructure and therefore have to cope with many (almost) insurmountable barriers such as complexity, resource sharing, crossing administrative (and even national) domains, handling IP and legal issues, dealing with sensitive data, working on interoperability, and facing the issue to expose every little detail of the underlying infrastructure services to the Grid application user. DEISA avoids most of these barriers by staying very focused: The main focus of DEISA is to provide the European supercomputer user with a flexible, dynamic, user-friendly supercomputing ecosystem (one could say Supercomputing Cloud, see next paragraph) for easy handling, submitting, and monitoring long-running jobs on the best-suited and least-loaded supercomputer(s) in Europe, trying to avoid the just mentioned barriers. In addition, DEISA offers application-enabling support. For a similar European funded initiative especially focusing on enterprise applications, we refer the reader to the BEinGRID project (2008), which consists of 18 so-called business experiments each dealing with a pilot application that addresses a concrete business case, and is represented by an end-user, a service provider, and a Grid service integrator. Experiments come from key business sectors such as multimedia, financial, engineering, chemistry, gaming, environmental science, and logistics and so on, based on different Grid middleware solutions, see (BEinGRID, 2008).

The DEISA Project

DEISA is the Distributed European Initiative for Supercomputing Applications, funded by the EU in Framework Programme 6 (DEISA1, 2004 – 2008) and Framework Programme 7 (DEISA2, 2008 – 2011). The DEISA Consortium consists of 11 partners, MPG-RZG (Germany, consortium lead), BSC (Spain), CINECA (Italy), CSC (Finland), ECMWF (UK), EPCC (UK), FZJ (Germany), HLRS (Germany), IDRIS (France), LRZ (Germany), and SARA (Netherlands). Further centers were integrated as associate partners: CEA-CCRT (France), CSCS (Switzerland), and KTH (Sweden).

DEISA developed and supports a distributed high performance computing infrastructure and a collaborative environment for capability computing and data management. The resulting infrastructure enables the operation of a powerful supercomputing Grid built on top of national supercomputing services, facilitating Europe's ability to undertake world-leading computational science research. DEISA is instrumental for advancing computational sciences in scientific and industrial disciplines within Europe and is paving the way towards the deployment of a cooperative European HPC ecosystem. The existing infrastructure is based on the coupling of eleven leading national supercomputing centers, using dedicated network interconnections (currently 10 GBs) of GÉANT2 and the NRENs.

DEISA2 developed activities and services relevant for applications enabling, operation, and technologies, as these are indispensable for the effective support of computational sciences in the area of supercomputing. The service provisioning model has been extended from one that supports a single project (in DEISA1) to one supporting Virtual European Communities (now in DEISA2). Collaborative activities are carried out with European and other international initiatives. Of strategic importance is the cooperation with the PRACE (2008) initiative which is preparing for the installation of a limited number of leadership-class Tier-0 supercomputers in Europe.

The DEISA Infrastructure Services

The essential services to operate the infrastructure and support its efficient usage are organized in three Service Activities: Operations, Technologies, and Applications:

Operations refer to operating the infrastructure including all existing services, adopting approved new services from the Technologies Activity, and advancing the operation of the DEISA HPC infrastructure to a turnkey solution for the future European HPC ecosystem by improving the operational model and integrating new sites.

Technologies cover monitoring of technologies in use in the project, identifying and selecting technologies of relevance for the project, evaluating technologies for pre-production deployment, and planning and designing specific sub-infrastructures to upgrade existing services or deliver new ones based on approved technologies. User-friendly access to the DEISA Supercomputing Grid is provided by DEISA Services for Heterogeneous management Layer (DESHL, 2008) and the UNiforme Interface for COmputing Resources (UNICORE, 2008).

Applications cover the areas 'applications enabling' and 'extreme computing projects', 'environment and user related application support', and 'benchmarking'. Applications enabling focuses on enhancing scientific applications from the DEISA Extreme Computing Initiative (DECI), Virtual Communities and EU projects. Environment and user related application support addresses the maintenance and improvement of the DEISA application environment and interfaces, and DEISA-wide user support in the applications area. Benchmarking refers to the provision and maintenance of a European Benchmark Suite for supercomputers.

In DEISA2, two Joint Research Activities (JRA) complement the portfolio of service activities. JRA1 (Integrated DEISA Development Environment) aims at an integrated environment for scientific application development, based on a software infrastructure for tools integration, which provides a common user interface across multiple computing platforms. JRA2 (Enhancing Scalability) aims at the enabling of supercomputer applications for the efficient exploitation of current and future supercomputers, to cope with a production infrastructure characterized by an aggressive parallelism on heterogeneous HPC architectures at a European scale.

DECI: DEISA Extreme Computing Initiative for Supercomputing Applications

The DEISA Extreme Computing Initiative (DECI, 2010) has been launched in May 2005 by the DEISA Consortium, as a way to enhance its impact on science and technology. The main purpose of this initiative is to enable a number of "grand challenge" applications in all areas of science and technology. These leading, ground breaking applications must deal with complex, demanding and innovative simulations that would not be possible without the DEISA infrastructure, and which benefit from the exceptional resources provided by the Consortium. The DEISA applications are expected to have requirements that cannot be fulfilled by the national HPC services alone.

In DEISA2, the single-project oriented activities (DECI) are qualitatively extended towards persistent support of Virtual Science Communities. This extended initiative benefits from and builds on the experiences of the DEISA scientific Joint Research Activities where selected computing needs of various scientific communities and a pilot industry partner were addressed. Examples of structured science communities with which close relationships are established are EFDA and the European climate community. DEISA2 provides a computational platform for them, offering integration via distributed services and web applications, as well as managing data repositories.

Applications Adapted to the DEISA Grid Infrastructure

In the following, we describe examples of application profiles and use cases that are well-suited for the DEISA supercomputing grid, and that can benefit from the computational resources made available by the DECI Extreme Computing Initiative.

International collaboration involving scientific teams that access the nodes of the AIX super-cluster in different countries, can benefit from a common data repository and a unique, integrated programming and production environment (via common global file systems). Imagine, for example, that team A in France and team B in Germany dispose of allocated resources at IDRIS in Paris and FZJ in Juelich, respectively. They can benefit from a shared directory in the distributed super-cluster, and for all practical purposes it looks as if they were accessing a single supercomputer.

Extreme computing demands of a challenging project requiring a dominant fraction of a single supercomputer. Rather than spreading a huge, tightly coupled parallel application on two or more supercomputers, DEISA can organize the management of its distributed resource pool such that it is possible to allocate a substantial fraction of a single supercomputer to this project which is obviously more efficient that splitting the application and distributing it over several supercomputers.

Workflow applications involving at least two different HPC platforms. Workflow applications are simulations where several independent codes act successively on a stream of data, the output of one code being the input of the next one in the chain. Often, this chain of computations is more efficient if each code runs on the best-suited HPC platform (e.g. scalar, vector, or parallel supercomputers) where it develops the best performance. Support of these applications via UNICORE (2008) which allows treating the whole simulation chain as a single job is one of the strengths of the DEISA Grid.

Coupled applications involving more than one platform. In some cases, it does make sense to spread a complex application over several computing platforms. This is the case of multi-physics, multi-scale application codes involving several computing modules each dealing with one particular physical phenomenon, and which only need to exchange a moderate amount of data in real time.

HPC APPLICATIONS IN THE CLOUD

With increasing demand for higher performance, efficiency, productivity, agility, and lower cost, since several years, Information Communication Technologies, ICT, are dramatically changing from static silos with manually managing resources and applications, towards dynamic virtual environments with automated and shared services, i.e. from silo-oriented to service-oriented architectures.

With sciences and businesses turning global and competitive, applications, products and services becoming more complex, and research and development teams being distributed, ICT is in transition again. Global challenges require global approaches: on the horizon, so-called virtual organizations and partner Grids will provide the necessary communication and collaboration platform, with Grid portals for secure access to resources, applications, data, and collaboratories.

One component which will certainly foster this next-generation scenario is Cloud Computing, as recently offered by companies like Amazon (2007 and 2010) Elastic Cloud Computing EC2, IBM (2008), Google (2008) App Engine and Google Group (2010), SGI (Cyclone, 2010), and many more. Clouds will become important dynamic components of research and enterprise infrastructures, adding a new 'external' dimension of 'elasticity' to them by enhancing their 'home' resource capacity whenever needed, on demand. Existing businesses will use them for their peak demands and for new projects, service providers will host their applications on them and provide Software as a Service, start-ups will integrate them in their offerings without the need to buy resources upfront, and setting up new social networks (Web 2.0 communities) will become very easy.

Cloud-enabling applications will follow similar strategies as with grid-enabling, as discussed in the previous paragraphs. Similarly challenging as with grids, though, are the cultural, mental, legal, and political aspects in the Cloud context. Building trust and reputation among the users and the providers will help in many scenarios. But it is currently difficult to imagine that users may easily entrust their corporate core assets and sensitive data to Cloud service providers. Today (in January 2011) the status of HPC Clouds seems to be similar to the status of Grids in the early 2000s: a few standard and well-suited HPC application scenarios run on Clouds, but many of the more complex and demanding HPC applications in research and enterprises will face barriers on Clouds which still have to be removed. For example, barriers may arise in the following context:

- The process of retrieving data from one cloud and move them into another cloud, and back to your desktop system, in a reliable and secure way.
- The fulfilment of (e.g. government) requirements for security, privacy, data protection, and the archiving risks associated with the cloud.
- The compliance with existing legal and regulatory frameworks and current policies (established far before the digital age) that impose antiquated (and sometimes even conflicting) rules about how to correctly deal with information and knowledge.
- The process of setting up a service level agreement.
- Migrating your applications from their existing environments into the cloud.

And for that matter…

- Do we all agree on the same security requirements; do we need a checklist, or do we need a federated security framework?
- Do our existing identity, access management, audit and monitoring strategies still hold for the clouds?
- What cloud deployment model would you have to choose: private, public, hybrid, or federated cloud?
- How much does the virtualization layer of the cloud affect application performance (i.e. trade-off between abstraction versus control)?
- How will clouds affect performance of high-throughput versus high-performance computing applications?
- What type of application needs what execution model to provide useful abstractions in the cloud, such as for data partitioning, data streaming, and parameter sweep algorithms?
- How do we handle large scientific workflows for complex applications that may be deployed as a set of virtual machines, virtual storage and virtual networks to support different functional components?
- What are common best practices and standards needed to achieve portability and interoperability for cloud applications and environments ?
- How can (and will) organizations like DMTF and OGF help us with our cloud standardization requirements?
- And last but not least, what if your cloud service provider fails?

One example of an early innovative Cloud system came from Sun Microsystems when in 2005 it truly built its SunGrid (Sun 2010) from scratch, based on the early vision that the network is the computer. As with other early technologies in the past, Sun paid a high price for being first and doing all the experiments and the evangelization. Its successor, Sun Network.com (Sun 2010), was popular among its few die-hard clients. This is because of an easy-to use technology (Grid Engine, Jini, JavaSpaces), but it's especially because of their innovative early users, such as CDO^2 (2008), a provider of innovative pricing and risk technology for organizations trading structured credit products.

It is interesting to observe how some of the earlier differences between Grids and clouds are fading away. While in the beginning of the Grid era, many Grid infrastructure prototypes were built and disappeared after a while, today we see many production Grids providing infrastructure, platform, and software services (almost) on demand, similar to the clouds, especially from an end-user point of view. One good example is the DEISA e-Infrastructure discussed in Chapter 5 above, with its DECI– DEISA Extreme Computing Initiative. Why is DECI currently so successful in offering millions of supercomputing cycles to the European e-Science community and helping scientists gain new scientific insights? Several reasons, in my opinion: because DEISA has a very targeted focus on specific (long-running) supercomputing applications and most of the applications just run on one – best-suited - system; because of its user-friendly access - through technology like DESHL (2008) and UNICORE (2008); because of staying away from those more ambitious general-purpose Grid efforts aiming at providing everything to everybody; because of its coordinating function which leaves the consortium partners (the 14 largest European supercomputer centers) fully independent; and – similar to network.com in the past – because of ATASKF (DECI, 2010), the application task force, consisting of application experts who help the users with porting their applications to the DEISA infrastructure. Because of the benefits of DEISA, the PRACE Consortium (PRACE, 2008) decided in 2010 to incorporate the DEISA Infrastructure into PRACE and provide access to the PRACE Petaflops systems via DEISA.

With this sea-change ahead of us, there will be a continuous strategic importance for sciences and businesses to support the work of the Open Grid Forum (OGF, 2008). Because only standards – recently also for clouds (OCCI, 2010) – will enable building e-infrastructures and grid- and cloud-enabled applications easily from different technology components and to transition towards an agile platform for federated services. Standards, developed in OGF, guarantee interoperation of different Grid and cloud components best suited for HPC applications, and thus reducing dependency from proprietary building blocks and services, keeping cost under control, and increasing research and business flexibility.

CLOUD CASE STUDIES: HPC APPLICATIONS ON AMAZON

Amazon Web Services (AWS) is Amazon's cloud computing platform, with Amazon Elastic Compute Cloud (EC2) as its central part, first announced as beta in August 2006. Users can rent Virtual Machines (VMs) on which they run their applications. EC2 allows scalable deployment of applications by providing a web service through which a user can boot an Amazon Machine Image (AMI) to create a virtual machine, which Amazon calls an "instance", containing any software desired. A user can create, launch, and terminate server instances as needed and paying by the hour for active servers. EC2 provides users with control over the geographical location of instances which allows for latency optimization and high levels of redundancy.

NAS Parallel Benchmark on Amazon EC2

In order to find out if and how clouds are suitable for HPC applications, Ed Walker (Walker 2008) run an HPC benchmark on Amazon EC2. He used several macro and micro benchmarks to examine the "delta" between clusters composed of state-of-the-art CPUs from Amazon EC2 versus an HPC cluster at the National Center for Supercomputing Applications (NCSA). He used the NAS Parallel Benchmarks (NAS 2010) to measure the performance of these clusters for frequently occurring scientific calculations. Also, since the Message-Passing Interface (MPI) library is an important programming tool used widely in scientific computing, his results demonstrate the MPI performance in these clusters by using the mpptest micro benchmark. For his benchmark study on EC2 he use the high-CPU extra large instances provided by the EC2 service.

The NAS Parallel Benchmarks (NPB 2010) comprise a widely used set of programs designed to evaluate the performance of HPC systems. The core benchmark consists of eight programs: five parallel kernels and three simulated applications. In aggregate, the benchmark suite mimics the critical computation and data movement involved in computational fluid dynamics and other "typical" scientific computation.

Research from Ed Walker (2008) about the runtimes of each of the NPB programs in the benchmark shows a performance degradation of approximately 7%–21% for the programs running on the EC2 nodes compared to running them on the NCSA cluster compute node.

Further results and an in-depth analysis showed that message-passing latencies and bandwidth are an order of magnitude inferior between EC2 compute nodes compared to between compute nodes on the NCSA cluster. Walker (2008) concluded that substantial improvements could be provided to the HPC scientific community if a high-performance network provisioning solution can be devised for this problem.

LINPACK Benchmark on Amazon Cluster Compute Instances

In July 2010, Amazon announced its Cluster Compute Instances (CCI 2010) specifically designed to combine high compute performance with high performance network capability to meet the needs of HPC applications. Unique to Cluster Compute instances is the ability to group them into clusters of instances for use with HPC applications. This is particularly valuable for those applications that rely on protocols like Message Passing Interface (MPI) for tightly coupled inter-node communication. Cluster Compute instances function just like other Amazon EC2 instances but also offer the following features for optimal performance with HPC applications:

- When run as a cluster of instances, they provide low latency, full bisection 10 Gbps bandwidth between instances. Cluster sizes up through and above 128 instances are supported.
- Cluster Compute instances include the specific processor architecture in their definition to allow developers to tune their applications by compiling applications for that specific processor architecture in order to achieve optimal performance.

The Cluster Compute instance family currently contains a single instance type, the Cluster Compute Quadruple Extra Large with the following specifications: 23 GB of memory, 33.5 EC2 Compute Units (2 x Intel Xeon X5570, quad-core "Nehalem" architecture), 1690 GB of instance storage, 64-bit platform, and I/O Performance: Very High (10 Gigabit Ethernet).

As has been benchmarked by the Lawrence Berkeley Laboratory team (2010), some applications can expect 10x better performance than on standard EC2. For the Linpack benchmark, they saw 8.5x compared to similar clusters on standard EC2 instances. On an 880-instance CC1 cluster, Linpack achieved a performance of 41.82 Tflops, bringing EC2 at #146 in the June 2010 Top 500 rankings.

MATLAB on Amazon Cluster Compute Instances

Another recent example for HPC on EC2 CCI comes form the MATLAB team at MathWorks (MATLAB 2010) which tested performance scaling of the backslash ("\") matrix division operator to solve for x in the equation $A*x = b$. In their testing, matrix A occupies far more memory (290 GB) than is available in a single high-end desktop machine—typically a quad core processor with 4-8 GB of RAM, supplying approximately 20 Gigaflops.

Therefore, they spread the calculation across machines. In order to solve linear systems of equations they need to be able to access all of the elements of the array even when the array is spread across multiple machines. This problem requires significant amounts of network communication, memory access, and CPU power. They scaled up to a cluster in EC2, giving them the ability to work with larger arrays and to perform calculations at up to 1.3 Teraflops, a 60X improvement. They were able to do this without making any changes to the application code.

Each Cluster Compute instance runs 8 workers (one per processor core on 8 cores per instance). Each doubling of the worker count corresponds to a doubling of the number of Cluster Computer instances used (scaling from 1 up to 32 instances). They saw near-linear overall throughput (measured in Gigaflops on the y axis) while increasing the matrix size (the x axis) as they successively doubled the number of instances.

Cloud User Scenario: Astronomic Data Processing on Amazon EC2

The following cloud user scenario has been taken from (Ahronovitz 2010): Gaia is a mission of the European Space Agency (ESA) that will conduct a survey of one billion stars in our galaxy (Gaia 2010). It will monitor each of its target stars about 70 times over a five-year period, precisely charting their positions, distances, movements, and changes in brightness. It is expected to discover hundreds of thousands of new celestial objects, such as extra-solar planets and failed stars called brown dwarfs.

This mission will collect a large amount of data that must be analyzed. The ESA decided to prototype a cloud-based system to analyze the data. The goals were to determine the technical and financial aspects of using cloud computing to process massive datasets. The prototype system contains the scientific data and a whiteboard used to publish compute jobs. A framework for distributed computing (developed in house) is used for job execution and data processing. The framework is configured to run AGIS (Astrometric Global Iterative Solution). The process runs a number of iterations over the data until it converges. For processing, each working node gets a job description from the database, retrieves the data, processes it and sends the results to intermediate servers. The intermediate servers update the data for the following iteration.

The prototype evaluated 5 years of data for 2 million stars, a small fraction of the total data that must be processed in the actual project. The prototype went through 24 iterations of 100 minutes each, equivalent to running a Grid of 20 Virtual Machines (VMs) for 40 hours. For the full billion-star project, 100 million primary stars will be analyzed along with 6 years of data, which will require running the 20 VM cluster for 16,200 hours. To evaluate the elasticity of a cloud-based solution, the prototype ran a second test with 120 high CPU extra large VMs. With each VM running 12 threads, there were 1440 processes working in parallel.

All of the VMs were running standard operating systems and none of the software used in the project is cloud-specific. The portability concern for this application would be the ability to migrate those VM images to another provider without having to rebuild or reconfigure the images.

The estimated cost for the cloud-based solution is less than half the cost of an in-house solution. That cost estimate does not include the additional electricity or system administration costs of an in-house solution, so the actual savings will be even greater. Storage of the datasets will be cloud-based as well.

CONCLUSION: GRIDS VERSUS CLOUDS FOR HPC

Time and again, people ask questions like "Will HPC codes move to the cloud?" or "Now that cloud computing is well accepted, are Grids dead?" or even "Should I now build my Grid in the cloud?" Despite all the promising developments in the Grid and cloud computing space, and the avalanche of publications and talks on this subject, many people still seem to be confused and hesitant to take the next step. A number of issues are driving this uncertainty, (Gentzsch, 2009), which are discussed in the following.

Grids didn't keep all their promises. Grids did not evolve into the next fundamental IT infrastructure for mainstream HPC, as had been anticipated by some experts. Because of the diversity of computing environments different middleware stacks (for department, enterprise, global, compute, data, sensors, instruments, etc.) had to be developed, and had to face different usage models with different benefits. HPC Grids are providing better resource utilization and flexibility, while global Grids are best suited for complex R&D application collaboration and resource sharing. For enterprise usage, setting up and operating Grids was often too complicated. For R&D experts this characteristic was seen to be a necessary evil: implementing complex HPC applications has never been easy.

Grid: the way station to the cloud. After 40 years of dealing with HPC, Grid computing was indeed the next big thing for the grand challenge, big-science researcher, while for the enterprise CIO, the Grid was a way station on its way to the cloud model. For the enterprise today, private and public clouds are providing all the missing pieces: easy to use, economies of scale, business elasticity up and down, and pay-as you go and thus getting rid of some capital expenditure (CapEx), but still concerned of removing the roadblocks mentioned above. And in cases where security matters, there is always the private cloud solution. In more complex HPC environments, with applications running under different policies, private clouds can easily connect to public clouds into a hybrid cloud infrastructure, to balance security with elasticity and efficiency.

Different policies, what does that mean? No HPC simulation job is alike. Jobs differ by priority, strategic importance, deadline, budget, IP and licenses. In addition, the nature of the code often necessitates a specific computer architecture, operating system, memory, and other resources. These important factors influence where and when a job is running. For any new type of job, a set of specific requirements decide on the set of specific policies that have to be defined and programmed into the scheduler, such that any of these jobs will run according to these policies. Ideally, this is guaranteed by a dynamic resource broker that controls submission to Grid or cloud resources, be they local or global, private or public.

Grids or clouds? One important question is still open: how do I find out, and then 'tell' the resource broker, whether my application should run on the Grid or in the cloud? The answer, among others, depends on the algorithmic structure of the compute-intensive part of the program, which might be intolerant of high latency and low bandwidth as they are often present in public clouds. This has been observed with benchmark results

(Walker, 2008). The performance limitations in clouds are exhibited mainly by parallel applications with tightly-coupled, data-intensive interprocess communication, running on hundreds or even thousands of processor cores.

The good news is, however, that many HPC applications do not require high bandwidth and low latency. Examples are parameter studies (sweeps) often seen in science and engineering, with one and the same application executed for a spectrum of parameters, resulting in many independent jobs, such as analyzing the data from a particle physics collider, identifying the solution parameter in numerical optimization, ensemble runs to quantify climate model uncertainties, identifying potential drug targets via screening a database of ligand structures, studying economic model sensitivity to parameters, simulating flow around an airplane wing with different angels of attach, and analyzing different materials and their resistance in crash tests, to name just a few.

HPC needs Grids and clouds. According to the DEISA Extreme Computing Initiative (DECI, 2010), there are plenty of complex grand challenge science and engineering applications that can only run effectively on the largest and most expensive supercomputers. Today, nobody would build an HPC cloud for these particular big-science grand-challenge applications. It simply isn't a profitable business: the "HPC market" is far too small and thus lacks economy of scale. In some specific science application scenarios, with complex workflows consisting of different tasks (workflow nodes), a hybrid infrastructure might make sense: cloud capacity resources combined with HPC capability nodes, providing the best of both worlds.

However, for a wide range of HPC applications like the parameter-sweeps mentioned above, clouds will be the way to go. We already see more and more HPC clouds today like Exa PowerFLOW (Exa, 2008), and Cyclone (SGI, 2010) which offers cloud services for engineering and scientific applications like BLAST, Gaussian, STAR-CCM+, and LS-DYNA.

CONCLUSIONS: TEN RULES FOR BUILDING SUSTAINABLE GRID AND CLOUD E-INFRASTRUCTURES FOR HPC APPLICATIONS

Grid-enabled applications require sustainable Grid infrastructures. It doesn't make any sense, for example, in a three-year funded Grid project, to develop or port a complex application to a Grid which will shut down after the project ends. We have to make sure that we are able to build sustainable Grid infrastructures which will last for a long time. Therefore, in the following, the author offers 'his' 10 rules for building a sustainable Grid or cloud infrastructure, originally presented in the OGF Thought Leadership Series (2008). These rules are derived from mainly four sources: research on major Grid projects published in a RENCI report (Gentzsch, 2007a), the e-IRG Workshop on "A Sustainable Grid Infrastructure for Europe" (Gentzsch, 2007b), the 2nd International Workshop on Campus and Community Grids at OGF20 in Manchester (McGinnis, 2007), and personal experience with coordinating the German D-Grid Initiative (D-Grid, 2008). The 10 rules are mainly non-technical, because we believe most of the challenges in building and operating a Grid are in the form of mental, cultural, legal and regulatory barriers. Although these rules have been derived originally for successfully building a sustainable Grid infrastructure, recent experience with cloud computing shows that most of these rules still hold for introducing, building or connecting to a cloud infrastructure.

Rule 1: Identify your specific benefits. Your first thought should be about the benefits for your users and your organization. What's in it for them? Identify the benefits which fit best: transparent access to and better utilization of resources; almost infinite compute and storage capacity; flexibility, adaptability and automation through dynamic and concerted interoperation of networked resources, in-house or from a public cloud; cost reduction through utility model; shorter time-to-market because of more simulations at the same time on the

Grid or in the cloud. Grid and cloud technologies help to adjust an enterprise's IT architecture to real business requirements (and not vice versa). For example, global companies will be able to decompose their highly complex processes into modular components of a workflow which can be distributed around the globe such that on-demand availability and access to suitable workforce and resources are assured, productivity increased, and cost reduced. Application of Grid and cloud technologies in these processes, guarantees seamless integration of and communication among all distributed components and provides transparent and secure access to sensitive company information and other proprietary assets, world-wide. Grid and cloud computing is especially of great benefit for those research and business groups which cannot afford expensive IT resources. It enables engineers to remotely access any IT resource as a utility, to simulate any process and any product (and product life cycle) before it is built, resulting in higher quality, increased functionality, and cost and risk reduction.

Rule 2: Evangelize your decision makers first. They are the ones who give you the money and authority for your project. The more they know about the project and the more they believe in it (and in you) the more money and time you will get, and the easier becomes your task to lead and motivate your team and to get things done. Present a business case (current deficiencies, specific benefits of the Grid and/or cloud (see Rule #1), how much will it cost and how much will it return, etc). They might also have to modify existing policies, top down, to make it easier for users (and providers) to cope with the challenges of and to accept and use these new services. For example, why would a researcher (or a department in an enterprise) stop buying computers when money continues to be allocated for buying it (CapEx)? This policy should be changed to support a utility model instead of an ownership model (OpEx). If you are building a national e-Infrastructure, for example, convincing your government to modify its research funding model is a tough task.

Rule 3: Don't re-invent wheels. In the early Grid days, many Grid projects tried to develop the whole software stack themselves: from the middleware layer, to the software tools, to grid-enabling the applications, to the portal and Web layer…and got troubled by the next technology change or by experts leaving the team. Today, so many Grid technologies, products and projects exist already that you first want to start looking for similar projects, select your favorite (most successful) ones which fit best your users' needs, and 'copy' what they have built: and that will be your prototype. Consider, however, that all Grids are different. For example, research Grids are mainly about sharing (e.g. sharing resources, knowledge, data) and collaboration, commercial enterprise Grids are about reducing cost and increasing productivity and revenue.

Rule 4: Keep It Simple. It took your users years to get acquainted with their current working environment, tools, and applications. Ideally, you won't change that. Try hard to stick with what they have and how they do things. Plan for an incremental approach and lots of time listening and talking. Social effects dominate in Grids and in clouds. Join forces with the system people to change/modify mainly the lower layers of the architecture. Your users are your customers, they are king. Differentiate between two groups of users: the end users who are designing and developing the company's products (or the research results) which account for all the earnings of your company (or reputation and therefore funding for your research institute), and the system experts who are eager to support the end users with the best possible services.

Rule 5: Evolution, not revolution. As the saying goes: "never change a running system". We all hate changes in our daily lives, except when we are sure that things will drastically improve. Your users and their applications deeply depend on a reliable infrastructure. So, whenever you have to change especially the user layer, only change it in small steps and in large time cycles. And, start with enhancing existing service models moderately, and test suitable utility models first as pilots. And, very important, part of your business plan has to be an excellent training and communications strategy.

Rule 6: Establish a governance structure. Define clear responsibilities and dependencies for specific tasks, duties and people during and after the project. An advisory board should include all stakeholders (e.g. your representatives of your end-users as well as application and system experts). In case of more complex projects, e.g. consisting of an integration project and several application or community projects, an efficient management board should lead and steer coordination and collaboration among the projects and the working groups. The management board (Steering Committee) should consist of leaders of the sub-projects. Regular face-to-face meetings are very important.

Rule 7: Money, money, money. Don't have unrealistic expectations that Grid and/or cloud computing will save you money from the start. In their early stage, Grid and cloud projects need enough funding to get over the early-adopter phase into a mature state with a rock-solid e-Infrastructure such that other user communities can join easily. In research grids, for example, we estimate this funding phase currently to be in the order of 2-3 years, with more funding in the beginning for the Grid infrastructure, and later more funding for the application communities. In larger (e.g. global) research grids, funding must cover Teams or Centers of Excellence, for building, managing and operating the e-Infrastructure, and for middleware tools, application support, training, and dissemination. Also, most of today's funding models in research and education are often project based and thus not ready for a utilitarian approach where resource usage is based on a pay-as-you-go approach. Old funding models first have to be adjusted accordingly before a utility model can be introduced successfully. For example, today's existing government funding models are often counter-productive when establishing new and efficient forms of utilitarian services (see Rule #2). In the long run, Grid and cloud computing will save you money through a much more efficient, flexible, reliable, and productive infrastructure.

Rule 8: Secure some funding for the post-project phase. Continuity especially for

Maintenance, support, and dissemination (the latter to attract more users) are extremely important for the sustainability of your Grid infrastructure. Make sure already at the beginning of your project that additional funding will be available after the end of the project, to guarantee service and support and continuous improvement and adjustment of the infrastructure.

Rule 9: Try not to grid-enable your applications in the first place. Adjusting your application to changing hardware and software technologies costs a lot of effort and money, and takes a lot of your precious time. Did you 'macro-assemble', vectorize, multitask, parallelize, or multithread your application yourself in the past? Then, grid-enabling such a code is relatively easy, as we have seen in this article before. But doing this from scratch is not what a user should do. Better to use the money to buy (lease, rent, subscribe to) software as a service or to hire a few experienced consultants who grid-enable your application and/or (even better) help you enable your Grid architecture to dynamically cope with the application and user requirements (instead vice versa). Today, in grids, or in Grid workflows, we are looking more at chunks of independent jobs, (or chunks of transactions). And we let our schedulers and brokers decide how to distribute these chunks onto the best-suited and least-loaded servers in the Grid or in the cloud, or let the servers decide themselves in an over-load situation to share the chunks with their neighboring servers automatically whenever they become available.

Rule 10: Adopt a 'human' business model. Don't invent new business models. This usually increases the risk for failure. Learn from the business models we have with our other service infrastructures: water, gas, telephony, electricity, mass transportation, the Internet, and the World Wide Web. Despite this wide variety of areas, there is only a handful of successful business models: on one end of the spectrum, you pay the total price, and the whole thing is yours (CapEx). Or you pay only a share of it, but pay the other share on a per usage basis. Or you rent everything, and pay chunks back on a regular basis, like a subscription fee or leasing. Or you pay just for what you use (OpEx). Sometimes, however, there are 'hidden' or secondary applications. For example, electrical power alone doesn't help. It's only useful if it generates something, e.g. light, or heat, or cooling. And this infrastructure is what creates a whole new industry of appliances: light bulbs, heaters, refrigerators, and so on. Back to Grids and clouds: providing the right (transparent) infrastructure (services) and the right (simple) business model will most likely create a new set of services which most probably will improve our quality of life in the future.

REFERENCES

Ahronovitz, M., et al. (2010). *Cloud computing use cases*. A white paper produced by the Cloud Computing Use Case Discussion Group. Retrieved from http://groups.google.com/ group/ cloud-computing-use-cases

Altintas, I., Berkley, C., Jaeger, E., Jones, M., Ludascher, B., & Mock, S. (2004). Kepler: An extensible system for design and execution of scientific workflows. *Proceedings of the 16th International Conference on Scientific and Statistical Database Management* (SSDBM), Santorini Island, Greece. Retrieved from http://kepler-project.org

Amazon Elastic Compute Cloud EC2. (2007). Retrieved from www.amazon.com/ec2

Badia, R. M., Labarta, J. S., Sirvent, R. L., Perez, J. M., Cela, J. M., & Grima, R. (2003). Programming Grid applications with GRID Superscalar. *Journal of Grid Computing*, *1*, 151–170. doi:10.1023/B:GRID.0000024072.93701.f3

Baker, S. (2007, December 13). Google and the wisdom of clouds. *Business Week*. Retrieved from www.businessweek.com/magazine/ content/07_52/ b4064048925836.htm

BEinGRID. (2008). *Business experiments in grids*. Retrieved from www.beingrid.com

Beltrame, F., Maggi, P., Melato, M., Molinari, E., Sisto, R., & Torterolo, L. (2006). SRB data Grid and compute Grid integration via the EnginFrame Grid portal. *Proceedings of the 1st SRB Workshop*, 2-3 February 2006, San Diego, USA. Retrieved from www.sdsc.edu/srb/Workshop / SRB-handout-v2.pdf

BIRN. (2008). *Biomedical Informatics Research Network*. Retrieved from www.nbirn.net/index.shtm

Buyya, R., Abramson, D., & Giddy, J. (2000). Nimrod/G: An architecture for a resource management and scheduling system in a global computational grid. *Proceedings of the 4th International Conference on High Performance Computing in the Asia-Pacific Region*. Retrieved from www.csse.monash.edu.au /~davida/nimrod/ nimrodg.htm

CCI. (2010). *Amazon cluster compute instances*. Retrieved from http://aws.amazon.com/ hpc-applications/

CDO^2. (2008). *CDOSheet for pricing and risk analysis*. Retrieved from www.cdo2.com

Chaubal, C. (2003). *Sun Grid engine enterprise edition—software configuration guidelines and use cases*. Sun Blueprints. www.sun.com/blueprints /0703/817-3179.pdf

D-Grid. (2008). Retrieved from www.d-grid.de/ index.php?id=1&L=1

DECI. (2010). *DEISA extreme computing initiative*. Retrieved from www.deisa.eu/science/deci

DEISA. (2010). *Distributed European infrastructure for supercomputing applications*. Retrieved from www.deisa.eu

DESHL. (2008). *DEISA services for heterogeneous management layer*. Retrieved from http:// forge.nesc.ac.uk/projects /deisa-jra7/

Dongarra, J., Foster, I., Fox, G., Gropp, W., Kennedy, K., Torczon, L., & White, A. (2003). *Sourcebook of parallel computing*. Morgan Kaufmann Publishers.

EnginFrame. (2008). *EnginFrame Grid and cloud portal*. Retrieved from www.nice-italy.com

Exa. (2008). *PowerFLOW on demand*. Retrieved from http://www.exa.com/ pdf/IBM_Exa_OnDemand _Screen.pdf

Foster, I. (2000). Internet computing and the emerging grid. *Nature*. Retrieved from www. nature.com/nature/ webmatters/grid /grid.html

Foster, I. (2002). *What is the grid? A three point checklist*. Retrieved from http://www-fp.mcs.anl. gov/ ~foster/Articles/ WhatIsTheGrid.pdf

Foster, I. Kesselman, & C., Tuecke, S. (2002). *The anatomy of the Grid: Enabling scalable virtual organizations*. Retrieved from www.globus.org/ alliance/ publications/papers/ anatomy.pdf

Foster, I., & Kesselman, C. (Eds.). (1999). *The Grid: Blueprint for a new computing infrastructure*. Morgan Kaufmann Publishers.

Foster, I., & Kesselman, C. (Eds.). (2004). *The Grid 2: Blueprint for a new computing infrastructure*. Morgan Kaufmann Publishers.

Fox, G., Williams, R., & Messina, P. (1994). *Parallel computing works!*Morgan Kaufmann Publishers.

Frey, J., Mori, T., Nick, J., Smith, C., Snelling, D., Srinivasan, L., & Unger, J. (2005). *The open Grid services architecture*, version 1.0. www.ggf. org/ggf_areas _architecture.htm

GAIA. (2010). *European space agency mission*. Gaia overview. Retrieved from http://www.esa. int/esaSC/ 120377_index_0_m.html

GAT. (2005). *Grid application toolkit*. Retrieved from www.gridlab.org/ WorkPackages/wp-1/

Gentzsch, W. (2004). Grid computing adoption in research and industry In Abbas, A. (Ed.), *Grid computing: A practical guide to technology and applications* (pp. 309–340). Charles River Media Publishers.

Gentzsch, W. (2004). Enterprise resource management: Applications in research and industry In Foster, I., & Kesselman, C. (Eds.), *The Grid 2: Blueprint for a new computing infrastructure* (pp. 157–166). Morgan Kaufmann Publishers.

Gentzsch, W. (2007a). *Grid initiatives: Lessons learned and recommendations*. RENCI Report. Retrieved from www.renci.org/publications / reports.php

Gentzsch, W. (Ed.). (2007b). *A sustainable Grid infrastructure for Europe*. Executive Summary of the e-IRG Open Workshop on e-Infrastructures, Heidelberg, Germany. Retrieved from www.e-irg. org/meetings /2007-DE/workshop.html

Gentzsch, W. (2008). *Top 10 rules for building a sustainable Grid*. Grid Thought Leadership Series. Retrieved from www.ogf.org/TLS/?id=1

Gentzsch, W. (2009). *HPC in the cloud: Grids or clouds for HPC?* Retrieved from http://www. hpcinthecloud.com /features/ Grids-or-Clouds-for-HPC-67796917.html

Gentzsch, W., Girou, D., Kennedy, A., Lederer, H., Reetz, J., Riedel, M., … Wolfrat, J. (2011). DEISA – Distributed European infrastructure for supercomputing applications. *Journal on Grid Computing*. Springer.

Gentzsch, W., Kennedy, A., Lederer, H., Pringle, G., Reetz, J., Riedel, M., et al. Wolfrat, J. (2010). DEISA: E-science in a collaborative, secure, interoperable and user-friendly environment. *Proceedings of the e-Challenges Conference e-2010*, Warsaw.

GEONGrid. (2008). Retrieved from www.geongrid.org

Goodale, T., Jha, S., Kaiser, H., Kielmann, T., Kleijer, P., & Merzky, A. … Smith, Ch. (2008). *A simple API for Grid applications* (SAGA). Grid Forum Document GFD.90. Open Grid Forum. Retrieved from www.ogf.org/documents / GFD.90.pdf

Google. (2008). *Google app engine*. Retrieved from http://code.google.com/appengine/

Google Groups. (2010). *Cloud computing*. Retrieved from http://groups.google.ca/ group/ cloud-computing

Grid Engine. (2001). *Open source project*. Retrieved from http://sourceforge.net/ projects/ gridscheduler/

GridSphere. (2008). Retrieved from www.gridsphere.org/ gridsphere/gridsphere

GridWay. (2008). *Metascheduling technologies for the Grid*. Retrieved from www.gridway.org/

Gustafson, J. (1987). Reevaluating Amdahl's law. *Communications of the ACM, 31*, 532–533. doi:10.1145/42411.42415

Jacob, B., Ferreira, L., Bieberstein, N., Gilzean, C., Girard, J.-Y., Strachowski, R., & Yu, S. (2003). *Enabling applications for Grid computing with Globus*. IBM Redbook. Retrieved from www.redbooks.ibm.com /abstracts/ sg246936.html?Open

Jha, S., Kaiser, H., El Khamra, Y., & Weidner, O. (2007). *Design and implementation of network performance aware applications using SAGA and Cactus*. 3rd IEEE Conference on eScience and Grid Computing, Bangalore, India, 10-13 Dec, (pp. 143-150).

Karonis, N. T., Toonen, B., & Foster, I. (2002). MPICH-G2: A Grid-enabled implementation of the Message Passing Interface. *Journal of Parallel and Distributed Computing, 63*, 551–563. doi:10.1016/S0743-7315(03)00002-9

Lederer, H. (2008). DEISA2: Supporting and developing a European high-performance computing ecosystem. *Journal of Physics, 125*. doi:10.1088/ 1742-6596/125/1/011003.

Lee, C. (2003). Grid programming models: Current tools, issues and directions In Berman, G. F., & Hey, T. (Eds.), *Grid computing* (pp. 555–578). USA: Wiley Press. doi:10.1002/0470867167.ch21

Luther, A., Buyya, R., Ranjan, R., & Venugopal, S. (2005). Peer-to-peer Grid computing and a. NET-based Alchemi framework. In M. Guo (Ed.), High performance computing: Paradigm and infrastructure. Wiley Press, USA. Retrieved from www.alchemi.net

MATLAB. (2010). *Amazon Web Services for high-performance cloud computing – MATLAB. Solving Ax=b*. Retrieved from http://aws.typepad.com/aws /2010/09/ high-performance-cloud-computing-nasa-matlab.html

McGinnis, L., Wallom, D., & Gentzsch, W. (Eds.). (2007). *2nd International Workshop on Campus and Community Grids*. Retrieved from http://forge.gridforum.org/ sf/go/doc14617?nav=1

MyGrid. (2008). Retrieved from www.mygrid.org.uk

NEESGrid. (2008). Retrieved from www.nees.org/

Neuroth, H., Kerzel, M., & Gentzsch, W. (Eds.). (2007). *German Grid initiative D-Grid*. Universitätsverlag Göttingen Publishers. Retrieved from www.d-grid.de/ index.php?id=4&L=1

NPB. (2010). *NAS parallel benchmark*. Retrieved from http://www.nas.nasa.gov/Resources /Software/npb.html

OCCI. (2010). *Open Cloud Computing Interface working group at OGF*. Retrieved 2010 from http:// forge.ogf.org/sf/ projects/occi-wg

OGF. (2008). *Open Grid forum*. Retrieved from www.ogf.org

P-GRADE. (2003). *Parallel Grid run-time and application development environment*. Retrieved from www.lpds.sztaki.hu /pgrade/

Perez, J.M., Bellens, P., Badia, R.M., & Labarta, J. (2007). CellSs: Programming the Cell/ B.E. made easier. *IBM Journal of R&D, 51*(5).

Portal, C. H. R. O. N. O. S. (2004). Retrieved from http://portal.chronos.org/ gridsphere/gridsphere

PRACE. (2008). *Partnership for advanced computing in Europe*. Retrieved from www.prace-project.eu/

Proactive. (2005). *Proactive manual*, rev.ed. 2.2. Proactive, INRIA. Retrieved from http://www-sop.inria.fr /oasis/Proactive/

Saara Väärtö, S. (Ed.). (2008). *Advancing science in Europe*. DEISA – Distributed European Infrastructure for Supercomputing Applications. EU FP6 Project. Retrieved from www.deisa.eu/ press/ DEISA-AdvancingScience InEurope.pdf

SAGA. (2006). *SAGA implementation homepage*. Retrieved from http://fortytwo.cct.lsu.edu:8000/ SAGA

Seymour, K., Nakada, H., Matsuoka, S., Dongarra, J., Lee, C., & Casanova, H. (2002). Overview of GridRPC: A remote procedure call API for Grid computing. *Proceedings of the Third International Workshop on Grid Computing* [Baltimore, MD: Springer.]. *Lecture Notes in Computer Science, 2536*, 274–278. doi:10.1007/3-540-36133-2_25

SGI. (2010). *Cyclone: HPC cloud results on demand*. Retrieved from http://www.sgi.com/ products /hpc_cloud/cyclone /index.htm

SIMDAT. (2008). *Grids for industrial product development*. Retrieved from www.scai.fraunhofer. de /about_simdat.html

Soh, H., Shazia Haque, S., Liao, W., & Buyya, R. (2006). Grid programming models and environments In Dai, Y.-S. (Eds.), *Advanced parallel and distributed computing* (pp. 141–173). Nova Science Publishers.

Streit, A., Bergmann, S., Breu, R., Daivandy, J., Demuth, B., & Giesler, A. … Lippert, T. (2009). UNICORE 6, a European Grid technology. In W. Gentzsch, L. Grandinetti, & G. Joubert (Eds.), High-speed and large scale scientific computing, (pp. 157-176). IOS Press.

Sun. (2010). *Sun Network.com, SunGrid, and Sun utility computing, now under Oracle*. Retrieved from www.sun.com/service/sungrid/

SURA Southeastern Universities Research Association. (2007). *The Grid technology cookbook. Programming concepts and challenges*. Retrieved from www.sura.org/cookbook/gtcb/

TAVERNA. (2008). *The Taverna workbench* 1.7. Retrieved from http://taverna.sourceforge.net/

TRIANA. (2003). *The Triana project*. Retrieved from www.trianacode.org/

UNICORE. (2008). *Uniform interface to computing resources*. Retrieved from www.unicore.eu/

Venugopal, S., Buyya, R., & Winton, L. (2004). A Grid service broker for scheduling distributed data-oriented applications on global grids. *Proceedings of the 2nd workshop on Middleware for Grid computing*, (pp. 75–80). Toronto, Canada. Retrieved from www.Gridbus.org/broker

Walker, E. (2008). *Benchmarking Amazon EC2 for high-performance scientific computing*. Retrieved from http://www.usenix.org/ publications/login/ 2008-10/openpdfs/walker.pdf

Chapter 3
Grid-Enabling Applications with JGRIM

Cristian Mateos
ISISTAN - UNCPBA, Argentina

Alejandro Zunino
ISISTAN - UNCPBA, Argentina

Marcelo Campo
ISISTAN - UNCPBA, Argentina

ABSTRACT

The development of massively distributed applications with enormous demands for computing power, memory, storage and bandwidth is now possible with the Grid. Despite these advances, building Grid applications is still very difficult. We present JGRIM, an approach to easily gridify Java applications by separating functional and Grid concerns in the application code, and report evaluations of its benefits with respect to related approaches. The results indicate that JGRIM simplifies the process of porting applications to the Grid, and the Grid code obtained from this process performs in a very competitive way compared to the code resulting from using similar tools.

INTRODUCTION

The Grid (Foster and Kesselman, 2003) is a distributed computing environment in which resources from dispersed sites are virtualized through specialized services to provide applications with vast execution capabilities. Just like an electrical infrastructure, which spreads over cities to convey and deliver electricity, the Grid offers a computing infrastructure to which applications can be easily "plugged" and efficiently executed by leveraging resources of different administrative domains. Precisely, "Grid" comes from an analogy with the electrical grid, since applications will take advantage of Grid resources as easily as electricity is now consumed.

Unfortunately, this analogy does not completely hold yet since it is difficult to "gridify" an application without rewriting or modifying it. A major problem is that most Grid toolkits provide APIs for merely implementing applications from scratch (Mateos et al., 2008a). Examples of

DOI: 10.4018/978-1-60960-603-9.ch003

such toolkits are JavaSymphony (Fahringer and Jugravu, 2005), Java CoG Kit (von Laszewski et al., 2003), GSBL (Bazinet et al., 2007), GAT (Allen et al., 2005) and MyCoG.NET (Paventhan et al., 2006). Hence, the application logic results mixed up with code for using Grid services, making maintainability, testing and portability to different Grid libraries and platforms somewhat hard. Furthermore, gridifying existing code requires to rewrite significant portions of it to use those APIs. These problems are partially addressed by tools that take an executable, along with user parameters (e.g. input arguments, CPU and memory requirements, etc.), and wrap the executable with a component that isolates the details of the Grid. Some tools falling in this category are GEMLCA (Delaittre et al., 2005), LGF (Baliś & Wegiel, 2008) and GridSAM (McGough et al., 2008). However, the output of these tools are coarse grained applications whose execution cannot be configured to make better use of Grid resources (e.g. parallelize and/or distribute individual application components). Overall, this represents a trade-off between ease of gridification versus flexibility to configure the runtime aspects of gridified applications (Mateos et al., 2008a).

To address these issues, we propose JGRIM, a novel method for porting Java applications onto service-oriented Grids, this is, based on Web Services. JGRIM minimizes the requirement of source code modification when gridifying Java applications, and provides simple mechanisms to effectively tune transformed applications. JGRIM follows a *two-step* gridification methodology, in which developers first implement and test the logic of their applications, and then Grid-enable them by undemandingly and non invasively injecting Grid services. Therefore, we conceive gridification as shaping the source code of an ordinary application according to few coding conventions, and then adding Grid concerns to it. In a previous paper (Mateos et al., 2008b), we reported preliminary comparisons between JGRIM and other approaches for gridifying software in terms of source code metrics. In this article we also report JGRIM execution performance on an Internet-based Grid, measuring execution time and network usage of two resource-intensive applications. The rest of the article analyzes the most relevant related works, describes JGRIM, and presents the experimental evaluations.

RELATED WORK

Motivated by the complex and challenging nature of porting conventional applications to the Grid (Gentzsch, 2009), research in tools and methods to easily gridify ordinary software is growing at an astonishingly rate. Besides providing APIs for developing and executing Grid applications, many of these tools actually materialize alternative approaches to support easy gridification of existing applications. For an exhaustive survey on technologies to port applications to the Grid, see (Mateos et al., 2008a). Below we describe a representative subset of such tools.

ProActive (Baduel et al., 2006) is a platform for parallel distributed computing that provides *technical services*, a support which allows users to address non-functional concerns (e.g. load balancing and fault tolerance) by plugging certain external configuration to the application code at deployment time. ProActive applications comprise one or more mobile entities whose creation, migration and lookup are performed by explicit code provisioning. Likewise, the JPPF (2008) framework supports distributed scheduling for CPU-intensive tasks on distributed environments. In both cases, after porting an application to a Grid, the application logic results mixed up with Grid-related code. Therefore, gridification as well as software maintenance thereafter become difficult. Furthermore, GridGain (GridGain Systems, 2008) attempts to minimize this problem by using Java annotations to seamlessly exploit distributed processors. However, GridGain does not target interoperability, and is not aimed at leveraging Grid services provided by other platforms.

JavaSymphony (Fahringer and Jugravu, 2005) provides a semi-automatic execution model that deals with migration, parallelism and load balancing of applications. The model also allows programmers to explicitly control such aspects through API primitives. Similarly, Babylon (van Heiningen, 2008) features weak mobility, remote object communication and parallelism in an uniform programming API. As JavaSymphony and Babylon are API-inspired gridification tools, they require developers to learn another API as well as to perform extensive modifications when gridifying their conventional applications.

Moreover, Ibis (van Nieuwpoort et al., 2005) is a Grid platform designed as an uniform and extensible communication facility on top of which a variety of distributed programming models are implemented. An interesting subsystem of Ibis is Satin (Wrzesinska et al., 2006), which allows developers to straightforwardly execute conventional divide and conquer codes in parallel on clusters and Grids. Similar to JPPF, Ibis offers limited support for using well-established Grid protocols such as WSDL and UDDI (Curbera et al., 2002). Consequently, interoperability is almost absent when using these tools to build Grid applications.

GMarte (Alonso et al., 2006) is a high-level API offering an object-oriented view on top of Globus (Foster, 2005). Developers can employ the API to compose and execute existing binary codes by means of a new Java application. GMarte also features metascheduling capabilities and fault-tolerance via custom checkpointing mechanisms. However, as GMarte treats these codes as black boxes, their structure cannot be altered to make better use of Grid resources, for example, parallelize or distribute portions of the codes. In addition, XCAT (Gannon et al., 2005) supports distributed execution of component-based applications on top of existing Grid platforms (mostly Globus). Application components can also represent legacy binary programs. XCAT provides an API to build workflow applications by assembling service and legacy components. Though this task can be performed with little programming effort, developers still have to manage component creation and linking in their programs. Besides, like GMarte, XCAT does not provide support for fine tuning components at the application level.

All in all, existing toolkits and frameworks for gridifying software can be grouped into two major categories (Mateos et al., 2008a): those that aim at separating application logic from Grid functionality, and those that do not. Our work aligns with the proposals in the former category. However, we believe that these efforts are some way off from being effective tools for gridifying applications. On one hand, those efforts that rely on an API-oriented approach to gridification require modifications to the input applications, which in turn requires developers to learn Grid APIs and negatively affects maintainability and portability. Nevertheless, developers have a deeper control of the internal structure of their applications. Conversely, tools based on gridifying by wrapping or composing existing applications (e.g. GEMLCA, GMarte, XCAT, LGF, GridSAM) simplify gridification, but prevent the usage of tuning mechanisms such as parallelization, mobility and distribution of individual application components. This represents a trade-off between ease of gridification versus true flexibility to configure the runtime aspects of gridified applications.

In this sense, JGRIM tackles this trade-off by avoiding excessive source code modifications when porting applications to the Grid, yet offering means to effectively tune these applications at a high level of abstraction once they have been transformed. Besides, developers are allowed to furnish application component with common Grid concerns such as parallelism and distribution at several levels of granularity. Moreover, JGRIM preserves the integrity of the application logic by allowing developers to seamlessly inject Grid concerns to their applications. This means that,

Figure 1. A layered view of JGRIM

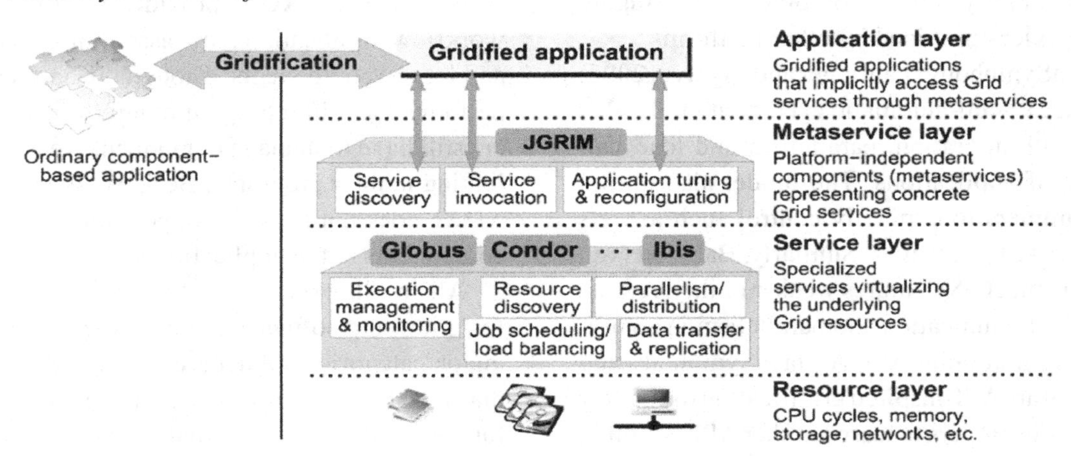

upon gridification, the ordinary application code does not get mixed with Grid-related code. This improves maintainability, testability and portability of the same source code to different Grid APIs and environments. In addition, unlike most of the aforementioned tools, the JGRIM API only have to be explicitly used when performing application tuning and, in such a case, the application logic is not affected. Finally, because of the component-based roots of its programming model, JGRIM is similar to using popular component development models for Java such as JavaBeans or EJBs. Given the widespread adoption of both Java and such models, our approach can benefit a large percentage of today's Java applications.

JGRIM

JGRIM is an approach for creating and deploying conventional applications on service-oriented Grids. Its goal is to allow applications to discover and efficiently use Grid services without requiring developers to provide code for it. JGRIM provides a layer whereby component-based Java applications are effortlessly transformed to applications that are furnished with specialized library components (see Figure 1). These components glue applications and the underlying Grid infrastructure by leveraging the services provided by existing Grid platforms. Conceptually, JGRIM is a software/hardware stack comprising the following layers:

- **Resource**: represents the physical infrastructure of the Grid (resources and transport protocols).
- **Service**: provides sophisticated services to applications (e.g. load balancing, brokering, parallelism, security, etc.) by means of existing Grid platforms (e.g. Ibis (van Nieuwpoort et al., 2005), ProActive (Baduel et al., 2006)) and resource management systems (e.g. Globus (Foster, 2005), Condor (Thain et al., 2003)). The Service layer is often the Grid entry point for gridification under most of the existing approaches, this is, gridified applications directly talk to Grid services.
- **Middleware**: comprises some metaservices that act as a glue between applications and the Grid. A metaservice is a representative of a set of related concrete services. Examples include service discovery, service invocation and application tuning.
- **Application**: contains applications consisting of a number of interacting compo-

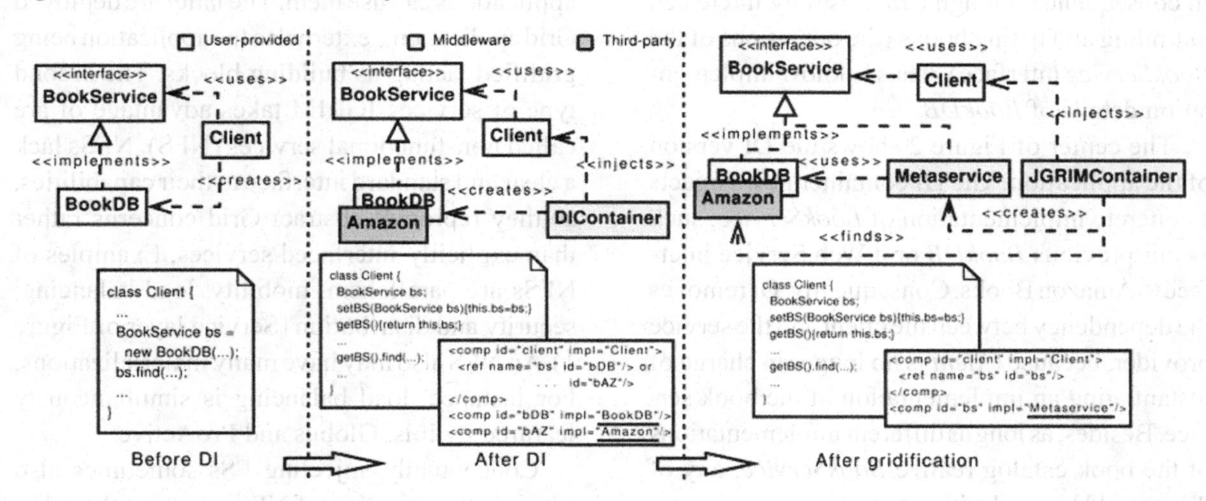

Figure 2. DI and Grid service injection

nents. During gridification, JGRIM alters some of them and their interactions by using metaservices, thus at runtime some operation requests originated by applications at this layer are handled by the Middleware layer.

JGRIM assumes that applications are properly componentized, which is the case of most Java applications. This allows JGRIM to treat an individual application as a collection of interacting components. The idea is to enhance these interactions by using metaservices, for example to add remoting, load balancing or security. In addition, individual components can be enriched too, for example to add parallelism, job scheduling, wrap as a Web Service, etc.

Central to JGRIM is the concept of Dependency Injection (DI) (Johnson, 2005). With DI, components providing services can be transparently injected into components that require these services. JGRIM exploits DI by allowing developers to inject metaservices into ordinary applications. Essentially, JGRIM targets the "ease of gridification versus flexible tuning" trade-off (Mateos et al., 2008a), minimizing the requirement of code modification when porting applications to the Grid, nonetheless providing useful mechanisms to tune Grid applications that give developers control over the way their gridified applications execute on a Grid.

Injecting Grid Services into Conventional Applications

DI achieves higher decoupling in component-based applications by having components described through public interfaces and reducing couplings by delegating the responsibility for component creation and linking to a DI *container* (Johnson, 2005). Put differently, components only know each other's interfaces, but it is up to the DI container to create and set (inject) into a client component an instance of another (provider) component implementing a required interface (center of Figure). A DI container is a runtime platform in charge of binding clients components to providers components.

Consider an application that includes a book catalog (*BookService*) and a client component (*Client*) accessing it (see Figure 2). The catalog may be implemented, for instance, by using a relational database (*BookDB*). Client has to setup a *BookBD* component by providing it with initialization parameters, specifically the location of the database, drivers, user name and password.

In consequence, though *Client* is only interested in finding and listing books (the operations of the *BookService* interface), it has to know implementation details of *BookDB*.

The center of Figure 2 shows the DI version of the application. The DI container now injects a concrete implementation of *BookService*, such as our previous *BookDB* or a Web Service interface to Amazon Books. Consequently, DI removes the dependency between the client and the service provider, because *Client* is no longer in charge of instantiating an implementation of the book service. Besides, as long as different implementations of the book catalog realize *BookService*, any of them could be used without modifying the source code of *Client*.

JGRIM takes DI a step further by introducing an indirection between software components to inject Grid metaservices (right of Figure 2). After gridification the container no longer injects a service implementation into the client but a metaservice, which is for example able to find the fastest service from several implementations residing in the Grid. The client interacts with the metaservice, which in turn interacts with an implementation of the required service. This indirection is transparent to the client: there is no need to change its code, since both the service implementation and the metaservice realize the same interface (*BookService*). Besides discovery, metaservices may add load balancing, fault tolerance, distribution, etc.

From an application perspective, after the metaservice finds a proper service implementation *S*, it becomes a proxy to *S*. A service such as the book catalog, for which many realizations may exist and access is mediated by interfaced metaservices, is called a *functional* service (FS). FSs are entities that expose their functionality through clear interfaces. Within Grids, they are often materialized as Web Services (Atkinson et al., 2005). FSs are categorized as internal or external. The former are parts of a complete application that, during gridification, are exposed so that other applications can use them. The latter are deployed Grid applications, external to the application being gridified, acting as building blocks. The second type of services JGRIM takes advantage of are called non-functional services (NFS). NFSs lack a clear and standard interface to their capabilities, as they represent abstract Grid concerns rather than explicitly-interfaced services. Examples of NFSs are parallelism, mobility, load balancing, security and distribution (Service layer of Figure 1). An NFS also may have many materializations. For instance, load balancing is simultaneously featured by Ibis, Globus and ProActive.

Conceptually, injecting FSs sometimes also requires the injection of NFSs, but not the other way around. For instance, this would be the case of using security mechanisms when contacting or invoking FSs.

Gridification Process

JGRIM prescribes a semi-automatic gridification process that developers have to follow to gridify their applications, which consists of the following steps (see Figure 3):

1. Developers identify application components and dependencies between components that will benefit from the Grid, or the **hot-spots** for gridification within their applications. Conceptually, hot-spot are the portions of an ordinary application to which one or more metaservices are associated.
2. Modification of the application code to obey some simple and standard object-oriented coding conventions, ensuring that application components defined in the previous step are implicitly linked through get/set accessors using the JavaBeans style. Any reference to a component *C* within the code must be done by calling a fictitious method *getC()*, instead of accessing it directly as *C.operation()*. For example, if an application reads data from a file component, it should

Figure 3. JGRIM: gridification process

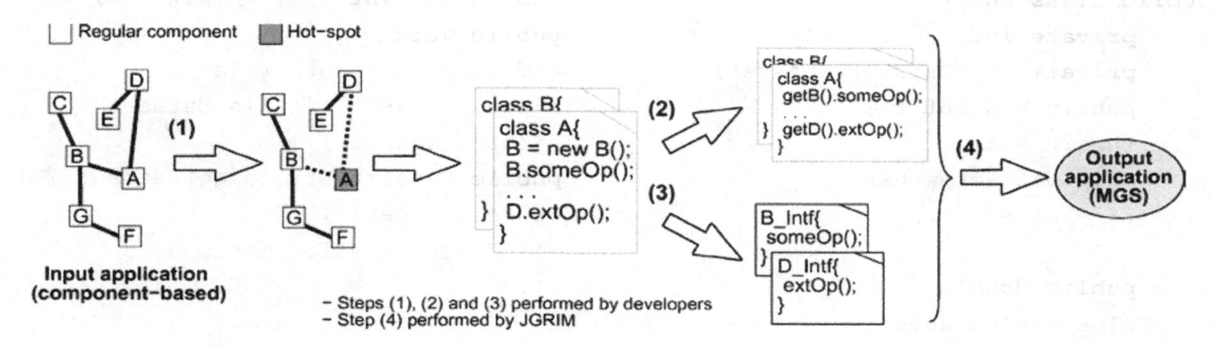

be accessed as *getFile().read()*. JGRIM modifies the application code to include the necessary instance variables and accessors. Since this style is commonplace in Java, this task often requires little or no effort.

3. Definition of internal and external interfaces. Involves separating what a component does from its implementation. Again, for the internal interfaces this is a common practice in Java. For the external interfaces it involves specifying the method signatures for either using third-party services or exporting application components as FSs.
4. Automatic assembling of the outputs of (2) and (3) with metaservices. Basically, the DI-enabled application code is injected with JGRIM API classes by using the Spring DI container (Walls and Breidenbach, 2005). The resulting application is a reactive mobile Grid service (MGS), a service capable of migrating its execution based on environmental conditions (Mateos et al., 2005) such as CPU load, storage availability, network latency, etc. More details on the JGRIM API and its implementation can be found in (Mateos, 2008).

The next section illustrates these steps through the gridification of a concrete application.

Gridifying the k-NN Classifier

The k-NN algorithm (Dasarathy, 1991) is a popular supervised learning technique for mining data. k-NN is computationally intensive, hence it is a suitable application for execution on a Grid. In the next paragraphs, we will gridify it with JGRIM.

k-NN classifies instances by placing them at a point of a multidimensional feature space. k-NN first partitions the space into regions according to class labels of several training samples, or dataset. Then, it assigns the class *C* to a point if *C* is the most frequent label among the *k* nearest training samples.

Let us suppose that the existing implementation of the k-NN algorithm consists of various helper classes plus a *KNN* class with three operations:

- **classifyInstance**: computes the label associated to an instance.
- **classifyInstances**: analogous to classifyInstance but operates on a list of instances.
- **sameClass**: tests whether two instances have the same label.

One of the helper classes provides access to a file-based dataset, which is accessed by these methods. Basically, the structure of the application code is:

```
public class KNN {
      private int k;
      private FileDataset dataset;
      public KNN(int k){
      this.k = k;
      this.dataset = new
      FileDataset();
      }
      public double
      classifyInstance(Instance
      instance) {...}
      public double[]classifyInstances
      (Instance[] instances) {...}
       public boolean
      sameClass(Instance instA,
      Instance instB) {...}
}
public class FileDataset {
      public Instance[] readItems(int
      rowStart, int rowEnd) {...}
      public int size() {...}
      public int dimensions() {...}
}
```

First, we must determine which classes (*KNN*) and interactions between components (*KNN* needs a data resource - the *KNN-FileDataset* interaction) to gridify. Then, we have to separate the implementation of the data resource from its interface, and replace all accesses to *dataset* by *getDataset()*. In the example, the valid operations of *dataset* were defined in *DatasetService*. Finally, we process the code with JGRIM, resulting in:

```
public interface DatasetService {
 public Instance[] readItems(int
 rowStart, int rowEnd);
 public int size();
 public int dimensions();
}
public class KNN extends jgrimapi.
MGS {
 private int k;
 private DatasetService dataset;
 public KNN(int k) { this.k = k; }
 public void
setdataset(DatasetService
injectedDataset) { this.dataset =
injectedDataset };
 public DatasetService getdataset() {
return dataset; }
 /** Classification methods */
 ...
}
```

Note that JGRIM added proper getter/setters for interacting with the dataset. Besides, the resulting source code is very clean, since it was not necessary to use any JGRIM API class for gridification purposes. Moreover, JGRIM generates an XML configuration file:

```
<beans>
 <bean id="knnComponent" class="KNN">
 <property name="dataset"
ref="datasetMetaService"/>
 </bean>
 <bean id="datasetMetaService"
class="jgrimapi.JGRIMServiceDiscov-
erer">
 <property name="requiredInterface"
value="DatasetService"/>
 </bean>
</beans>
```

The XML file links application components and JGRIM metaservices together through DI. Here, *KNN* is decoupled from the dataset implementation by linking it --via the *dataset* property-- with a component that provides runtime Web Service discovery. Currently, service discovery is based on the inspection of UDDI registries (Curbera et al., 2002). Consequently, *KNN* can use any external dataset service of the Grid provided it implements the *DatasetService* interface and is published to a UDDI registry.

So far we have decoupled the storage mechanism of the dataset from the *KNN* class. When the

KNN application is executed, JGRIM searches an appropriate dataset in the Grid and injects it into *KNN*. Besides, JGRIM mediates between these two components, hiding the actual location of the dataset and the communication details. Furthermore, the application is converted into a Grid service capable of transparently migrating its execution. Thus, the application becomes a callable entity that other applications can discover and use.

Now we will use JGRIM for executing KNN in multiple distributed threads to improve its performance. The *sameClass* operation classifies two instances and compares the results:

```
...
c1 = classifyInstance(instA);
c2 = classifyInstance(instB);
return (c1== c2);
```

The calls to *classifyInstance* are independent between each other, thus they can be computed concurrently. Let us exploit this by injecting parallelism into the *sameClass* operation. We have to define an interface for the *classifyInstance* operation:

```
public interface Classifier {
 public double
 classifyInstance(Instance instance);
}
```

After processing the code with JGRIM, a new component is added to the XML file:

```
<beans>
 <bean id="knnComponent" class="KNN">
 ...
 <property name="classifier"
ref="spawnerMetaService"/>
 </bean>
 ...
 <bean id="spawnerMetaService"
class="jgrimapi.JGRIMMethodSpawner">
 <property name="spawnableMethods"
value="Classifier"/>
 </bean>
</beans>
```

JGRIMMethodSpawner[1] parallelizes the invocations to the methods specified by the *Classifier* interface. Also, the programmer must replace the calls to *classifyInstance* by calls to a fictitious *getclassifier* method. Future uses of *c1* and *c2* will block the execution of *sameClass* until their values are computed by JGRIMMethodSpawner. This coordination is supported through Java futures, which are available in the java.util.concurrent package of the JVM since version 5.0. Behind scenes, JGRIM installs *classifyInstance* in several computers of the Grid and dynamically finds idle computers to execute one invocation per computer. This is, JGRIM not only parallelizes KNN, but also distributes its execution. For supporting parallelism on Grids, JGRIM relies on Satin (Wrzesinska et al., 2006), a subsystem of Ibis that is designed to execute embarrasingly parallel computations on distributed environments. Furthermore, a spawner based on raw, local threads is also available.

Besides parallelism, JGRIM allows developers to tune applications by using code mobility and policies based on environmental conditions. To briefly illustrate this mechanism, let us suppose our MGS is deployed on a Grid of several sites each hosting a replica of the dataset. Let us additionally assume that bandwidth between sites could drastically vary along time.

As KNN works by reading data blocks from the dataset and then performing computations on them, bandwidth indirectly affects response time. Particularly, accessing a replica through a busy network channel might decrease performance. JGRIM metaservices, unless otherwise indicated, assume that the best service instance is always the one offering the highest throughput. Through a policy, we can redefine what "best" means to an application, i.e. the highest transfer capabilities in our scenario.

To specify a policy for the dataset resource, we will attach to *DatasetInterface* a new class that implements four operations:

```
public class DatasetPolicy extends
jgrimapi.Policy {
 private boolean initialized = true;
 public String accessWith(String
methodA, String methodB){
 return jgrimapi.Constants.INVOKE;
 }
 public String accessFrom(String
siteA, String siteB){
 double trA = jgrimapi.Profiler.
instance().profile("bandwidth",
"localhost", siteA);
 double trB = jgrimapi.Profiler.
instance().profile("bandwidth",
"localhost", siteB);
 return (trA < trB) ? siteA: siteB;
 }
 public void before(){
 if (!this.initialized) {
 getOwnerMGS().move(jgrimapi.
Profiler.instance().idlestSite());
 this.firstEval = false;
 }
 }
 public void after(){...}
}
```

For simplicity, we have omitted the XML configuration that is generated by JGRIM to inject this policy into the application. The policy mechanism works as follows: upon each call to the dataset, *DatasetPolicy* is evaluated, which instructs KNN (through *accessWith* and *accessFrom* methods) to remotely contact the service replica which is hosted at the site that offers the best bandwidth (alternatively, the KNN application could be moved to that site). Methods *before* and *after* are used to perform initialization/disposal tasks before/after an individual evaluation of the policy takes place. For example, the policy causes KNN to migrate to the idlest site upon the first evaluation. Like any component, policies can maintain state (e.g. the initialized variable), and be associated just to single interface operations. Overall, by adding a simple policy, KNN is able to smartly interact with the dataset. Policy coding is not mandatory and, even more important to our work, it does not affect the application logic.

EVALUATION AND DISCUSSION

To provide empirical evidence about the practical soundness of our approach, we conducted a comparison between JGRIM, ProActive and Satin. In short, these tools were separately employed to gridify existing implementations of two different applications, namely the k-NN explained in past paragraphs, and an application for panoramic image restoration based on the enhancement algorithm proposed in (Tschumperlé and Deriche, 2003). After gridification, representative code metrics on the Grid-aware applications were taken to quantitatively analyze how difficult is to port a Java application to a Grid with either of the three alternatives. Besides, experiments were conducted to evaluate the performance of JGRIM applications with respect to the other two approaches.

The restoration application was originally implemented as a master component responsible for splitting/joining images, plus worker components for carrying out the CPU intensive processing, this is, running the actual restoration algorithm on individual portions of the whole panoramic image. Experiments were performed on a Grid comprising three Internet-connected clusters (see Figure 4). Each cluster hosted a replica of the k-NN dataset wrapped with a Web Service. For the sake of fairness, all gridified codes used the replicated datasets. Both the original codes of k-NN and the restoration application were implemented by an experienced Java programmer. On the other side, gridification was performed by a different developer with similar skills in Java programming

Figure 4. Grid used for the experiments

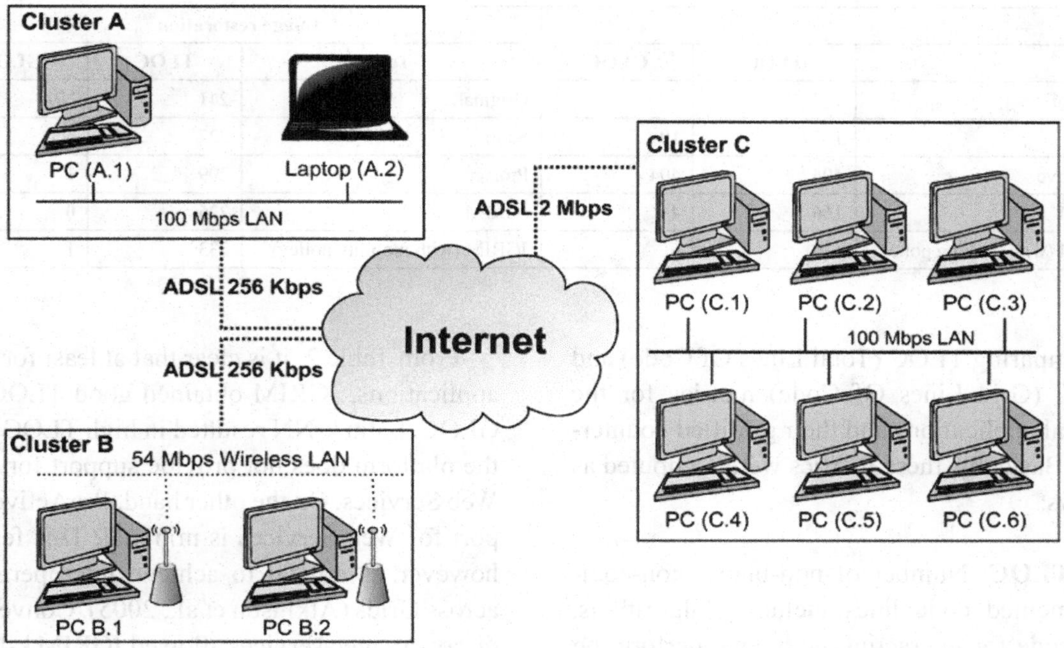

Table 1. Hardware specifications of the Grid machines

Cluster name	Machine name	CPU model	CPU frequency	Memory (MB)
A	A.1	AMD Athlon XP 2200+	1.75 Ghz.	256
	A.2	Intel Core2 T5600	1.83 Ghz. (per core)	1.024
B	B.1	AMD Sempron	1.90 Ghz.	512
	B.2	AMD Athlon 64 X2 Dual Core 3.600+	2.00 Ghz. (per core)	1.024
C	C.1	Intel Pentium 4	2.80 Ghz.	512
	C.2	Intel Pentium III (Coppermine)	852 Mhz.	256
	C.3	Intel Pentium III (Coppermine)	852 Mhz.	256
	C.4	Intel Pentium III (Coppermine)	852 Mhz.	384
	C.5	Intel Pentium III (Coppermine)	852 Mhz.	384
	C.6	Intel Pentium III (Coppermine)	798 Mhz.	256

but minimal background on JGRIM, ProActive and Satin. All experiments were performed during nighttime (from 11 P.M. to 8 A.M.), when the Internet traffic is low and the network latency has little variability.

Table 1 details the CPU and memory specifications of the nodes of the previous Grid setting. Machines were equipped with Ubuntu Linux (kernel version 2.6.20) and the Sun JDK 1.5.0. The reason of using such an heterogeneous hardware was to establish a realistic Grid testbed for the experiments.

We assessed the impact of gridification on the application code when employing the three tools

Table 2. Test applications: code metrics

k-NN			Image restoration		
Tool	**TLOC**	**GLOC**	**Tool**	**TLOC**	**GLOC**
Original	192	N/A	Original	241	N/A
Satin	1477	10	Satin	227	5
ProActive	404	404	ProActive	299	17
JGRIM	**166**	**4**	JGRIM	**226**	**0**
JGRIM (with caching policy)	179	6	JGRIM (with mobility policy)	233	1

by comparing TLOC (Total Lines Of Code) and GLOC (Grid Lines Of Code) metrics for the original applications and their gridified counterparts. Basically, these metrics were computed as follows:

- **TLOC**: Number of non-blank, non-commented code lines including algorithms, code for interacting with data, performing Grid exception handling, and parallelism. Note that this metric is closely related to the extra effort necessary to adapt the ordinary version of the applications to execute on our Grid.
- **GLOC**: Number of lines within the code of a gridified application that explicitly access the underlying Grid platform API. Intuitively, the larger the GLOC, the more the time a developer spends learning the API. In addition, greater GLOC means the application is more tied to a specific Grid library.

Before measuring, all codes were uniformly formatted with the help of the Eclipse SDK. Table 2 summarizes the obtained values for these metrics (lower values are better). Moreover, for the JGRIM applications we obtained two variants by implementing a caching policy for k-NN, which stores dataset accesses to reduce network traffic, and a mobility policy for the image application, which explicitly moves application components to reduce network latency.

From Table 2, it is clear that at least for these applications, JGRIM obtained good TLOC and GLOC. Satin k-NN resulted in high TLOC since the platform does not provide support for using Web Services. On the other hand, ProActive support for Web Services is minimal. This feature, however, is crucial to achieve interoperability across Grids (Atkinson et al., 2005). Conversely, discovery metaservices allowed JGRIM k-NN to delegate dataset discovery and access to the underlying platform, discarding the code for using a file-based dataset present in the original k-NN application. Moreover, achieving parallelism (i.e. classify several instances in parallel) with Satin and ProActive demanded more API code. Remarkably, unlike its competitors, the JGRIM API was only used for coding policies, not affecting the original codes. These facts suggest that using JGRIM may lead to more maintainable and portable Grid code, since JGRIM effectively pushes most of the code for handling Grid-specific concerns out of the application logic. Besides, the lower GLOC values of the JGRIM applications indicate that JGRIM is appropriate for users not proficient in JGRIM or even Grid technologies, as the amount of API functionality that is necessary to learn before using the tool is much less compared to employing Satin and ProActive.

To evaluate the performance and resource usage of the Grid-enabled codes, each gridified version of k-NN was used to classify several list of input instances with different sizes (5, 10, 15, 20 and 25 instances). For the image application

Figure 5. AET of the k-NN application (left) and the image restoration application (right)

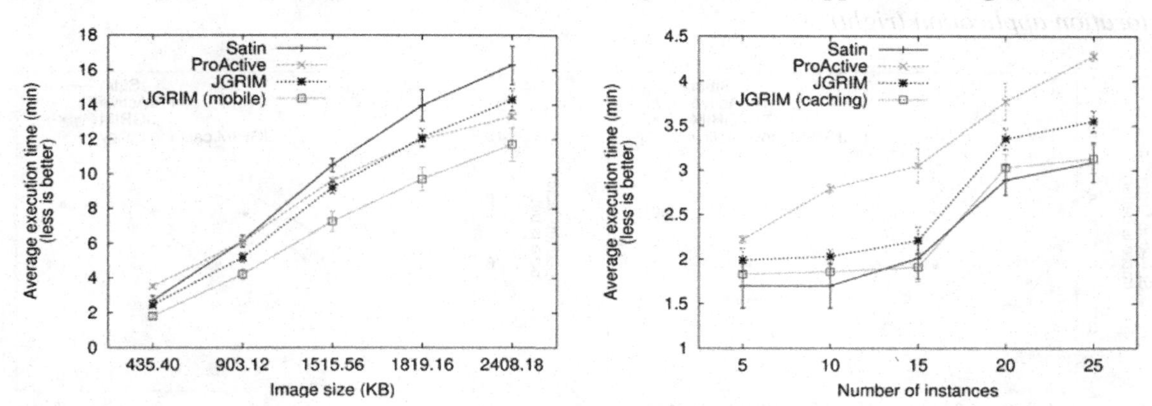

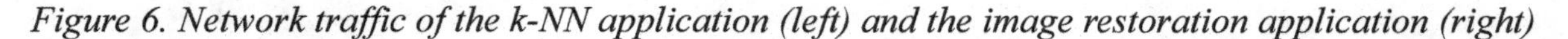

Figure 6. Network traffic of the k-NN application (left) and the image restoration application (right)

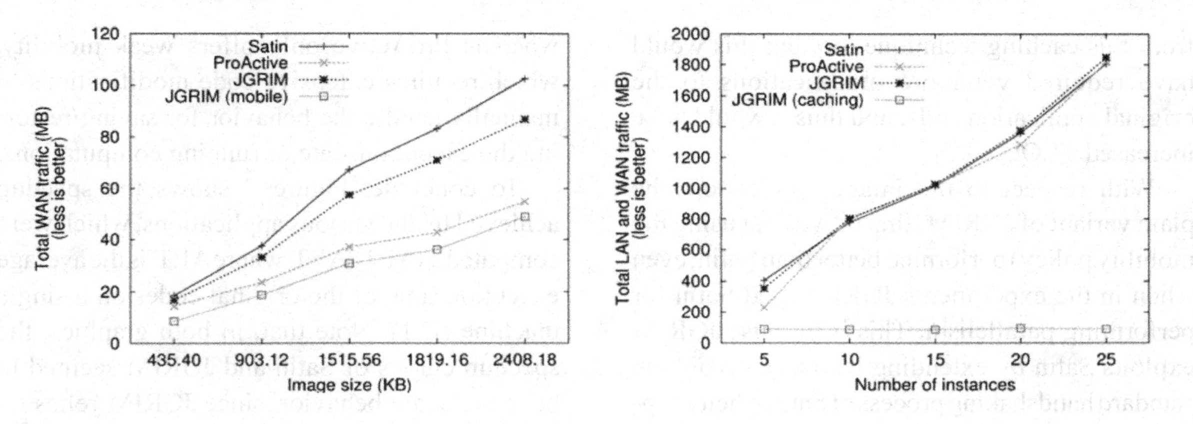

we used five pictures of various sizes (0.4 MB, 0.9 MB, 1.5 MB, 1.8 MB and 2.4 MB). We averaged the execution time (AET) and accumulated the network traffic for 10 executions per test (deviations were around 5%). Loopback network traffic was filtered out, as it does not consume bandwidth and it is negligible compared to LAN and WAN traffic. To capture network traffic, we used the *tcpdump*[2] network monitoring program. Figures 5 and 6 show the obtained results. As expected, JGRIM behaved similar to the alternatives. Besides, the use of JGRIM policies (caching and mobility) greatly improved both performance and network usage.

When not using the caching policy, the JGRIM variant of k-NN incurred in a performance overhead of 10-15% compared to its Satin counterpart. However, this overhead was associated to perform service discovery, a key Grid feature that is not present in Satin and ProActive. Besides, caching allowed JGRIM to continue using discovery -which intuitively translates into overhead- and at the same time to stay very competitive. Furthermore, ProActive k-NN performed poorly. Roughly, ProActive is strongly oriented towards simplifying the deployment of Grid applications, which contributes to make application setup slower. In principle, these results suggest that ProActive is not suitable for applications whose execution time is similar than their setup time. Moreover, caching significantly reduced network traffic, which is a consequence of performing less remote dataset accesses. It is worth noting that Satin and ProActive k-NN might have benefited

Figure 7. Speedups achieved by the grid-enabled versions of the k-NN application (left) and the image restoration application (right)

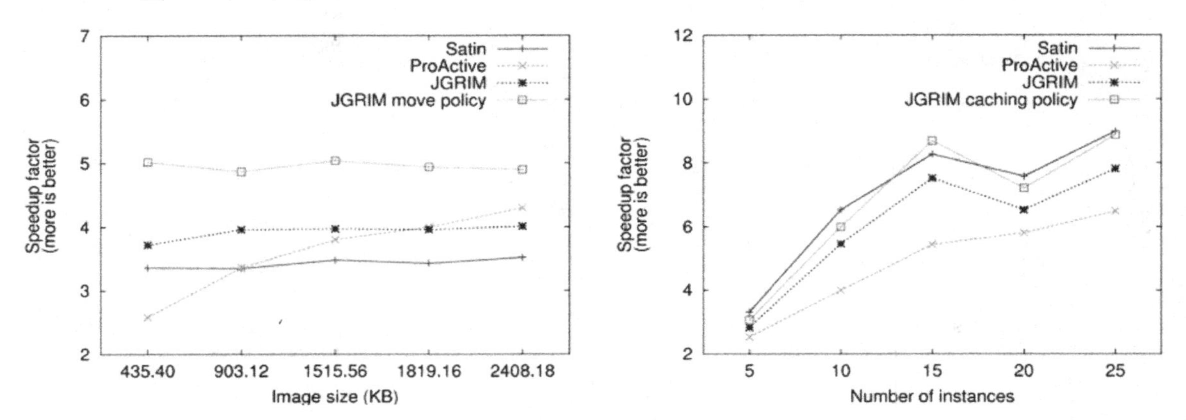

from this caching technique too, but this would have required yet more modifications to the original application code, and thus it would have increased TLOC.

With respect to the image application, the plain variant of JGRIM (this is, without using the mobility policy) performed better than Satin, even when in the experiments JGRIM used Satin for performing parallelism. This is because JGRIM exploits Satin by extending it so as to avoid the standard handshaking process of Satin when cooperatively executing applications. Furthermore, the ProActive version showed acceptable performance levels. In this case, unlike ProActive k-NN, the deployment times did not heavily impact in the performance, since these times were not significant with respect to the total execution times.

Moreover, ProActive generated the least amount of WAN traffic. Unlike Satin and therefore JGRIM, its job scheduling is not subject to random factors. Basically, the Satin platform is based on a load balancing algorithm by which each machine of the underlying Grid randomly asks other nodes for jobs to execute when it becomes idle. Nevertheless, injecting mobility allowed JGRIM to achieve higher performance and reduce this traffic. Again, the policy did not affect the original code. Unfortunately, Satin do not let developers to explicitly control mobility, whereas ProActive only offers weak mobility, which requires extensive code modifications to manually handle the behavior for saving/restoring the execution state of running computations.

To conclude, Figure 7 shows the speedup achieved by the various applications, which were computed as AET_s/AET, where AET_s is the average execution time of the original codes on a single machine (C.1). Note that, in both graphics, the speedup curves of Satin and JGRIM seemed to have the same behavior, since JGRIM relies on Satin for parallelism. This is, JGRIM inherits the job scheduling scheme of Satin. Due to the random nature of the Satin scheduler plus the heterogeneity of our Grid setting, for some experiments regarding the Satin and JGRIM applications, we obtained lower speedups for larger experiments. For example, note that for k-NN, there was a dip in the speedup for 20 instances (Figure 7 (left)). To a lesser extent, this effect was also present in the restoration application. Furthermore, the ProActive applications appeared to linearly gain speedup as the size of the experiments increased, but this trend should be further corroborated. In summary, the implications of the speedups are twofold. On one hand, the original codes certainly benefited from being gridified, thus they were representative Grid applications to experiment with. On the other hand, through the use of policies, JGRIM achieved very competitive speedups compared to both Satin and ProActive, while preserving the technical quality of the application code, as evidenced by the values for the code metrics discussed before.

CONCLUSION AND FUTURE WORK

In this article, we have presented JGRIM, a new approach to simplify the gridification of Java applications by hiding the complex nature of the Grid and its services. We showed the advantages of the approach through experimentation with two related approaches. A distinctive feature JGRIM is that it promotes a convenient model for developing Grid applications that is familiar to most Java programmers. JGRIM allows for a better separation of application logic and Grid code (e.g. for performing service discovery and invocation), and makes the task of consuming Grid services easier. Besides, custom decisions for tuning gridified applications can be specified separately from their logic through the use of policies, thus letting developers to seamlessly adapt the same application to different Grids and distributed environments. We experimentally showed that JGRIM simplifies gridification and produces better Grid code without resigning performance for the two aforementioned applications. However, we will conduct more experiments to further validate JGRIM. This will involve the gridification of more applications on larger Grids.

We are extending JGRIM in several directions. Since JGRIM applications can travel across different administrative domains looking for resources and services, security is crucial. A future research line is to incorporate security mechanisms into JGRIM. Another limitation arises from the assumptions made for gridifying applications, as JGRIM only accepts as input component-based applications, which does not likely hold for all applications. Fortunately, the problem of componentization of legacy object-oriented code has been addressed (Li & Tahvildari, 2006). Therefore, a similar approach could be employed to supply the JGRIM gridification process with an extra code transformation phase to ensure, prior to the first step of the current version of the process, that input applications are component-based. Similarly, it would be interesting to handle the case when the source code of ordinary applications is not available for gridification. We have implemented a tool that builds on the ideas presented in this article, but focuses on dynamically modifying Java bytecodes to run in parallel on a Grid. Basically, the tool takes advantage of the facilities provided by the Java Virtual Machine for altering classes at runtime to adapt ordinary bytecodes to transparently run on Satin clusters.

In addition, we are working on metaservices to leverage other state-of-the-art mechanisms for Grid resource discovery, such as those described in (Zhang et al., 2007), and more Grid execution services. With regard to the former, we are currently integrating JGRIM with GMAC (Gotthelf et al., 2008), a P2P protocol of our own that is designed for exchanging information between the hosts of a Grid in a scalable way. Specifically, GMAC will serve as a mean to efficiently gather information about the Grid resources available for executing applications, thus providing accurate metrics through the profiling interface to application programmers. With respect to the latter, we have already implemented a prototype integration with Condor that is based on a Java interface to Condor clusters[3]. Basically, this will allow JGRIM to smoothly delegate to Condor the execution of component operations representing resource-intensive, coarse-grained Grid jobs.

Also, since JGRIM is essentially a technology-agnostic gridification method, we are exploring the viability of materializing JGRIM in other programming languages besides Java. For example, languages such as C++ or Python are extensively employed for developing Grid applications. However, this will require to carefully study whether these new languages provide support for core features of JGRIM such as mobility, dependency injection and Web Service integration.

Finally, we are developing an Eclipse plug-in to supply developers with graphical tools to specify dependencies and associate metaservices. The plug-in is also expected to offer support for deploying, debugging and monitoring the execution of JGRIM applications. Basically, the goal of this line of research is to provide a full-fledged IDE for gridifying and running applications with JGRIM.

REFERENCES

Allen, G., Davis, K., Goodale, T., Hutanu, A., Kaiser, H., & Kielmann, T. (2005). The Grid Application Toolkit: Towards generic and easy application programming interfaces for the Grid. *Proceedings of the IEEE, 93*(3), 534–550. doi:10.1109/JPROC.2004.842755

Alonso, J., Hernández, V., & Moltó, G. (2006). GMarte: Grid middleware to abstract remote task execution. *Concurrency and Computation, 18*(15), 2021–2036. doi:10.1002/cpe.1052

Atkinson, M., DeRoure, D., Dunlop, A., Fox, G., Henderson, P., & Hey, T. (2005). Web Service Grids: An evolutionary approach. *Concurrency and Computation, 17*(2-4), 377–389. doi:10.1002/cpe.936

Baduel, L., Baude, F., Caromel, D., Contes, A., Huet, F., Morel, M., & Quilici, R. (2006). Grid computing: Software environments and tools. In *Programming, Composing, Deploying on the Grid,* (pp. 205-229)., Berlin, Heidelberg, and New York: Springer

Bartosz Baliś, M., & Wegiel, M. (2008). LGF: A flexible framework for exposing legacy codes as services. *Future Generation Computer Systems, 24*(7), 711–719. doi:10.1016/j.future.2007.12.001

Bazinet, A., Myers, D., Fuetsch, J., & Cummings, M. (2007). Grid Services Base Library: A high-level, procedural application programming interface for writing Globus-based Grid services. *Future Generation Computer Systems, 23*(3), 517–522. doi:10.1016/j.future.2006.07.009

Curbera, F., Duftler, M., Khalaf, R., Nagy, W., Mukhi, N., & Weerawarana, S. (2002). Unraveling the Web Services Web: An introduction to SOAP, WSDL, and UDDI. *IEEE Internet Computing, 6*(2), 86–93. doi:10.1109/4236.991449

Dasarathy, B. (1991). *Nearest neighbor (NN) norms: Nn pattern classification techniques*. IEEE Computer Society Press Tutorial.

Delaittre, T., Kiss, T., Goyeneche, A., Terstyanszky, G., Winter, S., & Kacsuk, P. (2005). GEMLCA: Running legacy code applications as Grid services. *Journal of Grid Computing, 3*(1-2), 75–90. doi:10.1007/s10723-005-9002-8

Fahringer, T., & Jugravu, A. (2005). JavaSymphony: A new programming paradigm to control and synchronize locality, parallelism and load balancing for parallel and distributed computing. *Concurrency and Computation, 17*(7-8), 1005–1025. doi:10.1002/cpe.840

Foster, I. (2005). Globus Toolkit version 4: Software for service-oriented systems. *In Network and Parallel Computing - IFIP International Conference, Beijing, China, 3779*, 2-13. Springer.

Foster, I., & Kesselman, C. (2003). *The Grid 2: Blueprint for a new computing infrastructure, chapter Concepts and Architecture* (pp. 37–63). San Francisco, CA, USA: Morgan-Kaufmann Publishers Inc.

Gannon, D., Krishnan, S., Fang, L., Kandaswamy, G., Simmhan, Y., & Slominski, A. (2005). On building parallel and Grid applications: Component technology and distributed services. *Cluster Computing, 8*(4), 271–277. doi:10.1007/s10586-005-4094-2

Gentzsch, W. (2009). Porting applications to grids and clouds. *International Journal of Grid and High Performance Computing, 1*(1), 55–77. doi:10.4018/jghpc.2009010105

Gotthelf, P., Zunino, A., Mateos, C., & Campo, M. (2008). GMAC: An overlay multicast network for mobile agent platforms. *Journal of Parallel and Distributed Computing, 68*(8), 1081–1096. doi:10.1016/j.jpdc.2008.04.002

GridGain Systems. (2008). *GridGain.* Retrieved October 16, 2008, from http://www.gridgain.com.

Johnson, R. (2005). J2EE development frameworks. *Computer, 38*(1), 107–110. doi:10.1109/MC.2005.22

JPPF. (2008). *Java Parallel Processing Framework.* Retrieved October 16, 2008, from http://www.jppf.org.

Li, S., & Tahvildari, L. (2006). JComp: A reuse-driven componentization framework for Java applications. *In 14th IEEE International Conference on Program Comprehension (ICPC'06),* (pp. 264-267). IEEE Computer Society.

Mateos, C. (2008). *An approach to ease the gridification of conventional applications. Doctoral dissertation.* Universidad del Centro de la Provincia de Buenos Aires, Argentina. Retrieved October 16, 2008, from http://www.exa.unicen.edu.ar/~cmateos/files/phdthesis.pdf.

Mateos, C., Zunino, A., & Campo, M. (2005). Integrating intelligent mobile agents with Web Services. *International Journal of Web Services Research, 2*(2), 85–103. doi:10.4018/jwsr.2005040105

Mateos, C., Zunino, A., & Campo, M. (2008a). A survey on approaches to gridification. *Software, Practice & Experience, 38*(5), 523–556. doi:10.1002/spe.847

Mateos, C., Zunino, A., & Campo, M. (2008b). JGRIM: An approach for easy gridification of applications. *Future Generation Computer Systems, 24*(2), 99–118. doi:10.1016/j.future.2007.04.011

McGough, S., Lee, W., & Das, S. (2008). A standards based approach to enabling legacy applications on the Grid. *Future Generation Computer Systems, 24*(7), 731–743. doi:10.1016/j.future.2008.02.004

Paventhan, A., Takeda, K., Cox, S., & Nicole, D. (2007). MyCoG.NET: A multi-language CoG toolkit. *Concurrency and Computation, 19*(14), 1885–1900. doi:10.1002/cpe.1133

Thain, D., Tannenbaum, T., & Livny, M. (2003). Condor and the grid . In Berman, F., Fox, G., & Hey, A. (Eds.), *Grid computing: Making the global infrastructure a reality* (pp. 299–335). New York, NY, USA: John Wiley & Sons Inc.

Tschumperlé, D., & Deriche, R. (2003). Vector-valued image regularization with PDE's: A common framework for different applications. *In IEEE Conference on Computer Vision and Pattern Recognition (CVPR '03), Madison, WI, USA, 1,* 651-656. IEEE Computer Society.

van Heiningen, W., MacDonald, S., & Brecht, T. (2008). Babylon: middleware for distributed, parallel, and mobile Java applications. *Concurrency and Computation, 20*(10), 1195–1224. doi:10.1002/cpe.1264

van Nieuwpoort, R., Maassen, J., Wrzesinska, G., Hofman, R., Jacobs, C., Kielmann, T., & Bal, H. (2005). Ibis: A flexible and efficient Java based Grid programming environment. *Concurrency and Computation, 17*(7-8), 1079–1107. doi:10.1002/cpe.860

von Laszewski, G., Gawor, J., Lane, P., Rehn, N., & Russell, M. (2003). Features of the Java Commodity Grid Kit. *Concurrency and Computation, 14*(13-15), 1045–1055. doi:10.1002/cpe.674

Walls, C., & Breidenbach, R. (2005). *Spring in action*. Greenwich, Connecticut, USA: Manning Publications Co.

Wrzesinska, G., van Nieuwport, R., Maassen, J., Kielmann, T., & Bal, H. (2006). Fault-tolerant scheduling of fine-grained tasks in Grid environments. *International Journal of High Performance Computing Applications*, *20*(1), 103–114. doi:10.1177/1094342006062528

Zhang, X., Freschl, J., & Schopf, J. (2007). Scalability analysis of three monitoring and information systems: MDS2, R-GMA, and Hawkeye. *Journal of Parallel and Distributed Computing*, *67*(8), 883–902. doi:10.1016/j.jpdc.2007.03.006

ENDNOTES

[1] This is an example of a metaservice representing an NFS (parallelism)

[2] tcpdump: http://www.tcpdump.org

[3] Condor Java API: http://staff.aist.go.jp/hide-nakada/condor_java_api/index.html

This work was previously published in International Journal of Grid and High Performance Computing (IJGHPC), Volume 1, Issue 3, edited by Emmanuel Udoh & Ching-Hsien Hsu, pp. 52-72, copyright 2009 by IGI Publishing (an imprint of IGI Global).

Section 2
Scheduling

Chapter 4
Moldable Job Allocation for Handling Resource Fragmentation in Computational Grid

Kuo-Chan Huang
National Taichung University of Education, Taiwan

Po-Chi Shih
National Tsing Hua University, Taiwan

Yeh-Ching Chung
National Tsing Hua University, Taiwan

ABSTRACT

In a computational Grid environment, a common practice is to try to allocate an entire parallel job onto a single participating site. Sometimes a parallel job, upon its submission, cannot fit in any single site due to the occupation of some resources by running jobs. How the job scheduler handles such situations is an important issue which has the potential to further improve the utilization of Grid resources, as well as the performance of parallel jobs. This paper adopts moldable job allocation policies to deal with such situations in a heterogeneous computational Grid environment. The proposed policies are evaluated through a series of simulations using real workload traces. The moldable job allocation policies are also compared to the multi-site co-allocation policy, which is another approach usually used to deal with the resource fragmentation issue. The results indicate that the proposed moldable job allocation policies can further improve the system performance of a heterogeneous computational Grid significantly.

DOI: 10.4018/978-1-60960-603-9.ch004

INTRODUCTION

Most parallel computing environments running scientific applications adopt the space-sharing approach. In this approach, the processing elements of a parallel computer are logically partitioned into several groups. Each group is dedicated to a single job, which may be serial or parallel. Therefore, each job has exclusive use of the group of processing elements allocated to it when it is running. However, different running jobs may have to share the networking and storage resources to some degree.

In a computational Grid environment, a common practice is try to allocate an entire parallel job onto a single participating site. However, this kind of allocation sometimes runs into a situation called resource fragmentation. The following is an example. Assume a Grid consisting of 4 computing sites each equipped with 32 processors. After a sequence of job allocations, at some moment the amounts of leftover processors for the four sites are 4, 2, 4, 6 in order. At the moment, a new job requiring 10 processors is submitted into the Grid. Apparently, there is no site being able to accommodate the job for immediate execution. It has to wait in queue. However, carefully inspecting the leftover processors reveals that some combinations among the four sites have a total amount of leftover processors larger than the requirement of the incoming job. For example, site 3 and site 4 add up to exactly 10 processors. Site 1, site2, and site3 together can make it, too. This is what we called *resource fragmentation* in Grid environments. This paper tries to deal with resource fragmentation through moldable job allocation.

Most current parallel application programs have the *moldable* property (Dror, Larry, Uwe, Kenneth, & Parkson, 1997). It means the programs are written in a way so that at runtime they can exploit different parallelisms for execution according to specific needs or available resource. Parallelism here means the number of processors a job uses for its execution. The moldable job allocation policies proposed in this paper take advantage of the moldable property of parallel programs to improve the overall system performance.

This paper develops moldable job allocation policies for both homogeneous parallel computers and heterogeneous computational Grid environments. The proposed policies require users to provide estimations of job execution times upon job submission. The policies are evaluated through a series of simulations using real workload traces. The effects of inexact runtime estimations on system performance are also investigated. The moldable job allocation policies are also compared to the multi-site co-allocation policy, which is another approach usually used to deal with the resource fragmentation issue. The results indicate that the proposed moldable job allocation policies are effective as well as stable under different system configurations and can tolerate a wide range of runtime estimation errors.

RELATED WORK

This paper deals with scheduling and allocating independent parallel jobs in a heterogeneous computational Grid. Without Grid computing local users can only run jobs on the local site. The owners or administrators of different sites are interested in the consequences of participating in a computational Grid, whether such participation will result in better service for their local users by improving the job turnaround time. A common load-sharing practice is allocate an entire parallel job to a single site which is selected from all sites in the Grid based on some criteria. However, sometimes a parallel job, upon its submission, cannot fit in any single site due to the occupation of some resources by running jobs. How the job scheduler handles such situations is an important issue which has the potential to further improve the utilization of Grid resources as well as the performance of parallel jobs.

Job scheduling for parallel computers has been subject to research for a long time. As for Grid computing, previous works discussed several strategies for a Grid scheduler. One approach is the modification of traditional list scheduling strategies for usage on Grid (Carsten, Volker, Uwe, Ramin, & Achim, 2002; Carsten Ernemann, Hamscher, Streit, & Yahyapour, 2002a, 2002b; Hamscher, Schwiegelshohn, Streit, & Yahyapour, 2000). Some economic based methods are also being discussed (Buyya, Giddy, & Abramson, 2000; Carsten, Volker, & Ramin, 2002; Rajkumar Buyya, 2002; Yanmin et al., 2005). In this paper we explore non economic scheduling and allocation policies with support for a speed-heterogeneous Grid environment.

England and Weissman in (England & Weissman, 2005) analyzed the costs and benefits of load sharing of parallel jobs in the computational Grid. Experiments were performed for both homogeneous and heterogeneous Grids. However, in their works simulations of a heterogeneous Grid only captured the differences in capacities and workload characteristics. The computing speeds of nodes on different sites are assumed to be identical. In this paper we deal with load sharing issues regarding heterogeneous Grids in which nodes on different sites may have different computing speeds.

For load sharing there are several methods possible for selecting which site to allocate a job. Earlier simulation studies in the literature (Hamscher et al., 2000; Huang & Chang, 2006) showed the best results for a selection policy called *best-fit*. In this policy a particular site is chosen on which a job will leave the least number of free processors if it is allocated to that site. However, these simulation studies are performed based on a computational Grid model in which nodes on different sites all run at the same speed. In this paper we explore possible site selection policies for a heterogeneous computational Grid. In such a heterogeneous environment nodes on different sites may run at different speeds.

In the literature (Barsanti & Sodan, 2007; John, Uwe, Joel, & Philip, 1994; Sabin, Lang, & Sadayappan, 2007; Srividya, Vijay, Rajkumar, Praveen, & Sadayappan, 2002; Sudha, Savitha, & Sadayappan, 2003; Walfredo & Francine, 2000, 2002) several strategies for scheduling moldable jobs have been introduced. Most of the previous works either assume the job execution time is a known function of the number of processors allocated to it or require users to provide estimated job execution time. In (Huang, 2006) without the requirement of known job execution time three adaptive processor allocation policies for moldable jobs were evaluated and shown to be able to improve the overall system performance in terms of average job turnaround time. Most of the previous work deals with scheduling moldable jobs in a single parallel computer or in a homogeneous Grid environment. In this paper, we explore moldable job allocation in a heterogeneous computational Grid environment. In addition to moldable job allocation, multi-site co-allocation (Sonmez, Mohamed, & Epema, 2010) is another approach usually used to deal with the resource fragmentation issue in computational Grid environments. We will compare the performance of these two approaches in this paper.

COMPUTATIONAL GRID MODEL AND EXPERIMENTAL SETTING

In this section, the computational Grid model is introduced on which the evaluations of the proposed policies are based. In the model, there are several independent computing sites with their own local workload and management system. This paper examines the impact on performance results if the computing sites participate in a computational Grid with appropriate job scheduling and processor allocation policies. The computational Grid integrates the sites and shares their incoming jobs. Each participating site is a homogeneous parallel computer system. The nodes within each site run at the same speed and are linked with a fast interconnection network that does not favor any specific communication pattern (Feitelson & Rudolph, 1995). This means a parallel job can be allocated on any subset of nodes in a site. The parallel computer system uses space-sharing and run the jobs in an exclusive fashion.

The system deals with an on-line scheduling problem without any knowledge of future job submissions. The jobs under consideration are restricted to batch jobs because this job type is dominant on most parallel computer systems running scientific and engineering applications. For the sake of simplicity, in this paper we assume a global Grid scheduler which handles all job scheduling and resource allocation activities. The local schedulers are only responsible for starting the jobs after their allocation by the global scheduler. Theoretically a single central scheduler could be a critical limitation concerning efficiency and reliability. However, practical distributed implementations are possible, in which site-autonomy is still maintained but the resulting schedule would be the same as created by a central scheduler (C. Ernemann, Hamscher, & Yahyapour, 2004).

For simplification and efficient load sharing all computing nodes in the computational Grid are assumed to be binary compatible. The Grid is heterogeneous in the sense that nodes on different sites may differ in computing speed and different sites may have different numbers of nodes. When load sharing activities occur a job may have to migrate to a remote site for execution. In this case the input data for that job have to be transferred to the target site before the job execution while the output data of the job is transferred back afterwards. This network communication is neglected in our simulation studies as this latency can usually be hidden in pre- and post-fetching phases without regards to the actual job execution phase (C. Ernemann et al., 2004).

In this paper we focus on the area of high throughput computing, improving system's overall throughput with appropriate job scheduling and allocation methods. Therefore, in our studies the requested number of processors for each job is bound by the total number of processors on the local site from which the job is submitted. The local site which a job is submitted from will be called the *home site* of the job henceforward in this paper. We assume all jobs have the moldable property. It means the programs are written in a way so that at runtime they can exploit different parallelisms for execution according to specific needs or available resource. Parallelism here means the number of processors a job uses for its execution. In our model we associated each job with several attributes. The following five attributes are provided before a simulation starts. The first four attributes are directly gotten from the SDSC SP2's workload log. The *estimated runtime* attribute is generated by the simulation program according to the specified range of estimation errors and their corresponding statistical distributions.

- **Site number**. This indicates the home site of a job which it belongs to.
- **Number of processors**. It is the number of processors a job uses according to the data recorded in the workload log.
- **Submission time**. This provides the information about when a job is submitted to its home site.
- **Runtime**. It indicates the required execution time for a job using the specified number of processors on its home site. This information for runtime is required for driving the simulation to proceed.
- **Estimated runtime.** An estimated runtime is provided upon job submission by the user. The job scheduler uses this information to guide the determination process of job scheduling and allocation.

The following job attributes are collected and calculated during the simulation for performance evaluation.

Table 1. Characteristics of the workload log on SDSC's SP2

	Number of jobs	Maximum execution time (sec.)	Average execution time (sec.)	Maximum number of processors per job	Average number of processors per job
Queue 1	4053	21922	267.13	8	3
Queue 2	6795	64411	6746.27	128	16
Queue 3	26067	118561	5657.81	128	12
Queue 4	19398	64817	5935.92	128	6
Queue 5	177	42262	462.46	50	4
Total	56490				

- **Waiting time**. It is the time between a job's submission and its allocation.
- **Actual runtime**. When moldable job allocation is applied, a job's actual runtime may be different from the runtime recorded in the workload log. This attribute records the actual runtime it takes.
- **Actual number of processors**. When the scheduler applies moldable job allocation, the number of processors a job actually uses for execution may be different from the value recorded in the workload log. This attribute records the number of processors actually used.
- **Execution site**. In a computational Grid environment, a job may be scheduled to run on a site other than its home site. The attribute records the actual site that it runs on.
- **Turnaround time**. The simulation program calculates each job's turnaround time after its execution and records the value in this attribute.

Our simulation studies were based on publicly downloadable workload traces ("Parallel Workloads Archive,"). We used the SDSC's SP2 workload logs[1] on ("Parallel Workloads Archive,") as the input workload in the simulations. The detailed workload characteristics are shown in Table 1.

In the SDSC's SP2 system the jobs in the logs are put into different queues and all these queues share the same 128 processors. In section 4, this original workload is directly used to simulate a homogeneous parallel computer with 128 processors. In section 5 the workload log will be used to model the workload on a computational Grid consisting of several different sites whose workloads correspond to the jobs submitted to the different queues respectively. Table 2 shows the configuration of the computational Grid according to the SDSC's SP2 workload log. The number of processors on each site is determined according to the maximum number of required processors of the jobs belonged to the corresponding queue for that site.

To simulate the speed difference among participating sites we define a speed vector, *e.g.* speed=(sp1,sp2,sp3,sp4,sp5), to describe the

Table 2. Configuration of the computational Grid according to SDSC's SP2 workload

	total	site 1	site 2	site 3	site 4	site 5
Number of processors	442	8	128	128	128	50

relative computing speeds of all the five sites in the Grid, in which the value 1 represents the computing speed resulting in the job execution time in the original workload log. We also define a load vector, *e.g.* load=(ld1,ld2,ld3,ld4,ld5), which is used to derive different loading levels from the original workload data by multiplying the load value ld_i to the execution times of all jobs at site *i*.

MOLDABLE JOB ALLOCATION ON HOMOGENEOUS PARALLEL COMPUTER

Moldable job allocation takes advantage of the moldable property of parallel applications to improve the overall system performance. For example, an intuitive idea is allowing a job to use a less number of processors than originally specified for immediate execution if at that moment the system has not enough free processors; otherwise the job has to wait in a queue for an uncertain period of time. On the other hand, if the system has more free processors than a job's original requirement, the system might let the job to run with more processors than originally required to shorten its execution time. This is called *moldable job allocation* in this paper. Therefore, the system can dynamically determine the runtime parallelism of a job before its execution through moldable job allocation to improve system utilization or reduce the job's waiting time in queue.

For a specific job, intuitively we know that allowing higher parallelism can lead to shorter execution time. However, when the overall system performance is concerned, the positive effects of raising a job's parallelism can not be so assured under the complex system behavior. For example, although raising a job's parallelism can reduce its required execution time, it might, however, increase other jobs' probability of having to wait in queue for longer time. This would increase those jobs' waiting time and in turn turnaround time. Therefore, it is not straightforward to know how raising a single job's parallelism would affect the overall system-level performance, *e.g.* the average turnaround time of all jobs. On the other hand, reducing a job's parallelism might shorten its waiting time in queue at the cost of enlarged execution time. It is not always clear whether the combined effects of shortened waiting time and enlarged execution time would lead to a reduced or increased overall turnaround time. Moreover, the reduced parallelism of a job would usually in turn result in the decreased waiting time of other jobs. This makes it even more complex to analyze the overall effects on system performance.

The above examples illustrate that the effects of the idea of moldable job allocation on overall system performance is complex and require further evaluation. In our previous work (Huang, 2006) we proposed two possible adaptive processor allocation policies. In this paper, we improve the two policies by requiring users to provide estimated job execution time upon job submission, just like what is required by the backfilling algorithms. The estimated job execution time is used to help the system determine whether to dynamically scale down a job's parallelism for immediate execution, *i.e.* shorter waiting time, at the cost of longer execution time or to keep it waiting in queue for the required amount of processors to become available. This section explores and evaluates the two improved moldable job allocation policies, which take advantage of the *moldable* property of parallel applications, on homogeneous parallel computers. The three allocation policies to be evaluated are described in detail in the following.

- No adaptive scaling. This policy allocates the number of processors to each parallel job exactly according to its specified requirement. The policy is used in this section as the performance basis for evaluating the moldable job allocation policies.
- Adaptive scaling down. If a parallel job specifies an amount of processors which

Figure 1. Performance comparison of moldable job allocation policies

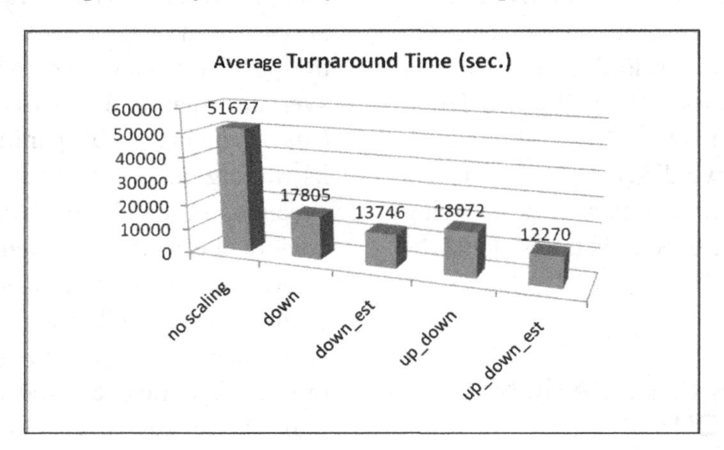

at that moment is larger than the number of free processors. The system has two choices for scheduling the job: scaling its parallelism down for immediate execution or keeping it waiting in queue. According to the estimated execution time of the job, the system can compute the job's enlarged execution time once scaling down its parallelism. On the other hand, based on the estimated execution time of each job running on the system, it is possible to predict how long it will take for the system to gather enough free processors to fulfill the original requirement of the job. Therefore, the system can compare the resultant performances of the two choices and choose the better one. We use a threshold variable to control the selection between the two choices. The system chooses to scale down the job's parallelism for immediate execution only if threshold × To > Tsd, where To is the predicted turnaround time if the job waits in queue until enough free processors are available and Tsd is the predicted turnaround time if the job run immediately with reduced parallelism.

- Conservative scaling up and down. In addition to the scaling down mechanism described in the previous policy, this policy automatically scales a parallel job's parallelism up to use the amount of total free processors even if its original requirement is not that large. However, to avoid a single job from exhausting all free processors, resulting in subsequent jobs' unnecessary enlarged waiting time in queue, the policy scales a parallel job's parallelism up only if there are no jobs behind it in queue. This is why it is called conservative.

Figure 1 shows the performance evaluation of various allocation policies where

- no scaling. No adaptive scaling.
- down. Adaptive scaling down without runtime estimation.
- down_est. Adaptive scaling down with runtime estimation.
- up_down. Conservative scaling up and down without runtime estimation.
- up_down_est. Conservative scaling up and down with runtime estimation.

For the adaptive policies with runtime estimation, we experimented with several possible threshold values and chose the best result to present in Figure 1. For the adaptive scaling down policy, the best threshold value is 2.1 and the conserva-

Figure 2. Effects of inexact runtime estimation under uniform distribution

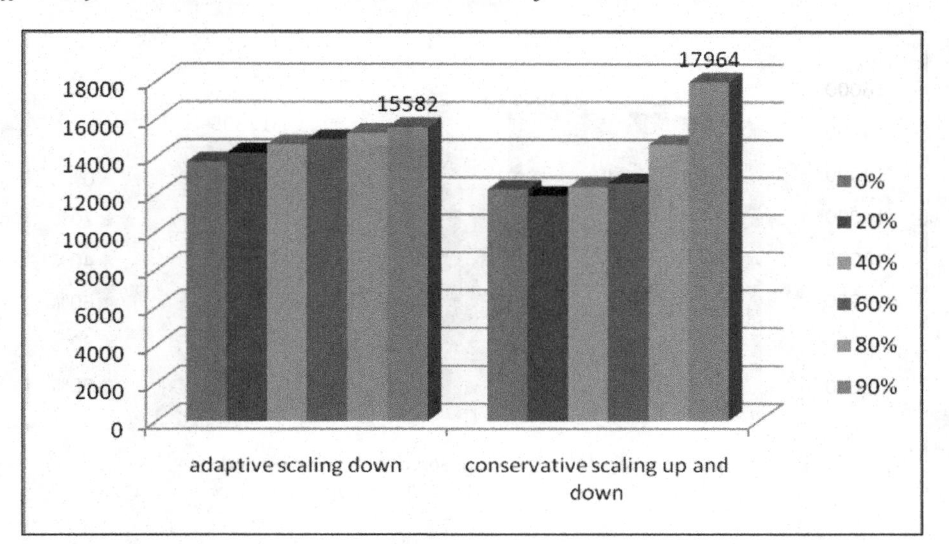

tive scaling up and down policy delivers the best performance when the threshold value is 2. Figure 1 shows that moldable job allocation in general can improve the overall system performance several times, compared to the traditional allocation policy sticking to a job's original amount of processor requirement. Moreover, the improved moldable job allocation policies presented in this paper can further improve the performance significantly with the aid of runtime estimation. For the original moldable job allocation policies, allowing scaling up parallelism cannot improve system performance further in addition to scaling down parallelism in terms of average turnaround time. However, for the improved moldable allocation policies, scaling up parallelism does improve the system performance delivered by the policy which scales down the parallelism only. Overall speaking, the conservative scaling up and down policy with runtime estimation outperforms the other policies.

The studies in Figure 1 assume that users always provide exact estimations of job execution times. However, this is by no means possible in real cases. Therefore, we performed additional simulation studies to evaluate the stability of the moldable job allocation policies when users provide only inexact estimations. The results are presented in Figure 2. The error range of estimation is relative to a job's actual execution time. Figure 2 shows that sometimes small estimation error might even lead to better performance than exact estimation such as the case of conservative scaling up and down with a 20% error range. In general, a larger error range results in degraded performance. However, up to 90% error range, the improved moldable job allocation policies with runtime estimation still outperform the original moldable allocation policies, compared to Figure 1. The results illustrate that the proposed moldable job allocation policies are stable and practical.

The simulations for Figure 2 assume the estimation errors conform to the uniform distribution. Figure 3 presents another series of simulations which evaluate the cases where the estimation errors conform to the normal distribution. The results again show that sometimes larger error ranges lead to better performances. Moreover, Figure 3 indicates that the moldable job allocation policies perform even more stably under the normal distribution of estimation errors, compared to Figure 2.

Figure 3. Effects of inexact runtime estimation under normal distribution

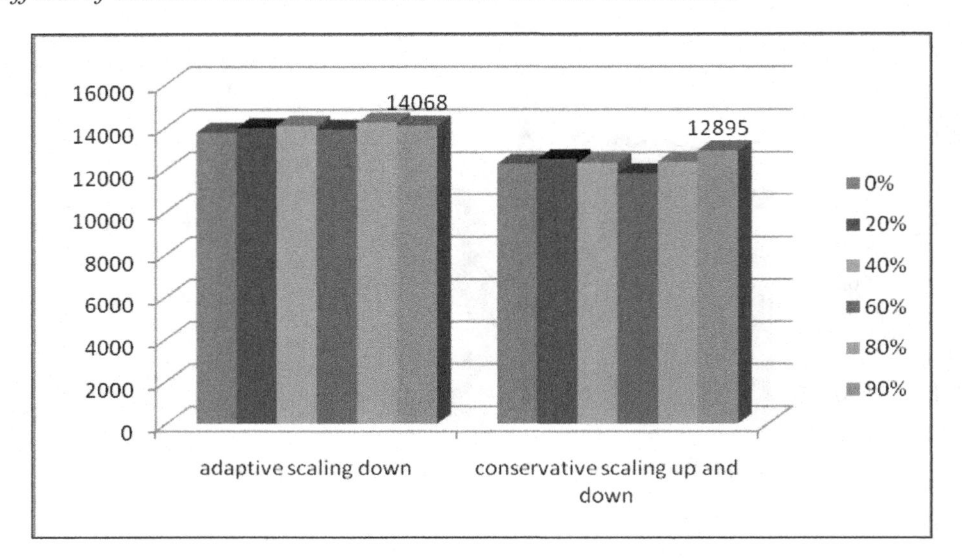

MOLDABLE JOB ALLOCATION IN HETEROGENEOUS GRID

In a computational Grid environment, a common practice is try to allocate an entire parallel job onto a single participating site. Sometimes a parallel job, upon its submission, cannot fit in any single site due to the occupation of some processors by running jobs. How the job scheduler handles such situations is an important issue which has the potential to further improve the utilization of Grid resources as well as the performance of parallel jobs. This section extends the moldable job allocation policies proposed in the previous sections to deal with the resource fragmentation issue in a heterogeneous computational Grid environment.

The detailed moldable job allocation procedure is illustrated in Figure 4. The major difference between the moldable job allocation procedures for a homogeneous parallel computer and for a heterogeneous Grid environment is the site selection process regarding the computation and comparison of computing power of different sites. A site's free computing power is defined as the number of free processors on it multiplied by the computing speed of a single processor. Similarly, the required computing power of a job is defined as the number of required processors specified in the job multiplied by the computing speed of a single processor on its home site.

In the following, we compare the performances of five different cases. They are independent clusters representing a non-Grid architecture, moldable job allocation without runtime estimation, moldable job allocation with exact runtime estimation, moldable job allocation with uniform distribution of runtime-estimation errors, moldable job allocation with normal distribution of runtime-estimation errors. Figure 5 presents the results of simulations for a heterogeneous computational Grid with speed vector (1,3,5,7,9) and load vector (10,10,10,10,10), where

- IC. Independent clusters.
- no estimation. Adaptive processor allocation without runtime estimation.
- exact estimation. Adaptive processor allocation with exact runtime estimation.
- uniform distribution. Adaptive processor allocation with uniform distribution of runtime-estimation errors.

Figure 4. Moldable job allocation procedure in heterogeneous Grid

Check if the waiting queue becomes empty after this job' s allocation

Yes

No

Pick up the site with the largest free computing power

Pick up the site whose free computing power is best close to the required computing power of the job

Predict if adaptive allocation would lead to better turnaround time than waiting in queue for more processors

Yes

Perform allocation

No

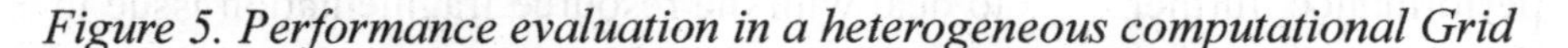

Figure 5. Performance evaluation in a heterogeneous computational Grid

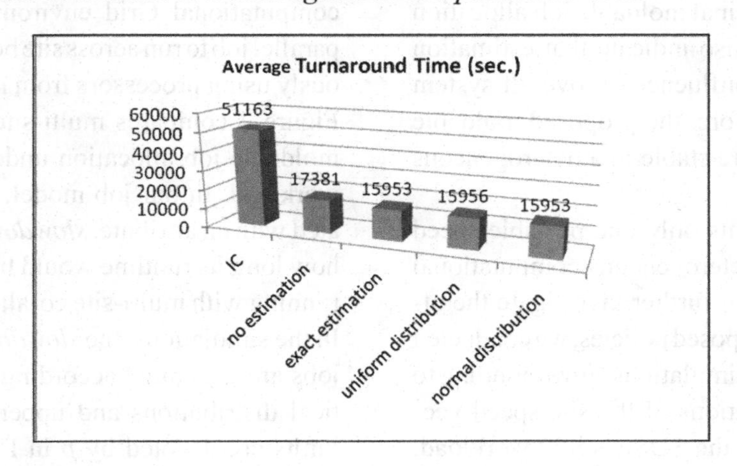

Figure 6. Average performance over 120 different speed configurations

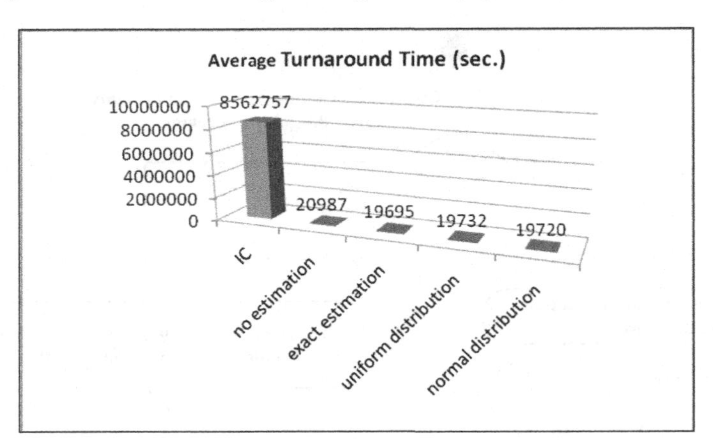

- normal distribution. adaptive processor allocation with normal distribution of runtime-estimation errors.

For the last two cases in Figure 5, we present their worst-case data within the estimation-error range from 10% to 100% with the step of 10%. The results in Figure 5 show that Grid computing with moldable job allocation can greatly improve the system performance compared to the non-Grid architecture. Moreover, the improved moldable job allocation policies with runtime estimation can improve the system performance further compared to the original moldable job allocation policy. The results also indicate that estimation errors lead to little influence on overall system performance. Therefore, the proposed moldable allocation policies are stable in a heterogeneous computational Grid.

Figure 5 represents only one possible speed configuration in a heterogeneous computational Grid environment. To further investigate the effectiveness of the proposed policies, we conducted a series of 120-case simulations corresponding to all possible permutations of the site speed vector (1,3,5,7,9) under the SDSC's SP2 workload. Figure 6 shows the average turnaround times over the 120 cases for the five allocation policies in Figure 5, accordingly. The results again confirm that the proposed moldable job allocation policies are stable and can significantly improve system performance. For the details, among all the 120 cases, the proposed moldable allocation policies with runtime estimation outperform the original moldable policy in 108 cases.

COMPARISON WITH MULTI-SITE CO-ALLOCATION

Multi-site co-allocation (Sonmez, Mohamed, & Epema, 2010) is another approach usually used to deal with the resource fragmentation issue in computational Grid environments. It allows a parallel job to run across site boundary, simultaneously using processors from more than one sites. Figure 7 compares multi-site co-allocation and moldable job allocation under the SDSC's SP2 workload. In our job model, each job is associated with an attribute, *slowdown*, which indicates how long its runtime would be extended to when running with multi-site co-allocation in the Grid. In the simulations, the *slowdown* values for these jobs are generated according to specified statistical distributions and upper limits. The upper limits are denoted by *p* in Figure 5. Two types of statistical distributions, uniform and normal distributions, are evaluated in the simulations. Results in Figure 5 show that the performance of multi-site co-allocation is greatly affected by the

Figure 7. Comparison under SDSC's SP2 workload for uniformly and normally distributed slowdown values

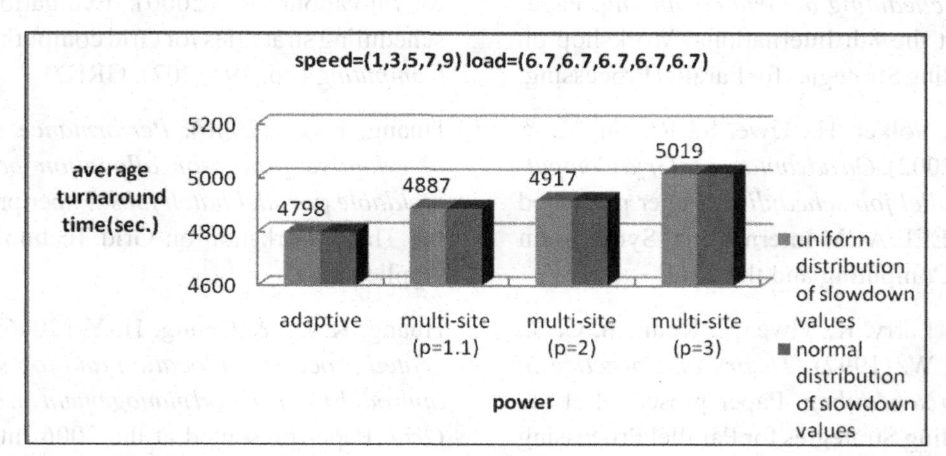

slowdown value which is determined by both the parallel program characteristics and underlying interconnection speed. On the other hand, performance of moldable job allocation is irrelative to the *slowdown* values and the results also indicate that moldable job allocation outperforms multi-site co-allocation in the simulations.

CONCLUSION

In the real world, a Grid environment is usually heterogeneous at least for the different computing speeds at different participating sites. The heterogeneity presents a challenge for effectively arranging load sharing activities in a computational Grid. This paper develops moldable job allocation policies based on the moldable property of parallel applications for heterogeneous computational Grids. The proposed policies can be used when a parallel job, during the scheduling activities, cannot fit in any single site in the Grid. The proposed policies require users to provide estimations of job execution times upon job submission. The policies are evaluated through a series of simulations using real workload traces. The results indicate that the moldable job allocation policies can further improve the system performance of a heterogeneous computational Grid significantly when parallel jobs have the moldable property. The effects of inexact runtime estimations on system performance are also investigated. The results indicate that the proposed moldable job allocation policies are effective as well as stable under different system configurations and can tolerate a wide range of estimation errors.

REFERENCES

Barsanti, L., & Sodan, A. (2007). Adaptive job scheduling via predictive job resource allocation. *Proceedings of the 12th Conference on Job Scheduling Strategies for Parallel Processing,* (pp. 115-140).

Buyya, R., Abramson, D., Giddy, J., & Stockinger, H. (2002). Economic models for resource management and scheduling in Grid computing. *Concurrency and Computation, 14*(13-15), 1507–1542. doi:10.1002/cpe.690

Buyya, R., Giddy, J., & Abramson, D. (2000). *An evaluation of economy-based resource trading and scheduling on computational power grids for parameter sweep applications.* Paper presented at the Second Workshop on Active Middleware Services (AMS2000), Pittsburgh, USA.

Carsten, E., Volker, H., & Ramin, Y. (2002). *Economic scheduling in Grid computing*. Paper presented at the 8th International Workshop on Job Scheduling Strategies for Parallel Processing.

Carsten, E., Volker, H., Uwe, S., Ramin, Y., & Achim, S. (2002). *On advantages of Grid Computing for parallel job scheduling*. Paper presented at the 2nd IEEE/ACM International Symposium on Cluster Computing and the Grid.

Dror, G. F., Larry, R., Uwe, S., Kenneth, C. S., & Parkson, W. (1997). *Theory and practice in parallel job scheduling*. Paper presented at the Job Scheduling Strategies for Parallel Processing Conference.

England, D., & Weissman, J. B. (2005). Costs and benefits of load sharing in the computational Grid. In *Proceedings of the Conference on Job Scheduling Strategies for Parallel Processing* (pp. 160-175).

Ernemann, C., Hamscher, V., Streit, A., & Yahyapour, R. (2002a). Enhanced algorithms for multi-site scheduling. In *Grid Computing* (pp. 219–231). GRID.

Ernemann, C., Hamscher, V., Streit, A., & Yahyapour, R. (2002b). On effects of machine configurations on parallel job scheduling in computational Grids. *Proceedings of International Conference on Architecture of Computing Systems, ARCS*, (pp. 169-179).

Ernemann, C., Hamscher, V., & Yahyapour, R. (2004). *Benefits of global Grid computing for job scheduling.* Paper presented at the Fifth IEEE/ACM International Workshop on Grid Computing, 2004.

Feitelson, D., & Rudolph, L. (1995). Parallel job scheduling: Issues and approaches. In *Proceedings of International Conference on Job Scheduling Strategies for Parallel Processing* (pp. 1-18).

Hamscher, V., Schwiegelshohn, U., Streit, A., & Yahyapour, R. (2000). Evaluation of job-scheduling strategies for Grid computing. In *Grid Computing* (pp. 191–202). GRID.

Huang, K.-C. (2006). *Performance evaluation of adaptive processor allocation policies for moldable parallel batch jobs*. Paper presented at the Third Workshop on Grid Technologies and Applications.

Huang, K.-C., & Chang, H.-Y. (2006). *An integrated processor allocation and job scheduling approach to workload management on computing Grid.* Paper presented at the 2006 International Conference on Parallel and Distributed Processing Techniques and Applications (PDPTA'06), Las Vegas, USA.

John, T., Uwe, S., Joel, L. W., & Philip, S. Y. (1994). *Scheduling parallel tasks to minimize average response time*. Paper presented at the fifth annual ACM-SIAM Symposium on Discrete algorithms.

Parallel Workloads Archive. (n.d.). Retrieved from http://www.cs.huji.ac.il/labs/ parallel/workload/

Sabin, G., Lang, M., & Sadayappan, P. (2007). Moldable parallel job scheduling using job efficiency: An iterative approach. In *Proceedings of the Conference on Job Scheduling Strategies for Parallel Processing* (pp. 94-114).

Sonmez, O., Mohamed, H., & Epema, D. (2010). On the benefit of processor coallocation in multicluster Grid systems. *IEEE Transactions on Parallel and Distributed Systems*, (June): 778–789. doi:10.1109/TPDS.2009.121

Srividya, S., Vijay, S., Rajkumar, K., Praveen, H., & Sadayappan, P. (2002). *Effective selection of partition sizes for moldable scheduling of parallel jobs*. Paper presented at the 9th International Conference on High Performance Computing.

Sudha, S., Savitha, K., & Sadayappan, P. (2003). *A robust scheduling strategy for moldable scheduling of parallel jobs*.

Walfredo, C., & Francine, B. (2000). *Adaptive selection of partition size for supercomputer requests*. Paper presented at the Workshop on Job Scheduling Strategies for Parallel Processing.

Walfredo, C., & Francine, B. (2002). Using moldability to improve the performance of supercomputer jobs. *Journal of Parallel and Distributed Computing*, *62*(10), 1571–1601.

Yanmin, Z., Jinsong, H., Yunhao, L., & Ni, L. M. Chunming, H., & Jinpeng, H. (2005). *TruGrid: A self-sustaining trustworthy Grid.* Paper presented at the 25th IEEE International Conference on Distributed Computing Systems Workshops, 2005.

ENDNOTE

1 The JOBLOG data is Copyright 2000 The Regents of the University of California All Rights Reserved.

Chapter 5
Speculative Scheduling of Parameter Sweep Applications Using Job Behaviour Descriptions

Attila Ulbert
Eötvös Loránd University, Hungary

László Csaba Lőrincz
Eötvös Loránd University, Hungary

Tamás Kozsik
Eötvös Loránd University, Hungary

Zoltán Horváth
Eötvös Loránd University, Hungary

ABSTRACT

The execution of data intensive Grid applications raises several questions regarding job scheduling, data migration, and replication. This paper presents new scheduling algorithms using more sophisticated job behaviour descriptions that allow estimating job completion times more precisely thus improving scheduling decisions. Three approaches of providing input to the decision procedure are discussed: a) single job description, b) multiple job descriptions, and c) multiple job descriptions with mutation. The proposed Grid middleware components (1) monitor the execution of jobs and gather resource access information, (2) analyse the compiled information and generate a description of the behaviour of the job, (3) refine the already existing job description, and (4) use the refined behaviour description to schedule the submitted jobs.

DOI: 10.4018/978-1-60960-603-9.ch005

INTRODUCTION

Resource management is one of the major tasks of Grid middleware. Resources include available computing power (i.e. CPUs), memory and secondary storage. The strategies implemented by the middleware fundamentally determine how early a job can finish its execution and provide the desired computing results. For data intensive parameter sweep applications the placement of data onto Storage Elements (SEs) and the selection of Computing Elements (CEs) have substantial impact on their completion time, therefore the combined efficiency of resource management and scheduling strategies significantly determine the performance of the Grid.

The resource management and scheduling algorithms may take into account the current state of the Grid, or statistics collected on the performance of the Grid components and applications. Some of the resource management strategies make use of sophisticated economy-based decision algorithms (Bell, Cameron, Carvajal-Schiaffino, Millar, Stockinger, & Zini, 2003), others focus chiefly on data replication, and present replica management Grid middleware (Laure, Stockinger, & Stockinger, 2005). Scheduling algorithms may apply statistical prediction methods (Gao, Rong, & Huang, 2005)(Nabrizyski, Schopf, & Weglarz, 2003), which can be used to rank the CEs by the estimated job completion time and select the optimal target CE.

Our resource management and scheduling approach is based on the realization that the completion time of a job on a CE can be determined exactly only after the given job has terminated. Furthermore, we could make perfect scheduling decisions if we were able to run the job on all possible CEs of the Grid one by one within the same circumstances, register the finishing times and run the job on the "best" CE. Obviously, such perfect decisions are not possible to be made, and we can only mimic the process of the selection of the best CE (Lőrincz, Kozsik, Ulbert, & Horváth, 2005).

In order to predict the completion time of the job the proposed scheduling strategies need to know the state of the Grid, the characteristics of the CEs and the expected resource access patterns of the job. For each job, the proposed Grid middleware services will (1) monitor the execution of the job and gather resource access information, (2) generate a compact description of the behaviour of the job, (3) use the job behaviour description to calculate the expected completion time of the job and schedule the job accordingly, and (4) refine the already existing behaviour description using the behaviour description reflecting its latest execution.

Our proposed scheduling strategies also take into consideration the effects of data replication and provide replication commands harmonising with the actual scheduling decision. For example, if the job accesses large chunks of data, it is most likely a good idea to schedule it to the Computing Element (or to a location in its neighbourhood) where the input files are available. However, if the job had to wait too long before it could be started on the chosen Computing Element, it would be worth copying the input files to another Grid component where the job can be executed earlier. In the case of jobs that are less data intensive (use less and smaller input files), the nearness of the files is not so important since the cost of the replication is very low. Furthermore, knowing the resource access patterns of the job the files can be replicated parallel to the execution of the job by fetching the necessary file fragments "just-in-time".

RELATED WORK

Our approach focuses on the resource access of jobs; the scheduling decisions are made based on the finishing time estimations exploiting the knowledge of the behaviour of jobs.

Nabrizyski et al. (Nabrizyski, Schopf, & Weglarz, 2003) gives an excellent overview of Grid resource management. Besides presenting a number of scheduling strategies (Ranganathan & Foster, 2003), in Chapter 16 W. Smith introduces new statistical prediction techniques for the execution times for applications. The first technique uses historical information of previous similar runs to form predictions. The similarity of runs are determined by categorising discrete characteristics of the submitted jobs. The second technique uses instance-based learning: a database of experiences is maintained and used to make predictions. Each experience consists of input and output features. The input feature is a simple job description (user name, job name, number of CPUs requested, requested operating system, etc.).

Similar to our approach, Y. Gao et al. (Gao, Rong, & Huang, 2005) introduces models for *estimating the completion time* of jobs in a service Grid and proposes scheduling algorithms minimising the average completion time of all jobs. The prediction of the completion time of an impending job is based on the number of jobs running on the Grid nodes and historical execution data of already completed jobs. In order to schedule a single job arriving at the node that shall take up the shortest time to execute the job an adaptive system-level job scheduling algorithm is used. To schedule multiple simultaneously arriving jobs genetic algorithms areapplied to minimise the completion time of all jobs.

In the context of workflow management systems Chervenak et al. (Chervenak, et al., September 2007) proposes improved data placement strategies based on the *knowledge of applications* and of expected data access patterns. Their research concentrates on the interplay between data placement services and workflow management systems. In order to improve performance pre-staging – using replication service and asynchronous data placement – is proposed; while the data placement operations are performed as the data sets become available – independently of the actions of the workflow management system.

The Data Intensive and Network Aware (DIANA) meta-scheduling approach (McClatchey, Anjum, Stockinger, Ali, Willers, & Thomas, March 2007) concentrates on the characteristics and *state of the hardware environment* when making scheduling decisions. Such characteristics are the data location and size, processing power and network bandwidth. The scheduler provides a global ranking of the computing resources based on their (changing) state and characteristics. Thereafter, the scheduling decision is made based on the global ranking and execution cost.

ARCHITECTURE OVERVIEW

Our scheduling solution has four keystone components. These are the *job behaviour description*, the *description repository service*, the *description generator*, and the *scheduler*. The relation of the components is depicted by Figure 1.

Each job may have a *behaviour description document*, which characterises the resource allocation and consumption strategy implemented and executed by the given job. A job may have at most one descriptor document. The job descriptions are stored and accessed through the *description repository service*. Besides storing the job descriptors the service is also capable of re-fining the descriptor of a job after it has been terminated using the descriptor relating to the latest execution. The *job description generator* monitors the execution of a job and creates the job description document relating to the actual job execution by analysing its resource access log. When a job is submitted to the Grid the *scheduler* queries its description document using the *description repository service* and selects the node on which the job must be executed.

Figure 2 depicts a proposed deployment scenario for the components. The *scheduler* is deployed on the entry-point of the Grid, which, in our case is the P-Grade portal (P-GRADE portal). The *description repository service* should be deployed in the vicinity of the *scheduler*, although it may be practical to use a different server machine.

Figure 1. Main components of the system

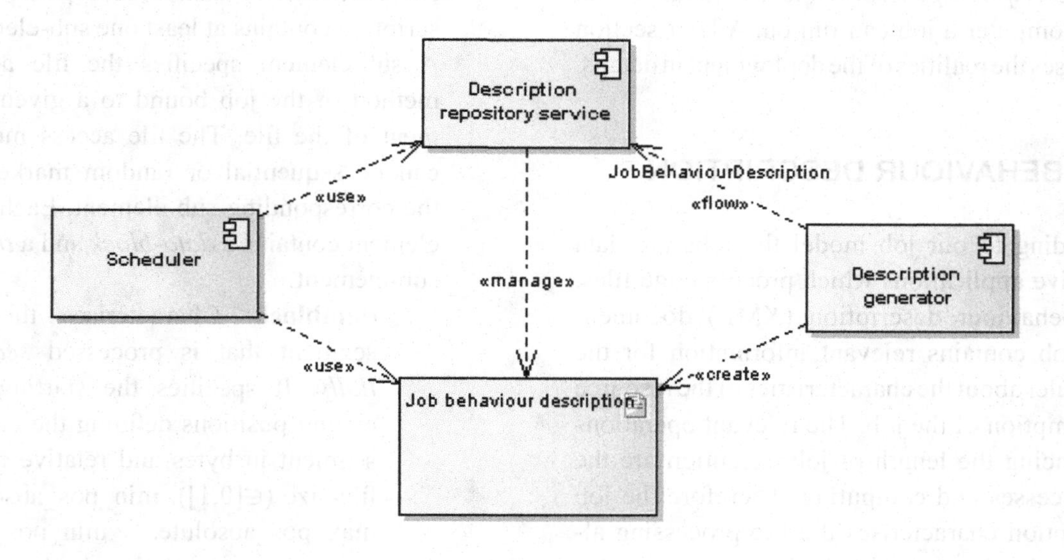

Figure 2. Deployment of the main system components

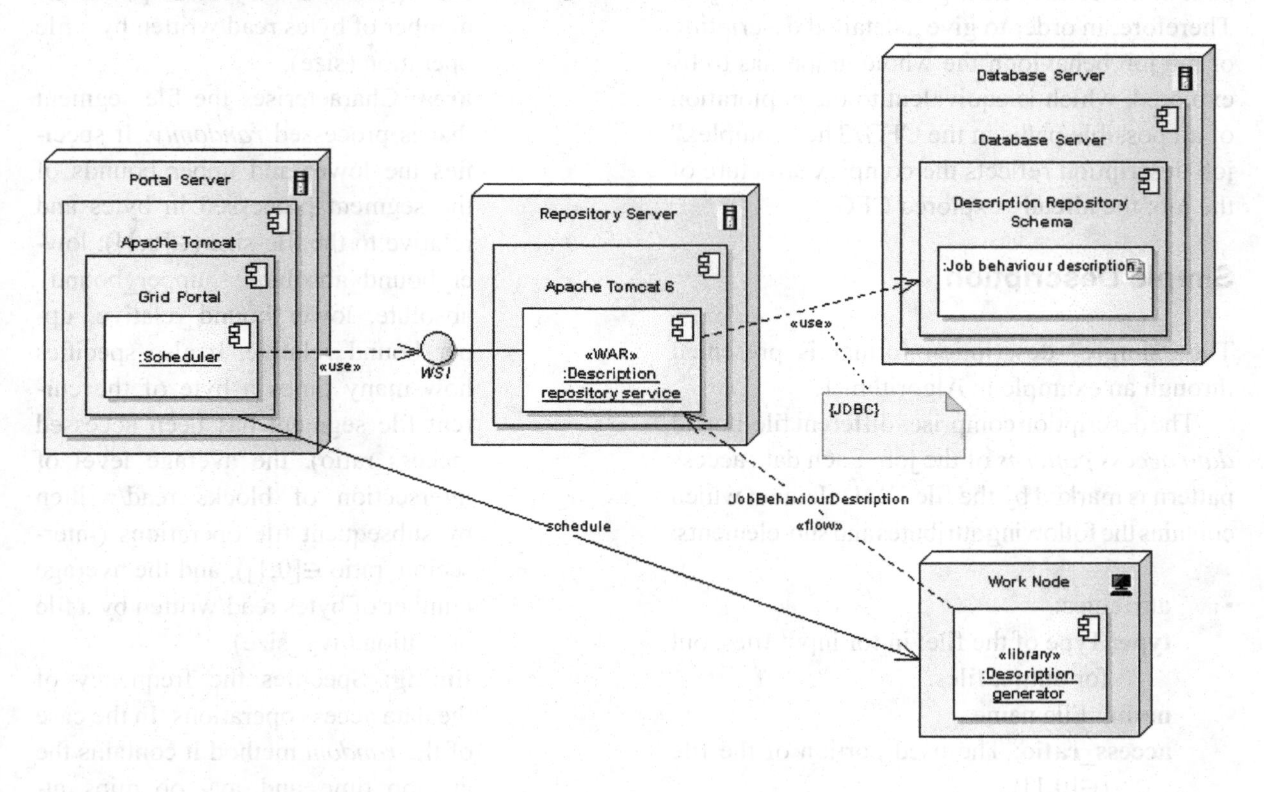

The *description generator* must be installed on each computer a job can run on. A later section discusses the realities of the deployment in details.

JOB BEHAVIOUR DESCRIPTION

According to our job model the jobs are data intensive applications which process huge files. The behaviour description (XML) document of a job contains relevant information for the scheduler about the characteristics of the resource consumption of the job. The relevant operations influencing the length of job execution are the file accesses and computing. Therefore the job description characterises the file processing algorithm implemented by the job.

After each execution of a job a "simple" description can be generated, which relates to a single path in the control flow graph (CFG) of the job. Therefore, in order to give a detailed description of the job behaviour, the whole graph has to be explored, which is equivalent to the exploration of all possible paths in the CFG. The "complex" job description reflects the complex structure of the job: the already explored CFG.

Simple Description

The "simple" description format is presented through an example in Algorithm 1.

The description comprises different file-bound *data access patterns* of the job. Each data access pattern is marked by the file XML element which contains the following attributes and sub-elements:

- attributes:
  type: Type of the file: in for input files, out for output files.
  name: File name.
  access_ratio: The used portion of the file (∈[0,1]).
  intersection_ratio: The file usage redundancy (∈[0,1]).
- sub-elements: A data access pattern description contains at least one sub-element. A sub-element specifies the file access method of the job bound to a given segment of the file. The file access method can be sequential or random marked by the corresponding sub-element. Each sub-element contains a *data-block* and a *timing* sub-element:
  - **datablock**: Characterises the file segment that is processed *sequentially*. It specifies the starting and ending positions defining the current segment in bytes and relative to the file-size (∈[0,1]): min_pos_absolute, max_pos_absolute, min_pos_relative, max_pos_relative. It also specifies the distance between starting positions of two successive data access operations in bytes (step) and the number of bytes read/written by a file operation (size).
  - **area**: Characterises the file segment that is processed *randomly*. It specifies the lower and upper bounds of the segment processed in bytes and relative to the file-size (∈[0,1]): lower_bound_absolute, upper_bound_absolute, lower_bound_relative, upper_bound_relative. It also specifies how many times a byte of the current file segment has been accessed (access_ratio), the average level of intersection of blocks read/written by subsequent file operations (intersecion_ratio ∈[0,1]), and the average number of bytes read/written by a file operation (avg_size).
  - **timing**: Specifies the frequency of the data access operations. In the case of the *random* method it contains the avg_op_time and avg_op_mips attributes: the average system time (in milliseconds) and CPU time (in mips)

Algorithm 1. Example job behaviour description

```
<file type="in" name="test1" access_ratio="1.47218" intersection_ratio="0.18">
  <sequential>
    <datablock
      min_pos_absolute="0" max_pos_absolute="24000"
      min_pos_relative="0" max_pos_relative="0.24"
      step="2000" size="1000" />
    <timing
      op_time="1" op_mips="4.341"
      avg_op_time="8" avg_op_mips="34.728" />
  </sequential>
  <sequential>
    <datablock
      min_pos_absolute="25000" max_pos_absolute="49000"
      min_pos_relative="0.25"  max_pos_relative="0.49"
      step="2000" size="2000" />
    <timing
      op_time="1" op_mips="4.341"
      avg_op_time="15" avg_op_mips="65.115" />
  </sequential>
  <random>
    <area
      lower_bound_absolute="50000" upper_bound_absolute="100000"
      lower_bound_relative="0.5"   upper_bound_relative="1"
      access_ratio="2.19436" intersection_ratio="0.36"
      avg_size="3300" />
    <timing avg_op_time="39" avg_op_mips="169.299" />
  </random>
</file>
```

between two consecutive operations. In the case of the sequential access method the timing specification also contains the minimum system and CPU time. The latter has significance in the case of dynamic scheduling and replication.

The job description example of Algorithm 1 depicts the data access pattern generated for an application that reads file "test1": in the first part the application reads sequentially blocks of 1000 bytes (skipping the following 1000 bytes); in the second part the application reads sequentially blocks of 2000 bytes; in the third part the application reads the blocks randomly.

Complex Description

A complex job description is a set of simple descriptions relating to the same job (see below). Each member description has a weight attribute, which specifies how many times the given member description reflected the actual job behaviour.

Structure of complex job behavior descriptions

```
<simple_description weight="...">
  ...
</simple_description>
```

GENERATING JOB DESCRIPTIONS

Simple job descriptions are generated by the *description generator* deployed on the computers the jobs will run on, the *complex* job descriptions are maintained by the *description repository service*. In the following the algorithms implemented by these components are introduced.

Generating Simple Job Descriptions

The "*simple*" description is generated during the job run. The *generator* monitors the activity of the job and re-fines the simple description whenever the job accesses a "relevant" resource. Such monitored activity is the computing (CPU usage) and file I/O (usage of secondary storage).

The analyser generates simple job descriptions by continuously processing the resource access information obtained by monitoring. For each file accessed by the job the analyser builds a *file access description*, which consists of one or more *file area access description(s)*. A *file area access description* presents the file access strategy used by the job when accessing a specific part of a file. Throughout the job execution, the analyser continuously keeps track of the file area access strategies applied by the job. The analyser recognises two kinds of *file area access methods*: *random* and *sequential*. The latter can be both *increasing* and *decreasing*.

Each of these methods is characterised by the following *behaviour parameters*:

- the average size of the blocks accessed by the individual file operations,
- the average time elapsed between two subsequent file operations working on the given file,
- the minimum and maximum file positions accessed by the job, and the number of times the job changes these positions.

When the analyser is called with a new activity, it refines the corresponding *file access description* by either refining the latest *file area access description* of the *file access description* or by adding a new *file area access description*. The changes in the applied file access methods are detected through the recalculation of the *behaviour parameters* and the comparison of the new values with the previous ones. If a parameter change is larger than a specified threshold value, the actual *file area access description* will be closed and a new one will be added to the *file access description*. For example, if the maximum file position would be needed to be updated in the case of a *decreasing sequential* method, the analyser will decide that the job stopped using the *decreasing sequential* method and it will try to determine the new method.

The detection of the behaviour changes is based on the *access log* which the analyser maintains for each file accessed by the job. An access log entry holds the position and size of the datablock accessed by the job and the time elapsed since the last file access. The size of the access logs is limited allowing the analyser to detect and determine the file access method changes in $O(1)$ time.

In order to determine the new file access method, the analyser resets all behaviour characterisation parameters and the access log. At this point, the file access method is *undetermined*. After the analyser has processed enough file access operations and has filled the access log, it determines the new method. Please note that the analyser actually detects changes of file access *behaviour*. This means that the *new* method is not necessarily a different *kind* of strategy but a file access method having different *behaviour parameters*. For example, if the job processes a file sequentially but from a certain point it will take much more (or less) time to process a data block and the analyser will decide that the strategy has been changed, the new method will be still *increasing sequential* but with different timing characteristics.

The file access method is determined in the following way:

- The method is *increasing sequential* if the maximum position has changed more times than a threshold value (e.g. if the access log size is 10, and the threshold is 7, the maximum position has to be updated 8 times after processing 10 file operations related to the given file).
- The method is *decreasing sequential* if the minimum position has changed more times than a threshold value.
- Otherwise the method is *random*.

After the method has been determined, whenever a new file operation is processed, the analyser updates the access log and the characteristic parameters and checks if the actual file access method has changed.

The analyser algorithm has several parameters, which determine how detailed the resulting *file access description* will be:

- access log size: Specifies how deeply the analyser can look into the past. The larger this parameter the less detailed the description is.
- progress detection threshold: Specifies how many times the maximum (minimum) position has to be changed in order to detect the increasing (decreasing) sequential access.
- behaviour parameter variation: Determines the scale by which the behaviour parameters can change.
- datablock log size: Determines how precise the access and intersection ratio will be. The access and intersection ratios are calculated by registering (per-file) the past few datablocks accessed by the job.

Generating Complex Job Descriptions

The generation of complex job descriptions is based on two different approaches. These will be presented in the following subsections along with the algorithm implemented by the *description repository service*, which combines them.

Single Generalized Description

The algorithm used by the *analyser sub-component* can be generalized to provide a refined description that conforms to all previous executions of the given job. According to the technique of the single generalized description, the job description cannot exclude an already completed sequence of operations. Therefore, the refinement of the description mostly will lead to the relaxation of the behaviour description. For example, if sometimes the job processes a file sequentially and other times the job processes it randomly, then the job description cannot state that the file is processed sequentially, because that would exclude the executions with random file processing. Therefore the description must state that the file access strategy is random. However, the parameters of the random behaviour description must not contradict with the parameters of the sequential behaviour (e.g. block size).

The algorithm of refining a *"simple"* job description is as follows. Let us presume that we have a job description that conforms to all previous job descriptions and reflects the job behaviour as close as possible. Let us also presume that after running the job again, the generator provides a new description that differs from the current one. The following derivation rules define the basic elements of job description refining:

1. If the new description contains parts referencing new files, add the corresponding description parts to the current description.
2. Skip those parts that exist in both the new and the current description and describe the same behaviour.
3. Modify those parts of the current description that exist in the new description, but describe different behaviours.
    a. If two sections of the part intersect according to the relative file positions, make a new section, which describes the intersection of the two sections.
    b. If the two intersecting sections are random, the new section will be random.
    c. If the two sections are sequential with the same directions, the new section will also be sequential.
    d. If the two sections are sequential with opposite directions, the new section will be random.
    e. If one of the sections is sequential and the other one is random, then the derived section will be random.

The attributes of the derived sections will comply with those of the originator sections. For example, if the originator sections are random, then the access_ratio of the derived section is the average of the access_ratio of the originator random sections.

This technique results in job descriptions that reflect the already visited control paths of a given job. However, the resulting description is globally less precise, as it is not able to give close descriptions of the individual control paths.

Multiple Descriptions

Instead of using the latest individual job description, according to the *multiple descriptions* approach, a complex and detailed job description is created by collecting *simple job descriptions* relating to different paths in the job's CFG. Besides this, the execution frequencies (weights) of the paths are also registered giving the probability of their execution.

The new complex job description must provide a more precise (compared to the *simple* job description) however non-redundant representation of the CFG of the job. In our case, redundancy means that the member job descriptions of the *composed job description* have to give significantly different completion time estimates. In order to generate the desired precise non-redundant composed job description:

1. the new job description is inserted into the old composed description, or
2. a similar job description is replaced by the new job description, or
3. the old composed job description is used.

The *similarity* of the new job description and the members of the old complex description determine which method is used to create the new job description. The similarity measure of the member job descriptions must be higher than a certain *threshold* value (i.e. the composed description cannot contain similar member descriptions).

After calculating the similarity of the newly generated individual job description (reflecting the behaviour of the job during its latest execution) and the member job descriptions, the complex job description is updated in the following way:

1. If the distances between the individual job description and the member job descriptions are greater than the threshold then the new description is *inserted* into the composed description. The absolute weight of the new member description will be 1.
2. If there is a member job description, which is closer to the new individual description than the similarity threshold value, but the diversity of the member descriptions would increase with the insertion of the new description, then the new description *replaces* the

"closest" member description. The absolute weight of the newly inserted member description will be the absolute weight of the description that was just replaced plus 1.
3. If at least one of the complex description members is closer to the new job description than the threshold and the diversity of the composed description would not increase with the insertion of the new individual description, than the *old composed description is used*, and the absolute weight of the "closest" member job description is increased by 1.

Many different similarity measures and threshold values can be defined. The similarity measure we have defined is based on the predicted execution time of jobs. The predicted job execution time is defined by the description of the job and the characterization of the Grid.

The Grid is characterised by its clusters, formally *grid profile* $g=\{c_1,c_2,\ldots,c_n\}$, where c_i is a *cluster profile*. Cluster profile $c_i=(mips,disk,net,k)$ describes the "typical" resource characteristics of a member cluster:

- *mips*: speed of a typical CPU in the cluster,
- *disk*: I/O bandwidth (CPU $\longleftrightarrow$ HDD communication),
- *net*: network bandwidth,
- $\kappa \in (0,1]$: weight of the given cluster in the Grid calculated as the number of hosts in the cluster divided by the total number of hosts in the Grid. Note that $\sum_{c\in g} c = 1$.

The composed job description d is described as the collection of its d_i member job descriptions: $d=\{d_1,d_2,\ldots,d_m\}$. The (d,i) weight of a d_i member job description is defined as the absolute weight of d_i divided by the sum of all *absolute* weights of description d. Note that $\sum_{i=1}^{n}(d,i) = 1$.

Let $C(c,d_i)$ denote the estimated execution time of the job running on cluster c behaving according to description d_i. Note that $C(c,d_i)$ can be easily calculated using cluster profile c, and the avg_op_*, datablock, area specifications of the job description. The similarity measure of two individual job descriptions d_i and d_j is defined by the following sum.

$$D(d_i,d_j) = \sum_{c\in g} |\, C(c,d_i) - C(c,d_j)\,|$$

The diversity of the job description is defined as the cumulative similarity of its member descriptions.

$$D(d) = \sum_{d_i,d_j\in d} D(d_i,d_j)$$

We have defined the similarity threshold as the half of the minimum distance of the member descriptions (the initial threshold is 0).

$$threshold = \min_{d_i,d_j\in d, i\neq j} D(d_i,d_j)\,/\,2$$

Compared to the usage of individual job descriptions the complex job description gives a more precise characterization of the behavior of the job. The complex job description offers therefore better job completion time estimates, which eventually result in better scheduling decisions.

Complex Descriptions with Mutation

According to the *multiple descriptions* approach, if the newly generated individual description is closer to an already existing description than a certain threshold, but adding the new description would increase the diversity, then the new description would replace the other one. However, this method unwillingly indicates that the new description is "better" than the description it replaces.

The complex job descriptions are generated with the *multiple descriptions with mutation* algorithm. The algorithm differs from the *multiple descriptions* approach, in that it considers, that in such cases if the sections of the new and the "to-be-replaced" description are the same and only their attributes differ, they presumably reflect the execution of the same sequence of operations. The different attributes indicate that the actual parameters were slightly different, however it cannot be said that either the new or the old description is closer to reality. Therefore, the to-be-replaced description should not be replaced but only *mutated*: the attributes of the sections have to be recalculated using the new attribute values (e.g. their average can be used) as determined by the algorithm presented by the *single generalized description* approach. The *mutation* operation is defined by the *single generalized description* approach.

After the execution of a given job the complex job description is updated as follows:

1. If the distance between the newly generated individual job description and the member job descriptions is greater than a threshold, the new individual description is *inserted* into the composed description. The absolute weight of the new member description will be 1.
2. If there is a member job description which is closer to the new individual description than the threshold value, and the diversity of the member descriptions would increase with the insertion of the new description, then the closest member job description is *mutated* using the algorithm presented in section Generating complex job descriptions. The absolute weight of the mutated member description is increased by 1.
3. If at least one of the complex description members is closer to the new job description than the threshold, and the diversity of the composed description would not increase with the insertion of the new individual description, then the *old composed description is used*, and the absolute weight of the "closest" member job description is increased by 1.

If the number of member job descriptions is *limited*, then mutation can be used to keep the number of member descriptions under the limit, and also to preserve the knowledge carried by the new individual job description. The algorithm resulting in complex job description with limited siye is as follows,

1. If the number of member job descriptions is less than the limit, the previously presented algorithm is used.
2. If the number of member job descriptions already reached the limit, this approach will *mutate* (using the algorithm presented in section Generating complex job descriptions) the member job description which is the closest to the new individual job description according to the similarity measure. The absolute weight of the mutated member job description is increased by 1.

SCHEDULING STRATEGIES

This section will present the proposed scheduling strategies that exploit the information stored by the job descriptions. The major difference between the scheduling strategies is that while the first, *static data feeder*, strategy prepares the input files before the job would be executed, the second, *dynamic data feeder*, strategy delivers the necessary files parallel to the execution of the job, in a just-in-time manner. The process of job scheduling and execution comprises of the following major steps.

1. The user submits the job and its description.
2. The system looks up the corresponding job behaviour description using the *description repository service*.

Algorithm 2.

```
void schedule(Job j, JobDescription d) {
    Map<CE, Long> m = new HashMap<CE, Long>();
    for (ClusterProfile c: g) {
        if (c.canRun(descr)) {
            m.put(c, estimate(c, d)); // calculate the est. finish time
        }
    }
    CE c = getOptimalCE(m); // get the optimal CE
    executeJob(j, c); // run job j on c
}
```

3. The *scheduler* applies the proposed scheduling algorithm, which – using the behaviour description and the information available on the current state of the Grid – calculates the estimated job finishing time for each Grid component, and schedules the job to the component where the job would be finished the earliest.
4. The job is executed on a computer belonging to the chosen Grid component. The resource consumption of the job is monitored, and after the job is terminated, the collected information is used by the description repository service to update the description repository with a refined description.
5. The output of the job (and the behaviour description of the job) is copied to the specified target node.

Static Data Feeder Strategy

The *static data feeder* strategy ranks each Computing Element (CE) by estimating the termination time of the submitted job on the given component. After the ranking of CEs the scheduler runs the job on the CE with the highest rank, i.e. the earliest completion time. The estimated job completion time depends on the job description and on the information collected from the GIS and the Replica Manager. The simplified code-snippet in Algorithm 2 presents the static data feeder algorithm.

The estimated execution time of a job described by *d* on cluster *c* is calculated as follows.

$$C(c,d) = \sum_{i=1}^{m} \lambda(d,i) * C(c,d_i)$$

The actual state of cluster *c* is obtained from the GIS. The *estimate(c,d)* estimated termination time of the given job on cluster *c* is the sum of the estimated job execution time *C(c,d)*, the "length" (measured by) of the job queue on that cluster (*Q*(*c*)), and the time necessary for preparing the input files (before running the job) and delivering result/output files (after the job is terminated):

estimate(c,d) = C(c,d) + Q(c) + fileTransferTime(c,d)

Please note that before running the job on the chosen cluster the necessary files are replicated by the Replica Manager (The DataGrid Project).

Dynamic Data Feeder Strategy

The basic idea behind the *dynamic data feeder* strategy is to download relevant parts of the input files (those parts that the job will presumably access) and to upload the output of the job to the specified destinations during *runtime*. Therefore, instead of dividing the execution of the jobs into three separate phases (download, run, upload), the execution of all steps is attempted at the same time: the input data is provided parallel to the running of the job.

The algorithm of the dynamic data feeder scheduler is similar to the algorithm of the static data feeder scheduler with two differences:

1. The estimated job execution time takes into account that the relevant parts of the necessary files may be delivered after the job is started (but before the job would access them). Therefore the calculation *fileTransferTime*(*c*,*d*) includes only the pre-run and post-run file transfer times, it does not include the transfer time of file segments that are copied parallel to the running of the job.
2. Replication commands are generated that allow the relevant file segments being copied parallel to the running of the job.

Please note that compared to the static data feeder strategy, the estimated completion time of a given job will be lower in most cases.

IMPLEMENTATION

The proposed architecture cannot be deployed completely in existing "production" Grid environments. Lack of administrative/authoritive credentials and missing services are among the most important reasons. We have chosen to extend the P-GRADE portal (P-GRADE portal) with our proposed components as it allowed us to implement an adopted version of the static scheduler. P-GRADE is a parallel application development system for Grid, which (among others) implements job scheduling, migration and checkpointing. P-GRADE supports the Globus (Globus Toolkit) and Condor(Condor Project) Grid environments.

Scheduler

The Portal runs a Java applet in the user's browser which communicates with the server layer. In order to implement the proposed components we needed to extend both the rich client and the server layer.

On the extended Portal interface the user can specify which scheduler algorithm should be used by the system. If our scheduler is selected the user also has to provide the job behaviour description.

Because the P-GRADE portal does not allow querying the size of input files directly, the implemented scheduler cannot consider it when estimating the finishing time of a job on a CE. Instead, the absolute file sizes contained by the job behaviour description are used. Moreover, the scheduler does not know the length of the wait queues of the CEs, therefore the maximum job running time estimates are used, which are specified by the job submitters.

Description Generator

The *Description generator* is implemented by a shared library, which monitors the resource access activity of jobs and prepares the job descriptions by analysing the pattern of activities.

File access monitoring is based on the interception of standard file handling operations defined in the *stdio.h*, *fcntl.h* and *unistd.h* libraries. In general, for a given file operation, the name of the operation, the file or stream descriptor, the name of the file, the opening mode flags, the amount of data read or written, or the new position in the stream are considered.

CPU usage information is collected between two consecutive file access operations. The */proc* - process information pseudo-filesystem (LinuxForum: Linux Filesystem Hierarchy, 1.10. /proc) - is used to access the kernel data structures containing the necessary CPU consumption information.

Because the component (for administrative reasons) cannot be deployed to all computers of

Algorithm 3.

```
universe            = vanilla
executable          = runjob
output              = stdout.log
error               = stderr.log
log                 = job.log
transfer_input_files = <executable>,descrgen.so
...
```

all CEs, it has to be sneaked in the target machine along with the job. The Condor classAD is prepared in this respect so that a *simple shell* script setting the LD_PREALOAD environment variable and running the job will be executed by the work node. The job and the shared object of the Description generator are transferred as input files of the job runner executable. The classAD fragment in Algorithm 3 demonstrates the technique.

Description Repository Service

For similar reasons which do not allow the permanent deployment of the *Description generator*, the *Description repository service* cannot be deployed inside the Grid either. Therefore, we have not implemented it in the current Grid environment supported by P-GRADE.

SIMULATION RESULTS

Simulations were conducted by using OptorSim v2.0 (Simulating data access optimization algorithms - OptorSim), which was extended with the proposed static data feeder and dynamic data feeder scheduler implementations. The extended OptorSim was configured to use the EDG topology specified by the configuration file shipped with the simulator. The CEs of the configuration were extended with MIPS values. One of the group of jobs submitted to the Grid (approx. 1/3 of the total jobs) was changed to simulate the single source shortest path searching algorithm in a graph. The job first parses the graph description loaded from a 300 MB input file then it starts to calculate the shortest path from the given parameter node to every other node in the graph. The jobs provided by OptorSim are using input files of 10 GB each. The number of jobs was also raised to 500 and 1000, to provide us with sufficient job queue sizes on the CEs. Before the simulation was performed OptorSim was supplied with the necessary complex job descriptions.

Due to the lack of support for querying some file related information from the current P-Grade portal, we have simulated mainly solutions that do not use such information during the scheduling process (the default schedulers in this scope in OptorSim are the Random and Queue Length strategies). The static data feeder strategy has been simulated both using and not using file information, while the dynamic data feeder strategy was simulated only with file information present. The benefits can be clearly seen. The static data feeder algorithm performs significantly better when the correct size of the files used by the jobs is known (Static DF) compared to the scheduling when information about expected file transfer times is absent (Static DF no FS info).

According to the simulation results (see Figure 3 for the mean job completion time values provided by OptorSim), using the static data feeder scheduler (Static DF no FS info) the mean job completion time of all jobs on Grid is about 3-4% lower than in the case of the schedulers which do

Figure 3. Simulation results – mean job completion time

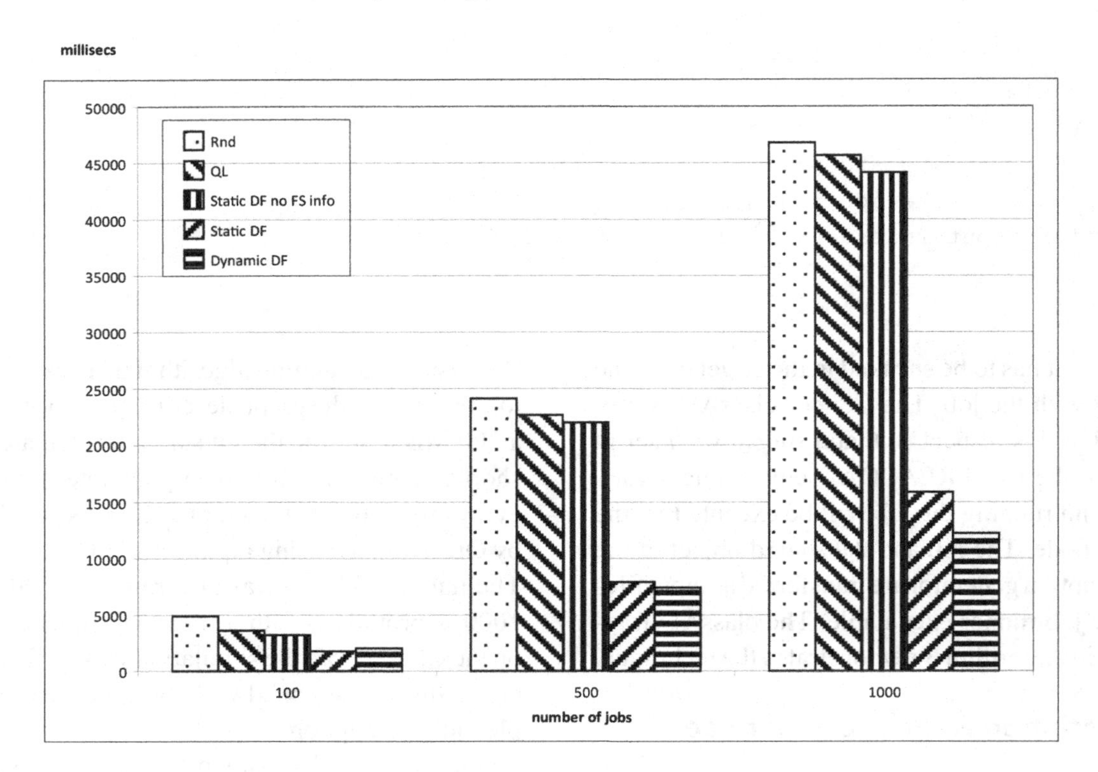

not use any file related information (Rnd, QL). As soon as file sizes are also considered (Static DF) the mean job time of all jobs on the Grid is about 40-60% lower than in the previous case.

Due to the more sophisticated file transferring approach, using the dynamic data feeder scheduler leads to even better (about 5-20% lower) mean job times. Besides, compared to the QL scheduler, the jobs scheduled by the dynamic strategy are finished 40-70% sooner. However the difference can further increase as CE queues would enlarge.

Another set of simulations was carried out for the Static Data Feeder strategy mainly for underlining the importance of refined job descriptions (see Figure 4). These measurements had been configured in such a way that the jobs monitored were consuming 10 times more CPU for the second execution than during their first run.

Four cases were compared: in the case of Static DF 1 the real execution of our jobs took 10 times longer than the values the scheduler was using during its calculations. There is an up to 4-5% speedup with the Static DF 2 strategy, which uses the real (multiplied) running times of the jobs during the scheduling process. Static DF 3 uses also the shorter execution time estimates during scheduling, while the real running times of the jobs were normal for about 50% of the jobs, and 10 times more for the other half of the jobs. Using a merged description (currently a 1-1 weighted average) from the two executions mentioned above (Static DF 4 strategy) will also reduce the mean job times with about 10% compared to the previous strategy.

Refining further these job descriptions with the execution of the monitored jobs can increase the credibility of the scheduling strategy, resulting

Figure 4. Performance of Static Data Feeder scheduler implementations

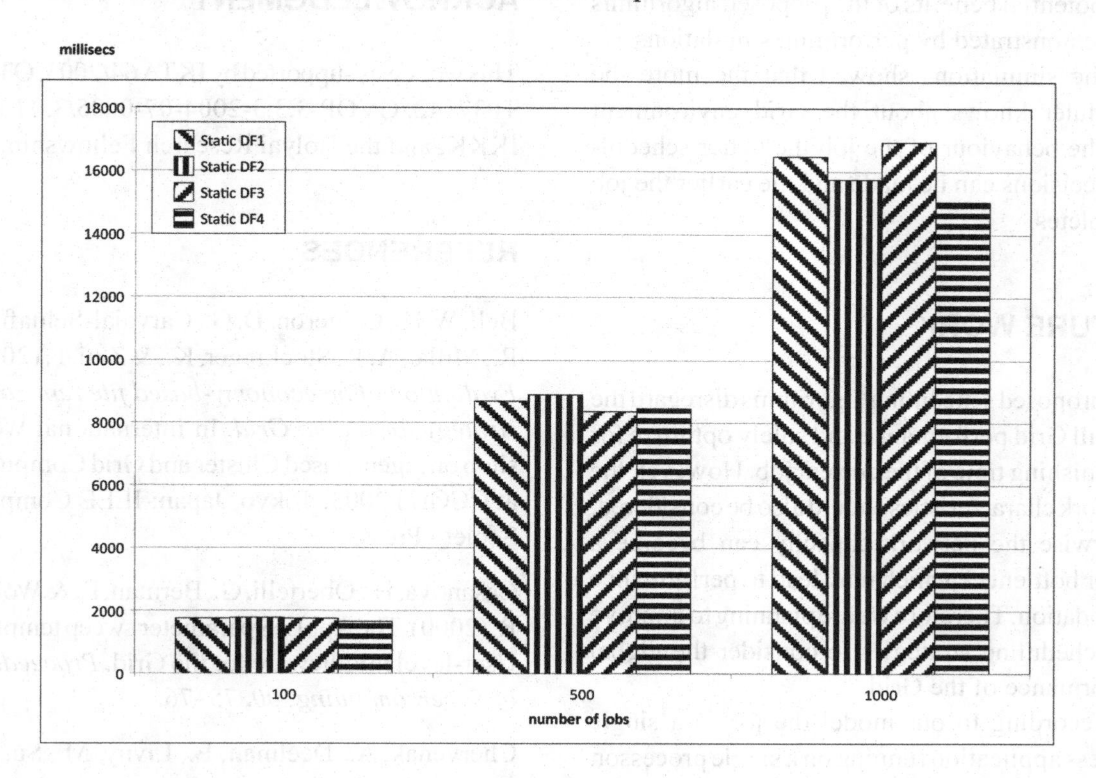

in more realistic assumptions, and lower execution times.

SUMMARY AND CONCLUSION

In this paper we have presented scheduling algorithms for parameter sweep applications in Grid. The scheduling algorithms estimate the job finishing times and select the target CE accordingly.

The key for the job completion time estimation is the description of the behaviour of the job. We have defined the job behaviour description so that it characterises the resource access of the job: the CPU consumption and secondary storage access. However, the description of a job alone is not enough to estimate its completion time; information about the characteristics and state of each CE is also required. Such information is the length of the job wait queue of the CE, the performance of the CPUs of the CE and the size and location of the files the given job would process.

We proposed algorithms for generating the job behaviour descriptions automatically after monitoring its resource access. The job behaviour descriptions generated after subsequent executions can be composed into a complex description. By using the complex description the proposed scheduling algorithms take into account that jobs can act in different ways when they process different files.

The scheduling of the job, the creation of its behaviour description, the refinement of the description and the maintenance of the complex description are supported by our proposed architecture. However, for various non-technical reasons, it is hard to implement the architecture in the proposed form in existing production Grid systems. Therefore, we could implement the components of the presented solution only partially.

The potential benefits of the proposed algorithms are demonstrated by performing simulations.

The simulations showed that the more the scheduler knows about the Grid environment and the behaviour of the job the better scheduling decisions can be made and the earlier the job completes.

FUTURE WORK

The proposed scheduling algorithms disregard the overall Grid performance and solely optimize for the finishing time of the current job. However, the network characteristics should also be considered, otherwise the network capacity can become a major bottleneck which may lead to performance degradation. Therefore we are planning to improve the scheduling strategies to consider the global performance of the Grid.

According to our model the job is a single process application running on a single processor. We would like to relax this limitation and extend the job behaviour description by including communication patterns for applications composed of parallel processes (e.g. PVM and MPI tasks). Accordingly we also intend to alter the scheduling (and estimation) algorithms to take the communication patterns into account.

The primary focus of our current work is data intensive applications and data Grids. We would like to generalize our approach and enable the scheduler to make efficient decisions in such cases when file access does not determine the execution time of the job significantly. The generalized approach should identify those operations which substantially influence the performance of the job. The job behaviour description and scheduling strategies should also be generalized to include and consider the relevant operations.

ACKNOWLEDGMENT

This work was supported by IKTA 64/2003, OTKA T037742, GVOP-3.3.3-2004-07-0005/3.0 ELTE IKKK, and the Bolyai Research Fellowship.

REFERENCES

Bell, W. H., Cameron, D. G., Carvajal-Schiaffino, R., Millar, A. P., Stockinger, K., & Zini, F. (2003). *Evaluation of an economy-based file replication strategy for a data Grid.* In International Workshop on Agent based Cluster and Grid Computing at CCGrid 2003. Tokyo, Japan: IEEE Computer Society Press.

Casanova, H., Obertelli, G., Berman, F., & Wolski, R. (2000). The AppLeS parameter sweep template: User-level middleware for the Grid. *Proceedings of Supercomputing, 00*, 75–76.

Chervenak, A., Deelman, E., Livny, M., Su, M.-H., Schuler, R., Bharathi, S., et al. (September 2007). Data placement for scientific applications in distributed environments. *Proceedings of the 8th IEEE/ACM International Conference on Grid Computing (Grid2007).*

Condor Project. (n.d.). Retrieved from http://www.cs.wisc.edu /condor/

Foster, I. (July 1998). *The Grid: Blueprint for a new computing infrastructure.* Morgan-Kaufmann.

Gao, Y., Rong, H., & Huang, J. Z. (2005). Adaptive grid job scheduling with genetic algorithms. *Future Generation Computer Systems, 21*, 151–161. doi:10.1016/j.future.2004.09.033

Globus Toolkit. (n.d.). Retrieved from http://www.globus.org/toolkit

Job Description Language Attributes. (n.d.). Retrieved from http://auger.jlab.org/jdl /PPDG_JDL.htm

Laure, E., Stockinger, H., & Stockinger, K. (2005). Performance engineering in data Grids. *Concurrency and Computation*, *17*(2-4), 171–191. doi:10.1002/cpe.923

LinuxForum. (n.d.). *Linux filesystem hierarchy,* 1.10. Retrieved from http://www.linuxforum.com /linux-filesystem/proc.html

Lőrincz, L. C., Kozsik, T., Ulbert, A., & Horváth, Z. (2005). A method for job scheduling in Grid based on job execution status. *Multiagent and Grid Systems - An International Journal 4 (MAGS) 1*(2), 197-208.

McClatchey, R., Anjum, A., Stockinger, H., Ali, A., Willers, I., & Thomas, M. (2007, March). Data intensive and network aware (DIANA) Grid scheduling. *Journal of Grid Computing*, *5*(1), 43–64. doi:10.1007/s10723-006-9059-z

Nabrizyski, J., Schopf, J. M., & Weglarz, J. (2003). Grid resource management: State of the art and future trends. In Nabrizyski, J., Schopf, J. M., & Weglarz, J. (Eds.), *International series in operations research and management*. Kluwer Academic Publishers Group.

OptorSim. (n.d.). *Simulating data access optimization algorithms.* Retrieved from http://edg-wp2.web.cern.ch/ edg-wp2/optimization/ optorsim.html

P-GRADE portal. (n.d.). Retrieved from http://www.lpds.sztaki.hu /pgrade/

Phinjaroenphan, P., Bevinakoppa, S., & Zeephongsekul, P. (2005). A method for estimating the execution time of a parallel task on a Grid node. *Lecture Notes in Computer Science*, *3470*, 226–236. doi:10.1007/11508380_24

Ranganathan, K., & Foster, I. (2003). Computation scheduling and data replication algorithms for data Grids. In Nabrzysk, J., Schopf, J., Weglarz, J., Nabrzysk, J., Schopf, J., & Weglarz, J. (Eds.), *Grid resource management: State of the art and future trends* (pp. 359–373). Kluwer Academic Publishers Group.

The DataGrid Project. (n.d.). Retrieved from http://eu-datagrid.web.cern.ch /eu-datagrid/

Chapter 6
A Security Prioritized Computational Grid Scheduling Model:
An Analysis

Rekha Kashyap
Jawaharlal Nehru University, India

Deo Prakash Vidyarthi
Jawaharlal Nehru University, India

ABSTRACT

Grid supports heterogeneities of resources in terms of security and computational power. Applications with stringent security requirement introduce challenging concerns when executed on the grid resources. Though grid scheduler considers the computational heterogeneity while making scheduling decisions, little is done to address their security heterogeneity. This work proposes a security aware computational grid scheduling model, which schedules the tasks taking into account both kinds of heterogeneities. The approach is known as Security Prioritized MinMin (SPMinMin). Comparing it with one of the widely used grid scheduling algorithm MinMin (secured) shows that SPMinMin performs better and sometimes behaves similar to MinMin under all possible situations in terms of makespan and system utilization.

INTRODUCTION

The grid, introduced in 1998, is an emerging field for compute-intensive tasks (Foster, Kesselman, Tsudik and Tuecke, 1998; Foster, Kesselman and Tuecke, 2001). A computational grid is a collection of geographically dispersed heterogeneous computing resources, providing a large virtual computing system to users. Idle computers across the globe can be utilized for such computations. Such an arrangement ultimately produces the

DOI: 10.4018/978-1-60960-603-9.ch006

power of expensive supercomputers which otherwise would have been impossible.

There are four factors behind the growing interest in grid computing: the evolution of key standards such as TCP/IP and Ethernet in networking; the ever-increasing bandwidth on networks reaching into the gigabit range; the increasing availability of idle megaflops on networked PCs, workstations and servers; and the emergence of Web services as a logical and open choice of software computing tasks (Prabhakar, Ribbens and Bora, 2002; Naedela, 2003). Grid scheduling software considers a job composed of tasks; finds suitable processors and other critical resources on the network; distributes the tasks; monitors their progress and reschedules any tasks that fail. Finally, the grid scheduler aggregates the results of the tasks so that the job is completed.

Grid computing has extensively supported collaborated science projects on the internet. Most of these projects have stringent security requirements. To a certain extent, the security may be provided by the application itself, but more usually it should be ensured and supported by the grid environment. The dynamic and multi-institutional nature of these environments introduces challenging security issues that demand new technical approaches for solutions. Scheduling algorithms play an important role in any distributed system. In an environment where security is of concern, responsibility is delegated to the scheduler to schedule the task on the resource that can meet the security requirement of the task. Such a scheduler is referred as the security aware scheduler (Jones, 2003; Tonelloto and Yahyapour, 2006). The goal of a security aware scheduler is to meet the desired security requirements as well as providing a high level of performance metric e.g. site utilization and makespan.

The most common public key authentication protocol used in the grid today is the Transport Layer Security (TLS) (Dierks and Allen, 2007; Apostolopoulos, Peris and Debanjan, 1999) protocol that was derived from the Secure Sockets Layer (SSL) (Freier, Karlton and Kocher, 1996). Different versions of SSL/TLS provide different level of security. Different version supports various cipher suites (security algorithms) for different security services like authentication, encryption and integrity. Thus it is the job of scheduler to allocate the tasks on the resources which supports the required security version and even supports required algorithm on a particular version to satisfy the demand.

Various grid scheduling models (algorithms) have been proposed in the past, but addressing little about security-aware scheduling. In this article, the thrust is security-aware scheduling model to optimize performance characteristics such as makespan (completion time of the entire job set) and site utilization along with the security demand of the task. The model is to consider the constraints exerted by both the job and the grid environment. In the proposed model, security prioritization is incorporated in MinMin scheduling strategy, resulting in renaming the model as Security Prioritized MinMin (SPMinMin).

The next section discusses the related work done in this field. Section 3 explains the proposed grid scheduling SPMinMin model. Section 4 shows some experiments and the observations over the results. Finally, section 5 concludes the work.

RELATED WORK

Often, grids are formed with resources owned by many organizations and thus are not dedicated to specific users. There are many important issues that a job scheduler should address for such a heterogeneous environment with multiple users. The grid resources have different security capability and computational power. The assignment of a task to a machine on which the task executes can significantly affect the overall performance. Resource contention should also be considered while scheduling tasks on grid resources with multiple users. Further, grid, being a non-dedicated networked system, has its own local jobs; i.e. it cannot provide exclusive services to remote jobs. Hence, scheduling algorithms need to address the performance measures of the jobs on non-dedicated network in the presence of multiple users.

Due to security heterogeneity, jobs that are dispatched to a remote site can possibly experience security and reliability problems. Scheduled grid tasks may have its security demand (SD) and the grid site offers a certain security level (SL). If security demand of the job (multiple tasks) is not met by the resource on which it is made to execute, the job may fail and is to be rescheduled on some other resources.

A security-aware scheduling algorithm need to satisfy the security constraints and at the same time has to optimize the performance parameters like site utilization (percentage of total task running time out of total available time on a given site), makespan (completion time of the entire job set), average response time (average value of all tasks' response time), average slowdown ratio (ratio of the task's response time to its service time).Therefore, multi-objective criteria have to be met. Some of the grids scheduling algorithms are discussed below. All these algorithms need prediction information on processor speed and the task length.

- **DFPLTF:** (Dynamic Fastest Processor to Largest Task First) gives the highest priority to the largest task but is not a security aware algorithm (Paranhos, Cirne and Brasileiro; 2003).
- **Suffer:** (Casonova, Legrand, Zagorodnov and Berman; 2000) allocates the processor to the task that would suffer the most if that processor is not assigned to it.
- **Round Robin:** (RR) proposed by Noriyuki Fujimoto and Kenichi Hagihara (2003) grid scheduling algorithm for parameter sweep applications which does not require prediction information regarding task length and processor speed. However RR does not consider security requirements.
- **MinMin:** gives highest priority to the task that can be completed first. In this, for each task the grid site that offers the earliest completion time is tagged and the task that has the minimum earliest completion time is allocated to the respective node. MinMin executes shorter task in parallel whereas longer task follows the shorter one (Freund et al., 1998).
- **MaxMin:** here the grid site that offers earliest completion time is tagged. Highest priority is given to the task with maximum earliest completion time. The idea behind max-min is overlapping long running task with short running ones. MaxMin executes many shorter tasks in parallel with the longer one (Freund et al., 1998).

MinMin and MaxMin are used in real world distributed resource management systems such as SmartNet (Freund et al., 1998). Both have time complexities of (mn^2) where m is the number of machines at the site and n is the number of tasks to schedule. They are suitable when the tasks to schedule are independent and compute intensive.

- **SATS**, suggested by Xie and Qin (2007), takes into account heterogeneities in security and computation. It provides a means of measuring overhead incurred by security services and quantitatively measuring quality of service (QoS) but it does not assure the desired security rather try to improve security and minimize computational overhead.
- **MinMin (Secure, Risky)** (Song, Kwok and Hwang; 2005), are secured version of MinMin. Secure mode allocates task to those sites that can definitely satisfy the security requirements. Risky mode allocates tasks to any available grid site and thus takes all possible risks at the resource

site. Merely imposing security demand in the MinMin degrades its performance which is discussed through cases in the later sections.

PROPOSED MODEL

This work proposes a guaranteed security aware scheduling model as Security Prioritized MinMin [SPMinMin]. It is a security aware scheduling model and assures security requirement of the job (multiple tasks) unlike algorithms like Suffer, DFPLTF, MinMin, MaxMin etc. MinMin gives highest priority to the task that can be completed first. Song et al. (2005) secured the MinMin by merely imposing security restrictions on it. It degrades the performance as security requirement act as the limiting factor rather than guiding factor. This work modified the Min-Min algorithm where the security demand of the task is the major guiding factor for scheduling decisions. SPMinMin allocates highest security demanding tasks first on faster resources. Tasks having same security requirement are then scheduled according to MinMin. Thus it never compromises with the benefits of MinMin but simply modifies it to work efficiently in a secured environment. Extensive experiments have been conducted over simulated grid environments for both MinMin and the SP-MinMin. The experimental test bed is divided on the basis of possible heterogeneous scenario that may exist in a real grid environment. The results obtained clearly indicates that the proposed model performs better and at the same time proves that in non-grid environment MinMin can outperform SPMinMin The results also reveal significant performance gain and better site utilization of SPMinMin over MinMin.

Terminologies Used

Grid is considered to be composed of number of non dedicated processing nodes and the node in turn can be a single processor or a group of heterogeneous or homogeneous processors. A grid job is comprised of "n" independent tasks. The aim is to find an optimum schedule for assigning the grid job (all the tasks) to the processing nodes that satisfies the security constraint. Following is the list of terminologies, used in this article.

$T_{complete}$ is the list of all the task of the given job that is to be scheduled.

T_{high} is the list of all the tasks with highest security requirement.

A task is characterized as $T_i = \{L_i, SD_i\}$ where, Li is the length (size) of the task (number of instructions in the task), and SD_i is the security level demand of the task.

A processing node is characterized as $N_j = (SP_j, SL_j, BT_j)$ where, SP_j is the execution speed of the processing node, SL_j is the maximum security level offered by the processing node, and BT_j is the begin time of the node (time to execute the tasks already assigned to the node).

$N_{qualified,i}$ is the list of processing nodes on which the ith task can be executed i.e. list of the nodes meeting the security demand of the ith task.

A schedule of the job is a set of n triplets $<P_j, T_i, CT_{ij}>$ where, P_j is j^{th} processing node, T_i is i^{th} task, CT_{ij} is completion time of ith task on j^{th} processing node.

$CT_{ij} = ET_{ij} + BT_j$ where, ET_{ij} is execution time of ith task on j^{th} processing node, and BT_j is the begin time of jth processor.

Earliest Completion Time (ECT) of a task is the minimum time amongst all the selected nodes taken to complete the task.

SPMinMin

In a Grid, tasks with different levels of security requests compete for the resources. It is assumed that a task with low security requirement can be executed on both types of resources; the one offering high security as well as the one offering low security. Thus, a task with a desired security level service can be executed only on a resource providing required or higher than the required security level.

As mentioned in section 2, MinMin is one of the most popular scheduling algorithms and is used in real world distributed resource management systems such as SmartNet (Freund et al., 1998). The original MinMin is not security aware and attempts have been made by Song et al. (2005) to make it secure. MinMin (secured) works as follows:

- Compute the completion time of all the tasks on all the nodes.
- Grid node that offers the minimum completion time while meeting the security demand is tagged for each grid task.
- Among all such task- node pair, the task which has the minimum completion time is allocated to the respective node.

MinMin was not designed to incorporate security as a scheduling parameter. The only guiding parameters for MinMin are size of the task and speed of the processor. Introducing security made it behave inefficiently especially under certain situations. Shorter tasks are scheduled on faster nodes at priority according to MinMin. In a typical situation where highly secured machines are the fastest and there are many shorter tasks with lower security requirements the performance of MinMin degrades significantly. The reason is, in the beginning shorter tasks even with low security requirement are unnecessarily assigned to the fastest node (highly secured) and at the end longer task also run on overloaded highly secured machines, as they cannot run on any other machines.

To overcome this shortcoming, we have modified the Min-Min algorithm to consider the security requirement as a guiding factor for the scheduling decisions. The modified Security Prioritized MinMin (SPMinMin) allocates highest security demanding tasks first on faster resources. Its working is as follows:

- Create a list of the tasks with highest security requirement (T_{high}) from the complete set of tasks ($T_{Complete}$).
- For each task of T_{high}, find the list of the nodes ($N_{qualified}$) which satisfies the security demand of the task.
- Compute the completion time for each task of T_{high} on its entire node list ($N_{qualified}$).
- For each task, tag the node(s) from the $N_{qualified}$ that offers minimum completion time.
- Among all such task-node pair, allocate the task which has the minimum completion time to the respective node.
- Remove the task from the T_{high} and $T_{Complete}$ list.
- Modify the begin time (BT) of the resource.
- Repeat the entire process till T_{high} list is empty. After all the tasks from T_{high} are allocated new T_{high} is generated and the entire process begins for the new T_{high}.

The Algorithm

The algorithm for the SPMinMin is given in Box 1.

EXPERIMENTAL EVALUATION

To validate and evaluate the performance, simulation experiments have been carried out. The experimental study considers the complete heterogeneous environment e.g. Security requirement of the tasks; security offered by the nodes, speed of the nodes and size of the task. Altogether, following possibility for the experimentation exists:

1. High speed nodes are more secured and heavy tasks require more security.
2. High speed nodes are more secured and heavy tasks require less security.
3. High speed nodes are more secured no dependency between length of task and security.

Box 1. The SPMinMin scheduling algorithm

```
do until (T_Complete != NULL)      //there are more tasks
  {
    Create T_high from T_Complete      // the tasks demanding high security
    do until (T_high, != NULL)      // there are tasks in T_high
        {
                    for each task i from T_high
                              {
                              Create N_qualified,i
                              for each node j from node list N_qualified,i
                                        compute CT_ij = ET_ij + BT_j
                              find the ECT(Earliest Completion Time)
                              for each task and its corresponding node.
                              Generate matrix ECT_task(i),node.
                              }
                    from the matrix ECT, find the task with
                    minimum ECT =( ECT_k,m)  ) // t^th task on m^th node
                    Schedule task t on node m,
                    Delete task t from T_high and T_complete
                    Modify BT_m = BT_m + CT_km    // begin time for the node m is
                    modified
                    }
}
```

4. High speed nodes are less secured and heavy tasks require more security.
5. High speed nodes are less secured and heavy tasks require less security.
6. High speed nodes are less secured and no dependency between length of task and security.
7. No dependency between speed of nodes and security and no dependency between length of task and their security requirement.

The parameters needed for the simulation to work in a secured grid environment are mentioned in Table 1.

Figure 1a to 1g shows the simulation results for the mentioned seven grid environments for MinMin and SPMinMin when the number of processing nodes for a grid environment is fixed to 16 and the number of task to be scheduled on them varies from 8 to 100. The aim is to study the performance of the two algorithms for different job sizes (number of tasks) when the grid size is kept constant. Figure 2a to 2g shows the simulation results when the tasks to be scheduled on a grid environment are fixed to 40 and the number of nodes varies from 4 to 24. This is to study the performance of the two algorithms for different grid size keeping job size fixed.

Also, the experiment was conducted to observe the time taken by both the algorithms under similar environment. Conspicuous is the fact that apart from offering better makespan the SPMinMin is better in terms of its own execution in comparison to MinMin. Figure 3 shows the comparison over the speed of the two algorithms. It has been observed that with the increase in the number of tasks the execution time of the algorithm improves exponentially.

Table 1. Parameters for simulations

Parameter	Value Range
No of nodes	fixed to 16 // **fig. 1a to 1g**
No of tasks	8 to 100 // **fig. 1a to 1g**
No of tasks	fixed to 40 // **fig. 2a to 2g**
No of nodes	4 to 24 // **fig. 2a to 2g**
Speed of the processing nodes (SP)	1, 2, 5, 10 (MIPS)
Security level of the processing node (SL)	4 / 6 / 9
No. of tasks	Up to 100
Size of tasks	10 to 100 (MB)
Security level demand of the grid task (SD)	4 / 6 / 9

OBSERVATIONS AND CONCLUSION

The present work proposes a security aware scheduling model for computational grid as an extension of MinMin model. It also compares the widely used MinMin algorithm with the proposed SPMinMin for performance metrics like makes-pan and site utilization. SPMinMin is a modified MinMin and it shows remarkable improvement where security demanded by the task cannot be compromised. According to MinMin, shortest task will be scheduled at priority on fastest node. This makes lighter and low security demanding tasks to be unnecessarily scheduled on faster high security nodes. This affects negatively on the site utilization and the entire jobs' (multiple tasks) makespan. To overcome this shortcoming, the Min-Min (secured) algorithm has been modified with the security demand of the task as a guiding factor to scheduling decisions. The modified Security Prioritized MinMin (SPMinMin) allocates highest security demanding tasks first on the faster resources. The highest security requiring tasks are then scheduled according to MinMin. Thus it never compromises the benefits of MinMin but very simply modifies it to work efficiently in a secured environment. Experimental results confirm our study.

The article has elaborated various possible situations in a grid environment based on the computational and security heterogeneity of grid resources and grid tasks. Experiments are conducted on simulated grid environments in two sets. In the first set, the number of nodes is fixed to 16 and the number of tasks varies from 8 to 100. In the second set number of tasks are fixed to 100 and number of nodes varies from 4 to 24. For all possible situations, experimental data is generated and the behavior of MinMin and SPMinMin for makespan and site utilization are studied and compared. The following observations are derived from the experiments:

Whenever heavy tasks demanded more security the proposed model always behaved much better than MinMin as shown in Figure 1a, 1d and 2a, 2d. The more the percentage of such tasks the more will be the improvement in performance of SPMinMin over MinMin. When there is no dependency between length of tasks and the security, even then SPMinMin outperforms MinMin as depicted from plots 3c, 3f and 4c, 4f. Similar results are obtained for both the sets; varying job size or varying grid size. Thus for any grid size or job size SPMinMin outperforms MinMin whenever heavy tasks are more security demanding.

In a grid situation when heavy tasks require more security and higher speed nodes are more secured, there is a significant improvement in the performance of SPMinMin over MinMin and this comes out to be the most favorable situation for

Figure 1. Makespan comparison for different grid environments when number of processing nodes is fixed to 16 and number of task varies from 8 to 100

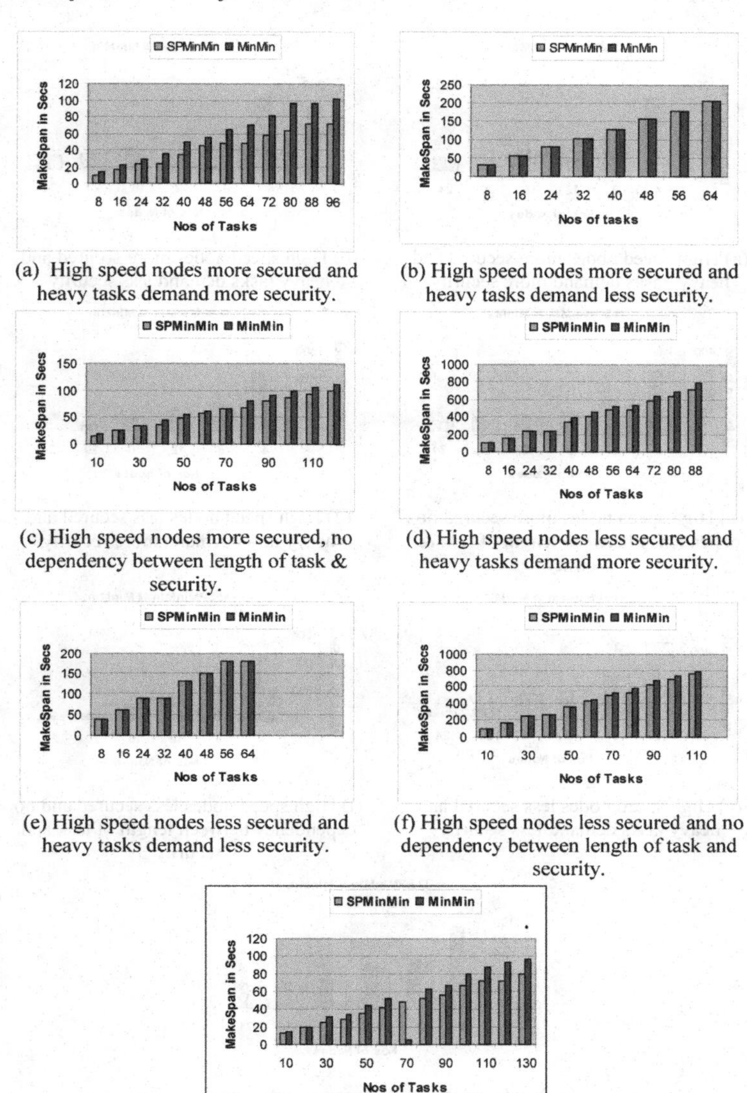

(a) High speed nodes more secured and heavy tasks demand more security.

(b) High speed nodes more secured and heavy tasks demand less security.

(c) High speed nodes more secured, no dependency between length of task & security.

(d) High speed nodes less secured and heavy tasks demand more security.

(e) High speed nodes less secured and heavy tasks demand less security.

(f) High speed nodes less secured and no dependency between length of task and security.

(g) No dependencies between speed of nodes and security, and no dependencies between length of task and their security demand.

SPMinMin. Under this grid environment, if we keep on increasing the tasks while keeping grid size constant, makespan of SPMinMin improves as shown in Figure 1a. For the same grid environment performance of SPMinMin is much better than MinMin for smaller grid as shown in Figure 2a.

In a situation when heavy task demand less security but high speed nodes are more secured, the two behaves similar for larger grid as shown in Figure 1b. For the same situation, it is also observed that for fewer grid nodes, SPMinMin behaves better than MinMin as depicted in Figure 2b.

Figure 2. Makespan comparison for different grid environments when number of tasks is fixed to 40 and the number of processing nodes varies from 4 to 24

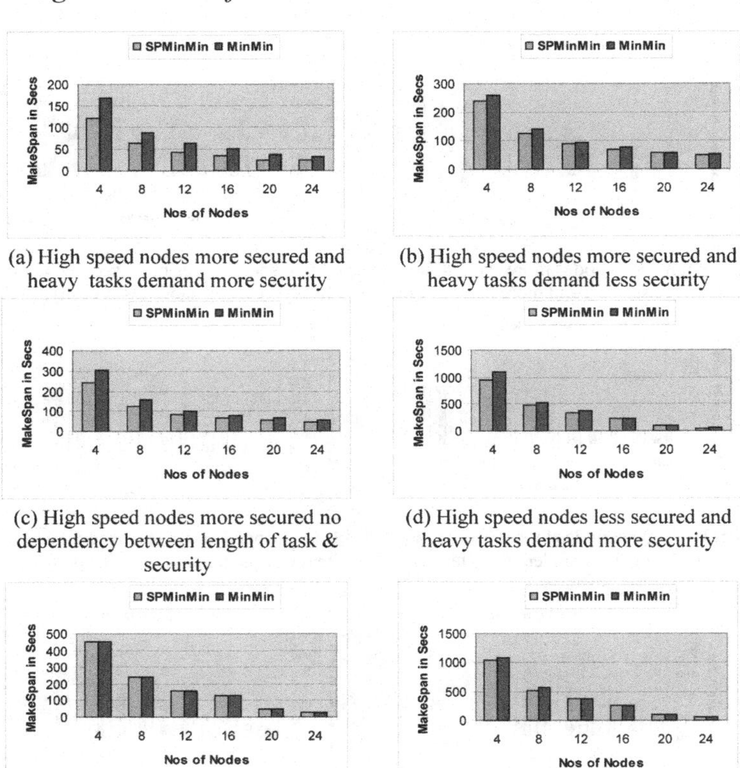

(a) High speed nodes more secured and heavy tasks demand more security

(b) High speed nodes more secured and heavy tasks demand less security

(c) High speed nodes more secured no dependency between length of task & security

(d) High speed nodes less secured and heavy tasks demand more security

(e) High speed nodes less secured and heavy tasks demand less security

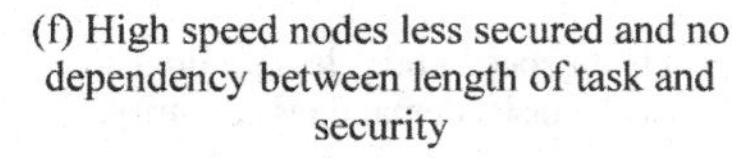

(f) High speed nodes less secured and no dependency between length of task and security

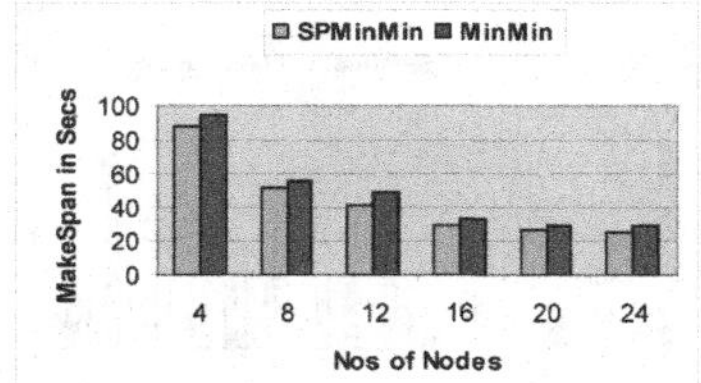

(g) No dependencies between speed of nodes and security, also no dependencies between length of task and their security demand

It has also been observed that whenever heavy tasks demands less security and high speed nodes are less secured, SPMinMin and MinMin behave identically as can be gleaned from Figure 1e and Figure 2e.

There are no dependencies between speed of nodes and security and between length of task and security demand. This is the most realistic grid situation and under most of the situation SPMinMin will perform better than MinMin giving a better makespan and site utilization as shown with the sample data above. Graph 3g and 4g confirm this.

Finally, it is concluded that under all possible situations where security of the task needs to be fulfilled, SPMinMin either outperforms MinMin or in the worst case behaves similar to MinMin. It

Figure 3. Comparison in speed of the two algorithms

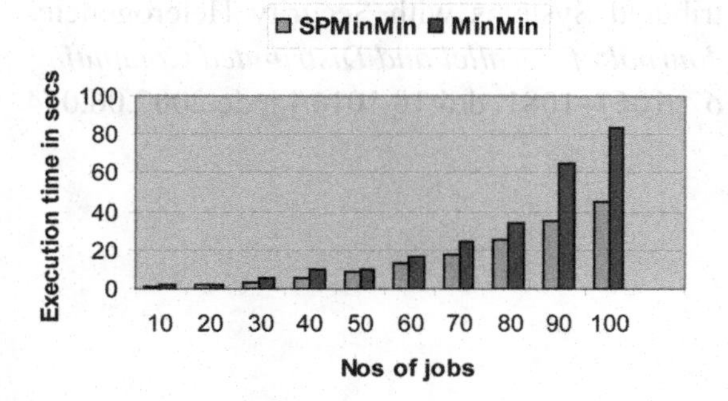

is also observed that SPMinMin is a faster algorithm than MinMin. Thus, SPMinMin is a prime candidate to be considered as a security aware scheduler on a computational grid.

REFERENCES

Apostolopoulos, G., Peris, V., & Debanjan Saha, D. (1999). Transport Layer Security: How Much Does it Really Cost. *Proceedings of the IEEE INFOCOM.* New York.

Casanova, H., Legrand, A., Zagorodnov, D., & Berman, F. (2000). Heuristics for scheduling parameter sweep applications in grid environments. *The Ninth IEEE Heterogeneous Computing Workshop (HCW)*,(pp. 349–363).

Dierks, T. (2007). *The Transport Layer Security (TLS) Protocol Version 1.2 Network Resonance, Inc.* Available at http://www.ietf.org /internet-drafts/draft-ietf-tls-rfc4346-bis-07.txt.

Foster, I., Kesselman, C., Tsudik, G., & Tuecke, S. (1998). *Security Architecture for Computational Grids*. ACM Conference on Computers and Security, (pp. 83-91).

Foster, I., Kesselman, C., & Tuecke, S. (2001). The Anatomy of the Grid: Enabling Scalable Virtual Organizations. *International Journal of High Performance Computing Applications*, 200–222. doi:10.1177/109434200101500302

Freier, A. O., Karlton, P., & Kocher, P. C. (1996). *Internet Draft: The SSL Protocol Version 3.0.* The Internet Engineering Task Force (IETF), Available at http://wp.netscape.com/eng/ssl3/draft302.txt,last accessed in November 2007.

Freund, R. F., Gherrity, R. M., Ambrosius, S., Campbell, M., Halderman, D., Hensgen, E., & Keith, T. Kidd, M. Kussow, Lima, J. D., Mirabile, F. L., Moore, L., Rust, B., & Siegel, H. J. (1998). Scheduling resources in multi-user, heterogeneous, computing environments with SMARTNET. *7th IEEE Heterogeneous Computing Workshop*, (pp. 184–199).

Fujimoto, N., & Hagihara, K. (2003). Near-optimal dynamic task scheduling of independent coarse-grained tasks onto a computational grid. *32nd Annual International Conference on Parallel Processing (ICPP-03)*, (pp. 391–398).

Jones, M. (2003). *Grid Security - An overview of methods used to create a secure grid.* Retrieved from http://www.cse.buffalo.edu/faculty/miller/Courses/Grid-Seminar/Security.pdf.

Naedele, M. (2003). Standards for XML and Web Services Security. *Computer, 36*(4), 96–98. doi:10.1109/MC.2003.1193234

Paranhos, D., Cirne, W., & Brasileiro, F. (2003). Trading cycles for information: Using replication to schedule bag-of-tasks applications on computational grids. *International Conference on Parallel and Distributed Computing (Euro-Par). Lecture Notes in Computer Science, 2790*, 169–180.

Prabhakar, S., Ribbens, C., & Bora, P. (2002). *Multifaceted web services: An approach to secure and scalable grid scheduling.* Proceedings of Euroweb, Oxford, UK.

Song, S., Kwok, Y. K., & Hwang, K. (2005). *Trusted Job Scheduling in Open Computational Grids: Security-Driven Heuristics and A Fast Genetic Algorithms*. Proceedings of International Symposium Parallel and Distributed Processing, Denver, Colorado.

Tonellotto, N., Yahyapour, R., & Wieder P. H.(2006). A Proposal for a Generic Grid Scheduling Architecture. *Core GRID TR-0025.*

Xie, T., & Qin, X. (2007). Performance Evaluation of a New Scheduling Algorithm for Distributed Systems with Security Heterogeneity. *Journal of Parallel and Distributed Computing*, *67*, 1067–1081. doi:10.1016/j.jpdc.2007.06.004

This work was previously published in International Journal of Grid and High Performance Computing (IJGHPC), Volume 1, Issue 3, edited by Emmanuel Udoh & Ching-Hsien Hsu, pp. 73-84, copyright 2009 by IGI Publishing (an imprint of IGI Global).

Chapter 7
A Replica Based Co-Scheduler (RBS) for Fault Tolerant Computational Grid

Zahid Raza
Jawaharlal Nehru University, India

Deo Prakash Vidyarthi
Jawaharlal Nehru University, India

ABSTRACT

Grid is a parallel and distributed computing network system comprising of heterogeneous computing resources spread over multiple administrative domains that offers high throughput computing. Since the Grid operates at a large scale, there is always a possibility of failure ranging from hardware to software. The penalty paid of these failures may be on a very large scale. System needs to be tolerant to various possible failures which, in spite of many precautions, are bound to happen. Replication is a strategy often used to introduce fault tolerance in the system to ensure successful execution of the job, even when some of the computational resources fail. Though replication incurs a heavy cost, a selective degree of replication can offer a good compromise between the performance and the cost. This chapter proposes a co-scheduler that can be integrated with main scheduler for the execution of the jobs submitted to computational Grid. The main scheduler may have any performance optimization criteria; the integration of co-scheduler will be an added advantage towards fault tolerance. The chapter evaluates the performance of the co-scheduler with the main scheduler designed to minimize the turnaround time of a modular job by introducing module replication to counter the effects of node failures in a Grid. Simulation study reveals that the model works well under various conditions resulting in a graceful degradation of the scheduler's performance with improving the overall reliability offered to the job.

DOI: 10.4018/978-1-60960-603-9.ch007

INTRODUCTION

Computational resources being scarce requires an efficient use of these resources. Resources may vary from specialized computational machines, storage machines to heterogeneous applications. Grid is the aggregation of the resources across the world seamlessly and enabling their use as, when and wherever desired rather than individual group investing heavily for high performance computational resources. In the era of high performance and high throughput computing, grid has emerged as an efficient means of connecting distributed computers or resources scattered all over the world for the purpose of collaborative computing thus essentially unifying various heterogeneous resources on a common platform while diminishing the administrative boundaries to provide a transparent access to a user. Essentially being a part of the grid means an infinite capability to execute and compute any kind of job anywhere by simply becoming its part. Therefore, even if the appropriate computational capabilities are not available with the user, the grid helps the job to be executed on the right resources thereby being efficient as well as cost effective.

Depending on the use grids can be classified as Computational grid, Data grid, Sensor grid, Biological grid etc. A computational grid emphasizes on the computing aspect thus scheduling the job to the grid resources by exploring the computational requirements of the job and effectively load balancing it. Scheduling can be based on various objectives like maximizing the reliability of job execution, minimizing the make span or maximizing the Quality of Service (QoS) for the job execution (Grid Computing Info centre, 2008; Baker, Buyya, & Laforenza, 2002; Tarricone & Esposito, 2005; Ernemann, Hamscher, & Yahyapour, 2002; Casanova, 2002; Vidyarthi, Sarker, Tripathi & Yang, 2009; Raza & Vidyarthi, 2008, 2009).

Execution of a job on the complex and dynamic grid poses number of challenges. One of these challenges is to ensure a reliable environment to the job so that it can cope with any kind of failure. Since the grid resources are heterogeneous in behavior and administrative control, introduction of fault tolerance in the system is very difficult. In addition, the jobs demanding execution on the grid themselves may be very complex and may take a long time to execute making them vulnerable to failures. Further, the resources are under the user control so even accidental damages or even a forced shutdown may fail the execution. Similar is true for the network failure also. These failures may range from hardware to software and to the network failures. The fault tolerant techniques can thus vary from proactive to reactive approaches to counter failure at any level (Dai, Xie, & Poh, 2002; Huda, Schmidt & Peake, 2005; Mujumdar, Bheevgade, Malik & Patrikar, 2008). In spite of these measures, the chances of failures cannot be overruled. The desired objective is to accept these failures and minimize their effect by gracefully degrading the system with continued job execution at the cost of a compromised overall performance. One of the popular mechanisms to handle failures is to introduce replication. This could be in the hardware form or the software form in which same application is executed or stored at more than one resources. Therefore, with the slight increase in the execution cost, replication increases the probability of the successful execution of the job, thus being fault tolerant.

Replication incurs a heavy cost but this cost can be minimized by adopting selective replication. The selection of nodes or job modules depends on certain parameters that can be decided by the system as per the scheduling requirements. The RBS works on the basis of replicating some of the modules allocated on a node with high failure rate on to those nodes with lesser failure rate. Therefore, it increases the fault tolerance of the system without severely affecting the performance.

This paper has six sections. Next section discusses the related work reported in the literature with the similar objective followed by a section elaborating the need and integration of RBS with a main scheduler. Working of the model using a suitable example is illustrated next along with the details of the results obtained from the simulation study. The chapter finally concludes detailing the achievements and drawbacks of the work.

RELATED MODELS

The grid being an aggregation of geographically distributed heterogeneous resources; the degree of unreliable behavior extends from the computational resources to the applications running to the network media. A reliable and fault tolerant scheduling has gained enough attention from the researchers and many models have been reported in the literature addressing these issues. A few models have been proposed to counter the effect of these failures by adopting proactive to reactive solutions. A reliability analysis of grid computing systems has been done in (Dai, Xie, & Poh, 2002). An agent oriented fault tolerant framework has been proposed in (Huda, Schmidt & Peake, 2005) to use agents to monitor the system and in case of any threat appropriate measures may be taken beforehand to prevent failures. A checkpoint-based mechanism has been adopted for recovery from failures from the last saved state as a reactive measure (Mujumdar, Bheevgade, Malik & Patrikar, 2008). Introduction of redundancy is a popular means to safeguard the application, as reported in many models in the literature. A study of the tradeoff between performance and availability has been carried out suggesting a file replication strategy (Zhang & Honeyman, 2008). The use of replication by determining the number of replicas required and then suggesting a scheduling strategy for the tasks submitted is reported in (Li & Mascagni, 2003). Another fault tolerant strategy using replication is proposed in (Liu, Wu, Ma, & Cai, 2008) whereas a model using database centric approach for static workload for data grid has been proposed in (Desprez & Vernois, 2007; Sathya, Kuppuswam & Ragupathi, 2006). Many more similar models are also available in the literature.

THE REPLICA BASED CO-SCHEDULER (RBS)

Replication can be applied in many ways for grid constituents to induce fault tolerance in the system. Depending on the requirements and availability it could be used at hardware or the software level. These techniques do well irrespective of the allocation strategy used by the scheduler but with the increased cost of execution both in terms of computational power and money. The degree and type of replication introduced, thus depends on the acceptable amount of failures the system can digest. Since grid is a heterogeneous environment, the failures may occur at many levels viz. the job may fail during the time of submission, the computational resource may fail while job is being scheduled or even after being scheduled, the network links may fail while the job is interacting with the user or within itself. Among all these failures, those accounting to failed resources or application before scheduling does not have a serious effect as they can be taken up again for scheduling. The problem is serious when the resources fail while executing the jobs. The most disastrous failure could be the node failure on which the job is getting executed. Robustness towards application failure and network failures is difficult to attain but the node failure can be handled a bit more easily if we have the information about the allocation of various modules (jobs) allocated on that node.

The proposed Replica Based Co-Scheduler (RBS) helps in the reliable execution of the modular job by replicating the modules allocated to the nodes with high failure rates (sick nodes) to the ones with a lower failure rates (healthy nodes). The reallocation is done only once for a module based on the random selection of nodes out of all the healthy nodes. This results in having duplicate copies of the modules on more than one node. In case of a node failure, the duplicate copies of the modules continues for the job execution. The duplicate copies are used only when a node fails otherwise the job is executed as per the originally scheduled allocation. The job of the RBS starts when the job of the main scheduler in allocating the job modules to various nodes has finished. It is then that the RBS takes control to provide robustness and fault tolerance to the cluster containing the computational resources. The RBS can be used along with any scheduler available in the grid middleware. The inclusion of RBS enables the grid to respond graciously to the node failures with the cost of compromising the performance of the grid, which is unavoidable since the replicated modules have an altered sequence of execution as compared to the original schedule. RBS strategy provides an important backup in absence of which the job needs to be scheduled afresh again resulting in consumption of computational energy that proves very costly for the high traffic environment such as grid. For the real time jobs the problem becomes much more severe as the failures may impact he grid performance thus hitting the financial prospects of the grid.

INTEGRATION OF RBS WITH TSM

To analyze the performance of the co-scheduler RBS it is essential to have a scheduler, which schedules the job submitted to the grid on appropriate resources based on certain optimization parameter. These parameters may vary e.g. turnaround time, reliability, security, Quality of Service (QoS) etc. Minimizing the turnaround time for the job submitted is often a desired parameter and has been addressed in the Turnaround Based Scheduling Model (TSM) for computational grids using Genetic Algorithm (GA) in [8]. The TSM model uses GA to schedule a modular job on a cluster based grid to suggest an allocation pattern in such a way that the turnaround time of the job is minimized. In the present work, the performance of the RBS has been analyzed by integrating it with a TSM scheduler.

The TSM model considers the grid as collection of many clusters, each with a specialization, consisting of a number of nodes for job execution. This is a multipoint entry grid in which the job can be fired at any node of the constituent clusters. The main scheduler (TSM) searches for the appropriate cluster matching the job's requirements and offering the minimum turnaround time to the job, on which the job is eventually scheduled. The job is submitted for execution along with its Job Precedence and Dependence Graph (JPDG) in which the position of each module of the job indicates its order of execution. It also depicts degree of parallelism and the interaction dependence of that module with the preceding modules in terms of the communication requirements.

The allocation status of the various jobs is maintained with each cluster in a data structure known as the Cluster Table (CT), which is updated periodically to reflect updated allocations. The CT consists of the following attributes

$$C_n (S_n, P_k, f_k, \lambda_{lt}, M_{ij}, T_{prkn})$$

Where C_n refers to the cluster under consideration with specialization S_n, number of nodes P_k, the clock frequency of each node f_k, failure rate of each node λ_{lt}, modules assigned on the nodes M_{ij} and the time to finish existing modules T_{prkn} on the nodes. As obvious, the CT provides the information regarding the cluster constituents e.g. the specialization of the cluster nodes to help allocating the jobs to appropriate resources as per its requirements and specifications, number of nodes in the cluster, their clock frequency, the failure rate of nodes, present allocation, and the time taken to finish the existing modules already allocated on the nodes. The main scheduler in this case is TSM but it can be any scheduler proposing a scheduling strategy for the modular job. Since the objective of the TSM is to minimize the turnaround

Table 1. Chromosome Structure

Node No. for Module1	Node No. for Module2	Node No. for Module3	Node No. for Module 4	Node No. for Module5

time of the job, the resultant allocation pattern corresponds to a chromosome suggesting the allocation of job modules on the appropriate nodes [8]. This information is helpful as it is eventually used by the RBS. Taking this allocation of the job modules as suggested by TSM as the prerequisite, RBS replicates the modules of the sick nodes to the healthy nodes as a precautionary measure to overcome the loss due to possible node failures thus increasing the fault tolerance of the system.

For the job submitted for execution, TSM generates a population of chromosomes populated randomly. This is done by dynamic generation of the chromosomes of size (number of genes) equal to the number of modules of the job such that each gene represents the allocation of a module to a node. Starting from the left hand side, the first gene corresponds to the node allocation for the first module, the second gene referring to the node allocation for the second module and so on till the last gene corresponding to the last module as shown in Table 1.

Table 2 presents an example of a job with five modules on a cluster with six nodes. The gene positions here can be read as module 1 being allocated to node 6, module 2 on node 2, module 3 on node 6, module 4 on node 1 and module 5 on node 5.

For the population, TSM uses GA to evolve towards a chromosome offering the minimum turnaround time using operators *selection*, *crossover* and *mutation*. This chromosome gives us the allocation pattern using which the job can be scheduled to minimize the turnaround time of the job. This process is done for all the clusters matching the specialization of the job resulting in a chromosome generated for each cluster offering the minimum turnaround time to the job. These costs are compared to select the cluster offering the least turnaround time corresponding to some allocation pattern responsible for it [8].

For any cluster of the grid, the allocation of modules to the individual nodes depends on three factors viz. processing speed of the node, time to finish execution of already allocated modules to a node and the communication cost in terms of the bytes exchange required between the modules. This cost becomes the fitness function for the allocation of a job with M modules and can be represented as

$$NEC_{kin} = \left[\sum_{i=1}^{M} \left(E_{ijkn}.x_{ijk} + \sum_{h=1}^{i-1} w\left(B_{ihj}.D_{kln}\right) x_{ijk}.x_{hjl} + T_{prkn} \right) \right] \quad \text{(i)}$$

Here E_{ijkn} represents the processing time of the node P_k under consideration calculated for node P_k for module m_i of size I_i of job J_j on cluster C_n as

Table 2. A Sample Allocation of Nodes to the Modules

6	2	6	1	5

$$E_{ijkn} = I_i * (1/f_k) + n * \alpha \qquad \text{(ii)}$$

x_{ijk} is the vector indicating the assignment of module m_i of job J_j on node P_k. It assumes a binary value. It is 1 if the module is allocated to the node and is 0 otherwise. T_{prkn} is the time to finish execution of the present modules on the node P_k. The factor $\sum_{h=1}^{i-1} w\left(B_{ihj}.D_{k\,ln}\right) x_{ijk}.x_{hjl}$ represents the communication cost between a module m_h with the previous modules m_i as per the JPDG, B_{ihj} being the number of bytes that need to be exchanged between modules m_i and m_h and D_{kl} is the hamming distance between nodes P_k and P_l involved in data exchange. w is the scaling factor to scale the term $\sum_{h=1}^{i-1}\left(B_{ihj}.D_{k\,ln}\right) x_{ijk}.x_{hjl}$ into time unit.

The reliability offered by the cluster of the grid, $ClusRel_{jn}$, as per the allocation pattern suggested by the chromosome can be written as shown in Box 1, where $ModRel_{ik}$ is the reliability offered by the grid when module m_i has been assigned on node P_k. Introduction of replicated modules increases the reliability of the job execution. At any time, the reliability offered to the job with replication, $ClusRelRep_{jn}$, can be written as

$$ClusRelRep_{jn} = ClusRel_{jn} + {}^{K}C_{I} * ClusRel_{jn} \qquad \text{(v)}$$

Here, $ClusRel_{jn}$ as stated in eq. (iii) is the reliability offered to the job J_j without node failure and ${}^{K}C_{I}$ accounts for the failure of 'I' nodes out of the available 'K' nodes on which original allocation has been made.

RBS Algorithm

The TSM essentially schedules the job on the cluster offering the minimum turnaround from a group of clusters with matching specialization of the job. Once the cluster is selected for job allocation, its Cluster Table (CT) is updated to accommodate the new job. The job of the RBS begins where the job of TSM finishes. For the cluster selected, the RBS evaluates the vulnerability of the nodes on which an allocation has been done by comparing their failure rates λ_{lt} with some threshold failure rate λ_{th} which depends on the domain knowledge of the cluster along with the acceptance level of the failures. Accordingly the nodes are judged as healthy and sick nodes. For the sick nodes, CT is referred to check for any allocations made. These modules are then duplicated on some healthy node, selected randomly. The algorithm for the same is shown in the box.

Now if a failure is detected the system does not fail completely as copies of the modules on the failed node are still available on some other nodes. The execution of the job still follows the JPG with the penalty of increase in the turnaround time. It is due to some nodes waiting for the pre-

Box 1.

$$ClusRel_{jn} = \prod_{i=1}^{M} ModRel_{ik} \qquad \text{(iii)}$$

$$ClusRel_{jn} = \prod_{i=1}^{M} \exp\left\{-\left[(\mu_{ij}+\lambda_{kn})\left[E_{ijkn}.x_{ijk}\right]+(\mu_{ij}+\xi_{kl})\left[\sum_{h=1}^{i-1} w(B_{ihj}.D_{kln})x_{ijk}.x_{hjl}\right]+\lambda_{kn}\left[T_{prkn}\right]\right]\right\} \qquad \text{(iv)}$$

Algorithm 1.

```
Replica (Job)
{
On the selected cluster Cn, for the submitted job Jj
 do
    Set the threshold failure rate λth.
   Get the failure rates λlt of each node on which allocation  has been made
   For each node, if λlt  > λth
    do
       {
        Get all modules Mij allocated on node Pk

   For each module
        do
         {
          Randomly allocate it to any processor with λlt > λth
         }
```

Table 3. Job J_0

Module (m_{ij})	Job Specialization (J_j)	Number of Instructions (I_i)
m_{00}	J_0	150
m_{10}	J_0	200
m_{20}	J_0	175
m_{30}	J_0	100
m_{40}	J_0	200

Table 4. Matrix B_{ih0} for Job J_0

	m_{00}	m_{10}	m_{20}	m_{30}	m_{40}
m_{00}	0	3	3	0	0
m_{10}	3	0	0	2	3
m_{20}	3	0	0	0	2
m_{30}	0	2	0	0	0
m_{40}	0	3	2	0	0

vious modules reallocated on other nodes to get executed. The RBS thus works as a supplement to the main scheduling algorithm by increasing the clusters fault appetite.

ILLUSTRATIVE EXAMPLE

To elaborate the working of the RBS an example has been illustrated using one of the results of the simulation study with other job as detailed in Table 3 and Table 4. The parameters taken are scaled down for the purpose of illustration. All the data values are generated randomly and conform to the similar studies. Table 5 and Table 6 represents the CT and the hamming distance between nodes respectively for cluster C_0, which is the selected cluster on which the job has been finally allocated by the TSM scheduler. Table 7 represents the processing time matrix for cluster

Table 5. Cluster table for C_0

Node Number (P_k)	Clock Frequency (f_k in MHz)	Specialization (S_n)	Time to finish (T_{prkn} in µS)	Modules allocated (M_{ij})	Node Failure rate (λ_{lt})
P_0	10	J_0	10	00	0.001
P_1	20	J_0	12	10	0.002
P_2	10	J_0	10	31	0.003
P_3	10	J_0	13	33	0.008
P_4	20	J_0	12	43	0.007

Table 6. Matrix D_{kl} for Cluster C_0

	P_0	P_1	P_2	P_3	P_4
P_0	0	1	2	3	2
P_1	1	0	3	2	3
P_2	2	3	0	1	2
P_3	3	2	1	0	1
P4	2	3	2	1	0

Table 7. E_{i0k0} on cluster C_0 for Job J_0

	m_{00}	m_{10}	m_{20}	m_{30}	m_{40}
P_0	15	20	17.5	10	20
P_1	7.5	10	8.75	5	10
P_2	15	20	17.5	10	20
P_3	15	20	17.5	10	20
P_4	7.5	10	8.75	5	10

C_0 for the given job. Final allocation of the job to the cluster selected considering allocation of individual modules by TSM is shown in Table 8.

Here, BEC_{ik} should be read as the Best Execution Cost for m_i module offered by node P_k and interpreted as the best turnaround time offered by a node to a module being considered for allocation. This becomes the best turnaround time offered by any node and results in allocation of the module to this node, which can then execute it in the minimum possible time.

The turnaround time for the above allocation is found to be

Turnaround Time

= max (NEC_{ikn}) for 'k' on which allocation has been made

= max (38, 29.75, 24, 13, 29.5) = 38

Therefore, the final allocation of the modules for cluster C_0 is represented in Table 9.

As can be seen from Table 5, the failure rates of the nodes are determined and the sick (S) and healthy nodes (H) are marked accordingly by comparing it with the threshold failure rate λ_{th}

Table 8. Allocation of modules for Cluster C_0

Module		P_0	P_1	P_2	P_3	P_4
m_{00}						
	E_{ijkn}	15	7.5	15	15	7.5
	T_{prkn}	10	12	10	13	12
	$\sum (B_{ihj} * D_{kl})$	0	0	0	0	0
	NEC_{ikn}	25	19.5	25	28	19.5 (BEC_{04})
m_{10}						
	E_{ijkn}	20	10	20	20	10
	T_{prkn}	10	12	10	13	19.5
	$\sum (B_{ihj} * D_{kl})$	6	9	6	3	0
	NEC_{ikn}	36	31	36	36	29.5 (BEC_{14})
m_{20}						
	E_{ijkn}	17.5	8.75	17.5	17.5	8.75
	T_{prkn}	10	12	10	13	29.5
	$\sum (B_{ihj} * D_{kl})$	6	9	6	3	0
	NEC_{ikn}	33.5	29.75 (BEC_{21})	33.5	33.5	38.25
m_{30}						
	E_{ijkn}	10	5	10	10	5
	T_{prkn}	10	29.75	10	13	29.5
	$\sum (B_{ihj} * D_{kl})$	4	6	4	2	0
	NEC_{ikn}	24	40.75	24(BEC_{32})	25	34.5
m_{40}						
	E_{ijkn}	20	10	20	20	10
	T_{prkn}	10	29.75	24	13	29.5
	$\sum (B_{ihj} * D_{kl})$	8	9	12	7	6
	NEC_{ikn}	38 (BEC_{40})	48.75	56	40	45.5

which is 0.005 in this case. This is shown in Table 10 along with the random replacements for the modules on the sick nodes.

So the new allocation becomes as shown in Table 11.

The node numbers shown in the brackets are the duplicate copies of the modules lying on the sick nodes, which becomes active as soon as the corresponding node fails. Assuming all the sick nodes fail, the new allocation becomes Table 12.

For the new allocation, now, the turnaround time can be calculated in the same way as shown in Table 8 as per equation (i). The new turnaround time calculation is shown in Table 13 for thresh-

Table 9. Final allocation of the job to the nodes

	Node on which allocation has been made
m_{00}	P_4
m_{10}	P_4
m_{20}	P_1
m_{30}	P_2
m_{40}	P_0

Table 10. Detection of sick and healthy nodes

	Node on which allocation has been made	Replacement Node
m_{00}	P_4(S)	P_2
m_{10}	P_4(S)	P_0
m_{20}	P_1(H)	P_1
m_{30}	P_2(H)	P_2
m_{40}	P_0(H)	P_0

Table 11. New allocation of the job to the nodes

	Node on which allocation has been made
m_{00}	$P_4(P_2)$
m_{10}	$P_4(P_0)$
m_{20}	P_1
m_{30}	P_2
m_{40}	P_0

Table 12. Modified allocation after node failure

	Node on which allocation has been made
m_{00}	P_2
m_{10}	P_0
m_{20}	P_1
m_{30}	P_2
m_{40}	P_0

old failure rate as λ_{th} =0.005. Since node P_3 and P_4 has been marked as a sick nodes these will not be considered for future allocation. Since the new allocation has suggested the allocation of modules in the order P_2, P_0, P_1, P_2 and P_0 for the modules m_{00}, m_{10}, m_{20}, m_{30} and m_{40}, the turnaround time offered for various modules by the corresponding nodes is as shown in Table 13.

The total turnaround time for the new allocation can thus be calculated as

Turnaround Time

= max (NEC_{ikn}) for 'k' on which allocation has been made

= max (58, 12, 65.5) = 65.5

As evident from Table 13, with node failures the turnaround time has increased from 38 to 65.5 but still the program overcomes the glitches of the node failures to finish the current execution and the job execution is guaranteed. Later these failed nodes can be eliminated from the CT of the respective clusters, which is C_0 in this case and will not be used for future allocation unless repaired.

Table 13. Turnaround time for the modified allocation for Cluster C_0

		P_0	P_1	P_2	P_3 (×)	P_4 (×)
m_{00}						
	E_{ijkn}			15		
	T_{prkn}	10	12	10		
	$\Sigma (B_{ihj} * D_{kl})$			0		
	NEC_{ikn}			25		
m_{10}						
		P_0	P_1	P_2	P_3 (×)	P_4 (×)
	E_{ijkn}	20				
	T_{prkn}	10	12	25		
	$\Sigma (B_{ihj} * D_{kl})$	6				
	NEC_{ikn}	36				
m_{20}						
		P_0	P_1	P_2	P_3 (×)	P_4 (×)
	E_{ijkn}		8.75			
	T_{prkn}	36	12	25		
	$\Sigma (B_{ihj} * D_{kl})$		9			
	NEC_{ikn}		29.75			
m_{30}						
		P_0	P_1	P_2	P_3 (×)	P_4 (×)
	E_{ijkn}			10		
	T_{prkn}	36	29.75	25		
	$\Sigma (B_{ihj} * D_{kl})$			4		
	NEC_{ikn}			39		
m_{40}						
		P_0	P_1	P_2	P_3 (×)	P_4 (×)
	E_{ijkn}	20				
	T_{prkn}	36	29.75	39		
	$\Sigma (B_{ihj} * D_{kl})$	2				
	NEC_{ikn}	58				

EXPERIMENTAL STUDY

Simulation experiments were carried out to observe the behavior of the model. Using equation (i), the turnaround time of the job is observed corresponding to the different number of failing nodes. For each allocation pattern, reliability with which the job can be executed is also calculated using equation (iii). The effect of the number of failed nodes on the turnaround time for varying number of modules of the job and cluster architecture is presented in Figure 1. For the same job and cluster architecture, effect on the reliability of the job execution is shown in Figure 2. Result of the experiment with 15 number of modules resulted in reliability values as low as 0.021. Since the value

Figure 1. Turnaround Time v/s Number of Failed Nodes

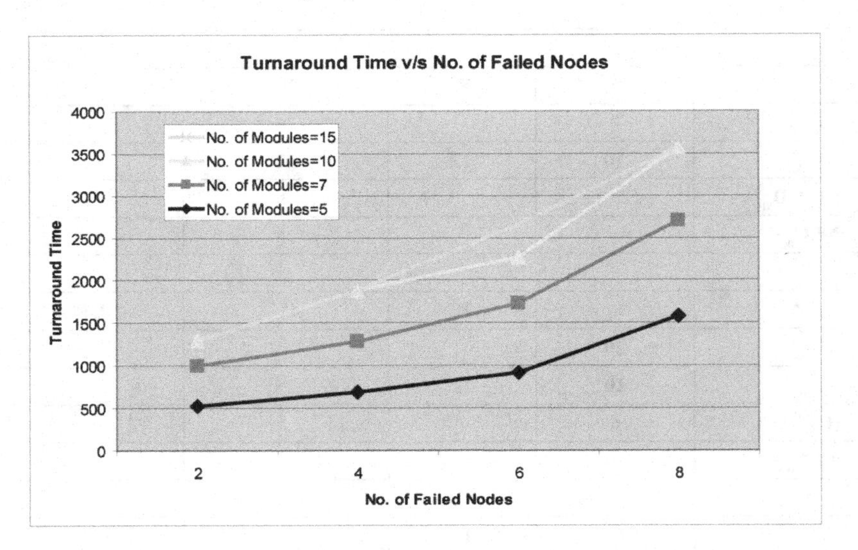

is too low, the result could not be accommodated along with other results shown in Figure 2. Figure 1 and Figure 2 corresponds to the experimental results without using any replication.

It is observed that the turnaround time keeps on increasing with increase in the number of failing nodes. In addition, replication of modules increases the reliability of job execution. This reliability is minimum when maximum number of nodes has failed and increases with reducing number of failing nodes. Thus, in spite of the node failures jobs gets executed with an increased turnaround time adhering to the purpose of the RBS.

Figure 3 and Figure 4 present the effect on grid reliability due to replication. It presents a comparison of the turnaround time and reliability obtained with and without node replication keeping the same grid environment and jobs. The experiments were set to have no node failure when no replication is there. The turnaround time and reliability values are observed. Also, experiment incorporated node failure feature along with module replication and the experiments were run again in the same grid environment for the same job. The turnaround time and reliability values with replication incorporated as reported here correspond to the ones with minimum nodes failed in each experiment.

It is evident from Figure 3 that the turnaround time increases with the introduction of replication owing to the cost of node failures resulting in execution of the job from the replicas. Thus the job gets executed though with some inflated turnaround time. Since, the job is getting executed due to the presence of replicas; it results in an increased reliability for the job as conspicuous in Figure 4. Thus the presence of replica ensures an increased reliability for the job execution. Same pattern, as reported in Figure 1 to Figure 4 is noticed in many more experiments validating the performance of the model.

CONCLUSION

The proposed Replica Based co-scheduler (RBS) helps in the reliable execution of the modular job by replicating the modules allocated to the nodes with high failure rates (sick nodes) to the ones with a lower failure rates (healthy nodes). In place of having full redundancy, partial redundancy has

Figure 2. Reliability v/s Number of Sick Nodes Failing

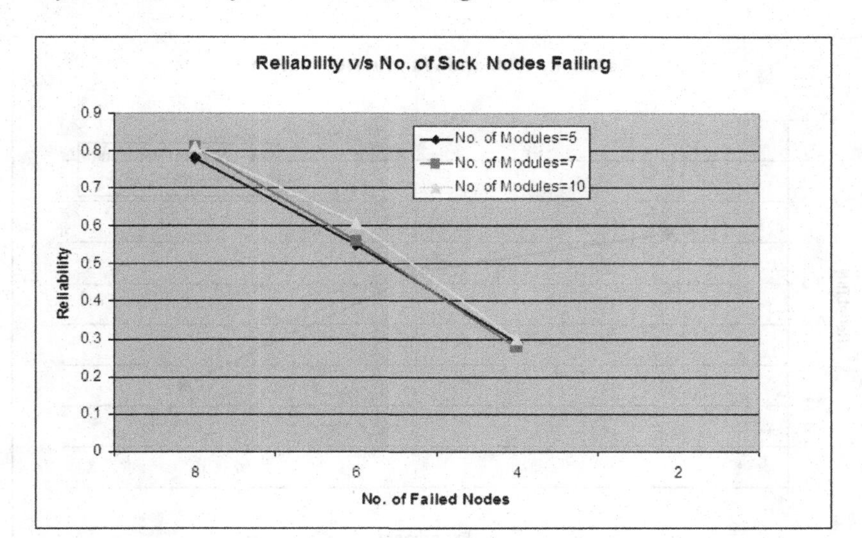

Figure 3. Turnaround Time With and Without Node Replication

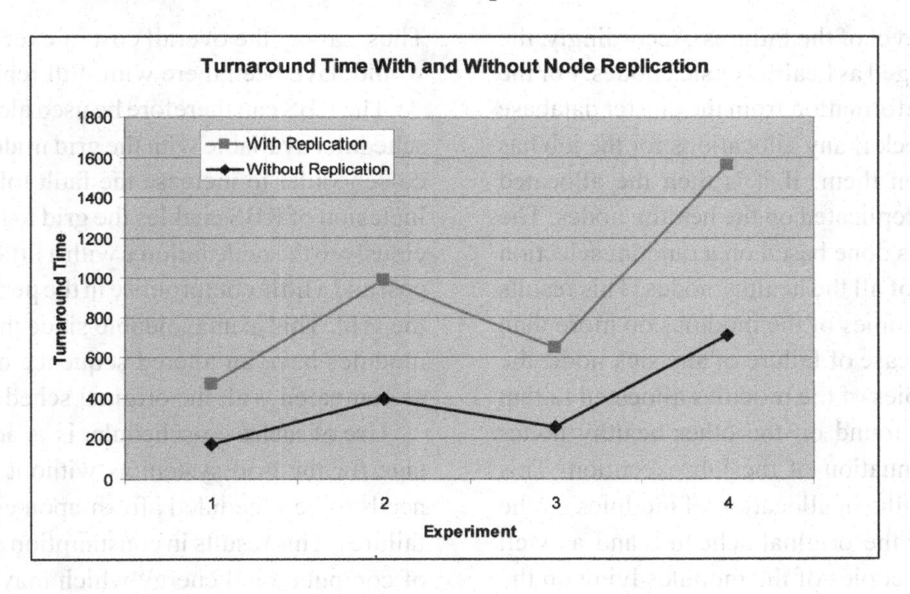

been introduced resulting in better fault tolerance on moderate cost. So a better cost-performance ratio is achieved.

The job of the RBS begins when the job of the main scheduler, responsible for selection of the cluster for job execution, finishes. Performance of RBS is analyzed by considering its integration with the TSM, which is a GA based scheduler proposing an allocation for the job modules which results in the minimum turnaround time offered to the job. For the cluster, selected by the TSM, the RBS evaluates the vulnerability of the nodes on which an allocation has been made by comparing their failure rates λ_{lt} with some threshold failure rate λ_{th}. Selection of the thresholds depends on the domain knowledge of the cluster along with the

Figure 4. Reliability With and Without Replication

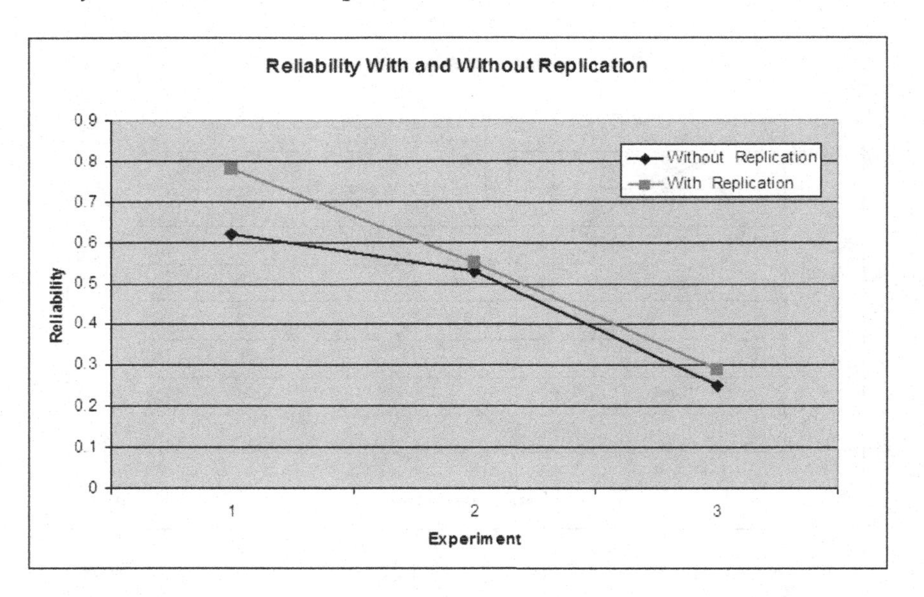

acceptance level of the failures. Accordingly, the nodes are judged as healthy or sick nodes. For the sick nodes, information from the cluster database is used to check if any allocations for the job has been made on them. if it is then the allocated modules are replicated on the healthy nodes. The reallocation is done based on a random selection of nodes out of all the healthy nodes. This results in duplicate copies of the modules on more than one node. In case of failure of any sick node, the duplicate copies of the modules allocated to that node can be found on the other healthy nodes for the continuation of the job execution. This operation results in allocation of modules on the nodes as per the original schedule and as well the duplicate copies of the modules lying on the failure prone nodes. Now if no failure occurs the job gets executed as planned but if node failures are detected, the system does not succumb to these failures rather it gracefully recovers with some additional computational cost. The model doesn't replicate all the modules of all the nodes rather only the modules on susceptible nodes. Thus, saving the overall cost of execution which would have been there with, full replication.

The RBS can therefore be used along with any scheduler available with the grid middleware as a co-scheduler to increase the fault tolerance. The inclusion of RBS enables the grid to respond graciously to the node failures with a little increase in cost and a little compromise in the performance of the grid. This is unavoidable since the replicated modules have an altered sequence of execution as compared with the original schedule.

Use of such a co-scheduler is an added advantage for the grid system as without this the job needs to be scheduled afresh upon encountering failures. This results in consumption and wastage of computational energy which may prove very costly for the high traffic environment like grid. For the real time jobs the problem becomes much more severe as the failures may impact the grid performance and thus hitting the financial prospects of the grid. The use of RBS does not affect the objective of the main scheduler allocating the job. Instead it helps it by providing necessary support towards failures. Experimental study reveals that the proposed RBS model works well under various conditions resulting in a graceful degradation of the grid performance.

ACKNOWLEDGMENT

The authors would like to thank Ms. Sayma Khan for her assistance in some of the experiments conducted for the study of the model.

REFERENCES

Baker, M., Buyya, R., & Laforenza, D. (2002). *Grids and Grid technologies for wide area distributed computing. SP&E.* John Wiley and Sons, Ltd.

Casanova, H. (2002). Distributed computing research issues in Grid computing. *ACM SIGACT News, 33*(3), 50–70. doi:10.1145/582475.582486

Dai, Y. S., Xie, M., & Poh, K. L. (2002). Reliability analysis of Grid computing systems. *Proceedings of the 2002 Pacific Rim International Symposium on Dependable Computing (PRDC'02), IEEE* (pp. 97-104).

Desprez, F., & Vernois, A. (2007). *Simultaneous scheduling of replication and computation for data-intensive applications on the Grid.* Kluwer Academic Publishers.

Ernemann, C., Hamscher, V., & Yahyapour, R. (2002). Benefits of global Grid computing for job scheduling. *Proceedings of the Fifth IEEE/ACM International Workshop on Grid Computing (GRID'04)* (pp. 374-379).

Grid Computing. (2008). *Info centre.* Retrieved from www.gridcomputing.com

Huda, M. T., Schmidt, W. H., & Peake, I. D. (2005). An agent oriented proactive fault-tolerant framework for Grid computing. *Proceedings of the First International Conference on e-Science and Grid Computing (e-Science'05), IEEE* (pp. 304-311).

Li, Y., & Mascagni, M. (2003). Improving performance via computational replication on a large-scale computational Grid. *Third IEEE International Symposium on Cluster Computing and the Grid (CCGrid'03), Tokyo, Japan* (pp. 442-448).

Liu, L., Wu, Z., Ma, Z., & Cai, Y. (2008). *A dynamic fault tolerant algorithm based on active replication.* Seventh International Conference on Grid and Cooperative Computing, China (pp. 557-562).

Mujumdar, M., Bheevgade, M., Malik, L., & Patrikar, R. (2008). *High performance computational Grids - fault tolerance at system level.* International Conference on Emerging Trends in Engineering and Technology (ICETET) (pp. 379-383).

Raza, Z., & Vidyarthi, D. P. (2008). *Maximizing reliability with task scheduling in a computational Grid.* Second International Conference on Information Systems Technology and Management(ICISTM), Dubai, UAE.

Raza, Z., & Vidyarthi, D. P. (2009). GA based scheduling model for computational Grid to minimize turnaround time. *International Journal of Grid and High Performance Computing, 1*(4), 70–90. doi:10.4018/jghpc.2009070806

Sathya, S. S., Kuppuswami, S., & Ragupathi, R. (2006). *Replication strategies for data Grids.* International Conference on Advanced Computing and Communications ADCOM, India (pp. 123-128).

Tarricone, L., & Esposito, A. (2005). *Grid computing for electromagnetics.* Artech house Inc.

Vidyarthi, D. P., Sarker, B. K., Tripathi, A. K., & Yang, L. T. (2009). *Scheduling in distributed computing systems*. Springer. doi:10.1007/978-0-387-74483-4

Zhang, J., & Honeyman, P. (2008). *Performance and availability tradeoffs in replicated file systems*. Eighth IEEE International Symposium on Cluster Computing and the Grid, Lyon, France (pp. 771-776).

Section 3
Security

Chapter 8
A Policy-Based Security Framework for Privacy-Enhancing Data Access and Usage Control in Grids

Wolfgang Hommel
Leibniz Supercomputing Centre, Germany

ABSTRACT

IT service providers are obliged to prevent the misuse of their customers' and users' personally identifiable information. However, the preservation of user privacy is a challenging key issue in the management of IT services, especially when organizational borders are crossed. This challenge also exists in Grids, where so far, only few of the advantages in research areas such as privacy enhancing technologies and federated identity management have been adopted.

In this chapter, we first summarize an analysis of the differences between Grids and the previously dominant model of inter-organizational collaboration. Based on requirements derived thereof, we specify a security framework that demonstrates how well-established policy-based privacy management architectures can be extended to provide the required Grid-specific functionality. We also discuss the necessary steps for integration into existing service provider and service access point infrastructures. Special emphasis is put on privacy policies that can be configured by users themselves, and distinguishing between the initial data access phase and the later data usage control phase. We also discuss the challenges of practically applying the required changes to real-world infrastructures, including delegated administration, monitoring, and auditing.

DOI: 10.4018/978-1-60960-603-9.ch008

INTRODUCTION

Using compute and storage services starts with selecting an appropriate IT service provider (SP). Within their terms of use and privacy statements, SPs define which information about a customer (and, if the customer is an organization, its users) they require in order to provide the selected service. It also must be specified for which purposes the collected data will be used, and how long it will be retained. Typically, customer and user information is required for accounting and billing purposes as well as for service personalization. Generally, it thus includes personally identifiable information (PII), i.e., data that can be used to uniquely identify a single person.

In order to prevent any misuse of such sensitive data, e.g., selling email addresses to marketing agencies, legislative regulations exist; they restrict how PII may be used on an organizational level and must be mapped to technical solutions, which often have been neglected in the past, resulting in potential vulnerabilities. Although privacy and data protection laws differ between countries and dedicated regulations exist for industrial sectors such as finance and healthcare, one classic and common principle is that data must only be used for purposes which the user has been informed about and agreed to.

As intra-organizational solutions so-called privacy management systems have successfully been implemented and deployed over the past few years. They are tightly coupled with the IT services used by the customers as well as with other management systems, such as billing and invoice management tools. Whenever a user's or customer's data is about to be accessed, rule sets are evaluated to determine whether the current access attempt is in accordance with the privacy policy the user has agreed to. Basically, such systems can be viewed as an extension of traditional access management systems in order to enforce the purpose limitation principle: They also take into consideration *for which specific purpose* someone is trying to access the data; formally specifying such policies requires extensive modeling of the involved roles, the acceptable purposes, and the available PII itself.

In inter-organizational service usage scenarios, such as Grid computing, privacy protection becomes an even more complicated issue, because multiple organizations – typically also located in different countries – are involved and SPs need to retrieve the required user data from the user's home organization in an automated manner.

Instead of a single organization's privacy policy, multiple heterogeneous demands must now be fulfilled regarding PII handling. For example, there usually will be Grid-wide privacy policies, such as those specified by a virtual organization (VO); they must often be adequately combined with SP-specific or user home organization specific policies, as well as policies eventually specified by the users themselves. Combining policies requires the handling of conflicting policy parts in a transparent manner.

In general, privacy management – intentionally with a strong focus on the user – becomes a two-tiered process: First, users must decide which of their data may be submitted to an SP at all, and second they must be able to monitor and control how their data is being used later on.

In the research areas of privacy enhancing technologies (PET) and federated identity management (FIM), various solutions to these issues have been suggested, with many of them already being used in production environments by commercial as well as academic SPs; a short overview will be given in the next section.

However, these solutions were originally not suitable for certain characteristics of Grid environments, such as the concept of VOs, and cover only the PII of the users themselves; thus, they neglect sensitive data submitted along with Grid jobs, such as medical records used as input data for those programs. In this article, we first discuss these differences of Grid environments and point out the relevant shortcomings of previous approaches regarding Grid-specific requirements.

Furthermore, we advocate that existing policy-based privacy management approaches can be adapted to provide the additional functionality required in Grids. Then, the architecture of our specifically privacy-aware security framework, which is based on the policy language XACML and intended to be applied by Grid architects and SPs' IT-security personnel, is presented. Afterwards, the integration of the discussed privacy management components into existing infrastructures along with its challenges in real-world projects are discussed. An outlook to our future research concludes the article.

PRIVACY MANAGEMENT IN LARGE-SCALE DISTRIBUTED SYSTEMS

The privacy management issues sketched above are, even on an inter-organizational level, neither a new nor a Grid-specific research issue. For this reason, we confine the following discussion of the state of the art to those approaches that are appropriate to build the base of a Grid-specific solution. To put the related work discussed below into the big picture, we simplify by stating that FIM provides an inter-organizational framework for the exchange of user data, while PET focus on the user-centric view of privacy management options; this means that PET puts the user in control of how her personal data is used by the involved organizations, which in turn use FIM protocols to actually exchange this data technically. In practice, FIM and PET must always go hand-in-hand due to regulatory requirements w.r.t. IT compliance.

From this legislative perspective, regulations regarding privacy and data protection become relevant as soon as personal data is being acquired by an organization, i.e., before the data is actually being used, e.g., for the personalization of the IT service ordered by the user. For distributed collaborative environments spanning several organizations, this implies that as a first step it must be decided which user PII is made available to which of the multiple involved organizations at all. Obviously, users must express their consent to such a distribution of their data adequately, i.e., either explicitly on a per-organization basis, or implicitly, e.g., based on a framework agreement or service contract. As an example, the acceptable use policies (AUP), which many Grid projects require their users to sign, typically include such consent. Privacy requirements have also been gathered for specific application domains, such as the use of Grids in medical research (see Manion, Robbins, Weems, & Crowley, 2009).

Once an organization has gained access to a user's data, there must be technical means to control and influence how the data may – or may not – be used in order to prevent the misuse of PII; this so-called *usage control phase*, which is typically parameterized with specific usage purposes along with the initial transmission of the data, ends with the deletion of the acquired data, e.g., after service usage as well as the accounting and billing processes have finished. These two phases will be discussed in the following subsections.

Managing Initial Data Access

All major federated identity management technologies, such as the Security Assertion Markup Language (SAML) (see Hughes & Maler, 2005), the Liberty Alliance (also known as the Kantara Initiative) specifications (see Wason, 2004), and the Web Services Federation Language (WS-Federation) (see Kaler & Nadalin, 2003), as well as several of the Grid middleware implementations that are currently in operation, use request-response-based protocols for the retrieval of information about the current user. As a consequence, decisions about which user data an SP is allowed to retrieve are often treated similarly to classic access control issues, and thus access control languages and suitable management tools for them are the most widely deployed solutions.

Figure 1. Managing data access at the home site / identity provider

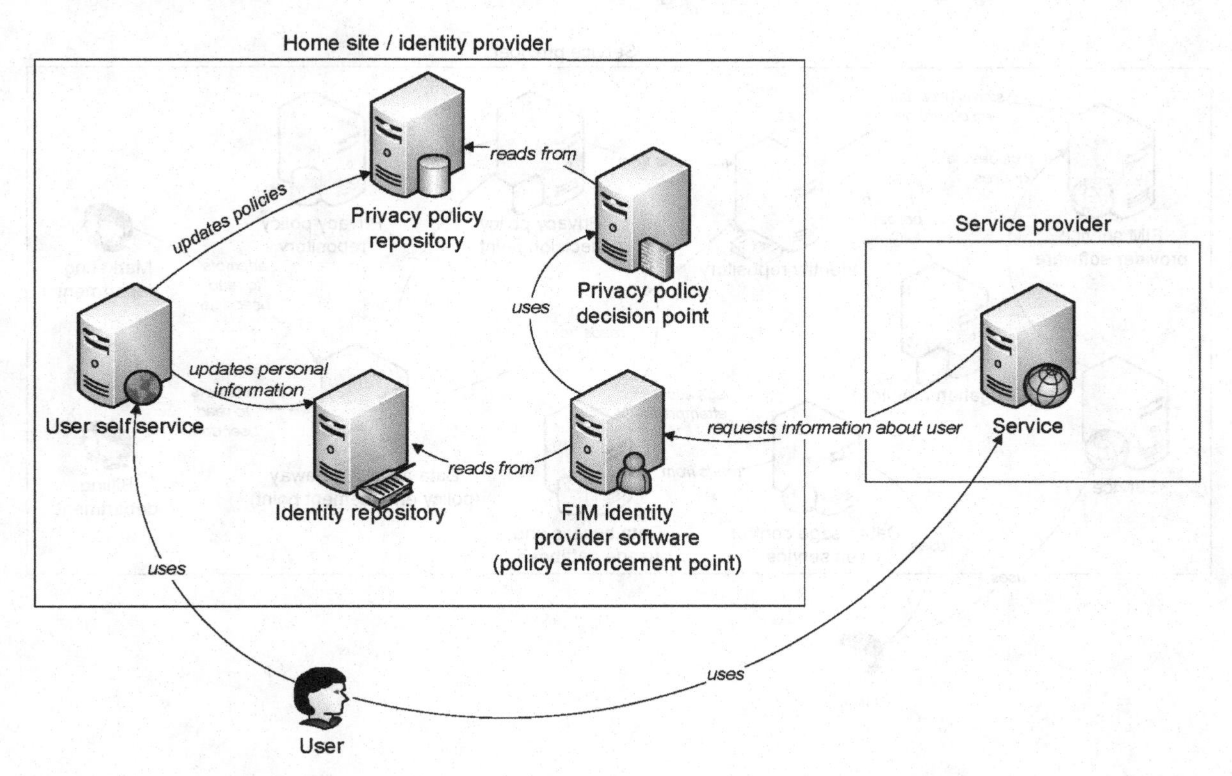

As shown in Figure 1, the user's home organization has the role of a FIM identity provider (IDP). All user data is stored in a local identity repository; this repository is usually realized as an LDAP-based enterprise directory, but for smaller deployments also relational database management systems (RDBMS) are being used in practice. A policy decision point (PDP) is used to determine which user attributes, such as name or email address, may be released to which service provider; this workflow has coined the terms attribute release policies (ARPs) and attribute release filtering (ARF).

Common to most current research approaches in this area is, in fact, the use of policy-based management. Thus, the technical architectures are quite similar and involve, among other components, policy repositories, policy decision points, and policy enforcement points (PEP). They differ, however, in the policy language that is actually being used: On the one hand, the language's expressiveness is relevant, e.g., whether and which usage purposes and obligations, for example concerning data retention limits, can be specified. On the other hand, arithmetical properties, such as efficiently calculating policy set intersections, are of major concern. Well-known approaches include Tschantz and Krishnamurthi (2006) and Spantzel, Squicciarini, and Bertino (2005), which put an emphasis on efficient negotiation handling and policy evaluation. A more detailed overview can be found in our previous work (Hommel, 2005a).

However, these approaches require the a priori definition of policies, which may be too complicated for many users. Thus, interactive solutions have been proposed by both research (e.g., Pfitzmann, 2002; Pettersson et al., 2005) and industry, e.g., the Liberty Alliance interaction service (Aarts, 2004). To enhance these approaches,

Figure 2. Controlling data usage at the service provider

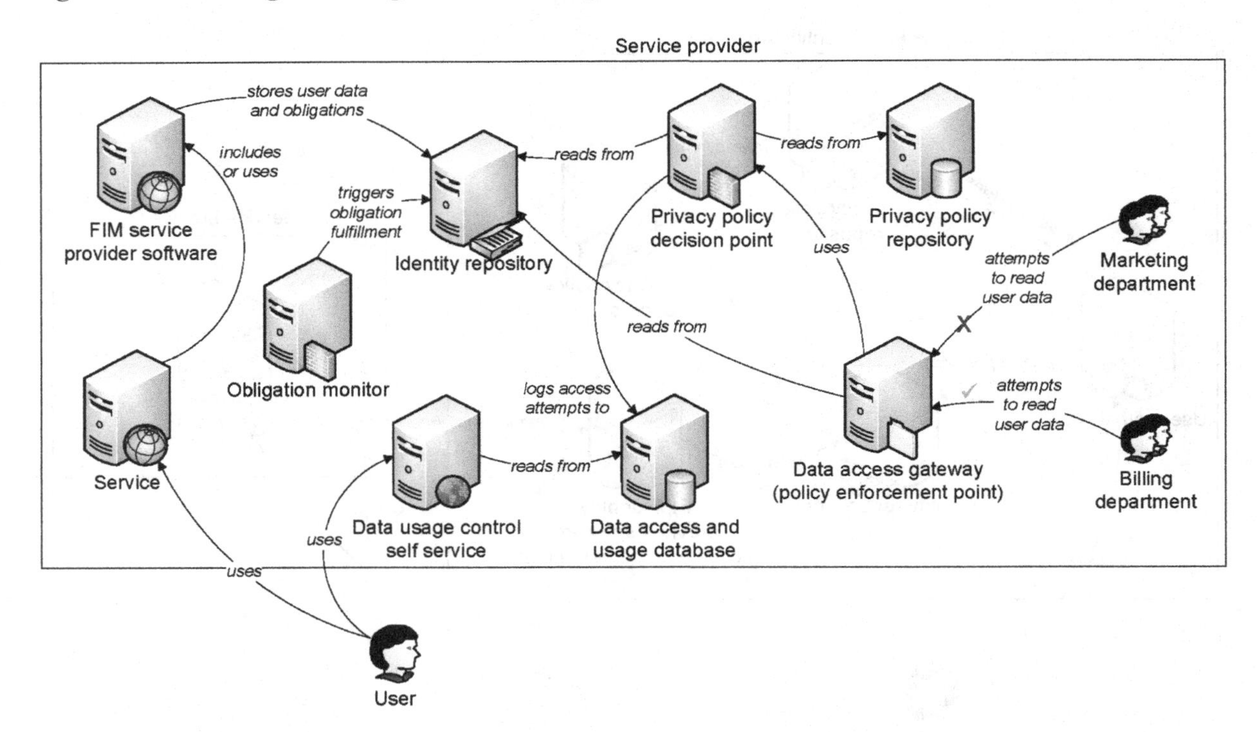

research focuses on usability issues, such as how to avoid that users grow tired of repeatedly giving their interactive consent to the transmission of their personal data to various service providers. Those usability aspects can be compared to how web browser users are asked about previously unknown server certificates for HTTPS access to web servers: The users must be made aware of security and privacy issues without harassing them when asking for their informed consent.

Controlling Data Usage

Complementary to the privacy management components on the IDP side, which have been described in the previous section, the SP, which retrieves the user data, also needs a privacy management infrastructure to ensure that such personal data is only used for the purposes agreed to by the user. Enhanced solutions additionally provide interfaces to the users, so that they can look up how their PII has been used. However, because an SP may not log all data access or eventually even lie about how the PII has been used, it is hard to reliably verify whether all privacy preferences have really been met from the user's perspective, which often remains a weak spot of technical implementations.

Privacy management systems, such as EPAL (Powers & Schunter, 2003), are typically also policy-based. Access to user data by any application is handled by a privacy PEP as shown in Figure 2. A PDP decides whether the application and its operator are allowed to access a particular user attribute for a given purpose. Thus, the key difference to traditional access control is the additional consideration of the purpose behind the data access. For example, an employee in the billing department may retrieve the user's postal address to send an invoice, while the marketing department must not access the address in order to avoid unsolicited advertisements.

Furthermore, so-called obligation monitors can be used to trigger the fulfillment of obligations which are part of privacy policies. Obligations can, among other goals, be used to restrict the PII data retention, so, e.g., all user data has to be deleted 90 days after the service usage has finished and all invoices have been settled. Some implementations also allow the users to specify obligations, e.g., to be notified by email whenever one's credit card is being charged by the SP, i.e., when the credit card detail attributes are being accessed for a purpose such as billing.

As an organization's privacy policies may change over time, it is vital for the privacy management system to keep track of which version of the policy was in use when a user signed up for a service. The sticky policy paradigm (Mont, Pearson, & Bramhall, 2003) glues the relevant policies to the user data so they cannot be separated anymore.

Protocols and log files of data access and usage are kept to support the organization's internal auditing processes, which are a mandatory part of legislative IT governance, risk management, and compliance regulations. Parts of this information can be made available to the user to prove that her data has only been used for the intended purposes. However, unless additional measures are taken, the usefulness and reliability of this information for the user is very limited, because malicious service providers could arbitrarily falsify the presented data. Thus, all recent approaches are based on certified software running on trusted computing platforms in order to guarantee the genuineness of the information given to the users (see Mont (2004)) as well as Bramhall and Mont (2005)). However, the complexity and costs of such solutions have so far impeded their wide-spread use. In Grids, trusted computing has already been applied to user management from the SP perspective (see Mao, Martin, Jin, & Zhang, 2009), but not yet vice versa to rate the SP trustworthiness from the users' perspective. Thus, having to trust SPs regarding their claims about what they use (or do not use) the PII for still remains a major challenge in research and in practice. For this reason, managing the initial data access phase and avoiding to transfer user data to untrusted SPs a priori is of high importance.

GRIDS AND THEIR REQUIREMENTS FOR PRIVACY MANAGEMENT

On the technical level, Grid computing is based on a Grid middleware which provides the required transparency layers and tools for submitting Grid jobs. Various Grid middleware implementations, such as the Globus Toolkit (Sotomayor & Childers, 2006), exist and are in practical use. In the first decade of Grid computing, the development of Grid middleware has focused on the core functionality. However, with increasing use in production environments and based on the goal of creating an environment that is also attractive to industry, the security and privacy properties finally get the required attention (see also Demchenko, de Laat, Koeroo, & Groep, 2008).

Because most of the organizations involved in Grid projects have identity management systems deployed nowadays, there is an increasing real-world demand to leverage the existing local infrastructure when participating in Grid projects. Concerning privacy management, however, this is not just a programming interface and implementation effort issue regarding the middleware. Grids have several characteristics and thus specific requirements which were not yet met by the approaches discussed in the previous section; we will discuss them next.

Starting with the technical aspects, which are – unlike the organizational issues discussed below – applicable to all Grids in general, it must be considered that using a Grid infrastructure differs from using other distributed systems and services in the concept of *Grid jobs*. When submitting a

Grid job, the user cannot only provide input data to a pre-defined service provided by an SP. Instead, the user lets own program code make use of the CPU and storage capacities provided by the SPs that are involved in the Grid.

This immediately leads to the consequence for privacy and data protection in Grids that any data related to a user's Grid job must be treated similarly to the user's PII:

- The Grid job's code, independent of whether it is being distributed in source or binary format, should be considered intellectual property of the Grid user. Especially in commercial Grid environments it must obviously be avoided that program code submitted by one user is redistributed by the service provider or made available to other users. However, this also affects whether an SP may modify the program code, e.g., in order to optimize it for the local computing architecture.
- Input data for the Grid job may contain sensitive data, e.g., when Grid-based data mining is performed on large sets of medical data. In this case, both the Grid user and the SP share a couple of responsibilities. On the one hand, the Grid user must have the permission to submit the data to the SP; this is a non-trivial organizational task because the utilized Grid service providers are, in general, unknown at the point in time when the input data is being collected. On the other hand, the SP to which a Grid job has been submitted is typically not allowed to make any use of the input data other than feeding it into the Grid job's code. Thus, similarly to the handling of PII, the user and the SP must agree on a set of purposes for which the data may be used. Obviously, there must be technical means to enforce this binding.
- On the SP side, the considerations for the input data must also be applied to the Grid job's output data. Depending on the Grid job, the output data may be even more sensitive than the input data. As an example, consider data mining on medical data which derives a set of potentially terminally ill patients. Thus, there must be an agreement about how the output data must be treated, both while the Grid job is running and after it has finished. This affects, for example, whether the output data has to be deleted from the service provider's systems after the user has retrieved it, or whether it should be kept, e.g., as input data for a subsequently submitted follow-up Grid job.

Additional aspects, such as whether the SP is allowed to backup or even archive these Grid job components, must also be taken into consideration. As an obvious resulting requirement, services which are shared by multiple or all organizations in the Grid, such as globally distributed file systems, must provide sufficient access control mechanisms to prevent organizations, which are not involved in a particular Grid job, from accessing its code, input data, and output data to achieve confidentiality and a separation of concerns on an organizational level (see also Cunsolo, Distefano, Puliafito, and Scarpa (2010)). In this context, it should be noted that encryption of input and output data would hardly increase security, as long as a potentially malicious SP runs the Grid job and thus gains access to the data in clear text.

Privacy and data protection settings may also vary with each Grid job, independent of the users' preferences regarding their own PII. As a consequence, the logical separation between PII and Grid job privacy management must be accounted for. This is not only relevant for Grid job execution engines, but also, e.g., for the design of (graphical) user interfaces.

Because the use of Grid middleware does not depend on the existence of an appropriate inter-organizational contractual framework, it is impossible to fully automate all privacy relevant decisions on the technical level. If the organizations involved in a Grid project decide to form a VO that becomes a legal entity, managing privacy preferences can be greatly simplified by treating the resulting Grid environment like a single organization. However, the technical approaches discussed in the previous section do not fully support the concept of VOs; a solution is discussed below.

Unless privacy-related contractual agreements can be arranged for all organizations participating in a Grid project, such as in VO scenarios, the vision of a Grid middleware offering total location transparency to the user is actually contradictory to the privacy management goal that users get to know exactly by whom their data is being processed. Thus, the traditional approach that users can define privacy preferences on a per-organization basis must be complemented by means to define what we call property-based privacy policies (PBPP). As an example, certain PII such as the user's email address should only be distributed to SPs which guarantee to only use it for contacting the users in case of technical problems, but not for other purposes such as sending marketing emails. Hence, this allows modeling the situation that it would not matter to the user which SP will actually execute the Grid job, as long as it is assured that all of the user's privacy preferences are met. In this regard, PBPPs can be seen as a contribution for attribute-based access control applied to organizations (cp. (Kuhn, Coyne, & Weil, 2010)).

We will discuss how previously established policy-based privacy management approaches need to be extended and enhanced to fulfill these new requirements in the next section.

ARCHITECTURE OF A POLICY-BASED SECURITY FRAMEWORK FOR PRIVACY-ENHANCING DATA ACCESS AND USAGE CONTROL IN GRIDS

The primary motivation for using a policy-based privacy management approach in Grids is to leverage existing identity and privacy management infrastructure components, which in turn is motivated by the goal to reduce the IT service management overhead and costs of solutions specific to the Grid domain. The basic suitability and applicability of policy-based approaches for privacy and data protection management has been pointed out by the previous work referred to above and is not discussed here, because the discussed Grid-specific requirements are by no means fundamental challenges to the policy-based management paradigm.

In this section, we motivate how policy-based privacy management can be used in Grids and demonstrate how the existing approaches can be extended and enhanced to fulfill the discussed Grid-specific requirements in a general manner, with the overall goal of protecting privacy relevant data from being misused by the SPs. The concrete application of this methodology to a selected privacy management architecture is discussed afterwards. As a first step, we need to consider that for any transmission of sensitive data, more than one policy may be relevant; in practice, there typically are four layers of policies:

1. Users can specify their personal privacy preferences, i.e., the conditions and obligations under which they are willing to share their data with an SP. This is also an effective way to delegate the management of dynamic policies to the users in order to reduce the overhead for home site and SP administrators. However, it also requires adequate, user-friendly management front-ends for policy creation, testing, and maintenance;

furthermore, trainings or introductory courses should be provided.

2. The user's home site (IDP) has privacy policies in place which typically provide default settings for all of its users. These defaults must be crafted carefully and are primarily intended to protect the privacy of the lesser privacy concerned users (see Berendt, Günther, and Spiekermann (2005) for an analysis of privacy-related user classifications). In general, these policies can be re-used for several Grid projects, VO memberships, and other external services.
3. Also each SP has its own privacy policies, which are not necessarily Grid-specific. For example, many academic supercomputing centers restrict access to their computing resources to users from selected countries. Thus, they can offer their service only to users whose nationality is revealed. If a user is unwilling to share her nationality, she will not be allowed to use the service. Similar to the home site policies, these SP policies can be re-used for external users from different Grid projects, VOs, or other inter-organizational collaborations.
4. Grid projects and VOs may have privacy policies which must be honored by all participating organizations and applied to all users (Schiffers et al., 2007), i.e., the implementation and management is delegated to the organizations participating in the project or VO.

In most approaches and implementations, the number of layers may vary with scenario-specific requirements, such as additional service-specific policies on top of SP-wide policies. There can be multiple policies in each layer, and it needs to be determined for each individual data request which policies are relevant. There may be conflicting policies, e.g., if an SP's privacy policy requires a user attribute such as the nationality when the user's personal privacy preference prohibits its release. In practice, sufficiently disjoint policies are ensured only on the same layer, usually by user-friendly management front-ends; thus, for example, administrators on the SP side are forced to formulate consistent SP policies. However, conflict resolution across the layers is often subject to a scenario-specific configuration, i.e., it cannot be defined in general whether, for example, user-specified policies override VO-wide policies or vice versa. Once such priorities have been defined, however, policy conflict resolution can be automated using PDP engines.

As discussed above, we must distinguish between privacy policies for PII and for Grid jobs on the user layer:

- The user's personal privacy preferences will usually stay the same over a certain period of time and are independent of the submitted Grid jobs to a certain (usually high) degree.
- While it must be possible to configure privacy policies for individual Grid jobs, there often is the situation that multiple Grid jobs belong to the same research project or are otherwise closely related. Thus, to reduce the management overhead, privacy policies must be applicable to groups of Grid jobs, which may arbitrarily be submitted sequentially or in parallel. Furthermore, if multiple Grid users are involved in the same research project, an additional Grid project policy layer contributes to simplifying the sharing of policies among all users submitting related Grid jobs.

However, the inter-organizational sharing of policies adds yet another layer of complexity and thus can often only be realized in later project stages. Enabling users to specify their privacy preferences locally at their home site usually is a good starting point.

Figure 3 shows the resulting modular privacy management architecture for the user's Grid

Figure 3. Privacy management architecture for the user's Grid home site

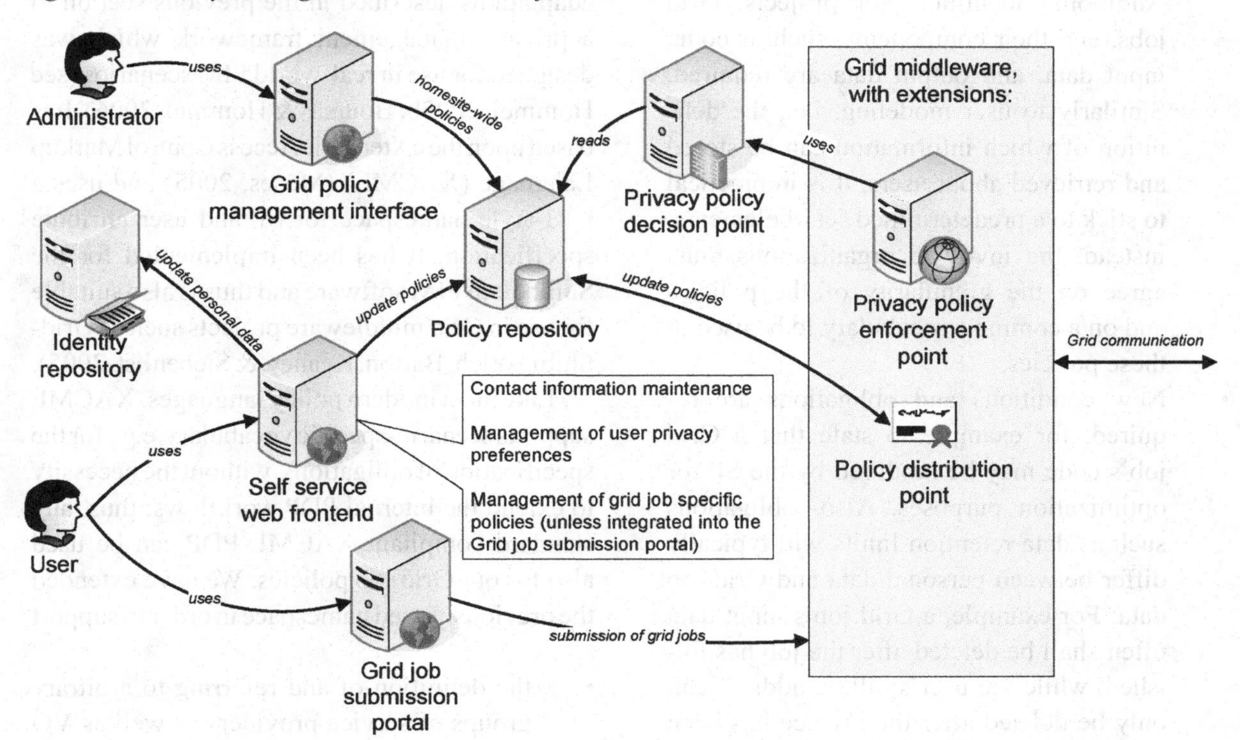

home site. Although each technical component is only shown once, high availability requirements can be fulfilled, e.g., based on hardware redundancy and clustering. Compared to previously used architectures as depicted in Figure 1, a logically separated management user interface is provided as part of the self services, which allows to configure project- and Grid-job-specific privacy policies. Furthermore, not only Grid-wide applicable policies must be exchanged between the involved organizations, but also the policies of those research projects whose users are spread among multiple organizations. The same policy distribution mechanisms are used for both use cases. However, it must be ensured that they provide metadata support to restrict a) to which organizations the policies are transferred to and b) which other users may access and modify them.

The components used in the architecture usually have multi-tenancy capabilities, i.e., they can be used for an arbitrary number of other services, Grid projects, VO memberships, and users, without requiring additional instances. They also often provide code hooks for site-specific extensions, so additional workflows can be triggered, e.g., in the policy evaluation process. At each home site, the Grid-specific components also can be combined with other security and privacy measures that are deployed locally.

The expressiveness of the used policy language is, in general, sufficient to handle the additional Grid job policies and groups thereof, so no in-depth modifications of PDPs and PEPs or other Grid-specific technology adaption are required. However, the syntactical basis for identifying and naming objects, often referred to as policy namespace, must be extended as follows:

- Instead of targeting a policy to a single SP, it must be possible to specify policies for arbitrary groups of organizations, up to a Grid environment such as a VO as a whole.

- Additional identifiers for projects, Grid jobs, and their components, such as code, input data, and output data are required. Similarly to user modeling, i.e., the definition of which information can be stored and retrieved about users, it is impractical to stick to a predetermined set of elements; instead, the involved organizations must agree on the granularity of the policies and on a common vocabulary to be used in these policies.
- New conditions and obligations are required, for example to state that a Grid job's code may be modified by the SP for optimization purposes. Also, obligations such as data retention limits will typically differ between personal data and Grid job data: For example, a Grid job's input data often shall be deleted after the job has finished, while the user's billing address can only be deleted after the invoice has been settled. Again, the complete definition of the necessary vocabulary is a task that is specific to each Grid environment, and standardization is required to provide a common subset of the vocabulary.

On the service provider side, no extensions to the privacy management architecture are required, with exception of support for any newly defined obligations. However, in practice so far only a limited number Grid SPs supports privacy management at all; the challenge of integrating the described privacy management components into Grid-specific workflows is discussed below.

APPLICATION OF THE SECURITY FRAMEWORK TO A XACML-BASED PRIVACY MANAGEMENT ARCHITECTURE

In order to show the feasibility of the presented approach, we have applied the extensions and adaptations described in the previous section to a privacy management framework which was designed for use in real-world FIM scenarios (see Hommel, 2005b; Boursas & Hommel, 2006). It is based upon the eXtensible Access Control Markup Language (XACML) (Moses, 2005) and uses a URI-style namespace for SP and user attribute specification. It has been implemented for the Shibboleth FIM software and thus is also suitable for use in Grid middleware projects such as GridShib (Welch, Barton, Keahey, & Siebenlist, 2005).

Like most modern policy languages, XACML supports scenario-specific vocabulary, e.g., for the specification of obligations, without the necessity to extend the internal PDP workflows; thus, any standard compliant XACML PDP can be used also for our Grid job policies. We have extended the previously used namespace in order to support

- the definition of and referring to arbitrary groups of service providers as well as VO identifiers (for VO management approaches, see Kirchler, Schiffers, & Kranzlmüller, 2009).
- the specification of Grid projects as groups of Grid jobs, the Grid jobs themselves, and their components; the granularity chosen for the components is *code*, *input*, and *output*. This granularity is a trade-off between very fine grained control and the implementation effort required at each involved SP.
- new conditions, such as *(allow/disallow) optimization (of code)* and *(allow/disallow) backup (of code, input, or output)*, as well as new obligations, e.g., *delete-after-execution (of input or code)*.

Figure 4 shows an example of a Grid job policy, which allows all Grid service providers to modify the code for the purpose of optimizations w.r.t. the local computer architecture. Note that XML namespaces have been omitted in the example to improve the readability of the XML fragment.

Figure 4. Example XACML Grid job policy to allow code optimization

```
<Policy id="GridJobPolicyExample1" RuleCombiningAlgorithm="first-applicable">
    <CombinerParameters>
        <CombinerParameter ParameterName='PolicyPriority'>
            100  <!-- Priority in case multiple policies are relevant for a request -->
        </CombinerParameter>
    </CombinerParameters>

    <Description> Grant access to code for optimization purpose </Description>

    <Rule id="ExampleRule1" effect="permit">
        <Target>
            <Resource> <!-- Specifies grid job data according to the chosen namespace -->
                https://org1.example.com/project/username/gridjobs/id/code
            </Resource>
            <Subject>  <!-- Grant access to all service providers within the VO, but... -->
                https://grid.example.com/members/VO
            </Subject>
            <Action>  <!-- ... restrict access based on the specified purpose. -->
                gridjobs/code-optimization
            </Action>
        </Target>
    </Rule>
</Policy>
```

Such a policy must be complemented by other policies for restricting the selection of suitable SPs and excluding other usage purposes in practice, which usually is a home site administrator task to be performed for all local users as a whole. Whether only few but complex, or many simple policies are used, depends on the management user interface; in real-world application, intuitive usability and the re-use of modular policies have so far proven to be of higher relevance than performance issues: Given the overall low number of policies and the average run-time of Grid jobs, evaluating the described policies does not cause any latency which the user would notice, and thus performance optimizations are currently not a priority, because more than sufficient scalability is already achieved.

INTEGRATION OF THE SECURITY FRAMEWORK'S PRIVACY MANAGEMENT COMPONENTS ON THE SERVICE PROVIDER SIDE

While the integration of privacy management components into the user's home site is straightforward, especially if an privacy-enhancing identity management system is already in use, the adaptation of Grid SPs is a challenging task. It also must be kept in mind that especially in scientific Grids, such as the European DEISA consortium (Niederberger & Alessandrini, 2004), often all involved organizations are both, home site and SP.

The use of FIM protocols, which are also typically being used for other aspects of user management, e.g., authentication and authorization, ensures that personal and Grid job data is only distributed to Grid SPs that are suitable from the privacy management perspective. Thus, privacy management on the SP side primarily pursues three goals:

Figure 5. Privacy management architecture for the Grid service provider

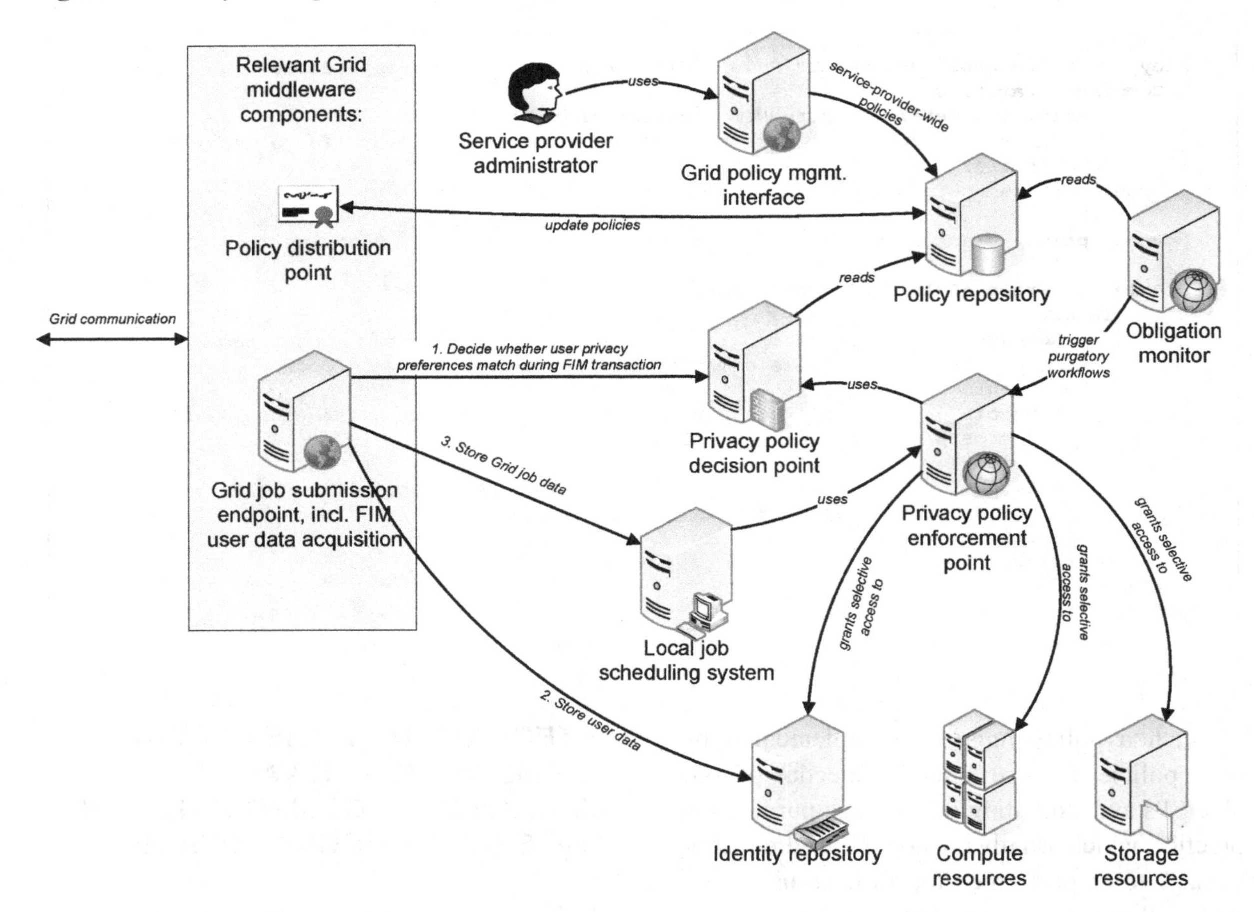

1. All personal and Grid job data may only be used in accordance with the privacy policies specified by the SP; it is safe to assume that these policies match the user's privacy preferences if all required data has been received via FIM protocols.
2. All user and Grid job specific obligations must be fulfilled. This necessitates the use of an obligation monitoring component.
3. It shall be possible for the user to verify whether the obligations have been fulfilled and that the user's PII has not been used for any other than the agreed purposes.

Figure 5 shows the resulting modular privacy management architecture with the required interfaces between the technical components and the Grid middleware. Clearly, protecting any personal and Grid job data from direct access by channelizing all data read, update, and delete attempts through the privacy PEP requires adequate hooks in the middleware on the data persistence layer, as well as additional error handling for privacy violation exceptions. Similarly to the home site architecture, the component's high availability can be achieved, e.g., through hardware redundancy.

For many supercomputing SPs, this also necessitates that the existing batch scheduling systems, which queue Grid as well as regular jobs to be run on the machines according to a local job execution policy, also contact the privacy PDP and honor the decision about whether the data may be accessed. Because many scheduling systems are proprietary or not available as open source, usually simple wrapping mechanisms have to be implemented; they can also be used to trigger obligation handling actions after the execution of a Grid job. To this extent, it is important to distinguish between the successful execution of a Grid job and errors (e.g., machine or job crash). The same mechanism can be used to extend the available components with site-specific additional workflows.

For the fulfillment of Grid-specific obligations, additional functionality is required within the local obligation monitor (OM). So far, the OMs in place have mostly been used to purge outdated user records from relational database management systems or enterprise directories, and to send emails to users or administrators to notify them about the status of their obligations. With privacy relevant data no longer stored only in databases and enterprise directories, additional workflow mechanisms are required to delete Grid job components from the involved compute platforms, including local as well as global or Grid-wide file systems. As this obligation handling typically requires site-specific implementation efforts, it is a good starting point to accept Grid jobs only without obligations first, and then add obligation support later on.

While the overall framework clearly has a preventive character, i.e., privacy policy violations shall be averted before they actually happen, there is also a demand for detecting irregularities and appropriately reacting to them. However, granting the users reliable insight into how their data has been used by the SP as a first step is still challenging: Grid users presently typically have terminal access via GridSSH or can manage their job files through Grid web portals. Both ways provide a suitable feedback channel, which can be used to make, for example, SP log file excerpts available to the user. However, there still is no guarantee that the logged information is sound and complete. The complexity to technically ensure that all data access is being logged and to prevent even administrators from tampering with the logs is incomparably higher than for single-SP services. Thus, until secure and trusted operating systems are used for Grid resources, the user's informational self-determination can already be supported, but the guaranteed enforcement of privacy policies cannot be verified in an absolute objective manner. Besides such information requests performed by the users themselves, there also must be an internal auditing and reporting process that checks the SP infrastructure for privacy policy violations on a regular basis in a pro-active manner. This process can often be supported and automated to a large degree with the available PMS, logfile correlation engines, or security information and event management systems. Reports should include, e.g., the number of successfully fulfilled privacy policies, detected policy conflicts, unfulfilled obligations, etc. The resulting figures are important feedback for different enterprise roles, such as privacy officers, policy writers, and service administrators. In general, selected events, such as policy violations, should also be used to trigger real-time alerting mechanisms. Policy violations and other undesired behavior should also be considered to serve as key performance indicators (KPIs) and, e.g., their maximum number per reporting period may become a service level parameter in contracts between home sites and SPs. They also should be used as a basis to identify and plan further security and privacy measures as a part of a continuous improvement process.

Given the number of additional components required at both the home sites and the SPs, suitable measures for ensuring the infrastructure availability and reliability must be taken. Because standard components are used on both sides, integration into existing monitoring systems is a tedious, but straight-forward task. For a better overview of the Grid-wide status, Grid Information Systems based monitoring solutions can be adopted as suggested by (Baur et al., 2009).

SUMMARY AND OUTLOOK

In this article, we have first motivated the necessity of privacy management in Grids. After sketching the state of the art, based on current research in the areas of federated identity management and privacy-enhancing technologies, we analyzed the characteristics of Grids, derived their specific requirements, and demonstrated that previous approaches fell short of fulfilling these requirements. We then presented in a security framework how policy-based privacy management can be adapted to Grid environments, and applied this methodology to a XACML-based management architecture. Finally, we discussed that the realization of a policy-based privacy management approach is a straight-forward task for Grid home sites, but very complex and challenging for Grid service providers.

Our ongoing work focuses on challenges with the practical application of the presented security framework, especially concerning its process-driven adaption to arbitrary SP infrastructures and long-term operational aspects, such as a tighter integration with the IT service management processes and an operational cost analysis. The research questions presented are also highly relevant for Cloud Computing infrastructures, which require an adaption of the solution components to Cloud technology, because they usually are not based on Grid middleware and target, e.g., virtual machines instead of high performance computing resources.

ACKNOWLEDGMENT

The authors wish to thank the members of the Munich Network Management (MNM) Team for helpful discussions and valuable comments on previous versions of this article. The MNM-Team, directed by Prof. Dr. Dieter Kranzlmüller and Prof. Dr. Heinz-Gerd Hegering, is a group of researchers of the University of Munich, the Technische Universität München, the University of the Federal Armed Forces Munich, and the Leibniz Supercomputing Centre of the Bavarian Academy of Sciences. The team's web-server is located at http://www.mnm-team.org/.

REFERENCES

Aarts, R. (Ed.). (2004). *Liberty ID-WSF interaction service specification*. Liberty Alliance document. Retrieved from http://www.projectliberty.org/

Baur, T., Breu, R., Kalman, T., Lindinger, T., Milbert, A., Poghosyan, G., … Rombert, M. (2009). An interoperable Grid Information System for integrated resource monitoring based on virtual organizations. *Journal of Grid Computing, 7*(3). Springer.

Berendt, B., Günther, O., & Spiekermann, S. (2005). Privacy in e-commerce. *Communications of the ACM, 48*(4). ACM Press.

Boursas, L., & Hommel, W. (2006). Policy-based service provisioning and dynamic trust management in identity federations. In [). IEEE Computer Society.]. *Proceedings of the IEEE International Conference on Communications, ICC*, 2006.

Bramhall, P., & Mont, M. (2005). Privacy management technology improves governance. In *Proceedings of the 12th Annual Workshop of the HP OpenView University Association*.

Cunsolo, V. D., Distefano, S., Puliafito, A., & Scarpa, M. L. (2010). GS3: A Grid storage system with security features. *Journal of Grid Computing, 8*(3). Springer.

Demchenko, Y., de Laat, C., Koeroo, O., & Groep, D. (2008). Re-thinking Grid security architecture. In *Proceedings of Fourth International Conference on eScience*. IEEE Computer Society.

Hommel, W. (2005a). Using XACML for privacy control in SAML-based identity federations. In *Proceedings of the 9th Conference on Communications and Multimedia Security (CMS 2005)*. Springer.

Hommel, W. (2005b). An architecture for privacy-aware inter-domain identity management. In *Proceedings of the 16th IFIP/IEEE Distributed Systems: Operations and Management (DSOM 2005)*. Springer.

Hughes, J., & Maler, E. (2005). *OASIS security assertion markup language (SAML), V2.0 technical overview*. OASIS Security Services Technical Committee Document.

Kaler, C., & Nadalin, A. (Eds.). (2003). *Web services federation language (WS-Federation)*. Web Services Specifications Document.

Kirchler, W., Schiffers, M., & Kranzlmüller, D. (2009). Harmonizing the management of virtual organizations despite heterogeneous Grid middleware – assessment of two different approaches. In *Proceedings of the Cracow Grid Workshop*.

Kuhn, D. R., Coyne, E. J., & Weil, T. R. (2010). Adding attributes to role-based access control. *IEEE Security*, June 2010.

Manion, F. J., Robbins, R. J., Weems, W. A., & Crowley, R. S. (2009). Security and privacy requirements for a multi-institutional cancer research data grid: An interview-based study. *BMC Medical Information and Decision Making, 9*(31).

Mao, W., Martin, A., Jin, H., & Zhang, H. (2009). Innovations for Grid security from trusted computing – protocol solutions to sharing of security resource. *LNCS 5087*. Springer. Mont, M., Pearson, S., & Bramhall, P. (2003). *Towards accountable management of identity and privacy: Sticky policies and enforceable tracing services*. (Report No. HPL-2003-49). Bristol, UK: HP Laboratories.

Mont, M. (2004). *Dealing with privacy obligations in enterprises*. (Report No. HPL-2004-109). Bristol, UK: HP Laboratories.

Moses, T. (Ed.). (2005). *OASIS eXtensible access control markup language 2.0, core specification*. OASIS XACML Technical Committee Standard.

Niederberger, R., & Alessandrini, V. (2004). DEISA: Motivations, strategies, technologies. In *Proceedings of the International Supercomputer Conference 2004*.

Pettersson, J. S., Fischer-Hübner, S., Danielsson, N., Nilsson, J., Bergmann, M., Clauss, S., et al. Krasemann, H. (2005). Making PRIME usable. In *Proceedings of the Symposium on Usable Privacy and Security (SOUPS)*. ACM Press.

Pfitzmann, B. (2002). Privacy in browser-based attribute exchange. In *Proceedings of the ACM Workshop on Privacy in Electronic Society (WPES 2002)*. ACM Press.

Powers, C., & Schunter, M. (2003). *Enterprise privacy authorization language*. W3C member submission. Retrieved from http://www.w3.org/Submission /2003/SUBM-EPAL-20031110/

Schiffers, M., Ziegler, W., Haase, M., Gietz, P., Groeper, R., Pfeiffenberger, H., et al. Grimm, C. (2007). Trust issues in Shibboleth-enabled federated Grid authentication and authorization infrastructures supporting multiple Grid middleware. In *Proceedings of IEEE eScience 2007 and International Grid Interoperability Workshop 2007 (IGIIW 2007)*. IEEE Computer Society.

Sotomayor, B., & Childers, L. (2006). *Globus toolkit 4 - programming Java services*. Morgan Kaufmann Publishers.

Spantzel, A., Squicciarini, A., & Bertino, E. (2005). *Integrating federated digital identity management and trust negotiation*. (Report No. 2005-46). Purdue University.

Tschantz, M. C., & Krishnamurthi, S. (2006). Towards reasonability properties for access-control policy languages. In *Proceedings of SACMAT 2006*. ACM Press.

Wason, T. (Ed.). (2004). *Liberty identity federation framework ID-FF architecture overview*. Liberty Alliance Specification. Retrieved from http://www.project-liberty.org/

Welch, V., Barton, T., Keahey, K., & Siebenlist, F. (2005). Attributes, anonymity, and access: Shibboleth and Globus integration to facilitate Grid collaboration. In *Proceedings of the Internet2 PKI R&D Workshop*.

Chapter 9
Adaptive Control of Redundant Task Execution for Dependable Volunteer Computing

Hong Wang
Tohoku University, Japan

Yoshitomo Murata
Tohoku University, Japan

Hiroyuki Takizawa
Tohoku University, Japan

Hiroaki Kobayashi
Tohoku University, Japan

ABSTRACT

On the volunteer computing platforms, inter-task dependency leads to serious performance degradation for failed task re-execution because of volatile peers. This paper discusses a performance-oriented task dispatch policy based on the failure probability estimation. The tasks with the highest failure probabilities are selected for dispatch when multiple task enquiries come to the dispatcher. The estimated failure probability is used to find the optimized task assignment that minimizes the overall failure probability of these tasks. This performance-oriented task dispatch policy is evaluated with two real world trace data sets on a simulator. Evaluation results demonstrate the effectiveness of this policy.

INSTRUCTION

Volunteer computing (Anderson, 2004) uses Internet-connected individual computers to solve computing problems. The pioneering research projects, including GIMPS (The Great Internet Mersenne Prime Search, http://www.mersenne.org), SETI@home (Anderson, 2004) and Distributed.net (http://www.distributed.net) are rather successful. GIMPS has already found a total of 9 Mersenne primes, each of which was the largest known prime number at the time of discovery. SETI@home has identified several candidate spots for extraterrestrial intelligence. Distributed.

DOI: 10.4018/978-1-60960-603-9.ch009

net has successfully provides the solutions of the DES, RC5-32/12/7 ("RC5-56"), and RC5-32/12/8 ("RC5-64") of the RSA secret-key challenge.

Nowadays, there are several well-known volunteer computing platforms such as Folding@home (http://folding.stanford.edu), BOINC (Berkeley Open Infrastructure for Network Computing, http://boinc.berkeley.edu), XtremWeb (Cappello, 2005), Entropia (Chien, 2003), Alchemi (Luther, 2005), and JNGI (Verbeke, 2005) to name a few. The volunteer computing platforms are providing more computing power than any supercomputers, clusters, or grid, and the disparity will grow over time. It is because of a large number of Internet-connected personal computers and latest generation game consoles. By November 2010, the most powerful volunteer computing platform - Folding@home achieved about 4 Petaflops computing power by connecting more than 5,700,000 CPUs (http:////fah-web.stanford.edu/cgi-bin/main.py?qtype=osstats). In contrast, the fastest supercomputer, Tianhe-1A achieves 2.566 Petaflops for the high-performance LINPACK benchmark (http://www.top500.org).

Despite the massive computing power offered by the existing volunteer computing platforms, they are lacking support for inter-task dependency. Our previous work solved this issue with a workflow management mechanism (Wang, 2007). However, inter-task dependency results in a status that none of the un-dispatched tasks can be dispatched, because these un-dispatched tasks require the results of one or several of the tasks that are being executed. This status may lead to serious performance degradation, because of the frequent task failures of volatile peers in volunteer computing platforms. Therefore, a redundant task dispatch policy (Wang, 2007) has been proposed to mitigate the performance degradation. Although the redundant task dispatch policy shown a significant performance improvement compared to the non-redundant one, it has a major limitation: the average failure rate model is not the best fit for the volunteer peers in the real world. Thus, this paper extends the policy so as to address the limitation.

This paper discusses a performance-oriented task dispatch policy for volunteer computing platforms. A heuristics-based mechanism for failure probability estimation is proposed based on a life cycle model of volunteer peers and the statistical data. The tasks with the highest failure probabilities are dispatched when multiple task enquiries come to the dispatcher. The estimated failure probability is used to find the optimized task assignment that minimizes the overall failure probability of these tasks. Once the optimized assignment is found, the dispatched tasks are sent to the workers. At the same time, the failure probabilities and other runtime information of the tasks are updated. While multiple types of workers exist in the real world, their different availability characteristics have to be considered. Thus, this work also studies the performance impact of identifying multiple worker types.

The rest of the paper is organized as follows. Section 2 reviews related work. Section 3 proposes a heuristics-based failure probability estimation method. Section 4 introduces the design of the least failure probability dispatch policy. Section 5 evaluates the proposed policy using a simulator, in terms of the total process time. Section 6 concludes and summarizes this paper.

RELATED WORK

The failure probability is estimated based on the analysis of peer availability data. The resource availability problem has been studied a lot for clusters, servers, PCs in a corporate network, grid, and volunteer computing systems.

Statistical Resource Availability Characterizing

There have been a large number of works on the problem of statistically characterizing resource availability.

Root Cause Analysis of Failures

Root cause analysis of failures has been studied in (Gray, 1990; Kalyanakrishnam, 1999; Oppenheimer, 2003; Schroeder, 2006). The software-related failure is reported to be around 20% (Oppenheimer, 2003), 50% (Gray, 1990, Kalyanakrishnam, 1999), and from 5% to 24% (Schroeder, 2006). The percentage of hardware-related failure is from 10% to 30% in (Gray, 1990; Kalyanakrishnam, 1999; Oppenheimer, 2003), and from 30% to over 60% (Schroeder, 2006). The network-related failure is significant in some of those works, while it accounts for around 20% (Kalyanakrishnam, 1999) and 40% (Oppenheimer, 2003) of the failures. Human errors also lead to 10% - 15% (Gray, 1990) and 14% - 30% (Oppenheimer, 2003) of the failures. These works reported different breakdown of failures, because of the different systems they studied.

Fitting Distribution to Empirical Availability Data

Some other works studied statistical distributions of empirical availability data such as *Time-to-Fail (TTF)* and *Down Time (DT)*. Such methods find the best fitted theoretical distribution for a given empirical data set, by estimating the parameters of the theoretical distributions with techniques such as Maximum Likelihood Estimation (MLE) (Aldrich, 1997). Several distributions have been used to model the peer availability, including log-normal, Weibull, exponential, hyper-exponential, and Pareto distributions. The detail of these distributions and their properties can be found in (Patel, 1976).

Exponential distribution and hyper-exponential distribution have been used to investigate the availability behaviors of software, operating system, workstation, and peer-to-peer file sharing system in (Goel, 1985; Iyer, 1985; Lee, 1993; Mutka, 1988; Plank, 1998; Tian, 2007). For the research such as process lifetime estimation (Harchol-Balter, 1997) and network performance (Paxson, 1997), Pareto distribution has been used a lot. Weibull distribution is another distribution widely used for modeling the resource availability. Xu et al.(1999) applied it to the modeling of network-connected PCs.

Several studies (Schoeder, 2006; Nurmi, 2005; Iosup, 2007; Nadeem, 2008) compared different distributions for the modeling. Nurmi (2005) and Brevik (2004) used exponential, hyper-exponential, Weibull, and Pareto distributions to model the *TTF* availability data gathered from student lab computers, a cycle-harvesting distributed computing system - Condor (Litzkow, 1988; Thain, 2005), and an early survey of Internet hosts (Long, 1995). Goodness-of-fit analysis indicated that hyper-exponential and Weibull distributions fit the empirical data more accurately. Schroeder et al.(2006) studied the distribution fitting of *TTF* in high-performance computing (HPC) systems with 4750 machines, using Welbull, lognormal, gamma, and exponential distributions. The results pointed out that Weibull distribution is a better fit. Iosup et al.(2007) found Weibull the best fitted among several distributions for *Mean Time Between Failure (MTBF)* and failure duration data of Grid'5000 (Bolze, 2006; http://www.grid5000.fr).

More recently, Nadeem et al.(2008) also applied several distributions to the analysis of grid resource availabilities. It introduced the class level modeling method by identifying three types of resources in the Austrian Grid (http://www.austriangrid.at). Based on the administration policy, it categorized the resources into three classes: dedicated resources, temporal resources and on-demand resources. The distribution fitting and goodness-of-fit tests are done separately for

Figure 1. Life cycle of a volunteer peer

Start — TTF → Failure — DT → Restart — TTF → Failure End

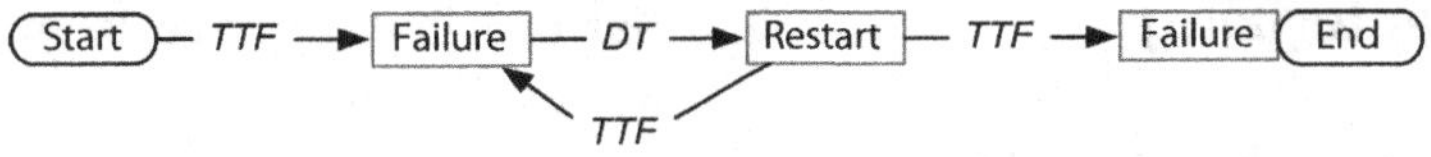

each class's availability (*TTF*) and unavailability (*Mean Time to Reboot (MTR)*) data. While other works found one or two best fitted distributions, this work found different best fitted distributions for different class.

Availability Prediction

Brevik et al. (2004) assumed a homogeneous environment, and proposed an availability prediction method on top of the found Weibull distribution. This method answered the question what is the largest availability duration for a given confidence value and a desired percentile. Iosup et al. (2007) proposed a resource availability model that considered the failure distribution among clusters, the *TTF* distribution, failure duration distribution, and the distribution of the failure size, which is the number of failed processors. This model is used to predict the failures in a multi-cluster grid system.

Some other works (Ren, 2006; Rood, 2007) utilized the availability pattern on weekdays and weekends to predict the availability. Nadeem et al. (2008) used Bayes Rule and Nearest Neighbor Rule to predict the resource availability. Mickens et al.(2006) proposed saturating counter predictors, state-based history predictors, a linear predictor, and a hybrid predictor that dynamically selects the best predictor. These predictors have been evaluated with trace data sets of distributed servers, peer-to-peer network, and corporation PCs.

A HEURISTICS-BASED FAILURE PROBABILITY ESTIMATION

The prediction methods of resource available status reviewed in Section 2 provide a different accuracy for their selected environments. Since this paper targets at finding optimized task assignment with estimated task failure probabilities, the distribution of empirical availability data can provide enough information. Here, a simple and straight heuristics-based failure probability estimation method is employed.

Life Cycle of a Volunteer Peer

The life cycle of a volunteer peer can be modeled as shown in Figure 1. *TTF* is the time between a peer's start/restart and the next failure/shutdown. *DT* is the time between a failure and the next peer restart. Given a statistical distribution of *TTF*, the *cumulative distribution function (CDF)* of this distribution's value at each uptime x is the probability that a peer's *TTF* is smaller than or equal to x, which equals to the failure probability at uptime x. The failure probability monotonously increases with time. Since none of a single distribution can characterize the resource availability accurately for any systems in large scale computing environments (Nurmi, 2005; Nadeem, 2008), a heuristics-based mechanism is proposed to estimate the failure probability at runtime with gathered *TTF* data.

Failure Probability Estimation

Volunteer computing platforms have two kinds of peers: dispatchers and workers. A task dispatcher is a specific server that controls a volunteer computing platform. Workers are volatile peers that compute tasks and send back the task results to the dispatcher. To estimate the failure probability of each worker, runtime *TTF* data are required. To gather such runtime data, a worker availability status list is maintained by the dispatcher. The list stores the start time of each worker. If a worker is currently unavailable, it is marked as *offline* in the list. The list is maintained as follows:

A Worker Goes Online

As shown in Figure 2(a), when a worker goes online, it sends an online notification message to the dispatcher. Once the notification is received, the dispatcher updates the worker availability status list as shown in Figure 2(b). The current time is stored as the start time of this worker.

Find Offline Worker

To gather the runtime *TTF* data, the dispatcher also checks the availability status of workers periodically. As shown in Figure 3(a), the dispatcher sends status checking messages to the workers that are marked online in the worker availability status list. Once the message is received by an alive worker, the worker sends a reply message back to the dispatcher as shown in Figure 3(b). If a worker is offline, it cannot reply the checking message. Then, it is marked as *offline* in the list. As an example, before the periodical status check, worker 4 in Figure 3 had been marked as *online* with a start time in the list, and then went offline. Thus, it does not reply the checking message. The dispatcher then updates the worker availability status list, and marks worker 4 to be *offline*. It also calculates the *TTF* of the worker 4's last online session. Given the current time and start time of the last online session, the *TTF* is 680 minutes (from 2010/11/23 3:10 to 2010/11/23 14:30).

With this simple periodical availability status checking mechanism, the runtime *TTF* data are gathered on the dispatcher. Thus, the *TTF* distribution can be found at runtime. Suppose the gathered *TTF*s are $\{ttf_1, ttf_2, ttf_3, ..., ttf_n\}$, where n is the number of gathered *TTF*s. The failure probability $F(x)$ of a worker (x is the time after a worker went online) can be estimated as shown in Equation (1):

$$F(x) = \frac{n_x}{n}, \tag{1}$$

where n_x is the number of *TTF*s that are less than or equal to x.

LEAST FAILURE PROBABILITY DISPATCH POLICY

With the failure probability estimation, this paper proposes a performance-oriented task dispatch policy - *Least Failure Probability Dispatch* (LFPD) for volunteer computing platforms. The assumptions are slightly different from the ones in our previous work (Wang, 2007). While the previous work assumes a homogeneous environment, this paper assumes that the volunteer computing platform is a heterogeneous environment, in which all the workers have different performances and different bandwidths to the dispatcher.

An Enhanced Workflow Management Mechanism

A workflow management mechanism has been proposed in our previous work (Wang, 2007). It is responsible for directing the workflow control and the task information update. It cannot fully satisfy the requirement of the LFPD, because it

Figure 2. Worker i goes online

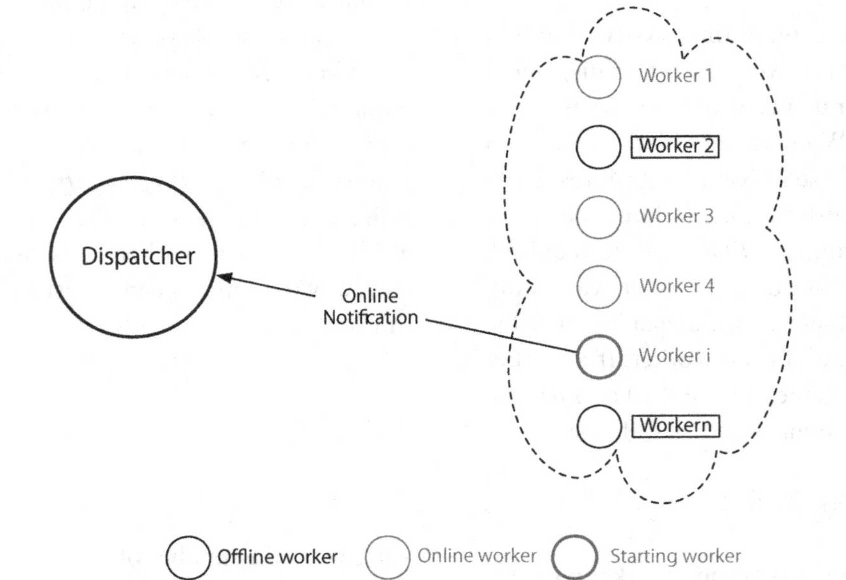

(a) Worker i goes online, sends an online notification to the dispatcher.

Worker Availability Status List

Worker ID	Start time
1	2010/11/22 7:20
2	offline
3	2010/11/21 5:50
4	2010/11/23 3:10
...	...
i	offline
...	...
n	offline

Received online notifi cation

Worker ID	Start time
1	2010/11/22 7:20
2	offline
3	2010/11/21 5:50
4	2010/11/23 3:10
...	...
i	2010/11/23 13:00
...	...
n	offline

Current time

(b) The status of worker i in the worker availability status list changed, after dispatcher received the notification.

assumes the same task failure probability for all the dispatched tasks. Thus, an enhanced workflow management mechanism is proposed to assist the LFPD.

To support the LFPD, the following information of each task *i* is stored and updated by the dispatcher.

- Status: ``undispatched", ``dispatched", and ``finished.
- Redundancy rate that records how many workers process the task *i* at the same time: RR_i.
- The list of worker IDs that process the task *i*: $workerID_i[RR_i]$.
- The list of estimated failure probability for each copy of task *i*: $EFPs_i[RR_i]$.

Figure 3. Checking worker availability status, gathering TTF data

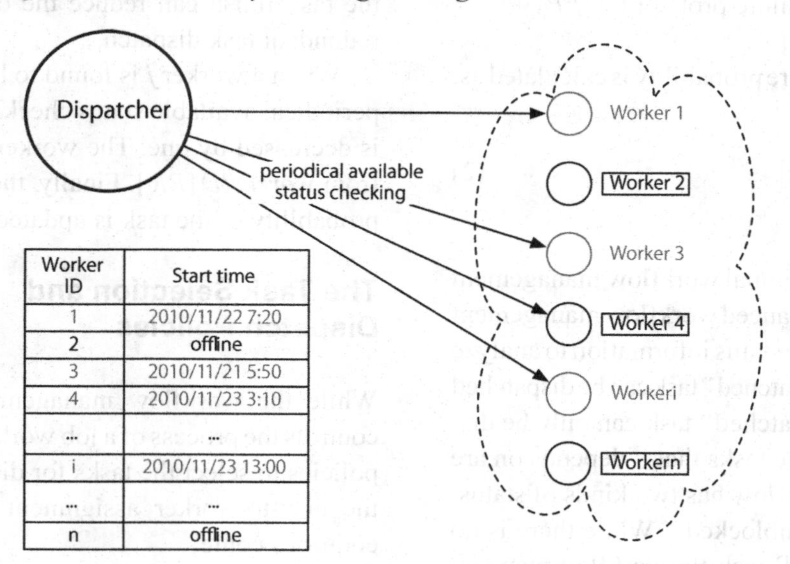

(a) Dispatcher periodically checks the availability status of workers that are marked online in the worker availability status list.

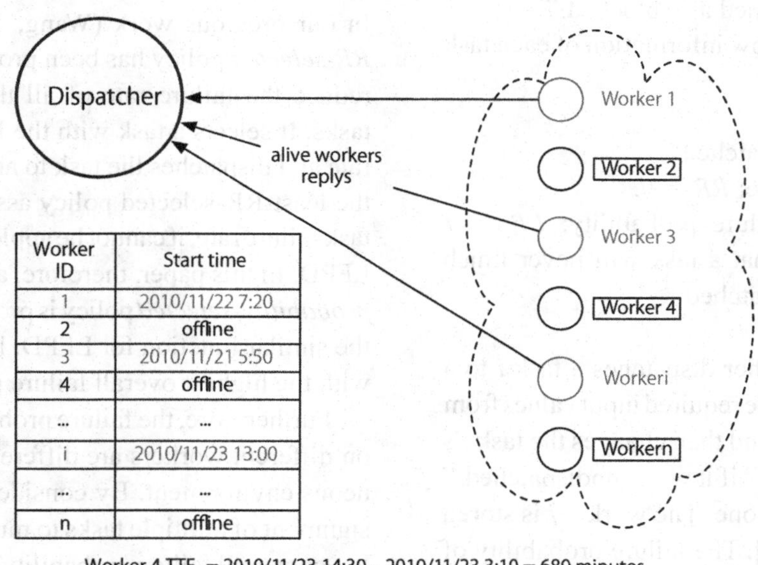

(b) Worker 4 is found to be offline, because it fails to reply the check message.

- The overall failure probability: FP_i.

The overall failure probability is calculated as:

$$FP_i = \prod_{k=1}^{RR_i} EFPs_i[k]. \quad (2)$$

Similar to the original workflow management mechanism, the enhanced workflow management mechanism uses the status information to analyze whether an ``undispatched'' task can be dispatched or not. An ``undispatched'' task can only be dispatched when all the tasks that it depends on are ``finished.'' A workflow has two kinds of status: ``blocked'' and ``unblocked.'' While there is no such ``undispatched'' task, the workflow management mechanism uses the redundant task dispatch to reduce the performance degradation. Such status of a workflow is defined as ``blocked.''

The initial workflow information of each task *i* is as follows:

- Status: ``undispatched.''
- Redundancy rate: $RR_i = 0$.
- The overall failure probability: $FP_i = 1$ which means that a task will never finish before it is dispatched

When the dispatcher dispatches a task *i* to a worker *j*, it provides the required input values from the preceding tasks, and then changes the task *i*'s status to ``dispatched'' if it was ``undispatched,'' and increases RR_i by one. The worker *j* is stored in the $workerID_i[RR_i]$. The failure probability of this assignment {task $i \rightarrow$ worker j } is estimated and stored in $EFPs_i[RR_i]$. The overall failure probability is calculated again.

When the dispatcher receives the result of a task *i* from a worker *j*, it changes the status of task *i* to ``finished'', and sends a ``cancel'' message to the workers in $workerID_i[RR_i]$, except worker *j*. The function to cancel duplicate copies after the task finish can reduce the overhead due to redundant task dispatch.

When a worker *j* is found to be offline by the periodical available status check, RR_i of this task is decreased by one. The worker ID is removed from $workerID_i[RR_i]$. Finally, the overall failure probability of the task is updated.

The Task Selection and Dispatch Policies

While the workflow management mechanism controls the process of a job workflow, it requires policies to select the tasks for dispatch, and find the task-to-worker assignment when the task enquiries come.

Task Selection Policy

In our previous work (Wang, 2007), the *least-RR-selected* policy has been proposed to equally reduce the failure rate of all the ``dispatched'' tasks. It selects a task with the least redundancy rate and dispatches the task to an idle worker. As the least-RR-selected policy assumes a constant task failure rate, it cannot be applied directly to the LFPD. In this paper, therefore, a *highest-failure-probability-selected* policy is proposed to provide the similar function for LFPD. It selects the task with the highest overall failure probability.

Furthermore, the failure probabilities of a task on different workers are different in a heterogeneous environment. By considering the task assignment of multiple tasks to multiple workers, a lower overall failure probability can be achieved. Thus, the idea of *dispatch window* is introduced. The dispatch window is the number of tasks that will be dispatched together. Given a window size *w*, the dispatcher waits for task enquiries from workers until the dispatch window is full, then it selects *w* tasks with the highest-failure-probability-selected policy.

Dispatch Policy

After getting w tasks to dispatch, the dispatcher finds the optimal task-to-worker assignment that minimizes the overall failure probability of the w tasks.

Suppose that each task i has its computation cost (cmp_i) and communication cost ($comm_i$) information, and each worker j has its performance ($Perf_j$) and bandwidth ($Band_j$) information. This information is available for the dispatcher. Given a task i and worker j, the estimated process time of task i on worker j is:

$$T_{i,j}^{EPT} = \frac{cmp_i}{Perf_j} + \frac{comm_i}{Band_j}. \qquad (3)$$

Thus, the estimated failure probability of this assignment is:

$$EFP_{\{i \to j\}} = F(T_{i,j}^{EPT} + CurrentTime - StartTime_j), \qquad (4)$$

where $F(x)$ is the CDF of *TTF*'s distribution; $StartTime_j$ is the start time of worker j in the worker availability status list.

Suppose that the window size is w, the selected tasks are $\{t_1, t_2, t_3, ..., t_w\}$, and the worker peers in the dispatch windows are $\{p_1, p_2, p_3, ..., p_w\}$. For each permutation of $\{p_1, p_2, p_3, ..., p_w\}$, there is an assignment. For example, the following assignment is for the permutation $\{p'_1, p'_2, p'_3, ..., p'_w\}$:

$$\begin{Bmatrix} t_1 \to p'_1 \\ t_2 \to p'_2 \\ t_3 \to p'_3 \\ \vdots \\ t_w \to p'_w \end{Bmatrix}. \qquad (5)$$

For each of the assignments, the estimated failure probability (EFP) of each task-to-worker pair is calculated with Equation (4). Then, the overall failure probability of the assignment is:

$$OFP = \prod_{k=1}^{w} EFP_{\{t_k \to p'_k\}}. \qquad (6)$$

There are $w!$ possible assignments. The dispatcher calculates each assignment's *OFP*, and compares them. The assignment with the least *OFP* is used for the task dispatch. The dispatcher sends tasks to the workers in the dispatch window, using the decided assignment. The workflow information is updated using the enhanced workflow management mechanism proposed in Section 4.1.

EVALUATION RESULTS

The effectiveness of the proposed LFPD is evaluated using a simulator that has been developed on a discrete event simulation environment - OMNeT++ (www.omnetpp.org). The purposes of this simulation are as follows:

1. To prove the effectiveness of the LFPD policy.
2. To verify the effect of the task dispatch window.
3. To study the effect of different parameters.
4. To analyze the effect of identifying multiple worker types.

Baseline Policies

To discuss the effectiveness of the LFPD policy, two baselines are used.

Simple Redundant Task Dispatch Policy

The window-size-1 is a special case of the LFPD policy, because there is only one (1!) task-to-work assignment. Thus, the window-size-1 *LFPD* policy can be considered as an extension of the original redundant task dispatch policy that uses the proposed heuristics-based failure probability estimation model. This simple redundant task dispatch policy is used as a baseline to discuss the effectiveness of the LFPD policy.

Greedy Dispatch Policy

The proposed LFPD policy selects a task-to-worker assignment with the least overall failure probability. Thus, the effectiveness of the LFPD policy highly depends on the estimation accuracy of the failure probabilities. If the dispatcher can predict task failures perfectly, it can eliminate all the task failures. In such case, an intensively optimized dispatch policy for volunteer computing platforms can be achieved. The comparison between such a dispatch policy and the LFPD policy can demonstrate the effectiveness of the LFPD policy. Therefore, in this paper, a greedy dispatch policy that can predict failure perfectly is used as another baseline in the following evaluation.

The greedy dispatch policy assumes that the dispatcher knows the perfect knowledge of the workers' future availability status. Using such knowledge, the dispatcher can perfectly predict whether a task can be finished on a worker without failure. The way to find the best task-to-worker assignment is similar to the LFPD policy. Instead of using the assignment with the least overall failure probability, the greedy dispatch policy uses the assignment with the least number of failures.

Using the LFPD policy, the computing power is wasted in some cases. These cases can be found in advance with the knowledge of the worker's future availability status. The greedy dispatch policy adopts new rules to handle such cases as follows:

1. A task copy will incur a failure on a worker. The computing power of this worker is wasted. The greedy dispatch policy does not dispatch such tasks that will incur failures. A ``sleep'' message is sent to the worker. The worker sleeps for a pre-defined period after receiving the message.
2. A task copy will finish on a worker. Multiple copies of this task are running on different workers. However, this copy will finish later than some other copies (larger estimated finish time). A task copy's estimated finish time can be calculated when it is dispatched as: $EFT = T^{EPT} + CurrentTime$. The computing power of this worker is also wasted, because it does not contribute to the process of the job. Therefore, instead of dispatching duplicate task copies, the greedy dispatch policy insures that there is only one copy of any task. This old copy of a task is continually replaced with a new copy that has a smaller *EFT* value, whenever a new task-to-worker assignment is found for the workers in the dispatch window. When an old task copy is replaced with a new one, the old copy is canceled on the worker that executes it. If the new task copy has a larger *EFT* value, a ``sleep'' message is sent to the worker.

The Simulator Configuration

The dispatcher and worker modules are implemented with the OMNeT++ to simulate the LFPD policy, the simple redundant task dispatch policy (the window-size-1 LFPD policy), and the greedy dispatch policy. To study the effectiveness of the policies in a real world environment, two sets of real world resource availability trace data are used to generate the worker failures. The Skype trace data set (Guha, 2006) has application-level resource availability data of 2,081 Skype supernodes

Table 1. Simulation parameters for LFPD policy and the greedy dispatch policy

	Skype Trace	Microsoft PCs Trace
Number of Workers	2,081	51,663
Number of Tasks	80,000	2,000,000
Mean Task Process Time	1250, 2500, 5000, 10000 seconds	
Number of Task Groups	5, 10, 20	
Idle Worker Inquire Interval	200 seconds	

for about 28 days. Skype is a peer-to-peer VoIP software that connects thousands of volatile peers. The Microsoft PCs trace data set (Bolosky, 2000) stores the availability data of 51,662 desktop PCs within the Microsoft corporation network for 35 days. The volatile peers in the peer-to-peer network and desktop PCs in the corporation network are two typical worker types for volunteer computing.

In a heterogeneous environment, the performance of each worker is different. To simulate such an environment, worker's performance parameters are generated with a power-law distribution. As this work focuses on the computation-intensive problems that satisfy ``*computation time ≫ data transfer time*,'' the communication cost is not considered in the simulation.

The simulation parameters are as follows:

- Number of Workers: the number of workers in the platform. It is the number of peers in the trace data sets.
- Number of Tasks: the number of tasks for the computing job.
- Mean Task Process Time: mean process time of a task. The process time of a task on a worker depends on the performance parameter of his worker.
- Idle Worker Inquire Interval: an inquire interval of idle workers that received the ``sleep'' message. The ``sleep'' message is only used in the greedy dispatch policy.
- Number of Task Groups: the number of task groups in the computing job. It is the factor of inter-task dependency.

The simulation parameters are listed in Table 1.

Since the same mean task process times are used to evaluate the two trace data sets, the different availability characteristics make it hard to compare the evaluation results of these two trace data sets. Thus, the Microsoft PCs trace data set is modified to have the same mean *TTF* as the Skype trace data set. The basic statistical properties of these two trace data sets are shown in Table 2.

Performance Evaluation

The two dispatch policies are evaluated with different parameters and different resource availability trace data sets. The total process time of the computing job for different combinations are compared and discussed. To simply the discussion, all the results are normalized with the correspond-

Table 2. Summary of the basic statistical properties of the data sets

	Skype Trace	Microsoft PCs Trace
Mean TTF (seconds)	55,125	55,125
Mean Down Time (seconds)	51,509	15,906
Average percentage of online node	33.15%	81.24%

ing total process time of the simple redundant task dispatch policy.

Figure 4 shows how the normalized total process time changes with the dispatch windows size and the mean task process time, using the Skype trace data set. The results with Microsoft PCs trace data set are shown in Figure 5.

Comparison with the Simple Redundant Task Dispatch Policy

The results indicate that the LFPD policy outperforms the simple redundant task dispatch policy (window-size-1 LFPD). The improvement is more significant for a larger number of task groups. A smaller mean task process time also leads to a slightly better improvement. For 20 task groups and the mean task process time of 1250 seconds, LFPD delivers up to 6% and 12% improvements for the Skype trace data set and Microsoft PCs trace data set, respectively.

The number of task groups is related to how many times a workflow is blocked during the process of the workflow. The ``blocked'' status introduces a serious performance overhead, because the computing power is used for re-execution of the failed tasks. The LFPD policy reduces the number of task failures, and thus mitigates this performance overhead. It explains the reason why the LFPD policy is more efficient for a larger number of task groups.

For a given trace data set, a larger mean task process time leaves less rooms for the LFPD dispatch to find a better assignment. As shown in Equation (4), the *EFP* of any task-to-worker assignment depends on the task process time, the current time, and the start time of the worker. Because the latter two values are given while finding a better task assignment, the *EFP* is decided only by the task process time. For example, there are two workers in the dispatch window, and two selected tasks. Given any assignment, a larger mean task process time leads to a longer task process time on both workers. Therefore, the *EFP* increases for both the two tasks. This *EFP* increment makes the overall failure probability (*OFP*) of both two possible assignments higher. The LFPD policy is designed to reduce the number of failures, by finding proper task-to-worker assignments. However, if all the assignments provide a high overall failure probability, the LFPD policy becomes less efficient. As a result, the LFPD policy delivers less improvement in the case of a larger mean task process time. In both of these two trace data sets, the mean *TTF* is 55125 seconds. The large mean task process time (10000 seconds) enlarges the tasks failure probability. Thus, the LFPD dispatch is less efficient, compared to the ones with a small mean task process time.

The results with two trace data sets are slightly different for their different availability characteristics. It is because of an overhead introduced by the dispatch window. When the dispatch window is not full, the workers that are waiting in the window are idle. Their computing power is wasted. Thus, the less time to fill a dispatch window, the better performance can be achieved. In the case of these two trace data sets, the average number of online workers in the Microsoft PCs trace is much larger. Thus the time of the Microsoft PCs trace data set to fill a dispatch window can be expected to be much shorter than that of the Skype data set. Therefore, the LFPD policy delivers a better performance improvement with the Microsoft PCs trace data set for all the parameter combinations.

Comparison with the Greedy Dispatch

As shown in both Figures 4 and 5, the greedy dispatch policy beats the LFPD policy for a large mean task process time (10000 seconds). The reason is that the greedy dispatch policy eliminates all the task failures with its perfect knowledge of the worker availability status. While both the simple redundant task dispatch policy and the LFPD policy suffer from the inefficiency for the high failure probabilities, the performance of the

Figure 4. Compare the LFPD policy and the greedy dispatch policy for different mean task process time (Skype Trace)

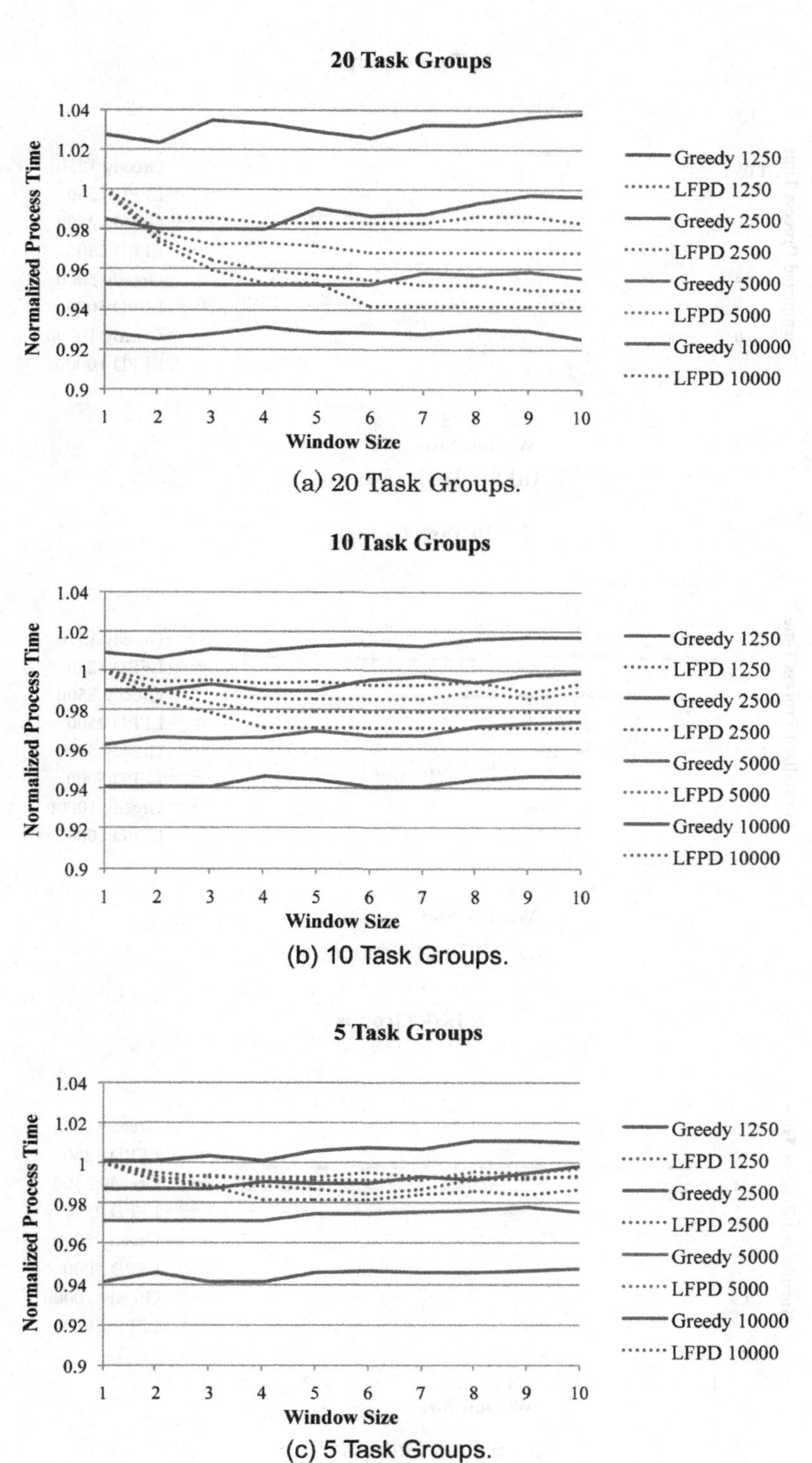

(a) 20 Task Groups.

(b) 10 Task Groups.

(c) 5 Task Groups.

Figure 5. Compare the LFPD policy and the greedy dispatch policy for different mean task process time (Microsoft PCs Trace)

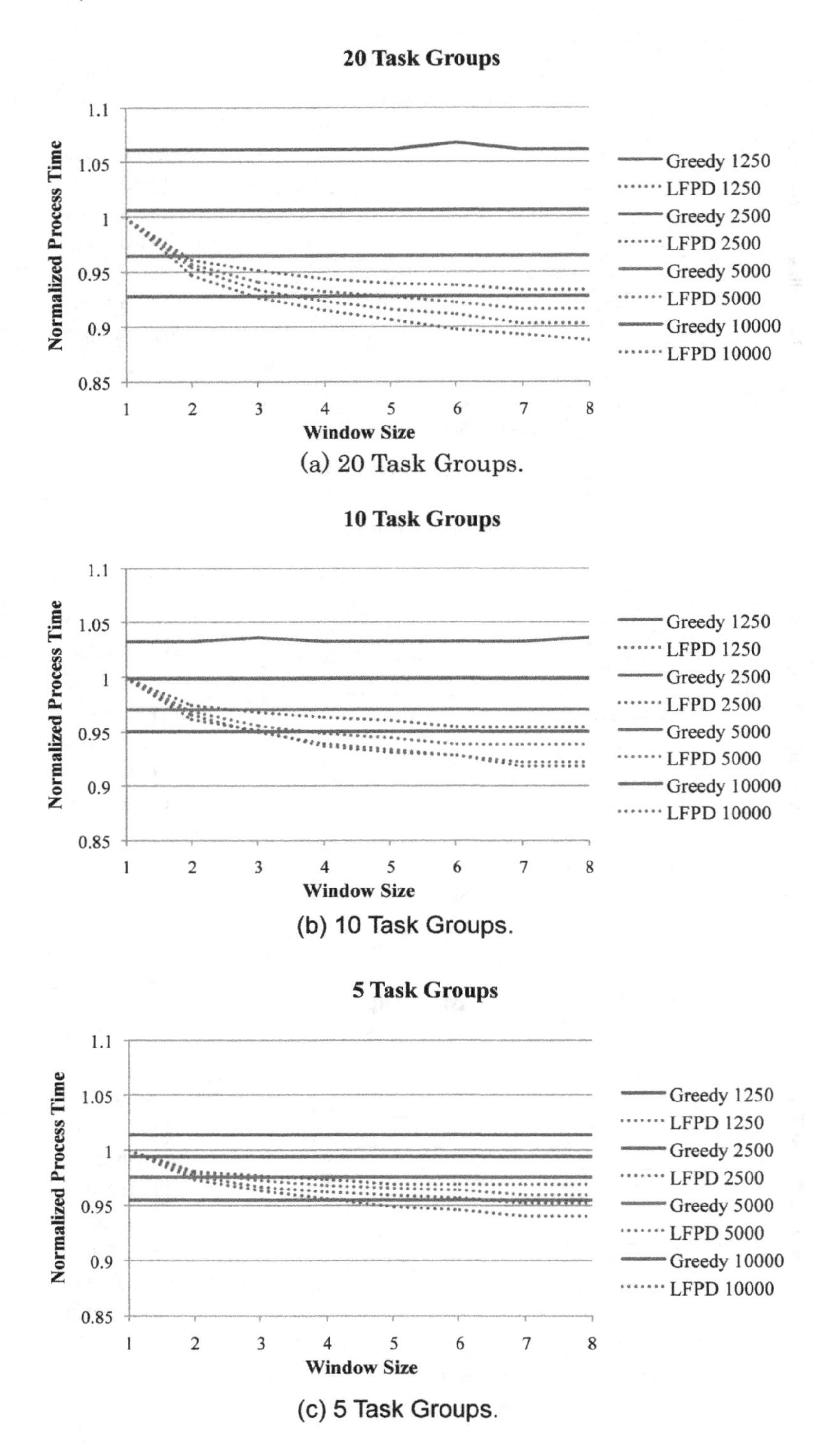

(a) 20 Task Groups.

(b) 10 Task Groups.

(c) 5 Task Groups.

greedy dispatch policy is not affected. Therefore, the normalized process time with the greedy dispatch policy decreases while increasing the mean task process time.

These two figures also show that the LFPD policy is more efficient than the greedy dispatch policy for a small mean task process time. It is because of the ``sleep" message used in the greedy dispatch policy. The greedy dispatch policy lets a worker sleep if it finds this dispatch to be a waste of computing power. It happens when the existing copy of a task has a smaller *EFT* than this new copy. This mechanism boosts the performance in most cases. However, it also introduces a possible overhead, for letting workers sleep even when the workflow is no longer ``blocked" and ``undispatched" tasks are available. When the mean task process time is small, the failures occur less frequently. Therefore, the greedy dispatch policy's advantage for eliminating task failures becomes less significant. In such cases, this particular overhead becomes more obvious and leads to worse efficiency.

Effects of Window Size on the Process Time

As shown in both Figures 4 and 5, a larger window size results in a shorter process time for the LFPD policy in most cases. This is because the LFPD policy is likely to find a better task-to-worker assignment with a larger window size, especially for the smaller mean task process time. As discussed earlier, a smaller mean task process time results in less frequent failures. Thus, the LFPD policy has a higher probability to find an assignment with less failures.

The overhead introduced by the ``blocked" status is not serious when the number of task groups is small. Thus, the improvement achieved with the LFPD policy is small. Therefore, while the window size increases, the overhead for the dispatch window becomes obvious. The overhead is more serious when the number of online workers is small. It explains why the process time with the LFPD increases when the window size exceeds a certain value in Figures 4(b) and 4(c). With a much larger number of online workers, the overhead for the dispatch window is not significant. Thus, the results with the Microsoft PCs trace are not affected by a small number of task groups and a large window size.

Improvement of the Performance by Identifying Worker Types

In the real world, multiple types of workers exist. A different type of workers has different availability characteristics. Nadeem et al.(2008) introduced the class level modeling method by pre-identifying three types of resources in the Austrian Grid (http://www.austriangrid.at). The *TTF* distribution of different types of resources is largely different across the three types. The heuristics-based failure estimation relies on the empirical distribution, and assumes that all the workers have similar availability behavior. Gathering multiple types of workers' *TTF* into a single *TTF* distribution leads to a low estimation accuracy. The low estimation accuracy will degrade the performance, because the LFPD policy cannot find the optimal task-to-worker assignments with the inaccurate failure estimations.

To improve the failure estimation accuracy, the worker type is considered. Two types of workers are selected from the two real world trace data sets. First, the two trace data sets are clustered into several types, using a K-Means clustering algorithm in the Weka toolkit (Witten, 2005). By extracting the *TTF* and the down time pair from the original trace data sets, two dimensional data are generated. The number of clusters is four, based on the assumption that four kinds of workers (diurnal, weekly, long *TTF*, and long downtime) exist. The clustering results are shown in Table 3. Each cluster shows different characteristics. Cluster 3 of Microsoft PCs trace shows a diurnal pattern, while Cluster 3 of Skype trace is highly volatile.

Table 3. Clustering results (node distribution, mean uptime/mean downtime)

	Skype Trace	Microsoft PCs Trace
Cluster 1	4.0%, 8 days/6.68 hrs	18.4%, 20.66 days/13.97 hrs
Cluster 2	6.3%, 9.49 hrs/4.40 days	5.7%, 28.68 hrs/4 days
Cluster 3	79.6%, 6.73 hrs/8.72 hrs	62.2%, 16.1 hrs/6.76 hrs
Cluster 4	10.1%, 2.75 days/6.50 hrs	13.7%, 7.97 days/8.99 hrs

5,000 peers are selected from each cluster of major clusters of both trace data sets to form a new 2-type trace data set. Thus, this 2-type trace data set is considered to consist of two types of workers. This trace data set is used in the simulation to study the effect of identifying worker types.

The simulation parameters are listed in Table 4. The LFPD policy with and without the ability to identify two worker types are simulated. If two types of workers are identified, each type of worker's *TTF* data is gathered separately. When failure probability estimation is needed for a task-to-worker assignment, the dispatcher selects the corresponding *TTF* distribution for each worker and then estimates the failure probability of the worker.

Figure 6 shows improvement achieved by identifying worker types. The vertical axis represents the normalized total process time with identifying worker types. These results are normalized by the total process time without identifying worker types. The results with the worker type identification show an average improvement of 0.7% (ranges from 0.1% to 1.5%). The results also indicate that the improvement is more significant for a larger mean task process time, a larger number of task groups, and a larger dispatch window size. As discussed in Section 5.3.1, a larger mean task process time leaves less rooms for the LFPD dispatch to find a better assignment. Therefore, the accuracy of failure estimation has a bigger impact on the performance. It has also been discussed that a larger number of task groups makes the performance degradation more serious. Thus, accurate failure estimation offers a higher improvement. The larger the dispatch window is, the more possible task-to-worker assignments exist. If failure estimation is not accurate, the LFPD policy cannot find the optimal assignment from these assignments. Therefore, the accuracy of the failure estimation is more critical.

CONCLUSION

The redundant task dispatch policy proposed in our previous work (Wang, 2007) has a major limitation: the average failure rate model is not the best fitted for the volunteer peers in the real world. To address this limitation, this paper has proposed a heuristics-based mechanism for failure probability estimation based on a life cycle model of volunteer peers and the statistical data. Then, the LFPD policy has been introduced. Instead of

Table 4. Simulation parameters for the 2-type trace data set

	2-type Trace Data
Number of Workers	5,000(Volatile) + 5,000(Diurnal)
Number of Tasks	400,000
Mean Task Process Time	1250, 2500, 5000, 10000 seconds
Number of Task Groups	5, 10, 20
Window Size	1, 2, 3, 4, 5, 6, 7, 8
Idle Worker Inquire Interval	200 seconds

Figure 6. The improvement archived by identifying multiple worker types

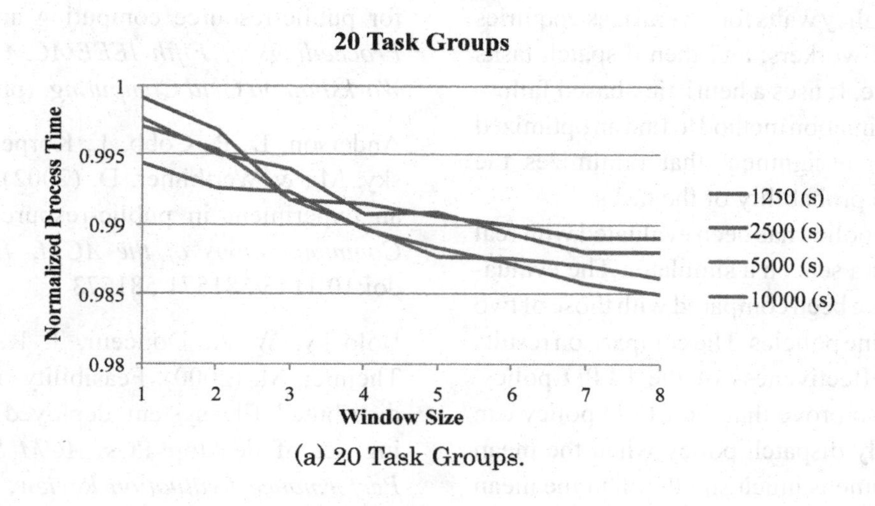

(a) 20 Task Groups.

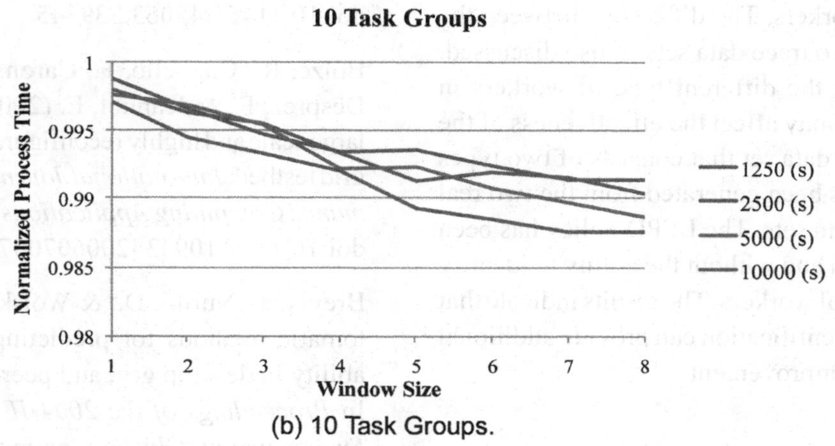

(b) 10 Task Groups.

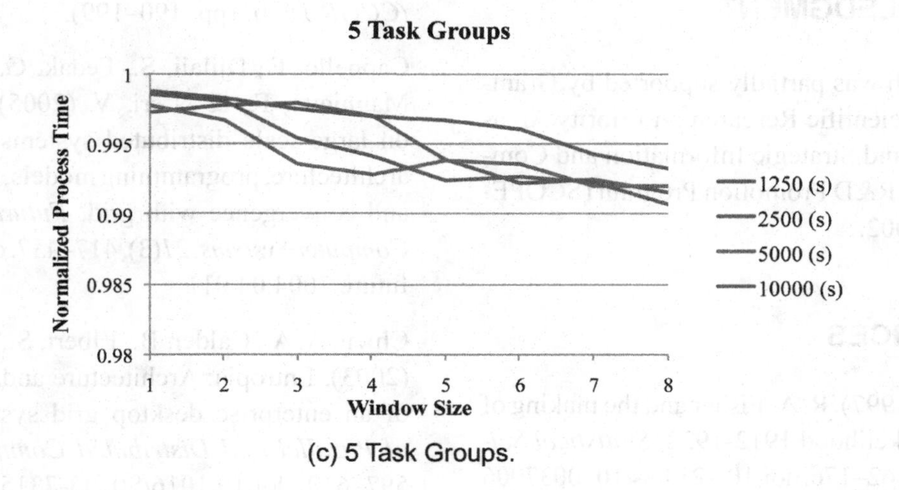

(c) 5 Task Groups.

dispatching a task whenever a task enquiry comes, this dispatch policy waits for several task enquiries from different workers, and then dispatch tasks to them at once. It uses a heuristics-based failure probability estimation method to find an optimized task-to-worker assignment that minimizes the overall failure probability of the tasks.

The LFPD policy has been evaluated with real world trace data sets on a simulator. The evaluation results have been compared with those of two selected baseline policies. The comparison results indicate the effectiveness of the LFPD policy. The results also prove that the LFPD policy can beat the greedy dispatch policy when the mean task process time is much smaller than the mean *TTF* of the workers. The difference between the results with two trace data sets is also discussed. To study how the different type of workers in the real world may affect the effectiveness of the LFPD, a trace data set that consists of two types of workers has been generated from the two real world trace data sets. The LFPD policy has been simulated with and without the ability to identify different type of workers. The results indicate that worker type identification can provide additional performance improvement.

ACKNOWLEDGMENT

This research was partially supported by Grant-in-Aid for Scientific Research on Priority Areas #18049003 and Strategic Information and Communications R&D Promotion Program (SCOPE-S) #061102002.

REFERENCES

Aldrich, J. (1997). R. A. Fisher and the making of maximum likelihood 1912-1922. *Statistical Science*, *12*(3), 162–176. doi:10.1214/ss/1030037906

Anderson, D. P. (2004). BOINC: A system for public-resource computing and storage. In *Proceedings of Fifth IEEE/ACM International Workshop on Grid Computing*, (pp. 4-10).

Anderson, D. P., Cobb, J., Korpela, E., Lebofsky, M., & Werthimer, D. (2002). Seti@home: an experiment in public-resource computing. *Communications of the ACM*, *45*(11), 56–61. doi:10.1145/581571.581573

Bolosky, W. J., Douceur, J. R., Ely, D., & Theimer, M. (2000). Feasibility of a serverless distributed file system deployed on an existing set of desktop PCs. *ACM SIGMETRICS Performance Evaluation Review*, *28*(1), 34–43. doi:10.1145/345063.339345

Bolze, R., Cappello, F., Caron, E., Dayd'e, M., Desprez, F., & Jeannot, E. (2006). Grid'5000: A large scale and highly reconfigurable experimental grid testbed. *International Journal of High Performance Computing Applications*, *20*(4), 481–494. doi:10.1177/1094342006070078

Brevik, J., Nurmi, D., & Wolski, R. (2004). Automatic methods for predicting machine availability in desktop grid and peer-to-peer systems. In *Proceedings of the 2004 IEEE International Symposium on Cluster Computing and the Grid (CCGRID04)*, (pp. 190–199).

Cappello, F., Djilali, S., Fedak, G., Herault, T., Magniette, F., & N'eri, V. (2005). Computing on large-scale distributed systems: Xtrem web architecture, programming models, security, tests and convergence with grid. *Future Generation Computer Systems*, *21*(3), 417–437. doi:10.1016/j.future.2004.04.011

Chien, A. A., Calder, B., Elbert, S., & Bhatia, K. (2003). Entropia: Architecture and performance of an enterprise desktop grid system. *Journal of Parallel and Distributed Computing*, *63*(5), 597–610. doi:10.1016/S0743-7315(03)00006-6

Goel, A. (1985). Software reliability models: Assumptions, limitations, and applicability. *IEEE Transactions on Software Engineering, 11*(12), 1411–1423. doi:10.1109/TSE.1985.232177

Gray, J. (1990). A census of tandem system availability between 1985 and 1990. *IEEE Transactions on Reliability, 39*(4), 409–418. doi:10.1109/24.58719

Guha, S., Daswani, N., & Jain, R. (2006). *An experimental study of the Skype peer-to-peer VoIP system.* In The 5th International Workshop on Peer-to-Peer Systems. Retrieved from http://saikat.guha.cc/pub /iptps06-skype.pdf

Harchol-Balter, M., & Downey, A. B. (1997). Exploiting process lifetime distributions for dynamic load balancing. *ACM Transactions on Computer Systems, 15*(3), 253–285. doi:10.1145/263326.263344

Iosup, A., Jan, M., Sonmez, O., & Epema, D. (2007). On the dynamic resource availability in grids. In *Proceedings of 8th IEEE/ACM International Conference on Grid Computing*, (pp. 26-33).

Iyer, R. K., & Rossetti, D. J. (1985). Effect of system workload on operating system reliability: A study on IBM 3081. *IEEE Transactions on Software Engineering, 11*(12), 1438–1448. doi:10.1109/TSE.1985.232180

Kalyanakrishnam, M., Kalbarczyk, Z., & Iyer, R. (1999). Failure data analysis of a LAN of Windows NT based computers. In *Proceedings of the 18th IEEE Symposium on Reliable Distributed Systems (SRDS99)*, (pp. 178-187).

Lee, I., Tang, D., Iyer, R., & Hsueh, M.-C. (1993). Measurement-based evaluation of operating system fault tolerance. *IEEE Transactions on Reliability, 42*(2), 238–249. doi:10.1109/24.229493

Litzkow, M., Livny, M., & Mutka, M. (1988). Condor - a hunter of idle workstations. In *Proceedings of the 8th International Conference of Distributed Computing Systems*, (pp. 104–111).

Long, D., Muir, A., & Golding, R. (1995). A longitudinal survey of internet host reliability. In *Proceedings of the 14th Symposium on Reliable Distributed System (SRDS95)*, (pp. 2-9).

Luther, A., Buyya, R., Ranjan, R., & Venugopal, S. (2005). *Alchemi: A. netbased enterprise grid computing system.* In International Conference on Internet Computing, (pp. 269-278).

Mickens, J. W., & Noble, B. D. (2006). Exploiting availability prediction in distributed systems. In *Proceedings of the 3rd Conference on Networked Systems Design & Implementation (NSDI06)*, (pp. 6-19).

Mutka, M. W., & Livny, M. (1988). Profiling workstations' available capacity for remote execution. In *Proceedings of the 12th IFIP WG 7.3 International Symposium on Computer Performance Modelling, Measurement and Evaluation*, (pp. 529–544).

Nadeem, F., Prodan, R., & Fahringer, T. (2008). Characterizing, modeling and predicting dynamic resource availability in a large scale multi-purpose grid. In *Proceedings of the 2008 8th IEEE International Symposium on Cluster Computing and the Grid (CCGRID08)*, (pp. 348-357).

Nurmi, D., Brevik, J., & Wolski, R. (2005). Modeling machine availability in enterprise and wide-area distributed computing environments. In *Proceedings of the 11th International Euro-par Conference*, (pp. 432-441).

Oppenheimer, D., Ganapathi, A., & Patterson, D. A. (2003). Why do internet services fail, and what can be done about it? In *Proceedings of USENIX Symposium on Internet Technologies and Systems (USITS 03)*, (p. 1).

Patel, J. K., Kapadia, C. H., & Owen, D. B. (1976). *Handbook of statistical distributions*. Marcel Dekker, Inc.

Paxson, V., & Floyd, S. (1997). Why we don't know how to simulate the Internet. In *Proceedings of the 29th Conference on Winter Simulation*, (pp. 1037–1044).

Plank, J., & Elwasif, W. (1998). *Experimental assessment of workstation failures and their impact on checkpointing systems*. Twenty-Eighth Annual International Symposium on Fault-Tolerant Computing, (pp. 48-57).

Ren, X., & Eigenmann, R. (2006). Empirical studies on the behavior of resource availability in fine-grained cycle sharing systems. In *Proceedings of 2006 International Conference on Parallel Processing*, (pp. 3-11).

Rood, B., & Lewis, M. (2007). Multi-state grid resource availability characterization. In *Proceedings of 8th IEEE/ACM International Conference on Grid Computing*, (pp. 42-49).

Schroeder, B., & Gibson, G. A. (2006). A large-scale study of failures in high-performance computing systems. In *Proceedings of the International Conference on Dependable Systems and Networks (DSN06)*, (pp. 249-258).

Thain, D., Tannenbaum, T., & Livny, M. (2005). Distributed computing in practice: The Condor experience. *Concurrency and Computation, 17*(2-4), 323–356. doi:10.1002/cpe.938

Tian, J., & Dai, Y. (2007). *Understanding the dynamic of peer-to-peer systems*. In Sixth International Workshop on Peer-to-Peer Systems (IPTPS2007).

Verbeke, J., Nadgir, N., Ruetsch, G., & Sharapov, I. (2002). Framework for peer-to-peer distributed computing in a heterogeneous, decentralized environment. In *Proceedings of Third International Workshop on Grid Computing*, (pp. 1-12).

Wang, H., Takizawa, H., & Kobayashi, H. (2007). A dependable peer-to-peer computing platform. *Future Generation Computer Systems, 23*(8), 939–955. doi:10.1016/j.future.2007.03.004

Witten, I. H., & Frank, E. (2005). *Data mining: Practical machine learning tools and techniques* (2nd ed.). Morgan Kaufmann.

Xu, J., Kalbarczyk, Z., & Iyer, R. (1999). Networked Windows NT system field failure data analysis. In *Proceedings of 1999 Pacific Rim International Symposium on Dependable Computing*, (pp. 178-185).

Chapter 10
Publication and Protection of Sensitive Site Information in a Grid Infrastructure

Shreyas Cholia
Lawrence Berkeley National Laboratory, USA

R. Jefferson Porter
Lawrence Berkeley National Laboratory, USA

ABSTRACT

In order to create a successful grid infrastructure, sites and resource providers must be able to publish information about their underlying resources and services. This information enables users and virtual organizations to make intelligent decisions about resource selection and scheduling, and facilitates accounting and troubleshooting services within the grid. However, such an outbound stream may include data deemed sensitive by a resource-providing site, exposing potential security vulnerabilities or private user information. This study analyzes the various vectors of information being published from sites to grid infrastructures. In particular, it examines the data being published and collected in the Open Science Grid, including resource selection, monitoring, accounting, troubleshooting, logging and site verification data. We analyze the risks and potential threat models posed by the publication and collection of such data. We also offer some recommendations and best practices for sites and grid infrastructures to manage and protect sensitive data.

DOI: 10.4018/978-1-60960-603-9.ch010

INTRODUCTION

Grid computing has become a very successful model for scientific collaborations and projects to leverage distributed compute and data resources. It has also offered the research and academic institutions that host these resources an effective means to reach a much larger community. As grid computing grows in scope, and as an increasing number of users and resources are plugged into the grid, there is an increasing need for metadata services that can provide useful information about the activities on that grid. These services allow for more sophisticated models of computing, and are fundamental components of scalable grid infrastructures. The scope of these services is fairly broad and covers a variety of uses including resource selection, monitoring, accounting, troubleshooting, logging, site availability and site validation. This list could grow, as grids evolve and other types of metadata become interesting to users and administrators. This means that it becomes important for a grid infrastructure to provide central collection and distribution points that can collate information gathered from multiple sources.

The typical publication model involves pushing data from site based informational end points to central collectors, using streaming feeds or periodic send operations. The central collectors then make this data available to interested parties using standard interfaces and protocols in the form of web services and database query engines. The usability of the grid depends on the widespread availability of this information. Given the increasingly open nature of grid computing these collectors and information services generally present publicly accessible front-ends.

Now consider the implications of this model for a site providing grid resources. Being included in a grid infrastructure means that a large amount of site information suddenly enters the public domain. This could include information deemed as sensitive or private from the perspective of the site, the user or the grid collaboration as a whole. It becomes very important then, to have controls on the access and flow of this data, so that the information sources can decide what data they want published and what data they want restricted. Since these models of informational flow are still evolving in today's grids, these controls are still in the process of being designed into the software infrastructure. As such, there isn't a standard way to control this flow of information. We think there is an urgent need to study the various vectors of information being provided by sites to grid infrastructures. This includes an analysis of the nature of the information itself, as well as the software publishing this information.

In our work, we use the Open Science Grid (OSG) ("Open Science Grid Consortium,") as a case study for this model of information flow, looking at the five major information collection systems within the OSG, and analyzing the security implications of this infrastructure. We also provide some recommendations on improving the current infrastructure to preserve the privacy and security of sensitive information.

THE OPEN SCIENCE GRID

The OSG offers a shared infrastructure of distributed computing and storage resources, independently owned and managed by its members. OSG members provide a virtual facility available to individual research communities, who can add services according to their scientists' needs.

It includes a wide selection of resource providers, ranging from small universities to large national laboratories. This broad range of sites results in a diverse set of security requirements. Reconciling these diverse security priorities is a challenge, and requires close interaction between the sites and the OSG managers. One approach to addressing this issue is to provide the necessary tools in the grid middleware stack, so that sites can configure security policies directly into the software. The OSG provides a software distribution called the Virtual Data Toolkit (VDT) ("Virtual Data Toolkit,"). This includes a packaged, tested and supported collection of middleware for participating compute and storage nodes, as well as a client package for end-user researchers.

The OSG also provides support and infrastructure services to collect and publish information from participating sites, and to monitor their resources. These services are provided by the OSG Grid Operations Center (GOC) ("OSG Grid Operations Center,"). The GOC provides a single point of operational support for the OSG. The GOC performs real time grid monitoring and problem tracking, offers support to users, developers and systems administrators, maintains grid services, and provides security incident responses. It manages information repositories for Virtual Organizations (VOs) and grid resources.

INFORMATION COLLECTION IN OSG

There are currently five major information collection systems in the OSG, which rely on information feeds from sites to centralized servers. The following is a description of each of these services, and an analysis of the information being published by them from a site security perspective.

Resource Selection Information

In the OSG framework, the Generic Information Provider (GIP) (Field, 2008) gathers site resource information. GIP aggregates static and dynamic resource information for use with LDAP-based information systems. Information published is based on the Glue Schema (Glue Working Group, 2007). The CEMon (Compute Element Monitor) (Sgaravatto, 2005) service is responsible for publishing this information to a central OSG information collector service called the CEMon Consumer. CEMon connections are authenticated and encrypted (using GSI). This information is then made public in two ways (Padmanabhan, 2007):

1. Class-ads are published to a Condor matchmaker service called the Resource Selection Service (ReSS), which allows Condor clients to select appropriate resources for job submission.
2. The Berkeley Database Information Index (BDII) collects this information for resource brokering. It tracks status of each participating cluster in terms of available CPUs, free CPUs, supported VOs, etc.

The Glue Schema provides a more detailed list of attributes supported in this scheme. For the purposes of this study, we concentrate on those attributes published by GIP that may be deemed sensitive by certain sites. This includes:

- Operating System version/patch information
- Authentication method (grid-mapfile, GUMS)
- Underlying job-manager and batch system information
- Internal system paths

In some sense, publication of this information is essential to a site's successful participation in the grid. However, a site must understand the implications of making this information public. Prior to joining the grid, much of this information was inherently under the control of the site, and limited to people under its own administrative domain. As such, administrators must be aware of any conflicts with the current site security policy and requirements that may have been drafted prior to participation in the grid.

Additionally, a site may only want to provide this information up to a desired level of detail. Since the GIP software will publish all available information in its default mode, a site may want to consider limiting, or overriding some of the attributes being published.

Another consideration is the public nature of this information, once it has been sent to the CEMon Consumers. Given that this information is only useful to actual users of the grid, it might be useful to provide some minimal restrictions so that the information is only accessible to current members of the OSG (or collaborating grids).

Accounting

The Gratia software provides the accounting framework for the OSG (Canal, Constanta, Green & Mack, 2007). Gratia consists of two components:

1. The Gratia probes that run on the site resource and interface with the site-specific accounting and batch systems. These probes extract resource usage information from the underlying infrastructure and convert it into a common Usage Record-XML (Global Grid Forum, 2003) based format. This is then sent to a central collector.
2. The Gratia collector is a central server operated by the OSG GOC that gathers information from the various probes, and internally stores this in a relational database. It makes this information publicly available through a web interface, in certain pre-defined views. The web interface also allows viewers to create their own reports and custom SQL queries against the usage data.

The Gratia records include information that might be considered sensitive by both the sites and the grid users. Specifically, we identified the following information as potentially sensitive:

- User account names
- User DN information
- Job file and application binary names

Given that this information can be accessed through a public SQL interface, all user activity on the OSG can be traced and analyzed in fairly sophisticated ways, by anyone with a web browser.

User account and DN information could be used by an attacker that has compromised an account on one site to query a list of sites with the same user account/DN, thus increasing the scope of the attack. It is not being suggested that masking this information will protect a site from a compromised account on another system. Certainly, once an account has been compromised, any other site that uses a common set of login credentials should be considered vulnerable. However, making this information less accessible to an attacker could mitigate the scope of the attack.

Job file or application names would be less useful to attackers, but could reveal information about the nature of the jobs being run. There is the potential for a rival project to gain valuable clues about the research being done from this information. A researcher may want to restrict this information to a limited set of people. On the other hand, from an accounting standpoint, the underlying file descriptions may not be as interesting as the actual resource consumption being measured. In most cases, the accounting software only needs to be able to uniquely identify a job, and doesn't care about the specifics of underlying job or application names.

For these reasons, it is recommended that access to this data be restricted along user and VO lines using grid certificates as the mechanism for controlling this. Sites can also mask sensitive information by modifying the probe software to apply filters to the records.

Logging

The OSG uses Syslog-ng ("Syslog-ng Logging System,") to provide centralized logging of user activity on the Grid. Syslog-ng is an extension to the Syslog protocol that provides more flexible support for distributed logging and richer content filtering options.

Currently OSG resources optionally log all information related to Grid processes using syslog-ng, and send this to a central collector managed by the GOC. The primary uses for this information are:

1. Troubleshooting – Being able to trace the workflow of a distributed job is very useful as a debugging tool for failures. It makes it significantly easier to detect how and why a job might be failing, especially when multiple sites are involved. The OSG GOC has a troubleshooting team to deal with such cases.
2. Security Incident Response – Having centralized logs available to the OSG security team, makes it very useful to be able to analyze the scope and extent of a security compromise. It allows the GOC to identify compromised sites or users, and to judge the nature of the compromise. Affected sites can then be notified for rapid incident response.

In the troubleshooting case, there is the need to protect failure modes from becoming publicly available, as this could reveal possible avenues for attack. For example, a poorly configured site may have vulnerabilities in the execution path. While not apparent through the standard client software, these may be exposed through syslog information. In general, logging information should only be available to authorized personnel within the OSG administrative domain, or to specific users when debugging problems. Another approach to this issue involves the level of logging performed by the site, so that only a minimal amount of information is logged by default. This translates to logging only the start and stop times for jobs and data transfers for a given user. In the event of a failure, the site can increase the level of logging, and work in conjunction with the troubleshooting team and the user to diagnose the specific problem.

Security incident information is perhaps even more sensitive, and syslog information revealing incident details must have tight access controls. Once again, this points to restricting the information to an authorized set of security personnel.

Syslog-ng allows for collectors on a per site basis (Tierney, Gunter & Schopf, 2007), which can then filter out the information getting passed to the OSG wide collector. This would allow sites to collect detailed information internally, while filtering the information sent to the OSG. Any information sent to the OSG GOC should be encrypted. As long as there is enough information being sent to identify a failure or compromise at a central level, the relevant sites can be notified of this. The sites can then address the specifics of the problem, and provide more information to the OSG GOC and security team, as necessary. This is the model that is expected to go into production for future OSG deployments.

Site Availability and Validation Data

The OSG GOC performs site availability and validity tests on participating compute and storage elements, and publishes these results online. These tests are run at regular intervals, either using a Perl script (site_verify.pl) or using a customizable set of probes called RSV (Resource and Service Validation) ("OSG Resource and Service Validation Project,"). The basic aim is to validate the services being advertised through the resource selection and monitoring modules (CEMon). Much of the information being collected here is analogous to CEMon information, and subject to the same issues. The RSV probes use a push model, similar to the Gratia service. The site_verify.pl script takes the form of a remote grid job run by the GOC at individual sites, relaying information back using the standard Globus data movement protocols

(GASS, GridFTP) ("Globus Toolkit,"). Possibly sensitive information being reported includes:

- Account Names
- Historical system availability information
- Currently running software information
- Internal System Paths

Given that site validation data is both being collected at regular intervals, and being archived, it offers the ability to track the state of a system over time. This may provide information about regular system downtimes, when a system may be in a transitional state and particularly susceptible to an attack.

Moreover, the archived nature of this information suggests that the site is subject to a "Google Hack" (Acunetix, "Google Hacking,"), even if system data is no longer been published. An attacker can use standard search-engine technology to scan the Internet for systems that match certain keywords. This can be used to scope out systems with known vulnerabilities based on advertised software levels. This is compounded by the fact that modern search-engines like Google do their own external caching and archiving of information, creating a situation where anything that is published on the web has the chance of persisting, despite a site no longer wishing to make that information publicly available. There are known methods to prevent a site form being listed in a search engine, and it is recommended to use these for this kind of data.

Monitoring

The OSG uses the CEMon software for monitoring sites. An analysis of this has already been included in the "Resource Selection Information" section.

The OSG also supports an optional package called MonALISA (MONitoring Agents using a Large Integrated Services Architecture) to monitor system availability and load. Sites using MonALISA send system information to a central MonALISA service, which allows general users to query site information from a web-based clickable map interface. It monitors the following information (Legrand, 2007):

- System information for computer nodes and clusters.
- Network information (traffic, flows, connectivity, topology) for WAN and LAN.
- Performance of applications, jobs and services.
- End user systems, and end-to-end performance measurements.

Since this includes performance and load information for systems and networks, it could be used to determine whether a machine is susceptible to a Denial-Of-Service attack. In other words, it could be used to target systems that are running close to their maximum capacity.

This type of information is, however, extremely useful to legitimate users of a grid - it helps them determine the optimal locations for their workloads. If possible, it should only be made available to grid users, without exposing it to the outside world.

SUMMARY OF SECURITY RISKS

So far we have identified the following pieces of information, that are published to the OSG, as being potentially sensitive to a site:

1. Operating system and software level information
2. Local account names
3. Supported grid user DNs
4. Underlying authentication methods
5. Job-manager / batch-system information
6. Internal system paths
7. Job names
8. Error and failure information
9. System load and performance information

10. User activity at the site
11. Historical system availability data

While much of this data is very important to users and VOs on the grid, and essential in creating a robust and flexible grid architecture, it is important to design the systems that publish this information such that they can support the desired level of protection for the data. In other words, information should be restricted to legitimate users of the grid, and sites should have ultimate control over what information they wish to publish, and at what level of detail.

RECOMMENDED GRID MIDDLEWARE CONFIGURATION

While software may evolve, and the specific methods for configuring software may change, the general goals for proper middleware configuration remain the same. The following recommendations will help provide some amount of control to sites that wish to protect sensitive data:

1. Override attributes that are considered sensitive with alternate values that can convey the equivalent information. For example the GIP allows named attributes to be overwritten by specifying them in a special file (alter-attributes.txt). This could allow a site to replace detailed software levels with more generic information.
2. Use site level collectors for multi-resource sites. This will allow the site to filter sensitive data at this level before forwarding it to OSG. Syslog-ng is designed with this sort of architecture in mind.
3. Turn down level of detail for the published information to the minimum required – during troubleshooting efforts, this can be turned up for more diagnostic information. This limits the overall exposure of the site.
4. Always use encrypted data streams and secure protocols to send information, instead of using clear text. Many OSG services, such as Gratia or Syslog-ng, offer both SSL and clear-text options to send data to their respective collectors. Sites should always use the former, when given a choice.

RECOMMENDATIONS FOR DATA PROTECTION

Additionally, it is in the best interest of the grid provider (OSG), to provide methods for protecting this data. This protection must happen in multiple ways:

1. All grid infrastructure software that transmits or collects data from public networks should support secure and encrypted communication protocols.
2. The software design should allow sites to override arbitrary attributes being published.
3. Information collectors should endeavor to authenticate the machines that publish site data – only machines whose identities can be verified should be allowed to publish their information. This prevents third parties from publishing fake or invalid data for a given site. GSI host certificates are an effective way to achieve this kind of authentication. CEMon already uses this, and the model could easily be extended to other OSG collection services.
4. Use of grid certificates to restrict access to data where possible. Web servers should attempt to verify the identity of the user before allowing access to grid resource information. Current technologies, (e.g. mod_gridsite ("Gridsite,") for Apache based web servers) provide the ability to control access based on the user certificates. Additionally, this information could be restricted along VO

lines, so that a VO is only authorized to access its own data.

5. Prevent indexing or caching of dynamic site information on web servers by search engines. This can be done by using files like robots.txt to prevent search engines from storing this information.
6. In the long run, there should be a concerted effort to consolidate software systems collecting similar information, so that site administrators and security officers have a single point of control for protecting such information. For example the Teragrid's Inca monitoring system consolidates resource validation, troubleshooting and monitoring functionality under a single engine ("Inca: User Level Grid Monitoring,"; "TeraGrid,").

Some of these features already exist in the OSG software, but there also needs to be a comprehensive effort to integrate these types of features across the middleware and collector infrastructure.

APPLICABILITY TO OTHER GRIDS

While our work has largely been a case study on the OSG, the general principles of securing site information are applicable to any major grid infrastructure. Collection and publication of resource information is a common feature across grids, and results in similar requirements and goals with respect to protection of such information.

Indeed, many of the discussed software systems are currently deployed in other grid infrastructures as well. e.g. CEMon and MonALISA at various EGEE sites ("MonALISA Repository for Alice,"; "Enabling Grids for E-Science,"). Other grids have their own information services providing equivalent functionality. The Teragrid uses the Inca monitoring system for resource availability, validation and monitoring purposes, collecting and publishing similar site information as that discussed in the "Information Collection in OSG" section. These systems face similar risks with respect to sensitive site information, and we expect the general techniques for protecting this information to be applicable as well.

There is an increasing trend towards interoperability among grids, with international collaborations and VOs driving usage and infrastructure requirements. There is a shift away from centralized grid providers, towards integrated VO architectures, where a given VO frames its own usage model. This points to cross-grid collection services that operate on a per-VO basis. Since VOs work in close collaboration with the major grid providers, many of the current technologies discussed have uses cases for such VO based services. For example, the ALICE VO uses MonALISA to provide integrated monitoring of its supporting resources. This means that VOs must also take site security requirements into consideration as they build their grid information frameworks.

FUTURE WORK

The focus of this work has been on the OSG, and its tools, infrastructure and metadata. It would be useful to extend this analysis to other major grid infrastructures such as the Teragrid or EGEE, to understand how they approach issues pertaining to sensitive site-related information. This would highlight common problems and solutions, and provide alternative approaches towards protecting site data.

Also, given that scientific collaborations are increasingly adopting the VO model of grid computing, where a VO maintains a certain amount of control over its own users and metadata, it would be interesting to analyze how VOs manage sensitive information, and how they publish and integrate this data across one or more grid infrastructures.

CONCLUSION

While a bulk of this article has been devoted to the importance of protecting information that might reveal weaknesses in a site's security infrastructure, this should not be taken as an endorsement of the "security by obfuscation" philosophy. We recognize that there is no substitute for hard security – regular fixing and patching of software, intelligent system monitoring, and strong security polices and practices are essential for a truly secure platform. However, practical security considerations demand that administrators account for the fact that not all vulnerabilities may be known at a given time. There may also be delays between the discovery and the patching of a vulnerability. Thus, it is prudent to minimize the amount of information available to a malicious entity and limit the extent of a compromise. While it is necessary to make certain kinds of information public for the success of open grid computing, it is also in the resource provider's best interest to understand the risks involved in doing so. Since grid architectures tend to be as generic as possible, some of the published information may be extraneous. The site must find a balance between how much information it seeks to publish about itself, and how much information it wishes to protect. It may also want to limit the consumers of this information to a controlled set of persons.

We believe that this article would serve as a useful tool for sites that wish to identify these channels of information, so that they can determine the appropriate level of protection they wish to apply to their published data. We also hope to motivate further study and discussion on the protection of site information across various grid infrastructure and middleware providers.

ACKNOWLEDGMENT

Supported by the U.S. Department of Energy under Contract No. DE-AC02-05CH11231.

REFERENCES

Acunetix. *Google Hacking*. from http://www.acunetix.com/websitesecurity/google-hacking.htm.

Canal, P., Constanta, P., Green, C., & Mack, J. (2007). *GRATIA, a resource accounting system for OSG*. CHEP'07, Victoria, British Columbia, Canada. Sep 2007. Enabling Grids for E-Science. from http://www.eu-egee.org/.

Field, L. (2008). *Generic Information Provider*. EGEE Middleware Support Group. from http://twiki.cern.ch/twiki/bin/view/EGEE/GIP.

Global Grid Forum. (2003). *Usage Record – XML Format*. Globus Toolkit. from http://globus.org.

Glue Working Group. (2007). *GLUE Schema Specification version 1.3 Draft 3*. Gridsite. from http://www.gridsite.org/.

Inca: User Level Grid Monitoring. from http://inca.sdsc.edu/drupal/.

Legrand, I. (2007). *MonALISA: An Agent Based, Dynamic Service System to Monitor, Control and Optimize Distributed Systems*. CHEP'07, Victoria, British Columbia, Canada. Sep 2007. MonALISA Repository for Alice. from http://pcalimonitor.cern.ch/map.jsp.

Open Science Grid Consortium. from http://www.opensciencegrid.org/.

OSG Grid Operations Center. from http://www.grid.iu.edu/.

OSG Resource and Service Validation Project. from http://rsv.grid.iu.edu/documentation/.

Padmanabhan, A. (2007). *OSG Information Services – A Discussion*. Presentation at OSG Site Administrators Meeting, Dec 2007.

Sgaravatto, M. (2005). *CEMon Service Guide*. from https://edms.cern.ch/document/585040.

Syslog-ng Logging System. from http://www.balabit.com/network-security/syslog-ng/.

Teragrid. from http://www.teragrid.org/.

Tierney, B. L., Gunter, D., & Schopf, J. M. (2007). *The CEDPS Troubleshooting Architecture and Deployment on the Open Science Grid*. J. Phys.: Conf. Ser. 78 012075, SciDAC 2007. Virtual Data Toolkit (VDT). from http://www.cs.wisc.edu/vdt/.

This work was previously published in International Journal of Grid and High Performance Computing (IJGHPC), Volume 1, Issue 2, edited by Emmanuel Udoh & Ching-Hsien Hsu, pp. 45-55, copyright 2009 by IGI Publishing (an imprint of IGI Global).

Chapter 11
Federated PKI Authentication in Computing Grids:
Past, Present, and Future

Massimiliano Pala
Dartmouth College, USA

Shreyas Cholia
Lawrence Berkeley National Laboratory, USA

Scott A. Rea
DigiCert Inc., USA

Sean W. Smith
Dartmouth College, USA

ABSTRACT

One of the most successful working examples of virtual organizations, computational Grids need authentication mechanisms that inter-operate across domain boundaries. Public Key Infrastructures (PKIs) provide sufficient flexibility to allow resource managers to securely grant access to their systems in such distributed environments. However, as PKIs grow and services are added to enhance both security and usability, users and applications must struggle to discover available resources-particularly when the Certification Authority (CA) is alien to the relying party. This chapter presents a successful story about how to overcome these limitations by deploying the PKI Resource Query Protocol (PRQP) into the grid security architecture. We also discuss the future of Grid authentication by introducing the Public Key System (PKS) and its key features to support federated identities.

DOI: 10.4018/978-1-60960-603-9.ch011

AUTHENTICATION IN VIRTUAL ORGANIZATIONS

Computational grids provide researchers, institutions and organizations with many thousands of nodes that can be used to solve complex computational problems. To leverage collaborations among entities, users of computational grids are often consolidated under very large *Virtual Organizations (VOs)*.

Participants in VOs need to share resources, including data storage, computational power and network bandwidth. Because these resources are valuable, access is usually limited, based on the requested resource and the requesting user's identity. In order to enforce these limits, each grid has to provide secure authentication of users and applications.

Erroneously granting access to unauthorized or even malicious parties can be dangerous even within a single organization---and is unacceptable in such large VOs.

Moreover, the dynamic nature of grid VOs requires the authentication mechanisms to be flexible enough to easily allow administrators to manage trust and quickly re-arrange resource-sharing permissions. Indeed, VOs are usually born from the aggregation of already existing organizations and constitute an umbrella that groups the participating organizations rather than replacing them. For example, large VOs like the ATLAS and CMS Large Hadron Collider collaborations may be distributed across multiple organizational and national boundaries. Authentication must allow individual organizations to maintain control over their own resources.

The Problem. When participating in a VO, an organization must solve the problem of securely identifying resource requesters that come from outside its boundaries. PKIs offer a powerful and flexible tool to solve the potential authentication nightmare. Nonetheless, grid and VO administrators are still striving to find an acceptable solution to address interoperability issues that originate from the way VOs differ in policies, infrastructures and resource control.

Consider the situation where access to grid resources is managed via a Web portal. The portal can use SSL to provide strong mutual authentication, between client and server, based on grid-approved PKI credentials. To do this, the portal administrator needs to set up the SSL Trust List to only allow credentials from approved CAs; the portal also needs to know how to validate the entire trust chain for that credential (that is, the end entity certificate presented, its issuer and the issuer's issuer, and so forth) up to the approved self-signed grid trust anchor.

To do this validation, the portal needs to know how to access services such as the location of the CA certificate and revocation data for each of these intermediate CAs. However, the portal cannot count on having pre-configured details for them. Even if it did—or if the information was packaged in each end entity certificate—this information may change over time, rendering this critical data stale. Having some way to dynamically discover service entry points of interest for grid-approved authorities (or indeed, the very authorities themselves) would solve a number of issues and would also provide for more flexible implementation options for the grid authorities, potentially lowering the costs of future service changes, and facilitating the future offering of additional services.

Our Solution Path. In order to help VOs to more efficiently address PKI interoperability issues we have started a collaboration with the TACAR project to foster the adoption of the *PKI Resource Query Protocol* (PRQP) which enables discovery of resources and services in inter-PKI and intra-PKI environments. Although PRQP provides a viable solution for immediate deployment, in this paper we extend this solution by advocating for the adoption of a Public Key System (PKS) in order to provide support for VO authentication over the Internet.

PAST AND PRESENT OF AUTHENTICATION IN GRIDS

According to Ian Foster, a *grid* is a system that "coordinates resources that are not subject to centralized control, using standard, open, general-purpose protocols and interfaces, to deliver nontrivial qualities of service" (Foster, 2002). In order for the grid computing model to be successful, users and VOs must access a wide variety of resources using a uniform set of interfaces. Given that most resource providers have their own security policies and schemes to begin with, grids must overcome the challenge of integrating a wide variety of authentication mechanisms to achieve this kind of resource sharing. Without a common authentication layer, Virtual Organizations and resource providers are forced to adopt ad hoc schemes to achieve integrated resource sharing. However, the adoption of arbitrary schemes discourages information sharing and collaboration among researchers, and essentially makes the grid model unworkable.

The Grid Security Infrastructure (GSI) has become the de facto security layer in scientific, research and academic grids. It provides applications, VOs and resource providers with a secure and standard means to perform authentication across organizational boundaries. GSI is built on top of a PKI layer and uses standard X509 v3 certificates for authenticating principals and granting access to local resources. Several major grid infrastructures, including Open Science Grid (OSG), European Grid Infrastructure, TeraGrid and Earth Systems Grid (ESG) rely on GSI for managing authentication between users and services.

In a distributed environment, it is important to maintain traceability back to the individual entity matching a given certificate. The task of identifying users is distributed across various grid CAs throughout the world. These CAs are accredited and audited by the International Grid Trust Federation and its three regional Policy Management Authorities. A list of accredited CAs is maintained by the IGTF and distributed to relying parties throughout the world.

Grid CAs issue users a PKI certificate, including a public key linked to the private key controlled by the grid subscriber. These certificates may either be long-lived (typically issued by classic grid CAs) or short-lived (typically issued by online CAs such as SWITCH (SWITCH, 2008) or MyProxy-based CAs (NCSA, 2008)) depending on the use case. The IGTF maintains different authentication profiles to manage CAs with different qualities of service, for the benefit of relying parties.

A resource provider or virtual organization relies on these CAs to be able to identify a given user. As such, if an end entity is able to present a valid certificate that is signed by a CA trusted by the relying party, the entity can be authenticated (of course, the end entity also needs to prove knowledge of the private key). GSI authentication is mutual (GLOBUS, 2008)—if a user wishes to access a service, both the user and the service must be able to present signed certificates to each other. The respective signing authorities must be trusted by the entity on each side of the transaction. Allowing the user and the service to have certificates signed by different CAs is the key to establishing cross-realm trust in grids. This also eases usability and scalability—the user need maintain only a single individual credential (single point of identity) no matter how many services she wishes to use. In order to improve usability, a user of grid services can sign a *Proxy Certificate (PC)* on his or her own behalf.

In general these proxies contain a slightly modified version of the user's identity (to indicate that it is a proxy certificate), a new public key, and a very short lifetime. These proxy credentials can then be used to access applications, or further delegated to application servers to perform actions on behalf of that user, without having to expose the user's original long-lived credential and private key—thus practicing the security principle of "least privilege."

Figure 1. Chain of Trust in grids environment. The usage of Proxy Certificates allows the user to delegate tasks without exposing her private key—since each Proxy Certificate has its own unique keypair

Most GSI-based grid applications can recognize PCs and will trust the credential as long as the chain of trust leads back to the original user and a trusted CA. A detailed scheme of the whole chain of certificates involved in identity verification is shown in Figure 1.

Additionally, grids and VOS may use special authorization services to handle fine-grained roles based access control. For example, OSG VOs use a *Virtual Organization Management Service (VOMS)* (Ciaschini, 2004) service to generate and sign an *Attribute Certificate* that contains one or more *Fully Qualified Attribute Name (FQAN)* strings, linked to the user's subject DN. This FQAN is embedded in the user's proxy certificate as an X.509v3 extension and defines that user's role within the VO. VOMS proxies can be used to manage roles and levels of access to resources, while using the same identity principal (user certificate) across the grid.

PKI RESOURCE DISCOVERY IN GRIDS

To use these more general PKIs, applications must be capable of finding and using services and data repositories provided by Certification Authorities. Unfortunately, even the retrieval of the list of revoked certificates (CRLs) is still a problem when dealing with CAs from different hierarchies or loosely coupled PKI meshes.

Grid PKIs can become rather complex, and the number of grid CAs accredited by the Policy Bodies (which are relatively young) is expected to grow in the near future. Indeed, as long as policies and common practices are established and well understood, the number of accredited CAs should increase in the number of hundreds, thus increasing the need for a standardized solution for a PKI resource discovery system.

Current Data Distribution. Currently, the mechanism for querying the trusted providers is fairly simple: administrators and users download a trusted CA distribution. This can either happen as part of a manual process, or it can be included within the grid software distribution (such as the Open Science grid software stack). This packaged data consists of a set of accredited CAs. (Accreditation is done by peer review in the various policy bodies.)

Because of the need to provide users and administrators with additional data besides the CA certificates, the downloaded package includes extra files. In particular, for a given CA, the package typically includes the following static information: the *CA certificate*, the *.info file*, a *CRL URL file*, a *namespaces file*, and a *signing policy file*.

The *.info file* contains general CA information along with contact information (including a URL). Applications can use information in the .info file to contact the CA. An example of a distributed .info file is shown in Figure 2. Some of the information distributed in this file (e.g. url, email or status) is required by applications and users to find details about the CA. The *CRL URL file* contains a URL pointer from where one would download the CRL. All accredited IGTF classic CAs provide this file. Sites and users build revocation lists by periodically querying the information in the CRL URL file and downloading

Figure 2. Example of distributed info file within grid communities. Notice how some of the distributed information have no equivalent pointers in standard X509 certificates

```
#
# @ (#) $Id: 1c3f2ca8.info,v 1.5 [...] $
# Information for CA DOEGrids
#   obtained from 1c3f2ca8 in DOEGrids/
alias = DOEGrids
url = http://www.doegrids.org/
crl_url = http://pki1.doegrids.org/CRL/1c3f2ca8.r0
email = trouble@es.net
requires = ESnet
status = accredited:classic
version = 1.16
sha1fp.0 = 2D:7C:01: [....] :F8:90
```

revocation lists from the CRL url for *each CA*. This means that many grid software installations in the world are downloading these large CRLs from the CA providers at regular intervals. From what we have seen, this has often created denial of service conditions for certain CAs.

The *namespaces file* defines the *Distinguished Names (DN)* namespace that the CA is authorized to use; the *signing policy file* defines the rules for the signing policy of that CA. The *namespaces file* and the *signing policy file* may contain overlapping information from a policy point of view (although only the *signing policy file* has an implementation in software). Although this information could be embedded into a CA's certificate, the need for updating this data periodically led to the creation of the .info file and bundling it together with the certificate.

TACAR (Terena Academic CA Repository) and *IGTF* register and distribute this information to users and sites as follows. The accredited CA sends the trust anchor information directly to the IGTF/TACAR through a TERENA officer or a TERENA TACAR trusted introducer.

The IGTF packages and distributes the official CA package. Relying parties download the IGTF package every time there is a new release (approximately once a month). Relying parties are encouraged to verify this against the TACAR repository. Then, based on the information within the downloaded package, relying parties download the CRL from the CRL URL on a daily basis.

Ultimately, in most cases, this relies on a very static "cron-based" process. There are several improvements to this that can be made by PRQP that would replace this type of static file and crontab based access with something more dynamic, and query driven.

Other Solutions. To publish pointers to data, a CA could use certificate extensions such as the *Authority Information Access* (AIA) and the *Subject Information Access* (SIA) (R. Housley, W. Polk, W. Ford, and D. Solo, 2002). Regrettably the lack of support built into applications and the difficulties in updating extensions in certificates clash with the need for flexibility required by today CAs.

To overcome the problem with updating the pointers, it is possible to use SRV records (A. Gulbrandsen, P. Vixie, and L. Esibov, 2000) in DNS (P. Mockapetris, 1987). Although interesting, the problem with this solution resides in the lack of correspondence between the DNS structure, which is built on a strictly hierarchical namespace, and PKIs where there are no requirements for the used namespace.

Other solutions are either overly complicated to solve our problem---e.g., Web Services (F. Curbera, M. Duftler, R. Khalaf, W. Nagy, N. Mukhi, and S. Weerawarana, 2002) uses *SOAP* (A. Karmarkar, M. Hadley, N. Mendolsohn, Y, Lafon, M.

Gudgin, J.-J. Moreau, H. Nielsen, 2007), *WSDL* (E. Christensen, F. Curbera, G. Meredith, and S. Weerawarana, 2001; R. Chinnici, M. Gudgin, J.-J. Moreau, and S. Weerawarana, 2005) and *UDDI* (L. Clement, A. Hately, C. von Riegen, and T. Rogers, 2004) or they are specifically targeted to local area networks---e.g., *Jini* (W. Edwards, 2000; K. Arnold, 2000) *UPnP* (UPnP Forum, 2008; M. Jenronimo and J. Weast, 2003) or *SLP* (E. Guttman, C. Perkins, and J. Kempf, 1999; E. Guttman, 1999).

TRUST AND CERTIFICATION POLICIES

The use of a standardized and well-established technology such as public key certificates has enabled applications such as browsers to facilitate ease of use within grids. However, especially when integrating credentials from different authorities, an important aspect to consider is the policies under which those credentials have been issued. Although a PKI potentially provides the benefit of strong binding of identities to public keys, the strength of that binding is really dependent on the policies and practices followed by the issuing authority, and the subscribers.

A CA is a trusted third party entity which issues digital certificates for use by relying parties. In a certificate, the CA attests that the public key matches the identity of the owner of the corresponding private key, and also that any other data elements or extensions contained in the certificate match the subject of the certificate. The obligation of a CA (and its registration authorities) is to verify an applicant's credentials, so that relying parties can trust the information contained in the certificates it issues. If a relying party trusts the CA and can verify the CA's signature, then it can also verify that a certain public key does indeed belong to whoever is identified in the certificate (as long as they accept this, the end entity is fulfilling its responsibilities with respect to protecting the private key). If the CA can be subverted, then the security of the entire system is lost; likewise, if an end entity is negligent, then the security and trust associated with their particular credential could be lost.

The degree to which a relying party can trust the binding embodied in a digital certificate depends on several factors. These factors can include the practices followed by the certification authority in authenticating the subject; the CA's operating policy, procedures, and security controls; the scope of the subscriber's responsibilities (for example, in protecting the private key); and the stated responsibilities and liability terms and conditions of the CA (e.g warranties, disclaimers of warranties, and limitations of liability). The processing of information contained in these multiple complex documents for the purpose of making a trust decision about each PKI involved is too onerous for the average user. Relying parties therefore usually accept recommendations from trusted accreditation bodies about the relative trustworthiness and suitability of credentials being issued by a particular CA. For grids, those accreditation bodies are the three regional PMAs that constitute the IGTF. *TAGPMA* is the accreditation authority for the Americas (covering a geographical region from Canada to Chile).

TAGPMA conducts peer reviews of grid CA operations. A grid CA can be accredited as a grid credential issuer after TAGPMA reviews their *Certificate Policy (CP)* and *Certification Practices Statement (CPS)* to ensure that the practices implement the policies and that the policies are equivalent to standard approved grid profiles. Once approved, the CA and associated information is packaged for official distribution for IGTF relying parties. Re-review of a CA is conducted on a periodic basis to ensure they are still compliant with the standard grid profiles.

Not all grid CA accreditation applicants are able to map their existing policies and practices to an approved IGTF profile. However, a relying party may still wish to accept the credentials of such a CA operator based upon their own assessment of trustworthiness of the CA. In order for the relying party to make a local trust decision, they should consider the statements by the CA published in their CP and CPS and also review any other relevant security or trust-related documentation. Currently this information is generally not readily available to a relying party from the CA's certificate, nor can a relying party or potential subscriber easily find the URI for the application or revocation of credentials from such CAs. A mechanism for publishing and updating this information would greatly enhance the flexibility, and usability of potential grid PKIs. The PRQP is a perfect candidate for providing such functionality.

INTEROPERABLE GRID PKIS: FIRST STEPS

Effective authentication frameworks that make use of certificates potentially require many different services such as OCSP servers, CRL repositories, or timestamping to validate certificates issued by accredited CAs. As a consequence, certification authorities need to be able to provide these services and to enable applications to discover them.

Because the need to distribute PKI related data and pointers to services is of primary concern in grids, each grid environment defines its own specific format and solution. Although this might temporarily solve specific issues within a specific grid community, it does not encourage the exchange of information and interoperability with other organizations.

It is to be noted that because of the customized nature of current solutions, specific extensions to applications must be developed in order to be able to operate in such environments.

The PKI Resource Discovery Protocol. The notion of a discovery protocol for PKIs first appeared in our earlier paper (M. Pala and S. W. Smith, 2007), which proposed the *PKI Resource Query Protocol (PRQP)*[1] to provide pointers to any available PKI resource from a particular CA.

The PRQP (M. Pala, 2008) has been already discussed in the IETF PKIX working group. The updated version of the PRQP specification, which includes grid-specific enhancements proposed in this paper, is published as an Experimental-Track Internet Draft. In PRQP, the client and a *Resource Query Authority (RQA)* exchange a single round of messages where the client requests a resource token by sending a request to the server. The server replies back by sending a response to the requesting entity.

The client can request the address of one or more specific services by embedding one or more Object Identifiers (OIDs) into the request. The resources might be items that are (occasionally) embedded in certificates today—such as URLs for CRLs, OCSP, SCVP or CP/CPS locations---as well as other items, such as addresses for the CA website, the subscription service, or the revocation request.

Alternatively, the client may ask for the location of all the services provided by a CA by not specifying any identifier in the request.

The Resource Query Authority. In PRQP, the server is called the *Resource Query Authority* (RQA). An RQA can play two roles. First, a CA can directly delegate an RQA as the party that can answer queries about its certificates, by issuing a certificate to the RQA with a unique value set in the *extendedKeyUsage* (i.e. prqpSigning). The RQA will provide authoritative responses for requests regarding the CA that issued the RQA certificate. Alternatively, an RQA can act as *Trusted Authority (TA)* ("trusted" in the sense that a client simply chooses to trust the RQA's recommendations and assertions). In this case, the RQA may provide responses about multiple CAs without the need to have been directly certified by them.

In this case, provided responses are referred to as *non-authoritative*, meaning that no explicit trust relationship exists between the RQA and the CA. To operate as a TA, a specific extension (*prqpTrustedAuthority*) should be present in the RQA's certificate and its value should be set to TRUE. In this configuration the RQA may be configured to respond for different CAs which may or may not belong to the same PKI as that of the RQA.

Security Considerations. The PRQP provides URLs to PKI resources, therefore it only provides locators to data and services, and not the real data. It still remains the client's job to access the provided URLs to gather the needed data, and to validate the data (e.g., via signatures or SSL).

Because of this consideration, both the NONCE and the signature are optional in order to provide flexibility in how requests and responses are generated.

Also, it is then possible to provide pre-computed responses in case the NONCE is not provided by the client. If an authenticated secure channel is used at the transport level between the client and the RQA (e.g. HTTPS or SFTP) signatures in requests and responses can be safely omitted.

Distribution of RQA addresses. The distribution of the RQA's address to clients is still an open issue. There are four possible approaches. A first option would be to use the AIA and SIA extensions to provide pointers to RQAs. We believe that by using these extensions in certificates to locate the RQA, one could provide an easy way to distribute the RQA's URL. The size of issued certificates would be smaller than embedding all the pointers for CA's resources, thus providing a more space efficient solution.

The second option is applicable mostly for LANs, and consists of providing the RQA's address by means of DHCP. This method would be mostly used when a trusted RQA is available on a local network. These two techniques can then be combined together. Although the service number for DHCP and DHCPv6 for PRQP have not yet been assigned by IANA, the official protocol draft describes how to provide local RQAs addresses via dynamic host configurations.

The third option—which could be successfully applied in special-purpose application environments like Grid Computing—is to embed the RQA's address directly into application software distributions. This approach could be adopted in grids and VOs where a centralized software distribution system is in place. At each software update, the RQA network address can be updated as well. If the distributed software is not digitally signed by a trusted authority, this approach could be subject to serious security threats, e.g. distribution of an altered package by a malicious attacker where the configured RQAs are not the "official" ones. Besides the security considerations already discussed above, the trust level in the application's RQA configuration should be not less then that put into browser or operating system certificate stores.

Ultimately, the RQA address can be retrieved by querying the DNS for specific service records. The SRV records—or Service records—technique was meant as a way to provide pointers to servers directly in the DNS. As defined in RFC 2782 (A. Gulbrandsen, P. Vixie, and L. Esibov, 2000), the introduction of this type of record allows administrators to perform automatic discovery for local network services. The core idea behind SRV records is to have the client query the DNS for a specific SRV record. For example if an SRV-aware OCSP client wants to discover an OCSP server for a certain domain, it performs a DNS lookup for ocsp.tcp.example.com (the " tcp" means the client requesting a TCP enabled OCSP server). The returned record contains information on the priority, the weight, the port and the target for the service in that domain.

Although used for many different network-related configurations (e.g., printing services), this approach has not been successfully deployed for PKI-related services. Besides the issues related to relying on not authenticated services for discovering the network addresses of specific resources, the main issues are related to the fact that there is no correspondence between DNS structure and data contained in the certificates. The only exception being when the Domain Component (DC) attributes are used in the certificate's Subject. Fortunately, with the recent deployment of DNSSEC (Arends, R.; Austein, R.; Larson, M.; Massey, D. & Rose, S., 2005; Weiler, S. & Ihren, J., 2006) services and their integration with current OSes, some of the trust considerations related to the local service discovery via DNS records will be soon solved.

However, this approach can be successfully adopted in VOs where the centralized policy body authority could provide the RQA configurations on behalf of the whole VO.

Finally, we want to point out that other mechanism will be available to discover LAN provided services in IPv6 (Deering, S. & Hinden, R., 1998) based on simple ping of reserved IP addresses in the local segment.

INTEGRATING PRQP INTO GRIDS

In our work toward a dynamic discovery of PKI-related services for Computing Grids, we analyzed the security requirements and the current challenges in distributing pointers to authentication data. To ease the administrators' burden and to provide a more efficient way to distribute resource locators, we extended the PRQP specification with grid-specific support. In particular, these extensions provide an interoperable method to distribute information about provided services. Although some solutions already exist in the computing grid environment (e.g. the monthly IGTF/TACAR update), our work addresses the problem by providing a more standardized solution that would allow for better interoperability between organizations (as discussed earlier).

OpenCA's LibPKI (OpenCA, 2008a) provides an updated implementation of the full PRQP protocol. At present, a PRQP server is also available as a stand-alone application (OpenCA Labs, 2008c) and freely downloadable[2]. The GSI based security layer, used across several major grids and VOs, is built on top of the OpenSSL library, a widely used open-source library. Since GSI is based on standard PKI mechanisms, it plugs nicely into the PRQP model. A PRQP client can be implemented at the GSI layer using callouts – we plan to implement this in the future.

Grid-Specific Resources. In order to better leverage PRQP in the Grid environment, we defined a set of object identifiers (OIDs) that enhance PRQP with the ability to provide grid-specific data distribution. Because grid communities organize themselves in VOs that accept common authentication profiles (such as those of the IGTF), it has been easy to analyze the requirements and identify the needed enhancements to PRQP.

Besides identifying the OIDs for general PKI operations (e.g., HTTP based or browser-specific services, CA "communication gateways", etc.)[3], we also defined some Grid-specific pointers (see Table 1).

The *accreditationBody* and the *accreditationPolicy* pointers can be used to specify the bodies and the policies (or profiles) under which a CA has been accredited. In addition to these, we also defined the *commonDistributionUpdate* and the *accreditedCACertificates* OIDs. These identifiers can carry information about pointers to the most recent Grid distribution data (the former) and to the set of accredited CA certificates (the latter).

One interesting feature of PRQP is its flexibility. It can provide CA management with a dynamic model to add services or, if needed, to switch to newer and more efficient ones. This feature becomes of primary concern in grids where currently grid-specific services have not been standardized yet.

CAs can leverage PRQP flexibility properties in order to provide dynamically updated information about its accreditation status to applications via the *accreditationStatus* pointer. The set of grid-specific pointers we introduced facilitates more flexible trust options from the VO's perspective, in the set of CAs it chooses to trust. For instance, besides the generally accepted IGTF distribution,

Table 1. Newly Identified OIDs for Grid Operations. Of particular interest are the Grid specific pointers that enable an RQA to provide Grid specific information to applications. It is also to be noted that some of the proposed PKIX Identifiers refer to services that are not yet standardized

	OID	Text	Description
PKIX	{id-ad 1}	ocsp	OCSP Service
	{id-ad 2}	caIssuers	CA Information
	{id-ad 3}	timeStamping	TimeStamping Service
	{id-ad 10}	dvcs	DVCS Service
	{id-ad 11}	scvp	SCVP Service
General PKI operations	{id-ad 50}	certPolicy	Certificate Policy (CP) URL
	{id-ad 51}	certPracticesStatement	Certification Practices Statement (CPS) URL
	{id-ad 60}	httpRevokeCertificate	HTTP Based (Browsers) Certificate Revocation Service
	{id-ad 61}	httpRequestCertificate	HTTP Based (Browsers) Certificate Request Service
	{id-ad 62}	httpRenewCertificate	HTTP Based (Browsers) Certificate Renewal Service
	{id-ad 63}	httpSuspendCertificate	Certificate Suspension Service
	{id-ad 40}	cmsGateway	CMS Gateway
	{id-ad 41}	scepGateway	SCEP Gateway
	{id-ad 42}	xkmsGateway	XKMS Gateway
	{eng-ltd 3344810 10 2}	webdavCert	Webdav Certificate Validation Service
	{eng-ltd 3344810 10 3}	webdavRev	Webdav Certificate Revocation Service
Grid Specific	{id-ad 90}	accreditationBody	Accreditation Body URL
	{id-ad 91}	accreditationPolicy	Accreditation Policy
	{id-ad 92}	accreditationStatus	Accreditation Status Document
	{id-ad 95}	commonDistributionUpdate	Grid Distribution Package
	{id-ad 96}	accreditedCACertificates	Certificates of Currently Accredited CAs

these pointers also allow a VO to specify a set of additional CAs that the VO wishes to trust locally (that the VO has vetted itself for use within the community), by simply specifying an additional local distribution maintained by the VO or any entity it delegates this responsibility to (e.g. refer to the additional non-IGTF accredited CAs that are accepted by TeraGrid).

PRQP AND TACAR: A REAL WORLD DEPLOYMENT

An interesting aspect of the grid trust model is the presence of a central authority, often embodied by the grid policy management authority. Usually this authority is represented by a federation of authentication providers and relying parties responsible for accreditation of CAs willing to participate in the organization.

The presence of such an authority eases the deployment of PRQP in that it provides a central point where the RQA can be deployed. In this section, we discuss the real-world experience in deploying the PRQP service for the TACAR project. To speed up the service deployment and ease CA administrators from running an additional service, we deployed a centralized RQA service that serves the entire grid community.

Trusting a Central RQA. In the TACAR PRQP deployment, we adopted a trust model that utilizes a centralized Resource Query Authority which serves all the organizations participating in the grid community.

This model is easily applicable when the VOs and grids share the same set of accredited Certification Authorities. In this case, the client application queries the central RQA to discover the needed information about CAs participating in the VO. For this model to work, the central RQA must know the pointers for each and every CA that is recognized by the VO. In this case, the RQA is to be trusted by all the participating parties. The RQA can be configured to act as a trusted responder or, if every participating CA is willing to certify the RQA's key pair, as an authoritative responder.

It may be unrealistic to expect a policy authority (like IGTF) to operate a central RQA which would require 24x7 support; however, the operation could be easily delegated by the policy authority to one of the more prominent accredited CA sites that are already geared for 24x7 services. The policy body would then simply need to require periodic assertions (or audits) to confirm that the central service was operated precisely and integrally.

In our PRQP deployment for TACAR we adopted a delegated model where the central RQA service is run by one of the accredited CAs. Moreover, in order to facilitate the update of the pointers provided by each CA to the RQA administrators, we provided a web-based configuration tool (integrated with the TACAR control panel) that allows CA administrators to easily update/add URL pointing to the provided services. The configuration is then pushed to the RQA server and automatically deployed at regular intervals during the day.

TOWARDS GLOBAL GRID AUTHENTICATION

Our experience with PRQP provided us with the idea that an Internet-wide service aimed at enhancing trust-infrastructures deployment and interoperability is both needed and soon deployable. In particular, we started working at the deployment of a distributed support system for trust infrastructures suitable for Internet-scale deployment and dynamic federation management, namely the Public Key System (PKS). In order to ease roll over between isolated PKI islands to globally available and locally configurable PKI services, this infrastructure will allow smooth co-existence and progressive integration with existing infrastructures.

The PKS we first designed in (M. Pala, 2010) and that we plan to develop and deploy for Grid authentication purposes first, is composed of three main parts: a DHT-based overlay network, a unified message format, and the support for federated identities.

The PKS uses a peer-to-peer overlay network to route messages to the target CAs and federation authorities. In particular, we use a simplified version of the Chord protocol based on the PEACH (M. Pala and S. W. Smith, 2008) system. We selected this routing algorithm for two reasons. First, it already provides support for node identifiers based on public key certificates. Secondly, the PEACH protocol is easy to support from the developers point of view: other protocols like Kademilia (Maymounkov, P., Mazieres, D., 2002) or P-Grid (Aberer, K., Mauroux, P.C., Datta, A., Despotovic, Z., Hauswirth, M., Punceva, M., Schmidt, R., 2003) might provide additional features at a greater implementation costs.

A collaborative Approach. In our previous work, we designed and prototyped a scalable system for PKI resources look-up. In (M. Pala and S. W. Smith, 2008) we introduced a new peer-to-peer overlay network that makes use of a Distributed Hash Table routing protocol (namely, Peach). Results from this work have demonstrated that PKIs can make effective use of peer-to-peer technologies and have laid the path for the next steps in this new field. Building on our previous work, we extended this approach to provide a support system for Public Key trust infrastructures deployment. In particular, we enhanced the peer-to-peer protocol to support (1) interoperable PKI message exchange among CAs, and (2) usable federated identities deployment. The most noticeable addition to the PEACH network infrastructure is introduction of a new type of nodes, the PK Federation Authorities.

In the PKS model, network administrators deploy local PKS responders. As such, the PKS is similar to the DNS where caching servers are deployed in LANs to ease client configurations. The PKS responders, in this case, act as a PKI proxy for applications. In case the local organization also deployed it's own CA, the local PKS node will reply to PKI requests for the local PKI in addition to forwarding requests that are addressed for external CAs.

In order to locate available CAs efficiently on the PKS network, we use unique node identifiers for each CA. We leverage the availability of the CAs' digital certificates by deriving the node's identifier from the fingerprint of the CA certificate itself. For example, if CA1 wants to participate in the PKS network, it will setup a PKS node and issue a certificate that identifies it as the authoritative PKS responder.

When joining the PKS network, the PKI gateway will present its own certificate together with its issuing CA's certificate. The node identifier, that is the identifier that will enable the node to be found on the network, is calculated by using the fingerprint of the CA's certificate. To validate the identity of the joining node, a simple validation of the presented certificate chain will guarantee that the joining node has been authorized as a PKS responder for that particular CA. This approach guarantees high scalability, provides a simple approach to PKS responders deployment, and is logically compatible with the Peer Name Resolution Protocol (Microsoft) already available in the Windows operating system (although available only over IPv6).

Ultimately, we notice that the PKS network can support any type of public key identifiers, not only X.509 certificates. This feature stems from the use of the output of the hash function to link a node on the PKS network to an identity (e.g., a CA or a PK-FA). Although our work primarily focuses on X.509 certificates, the PKS overlay network is capable of supporting multiple types of public key based identifiers.

Two-Tier Approach. To ease the deployment of PKS, applications such as browsers or email clients, access the PKS by querying the local PKS server. The local PKS responder is responsible of discovering if the responder authoritative for the CA requested by a client is available on the PKS network and, if so, it forwards the application's request to the target node. The response is then routed back to the client through the same local PKS responder.

In other words, applications use only one simple transport protocol for all PKI-related queries (e.g., OCSP, CMM, SCEP, etc.) and do not need to implement any of the peer-to-peer overlay network operations (e.g., *join*() or *lookup*()).

The Quest for Federated Identities. One of the urgent needs in today's on-line communities is the possibility to demonstrate one's participation to one or more federations. In the case of Computing Grids, these federations are identified by saccreditation bodies. These authorities decide the policies (or rules) that an organization must follow in order to be accredited. They also perform audits to check on the compliance of an accredited organization with the policy of the VO. Therefore, being the authority recognized by every member participating to the VO, the policy body is the authoritative source of isnformation about the VO membership. Regrettably, there is no standardized way to dynamically provide that information to applications.

To accommodate the need to federate existing organizations, the PKS supports PK Federation Authorities (PK-FA) nodes. These nodes provide information about the deployed federations by indicating if a particular entity is part of a specific federation or not. The protocol we designed in

(M. Pala, 2010) allows clients to sensibly reduce the list of trust anchors (or Trusted Certification Authorities). In particular, by trusting the PK-FA certificate, a client can dynamically discover if a CA is part of the trusted federation, and, if so, can use the PKS to correctly route the requests about the provided PKI services.

Since the source of trust is the PK-FA, the trust is built by combining the PK-FA response with the usual certificate validation of the certificate that is being verified. The use of dynamically generated PK-FA responses allows infrastructures to dynamically join or leave federations. In fact, although that there is no direct certification link between the PK-FA (the trusted entity) and the certificate to be verified, the trust (from a federation point of view) flows from the signed PK-FA response as it identifies the certificate issuing CA as part of the trusted federation. In other words, the PK-FA provides a source of technical bridge that allows to verify (from an application standpoint) the compliance of an organization to a well-known policy without the need of cross certification among trust infrastructures.

This allows applications to implement user-friendly trust anchor management systems based on the idea of federation (e.g., the Banking Federation, the Credit Cards Association, etc.).

CONCLUSION

In our work we provide a description of the grid authentication layer. We also provide an overview of the issues that grids and virtual organizations face every day in distributing crucial information that enables the usage of digital certificates.

Our work also analyzes the current status of the PKI Resource Query Protocol and describes the TACAR experience in integrating the protocol into an existing infrastructure.

We believe that PRQP can provide an effective solution to the PKI services pointer distribution issue, especially in virtual organizations where a common authentication layer exists. The PRQP introduces a new layer of indirection that allows mapping of PKI resource discovery to network addresses. Today, no existing software provides such a flexible service. In fact, no deployed infrastructure exists that provides an efficient and interoperable PKI resource-discovery service.

Building on top of our experience with PRQP deployment, we focused on allowing for improved interoperability among trust infrastructures by introducing the Public Key System (PKS) and its promising characteristic toward an Internet-wide support infrastructure for federated identities.

ACKNOWLEDGMENT

The authors would like to thank the IGTF members for their contribution and inspiring suggestions. This work was supported in part by CISCO; the NSF (under Grant CNS-0448499); the U.S. Department of Homeland Security (under Grant Award Number 2006-CS-001-000001); and the Director, Office of Science, Office of Advanced Scientific Computing Research of the U.S. Department of Energy (under Contract No. DE-AC02-05CH11231). The views and conclusions contained in this document are those of the authors and should not be interpreted as necessarily representing the official policies, either expressed or implied, of any of the sponsors. A preliminary version of this work appeared as Pala et al, "Extending PKI Interoperability in Computational Grids," 8th *IEEE International Symposium on Cluster Computing and the Grid.*

REFERENCES

Aberer, K., Mauroux, P. C., Datta, A., Despotovic, Z., Hauswirth, M., Punceva, M., & Schmidt, R. (2003). P-Grid: A self-organizing structured P2P system. *SIGMOD, 32*. ACM.

Arends, R., Austein, R., Larson, M., Massey, D., & Rose, S. (2005). *DNS security introduction and requirements. RFC 4033. Internet Engineering Task Force*. IETF.

Arnold, K. (2000). *The Jini specification* (2nd ed.). Addison-Wesley.

Chadwick, D. W. (2007). *Use of WebDAV for certificate publishing and revocation. Internet Engineering Task*. IETF.

Chinnici, R., Gudgin, M., Moreau, J.-J., & Weerawarana, S. (2005). *Web services description language (WSDL) version 2.0 part 1: Core language.* Retrieved from http://www.w3.org/TR/wsdl20

Christensen, E., Curbera, F., Meredith, G., & Weerawarana, S. (2001). *Web services description language (WSDL) 1.1.* Retrieved from http://www.w3.org/TR/2001/NOTE-wsdl-20010315

Ciaschini, V. (2004). *A VOMS attribute certificate profile for authorization.* Retrieved from http://grid-auth.infn.it/docs/AC-RFC.pdf

Clement, L., Hately, A., von Riegen, C., & Rogers, T. (2004). *UDDI version 3.0.2.* Retrieved from http://uddi.org/pubs/uddi v3.htm

Curbera, F., Duftler, M., Khalaf, R., Nagy, W., Mukhi, N., & Weerawarana, S. (2002). Unraveling the Web services Web: An introduction to SOAP, WSDL, and UDDI. *IEEE Internet Computing, 6*, 86–93. doi:10.1109/4236.991449

Deering, S., & Hinden, R. (1998). *Internet protocol, version 6 (IPv6) specification. Internet Engineering Task Force*. IETF.

Edwards, W. (2000). *Core Jini* (2nd ed.). Prentice-Hall.

Foster. (2002). What is the Grid? A three point checklist. *GRIDtoday, 1*(6).

GLOBUS. (2008). *Overview of the Grid security infrastructure.* Retrieved from http://www.globus.org/security/overview.html

Gulbrandsen, A., Vixie, P., & Esibov, L. (2000). *A DNS RR for specifying the location of services (DNS SRV). RFC 2782. Internet Engineering Task Force*. IETF.

Guttman, E. (1999). Service location protocol: Automatic discovery of IP network services. *IEEE Internet Computing, 3*(4), 71–80. doi:10.1109/4236.780963

Guttman, E., Perkins, C., & Kempf, J. (1999). *Service templates and schemes. Internet Engineering Task Force*. IETF.

Guttman, E., Perkins, C., Veizades, J., & Day, M. (1999). *Service location protocol, version 2. Internet Engineering Task Force*. IETF.

Housley, R., Polk, W., Ford, W., & Solo, D. (2002). *Certificate and certificate revocation list (CRL) profile*. RFC 3280. Internet Engineering Task Force (IETF).Jenronimo, M., & Weast, J. (2003). *UPnP design by example: A software developer's guide to universal plug and play. Intel Press., ISBN-13*, 978–0971786110.

Karmarkar, A., Hadley, M., Mendolsohn, N., Lafon, Y., Gudgin, M., Moreau, J. J., & Nielsen, H. (2007). *SOAP version 1.2 part 1: Messaging framework* (2nd ed.). Retrieved from http://www.w3.org/TR/2007/REC-soap12-part1-20070427/

Maymounkov, P., & Mazieres, D. (2002). *Kademlia: A peer-to-peer Information System based on the XOR metric*.

Microsoft. (n.d.). *Peer name resolution protocol.* Retrieved from http://technet.microsoft.com/en-us/library/bb726971.aspx

Mockapetris, P. (1987). *Domain names - implementation and specification. RFC 1035. Internet Engineering Task Force*. IETF.

Myers, M., & Schaad, J. (2007). *Certificate management over CMS (CMC) transport protocols. Internet Engineering Task Force*. IETF.

NCSA. (2008). *MyProxy credential management service.* Retrieved from http://grid.ncsa.uiuc.edu/myproxy/ca/

Open, C. A. (2008a). *LibPKI: The easy PKI library*. Retrieved from http://www.openca.org/projects/libpki/

Open, C. A. (2008b). *OpenCA-NG: The next generation CA*. Retrieved from http://www.openca.org/projects/ng/

Open, C. A. Labs. (2008c). *OpenCA's PKI resource discovery package*. Retrieved from http://www.openca.org/projects/prqpd/

Pala, M. (2008). *PKI resource discovery protocol (PRQP). Internet Engineering Task Force*. IETF.

Pala, M. (2010). *A proposal for collaborative Internet-scale trust infrastructures deployment: The public key system*. 9th Symposium on Identity and Trust on the Internet (IDTrust 2010). Gaithersburg, MD: NIST.

Pala, M., & Smith, S. W. (2007). AutoPKI: A PKI resources discovery system. *Public Key Infrastructure: EuroPKI 2007.* [Springer-Verlag.]. *LNCS, 4582*, 154–169.

Pala, M., & Smith, S. W. (2008). PEACHES and peers. *5th European PKI Workshop: Theory and Practice. LNCS 5057*, (pp. 223-238). Springer-Verlag.

SWITCH. (2008). *SWITCHpki, an X.509 public key infrastructure for the Swiss higher education system.* Retrieved from http://www.switch.ch/pki/

UPnP forum. (2008). *Universal plug and play specifications.* Retrieved from http://www.upnp.org/resources/

Weiler, S., & Ihren, J. (2006). *Minimally covering NSEC records and DNSSEC online signing. Internet Engineering Task Force*. IETF.

ENDNOTES

1. The subsequent description here of the PRQP protocol is derived from our earlier paper (M. Pala and S.W. Smith 2007).
2. http://www.openca.org/projects/prqpd/
3. A more complete explanation of the non grid-specific pointers is currently submitted for publication.

Chapter 12
Identifying Secure Mobile Grid Use Cases

David G. Rosado
University of Castilla-La Mancha, Spain

Eduardo Fernández-Medina
University of Castilla-La Mancha, Spain

Javier López
University of Málaga, Spain

Mario Piattini
University of Castilla-La Mancha, Spain

ABSTRACT

Mobile Grid includes the characteristics of the Grid systems together with the peculiarities of Mobile Computing, with the additional feature of supporting mobile users and resources in a seamless, transparent, secure, and efficient way. Security of these systems, due to their distributed and open nature, is considered a topic of great interest. We are elaborating a process of development to build secure mobile Grid systems considering security on all life cycles. In this chapter, we present the practical results applying our development process to a real case, specifically we apply the part of security requirements analysis to obtain and identify security requirements of a specific application following a set of tasks defined for helping us in the definition, identification, and specification of the security requirements on our case study. The process will help us to build a secure Grid application in a systematic and iterative way.

INTRODUCTION

Grid computing has emerged to cater the need of computing-on-demand (Jana, Chaudhuri, & Bhaumik, 2009) due to the advent of distributed computing with sophisticated load balancing, distributed data and concurrent computing power using clustered servers. The Grid enables resource sharing and dynamic allocation of computational resources, thus increasing access to distributed data, promoting operational flexibility and collaboration, and allowing service providers to scale efficiently to meet variable demands (Foster & Kesselman, 2004).

DOI: 10.4018/978-1-60960-603-9.ch012

Mobile computing is pervading our society and our lifestyles with a high momentum. Mobile computing with networked information systems help increase productivity and operational efficiency. This however, comes at a price. Mobile computing with networked information systems increases the risks for sensitive information supporting critical functions in the organization which are open to attack (Talukder & Yavagal, 2006).

At first glance, it seems that the marriage of mobile wireless consumer devices with high-performance Grid computing would be an unlikely match. After all, Grid computing to date has utilised multiprocessors and PCs as the computing nodes within its mesh. Consumer computing devices such as laptops and PDAs are typically restricted by reduced CPU, memory, secondary storage, and bandwidth capabilities. However, therein lies the challenge. The availability of wirelessly connected mobile devices has grown considerably within recent years, creating an enormous collective untapped potential for resource utilisation. To wit, recent market research shows that in 2008, 269 million mobile phone and 36 million smartphone (Gartner, 2009) were sold worldwide, and that in 2006, 17 million PDAs (Gartner, 2007) were sold worldwide. Although these individual computing devices may be resource-limited in isolation, as an aggregated sum, they have the potential to play a vital role within Grid computing (Phan, Huang, Ruiz, & Bagrodia, 2005).

Mobile Grid, in relevance to both Grid and Mobile Computing, is a full inheritor of Grid with the additional feature of supporting mobile users and resources in a seamless, transparent, secure and efficient way (Litke, Skoutas, & Varvarigou, 2004). Grids and mobile Grids can be the ideal solution for many large scale applications being of dynamic nature and requiring transparency for users.

Security has been a central issue in grid computing from the outset, and has been regarded as the most significant challenge for grid computing (Humphrey, Thompson, & Jackson, 2005). The characteristics of computational grids lead to security problems that are not addressed by existing security technologies for distributed systems (Foster, Kesselman, Tsudik, & Tuecke, 1998; Welch et al., 2003). Security over the mobile platform is more critical due to the open nature of wireless networks. In addition, security is more difficult to implement into a mobile platform due to the limitations of resources in these devices (Bradford, Grizzell, Jay, & Jenkins, 2007).

The reasons that led us to focus on this topic are several: Firstly, the lack of adequate development methods for this kind of systems since the majority of existing Grid applications have been built without a systematic development process and are based on ad-hoc developments (Dail et al., 2004; Kolonay & Sobolewski, 2004), suggests the need for adapted development methodologies (Giorgini, Mouratidis, & Zannone, 2007; Graham, 2006; Jacobson, Booch, & Rumbaugh, 1999; Open Group, 2009). Secondly, due to the fact that the resources in a Grid are expensive, dynamic, heterogeneous, geographically located and under the control of multiple administrative domains (Bhanwar & Bawa, 2008), and the tasks accomplished and the information exchanged are confidential and sensitive, the security of these systems is hard to achieve. And thirdly, because of the appearance of a new technology where security is fundamental together with the advances that mobile computation has experienced in recent years that have increased the difficulty of incorporating mobile devices into a Grid environment (Guan, Zaluska, & Roure, 2005; Jameel, Kalim, Sajjad, Lee, & Jeon, 2005; Kumar & Qureshi, 2008; Kwok-Yan, Xi-Bin, Siu-Leung, Gu, & Jia-Guang, 2004; Sajjad et al., 2005).

In this paper, we will apply the activity of security requirements analysis for obtaining a set of security requirements on a mobile grid environment for a case study of media domain where the mobile devices participate as actives resources. Using misuse cases and security use cases we obtain a vision about the threats and risks of the system and about the security requirements and mechanisms that we must use to protect to our mobile grid system.

The rest of paper is organized as follows: First, we present the related work with this topic. Next, we will describe some of the security requirements most important on grid environments and will identify the common attacks that can appear on a mobile grid system. Later, we give a brief overview of our development process for mobile grid systems, we will describe the analysis activity and we will study one of the tasks of this activity, the "Identifying secure Mobile Grid Use Cases" task. After, we will present a case study and we will apply the task of identifying security requirements for obtaining a set of security requirements for our real application. Finally, we will finish by putting forward our conclusions as well as some research lines for our future work.

BACKGROUND

There are numerous approaches related to secure development processes but here we present some of those that we believe to be most interesting and that consider security as an important factor for success and application in Mobile Grid environments. Rational Unified Process (RUP) (Kruchten, 2000) describes how to effectively deploy commercially proven approaches to software development for software development teams, although it does not specifically address security. One extension of the Unified Process is defined in (Steel, Nagappan, & Lai, 2005) in which the authors present a methodology for the integration of security into software systems which it is called the Secure Unified Process (SUP). SUP establishes the pre-requirements to incorporate the fundamental principles of security. It also defines an optimized design process of security within the life cycle of software development. The problem is that it is a very general approach that has to be adapted for each specific application that we wish to develop. The specific aspects of Mobile Grid systems necessitate the definition of new activities, artefacts, roles, techniques and security disciplines which are not considered in Secure UP. Another recent approach proposes the integration of security and systems engineering by using elements of UML within the Tropos methodology (Castro, Kolp, & Mylopoulos, 2001; Mouratidis & Giorgini, 2006). Secure Tropos (Mouratidis, 2004) is an extension of the Tropos methodology (Bresciani, Giorgini, Giunchiglia, Mylopoulos, & Perin, 2004) and has been proposed to deal with the modelling and reasoning of security requirements and their transformation to design that satisfies them. There are many security aspects that cannot be captured as a result of the dynamic behaviour and mobile considerations of Mobile Grid systems.

Several approaches for the integration of the security in the development process for specific domains appear in the relevant literature. For example, in (Fernández-Medina & Piattini, 2005), the authors propose a methodology with which to build multilevel databases, taking into consideration aspects of security (with regard to confidentiality) from the earliest stages to the end of the development process. SEDAWA (Trujillo, Soler, Fernández-Medina, & Piattini, 2009) is another approach that proposes a comprehensive methodology with which to develop secure Data Warehouses based on the MDA framework. Approaches which integrate security in the development process for generic applications and systems also exist, such as for example, (Georg et al., 2009) which proposes a methodology based on aspect-oriented modelling (AOM) with which to incorporate security mechanisms into an application, and (Fernández-Medina, Jurjens, Trujillo, & Jajodia, 2009), whose authors explore current research challenges, ideas and approaches for employing Model-Driven Development to integrate security into software systems development through an engineering-based approach, avoiding the traditional ad hoc security integration. None of these approaches are defined and designed for Grid computing and none of them support mobile nodes.

A further approach (Jurjens, 2001, 2002) concentrates on providing a formal semantics for UML to integrate security considerations into the software design process. The approach presents UMLsec (Jan Jürjens, 2005) which is an extremely interesting approach which incorporates security properties into the UML model. UMLsec has been applied in security-critical systems and in the industrial context of a mobile communication system (J. Jürjens, Schreck, & Bartmann, 2008; Popp, Jürjens, Wimmel, & Breu, 2003), and the security aspects of this kind of systems has been analyzed, but it has not been applied in Grid environments with specific security aspects. UMLsec is a perfect candidate to model the mobile security aspects within the diagrams of deployment, activity, classes, collaboration, etc., which complement to the use cases and describe the complete behavior of detailed way. Our approach models mobile Grid security aspects in use cases diagrams, so that our approach and UMLsec can work together to capture, between other things, the mobile security requirements in the different UML diagrams used in the analysis. A model driven architecture approach towards security engineering, called Model Driven Security, is introduced in reference (Basin, Doser, & Lodderstedt, 2003). This approach, called SecureUML (Basin & Doser, 2002), integrates role-based access control policies into a UML-based model-driven software development process, but is not focused on Grid systems. The Comprehensive, Lightweight Application Security Process (CLASP) is a life-cycle process that suggests a number of different activities throughout the development life cycle in an attempt to improve security (Graham, 2006). Finally, AEGIS (Flechais, Sasse, & Hailes, 2003) is the only approach found in which the authors attempt to apply the methodology to Grid systems, although they do not explain how to do this, and do not define guidelines and practices with which to capture specific security aspects in Grid systems. This approach should be adapted to the necessities and features of Grid systems.

We conclude that the existing proposals are not specific enough to provide a complete solution of security under a systematic development process for Mobile Grid environments. This is due to the fact that none of the approaches defines a systematic development process for this specific kind of systems that incorporates security from the earliest stages of the development. The approaches which provide security to the software development processes for Mobile Grid systems are scant or nonexistent, because the secure development approaches are not focused on Grid systems and they do not take into account mobile devices. Thus, reflected the need to advance in the study of new contributions to the secure systematic development of Grid systems incorporating mobile devices.

SECURITY REQUIREMENTS AND ATTACKS ON A MOBILE GRID SYSTEM

Defining Security Requirements

The special security requirements of Grid applications derive mainly from the dynamic nature of Grid applications and the notion of virtual organization (VO), which requires the establishment of trust across organizational boundaries. In this kind of environment, security relationships can be dynamically established among hundreds of processes spanning several administrative domains, each one with its own security policies. As a result, the Grid security requirements are complex and pose significant new challenges.

The most common general security requirements and challenges associated with Grids and Mobile systems (Bellavista & Corradi, 2006; Foster & Kesselman, 2004; Nagaratnam et al.,

2003; Open Grid Forum, 2006; Vivas, López, & Montenegro, 2007) are presented below:

- *Authentication.* Authentication mechanisms and policies are supposed to constitute the basis on which local security policies can be integrated within a VO. Difficult issues with respect to authentication in Grids are scalability, trust across different certification authorities, revocation, key management, and delegation.
- *Confidentiality.* The nature of Grids forces data to be stored in accessible online databases. Confidential code may be requested to execute on a remote host, and confidential data may need to be used at remote locations. Data may also need to be replicated at multiple sites, and thus should be stored in an encrypted form and remain consistent throughout.
- *Integrity.* Many applications have strong code or data integrity concerns. The trust status of remote resources is important when data arises from remote processing as the accuracy of results can be trusted only to the extent that the remote host generating the data is trusted.
- *Authorization and access control.* Authorization refers to the ability to control the level of access that individuals or entities have to a wireless network or resource and how much information they can receive. In Grids local access mechanisms should be applied whenever possible, and the owner of a resource should be able to enforce local user authorization.
- *Revocation.* Revocation is crucial for authentication in case of a compromised key and for authorization when a VO is terminated or a user or mobile user proves untrustworthy.
- *Distributed trust.* Trust is a complex theoretical issue. A Grid must be constructed in a dynamic fashion from components whose trust status is hard to determine. Determining trust relations between participant entities in the presence of delegation is important, and delegation mechanisms must rely upon stringent trust requirements.
- *Freshness.* Freshness is related to authentication and authorization and is important in many Grid applications. Validity of a user's proof of authentication and authorization is an issue when user rights are delegated and the duration of a job may span several weeks.
- *Scalability.* A Grid must be easy to extend and capable of progressive replacement in mobile environments. Fault recovery and dynamic optimization should be usually possible, and degradation should happen gracefully.
- *Trust.* Trust refers to the assured reliance on someone or something. Since VOs can span multiple security domains, trust relationships between domains are of paramount importance. Sites in a Grid must be able to enter into trust relationships with Grid users, mobile users and maybe other Grid sites as well. In a Grid environment trust is usually established through exchange of credentials, either on a session or a request basis.
- *Single sign-on.* A user should be able to authenticate only once, whereupon he may acquire, use and release resources without further authentication in different domains of the Grid. Users may want to initiate computations running for long periods of time without needing to remain logged on all the time.
- *Delegation.* Privilege delegation for operations executed by a proxy is a basic requirement for Grid environments, among other reasons in order to satisfy the single sign-on requirement. Delegation of user rights depends upon the security requirements of the application.

- *Privacy*. Privacy is the ability to keep information from being disclosed to determined actors. Privacy can be important in many Grid applications, for instance in medical and health Grids (Herveg, Crazzolara, Middleton, Marvin, & Poullet, 2004). It is also very important in mobile devices with limited memory and whose access is through wireless networks.
- *Non-repudiation*. Non-repudiation refers to the inability to falsely deny the performance of some action. It is especially important in e-commerce involving money transactions and mobile environments. With the advent of Enterprise Grid this requirement becomes very important.
- *Credentials*. Interdomain access requires a uniform way of expressing the identities of users or resources, and must thus employ a standard for the encoding of credentials. In many scenarios, a job initiated by a user may take longer than the life span of the user's initially delegated credential. In those cases, the user needs the ability to be notified prior to expiration of the credentials, or the ability to refresh those credentials such that the job can be completed.
- *Exportability*. Code is required to be exportable and executable in multinational testbeds. As a result, bulk encryption cannot be required.
- *Secure group communication*. Authenticated communications for dynamic groups is required since the composition of a process group may change dynamically during execution.
- *Multiple implementations*. It should be possible to enforce security requirements with distinct security technologies and mechanisms.
- *Interoperability*. In the context of mobile Grids, interoperability means that services within a single VO must be able to communicate across heterogeneous domains. Interoperability guarantees that services located in different administrative domains are able to interact at multiple levels.
- *Interoperability with local security solutions*. Access to local resources is normally enforced by local security policies and mechanisms. Interoperability between sites and domains with different local policies is necessary in a mobile Grid environment. In order to accommodate interdomain access, one or several entities in a domain may act as agents of external entities for local resources.
- *Integration*. In order to allow the use of existing services and resources, integration requirements call for the establishment of an extensible architecture with standard interfaces. Security integration is facilitated by the use of existing security mechanisms.
- *Uniform credentials and certification infrastructure*. A common way of expressing identity, e.g. by a standard such as X.509, is necessary for interdomain access.
- *Policy exchange*. Allow service requestors and providers to exchange dynamically security (among other) policy information to establish a negotiated security context between them.
- *Secure logging*. Provide all services, including security services themselves, with facilities for time-stamping and securely logging any kind of operational information or event in the course of time - securely meaning here reliably and accurately, i.e. so that such collection is neither interruptible nor alterable by adverse agents.
- *Assurance*. Provide means to qualify the security assurance level that can be expected of a hosting environment.
- *Manageability*. Explicitly recognize the need for manageability of security functionality within the OGSA security model. For example, identity management, policy

management, key management, and so forth.

- *Firewall traversal.* A major barrier to dynamic, cross-domain Grid computing today is the existence of firewalls. As noted above, firewalls provide limited value within a dynamic Grid environment. However, it is also the case that firewalls are unlikely to disappear anytime soon.
- *Anonymity.* Anonymity is the state of being not identifiable within a set of principles (Pitzmann & Köhntopp, 2001). Preserving anonymity is of greater concern in mobile systems for several reasons. Mobile systems yield more easily to eavesdropping and tapping, compared to fixed networks, making it easier to tap into communication channels and obtain user information.
- *Mobility.* Because mobile devices come with many capabilities, mobile applications must run on a wide variety of devices, including the devices embedded in various environments and devices carried by users. Applications must also support varying levels of network connectivity.
- *Self-organization.* The wireless networks topology must be adapted in case of node or system compromise and failure. If a malicious node discloses the network topology, routing establishment paths may be affected as well.

All these security requirements must be identified and analyzed in the analysis activity of our development process from the mobile grid security use cases defined in this activity and that we will explain further on.

Defining Attacks on Mobile Grid Environments

According to (Enterprise Grid Alliance Security Working Group, 2005), the following include some of the threats and risks based on the unique characteristics of an enterprise Grid:

- *Access control attacks*: defines risks with unauthorized entities, as well as authorized entities, bypassing or defeating access control policy.
- *Mobile colluding attackers*: adversaries having different levels of attacking ability can collaborate through separate channels to combine their knowledge and to coordinate their attacking activities. This realizes the strongest power at the adversary side.
- *Defeating Grid auditing and accounting systems*: includes threats to the integrity of auditing and accounting systems unique to an enterprise Grid environment. This may include false event injection, overflow, event modification, and a variety of other common attacks against auditing systems.
- *Denial of Service (DoS)*: this describes an attack on service or resource availability. As an enterprise Grid is often expected to provide a better availability compared to a non-Grid environment, the following DoS threats must be considered as part of a risk assessment:
    - DoS attack against the Grid component join protocol to prevent new authorized Grid components/users from successfully joining.
    - Authorized Grid component or user is "forced" to leave the grid.
    - User or service attempts to flood the Grid with excessive workload which may cause compute, network and/or storage components to become exhausted, or the latency to access those resources significantly impacts other Grid users.
    - Altering scheduling (or other Quality of Service) priorities that have been defined for Grid components to un-

fairly prioritize one application/service over another.

- *Malicious code/"malware"*: this describes any code that attempts to gain unauthorized access to the Grid environment, to subsequently elevate its privileges, hide its existence, disguise itself as a valid component, or propagate itself in clear violation of the security policy of the enterprise Grid.
- *Masquerading attacks*: describes a class of attacks where a valid Grid component may be fooled into communicating or working with another entity masquerading as valid Grid component. Such an attack could permit the disclosure or modification of information, the execution of unauthorized transactions, etc.
- *Mobile eavesdropper and traffic analyst*: such an adversary can at least perform eavesdropping and collect as much information as possible from intercepted traffic. It is mobile and equipped with GPS to know its exact location. The minimum traffic it can intercept is the routing traffic from the legitimate side. An eavesdropper with enough resources is capable of analyzing intercepted traffic on the scene. This ability gives the traffic analyst quick turnaround action time about the event it detects, and imposes serious physical threats to mobile nodes.
- *Mobile node intruder*: if adequate physical protection cannot be guaranteed for every mobile node, node compromise is inevitable within a long time window. A successful passive node intruder is protocol compliant, thus hard to detect. It participates in collaborative network operations (e.g., ad hoc routing) to boost its attack strength against mobile anonymity; thus it threatens the entire network including all other uncompromised nodes. This implies that a countermeasure must not be vulnerable to a single point of compromise.
- *Object reuse*: this describes how sensitive data may become available to an unauthorized user, and used in a context other than the one for which it was generated. In the enterprise grid context, this is a risk if a Grid component is not properly decommissioned.
- *Sniffing/snooping*: involves watching packets as they travel through the network. An enterprise Grid potentially introduces additional network traffic between applications/services, the system and grid components that should be protected. Failure to address this threat may result in other types of attacks including data manipulation and replay attacks.

In addition to these, it is also necessary to adopt the general security mechanisms applicable in any enterprise scale IT infrastructure, and includes physical security to protect against threats from humans (either malicious or accidental) as well as man-made and natural catastrophes.

OVERVIEW OF OUR PROCESS

A. Process of Development

The process is designed for building software systems based on Mobile Grid computing with security aspects. It is a process which builds, from initial requirements and needs of Mobile Grid systems, a secure executable software product. It is not a process for including only security in a development process but it is a development process in itself incorporating security aspects during all the process.

Our systematic process of development (Rosado, Fernández-Medina, López, & Piattini, 2008) is an iterative and incremental process. An iterative approach refers to the cyclic nature of the process in which activities are repeated in a structured manner and proposes an understanding of the problem through successive refinements, and an incremental growth of an effective solution through several versions. Thus, in each iteration of the process, new and necessary characteristics can be added and extended so that a complete final design is obtained. Also, it is a reusable process in the sense of the utilization of artifacts built in others executions of this process or in previous iterations which have been validated and tested and that improve the quality of the new artifacts built and save developers' time and effort.

The structure of the process which we propose follows the classical cycle, in which we find a planning phase, a development phase including analysis, design and construction and finally a maintenance phase. The phases of planning and maintenance are common phases which any development of information systems has to define, so we move on a generic development process to carry out the activities and tasks of these phases. Thus, our work focuses on defining what is really specific and differentiating in developing systems based on Grid computing, the development phase. This phase consists of three activities, analysis, design and construction, and each of them defines the specific tasks necessary, the artifacts to be used, and the steps to take to analyze, design and build specific information systems as Mobile Grid systems are.

Therefore, the main block of this process consists of a requirements analysis activity driven by use cases (Rosado, Fernández-Medina, López, & Piattini, 2010a), a design activity that focuses on architecture (Rosado, Fernández-Medina, López, & Piattini, 2011), and construction activity oriented to implementation. All these activities are supported by a repository where different reusable elements which can be used in the different activities and tasks of the process are stored. These reusable elements are use cases and security use cases diagrams oriented to Grid systems to be reused in the analysis activity to capture the security requirements (Rosado, Fernández-Medina, & López, 2009a, 2009b, 2009c; Rosado, Fernández-Medina, López, & Piattini, 2010b); a reference security architecture (Rosado, Fernández-Medina, & López, 2011b) where we define security services for Mobile Grid environments reused in the design activity which guarantees that the system is built under a secure environment and meets all the requirements and security needs of the system; and implemented interfaces based on Grid tools and platforms (as Globus) to be reused in the construction activity (See Figure 1).

In this paper, we study one of the tasks of the secure mobile grid system analysis activity, the Identification of secure Mobile Grid Use Cases task whose steps can be seen in Figure 3. In this task we identify threats and risks related to mobile grid environments which attack assets that we want to protect, and we build the diagrams of security use cases and misuse cases for mobile grid environments considering these assets, threats and attacks.

B. Secure Mobile Grid System Analysis Activity

The analysis activity is based on use cases in which we define the behaviour, actions and interactions with those implied by the system (actors) to obtain a first approach to the needs and requirements (functional and non-functional) of the system to be constructed. This activity is supported by repositories in which several types of elements appear: Firstly, the elements that have been developed in earlier stages; secondly, those that have been built at the beginning of the process and finally, those that come from other executions of the process from which we have obtained elements that can be reused by other applications. Reuse is appropriate here thanks both to the common features of applications based on Grid computing (CPU intensive, data intensive, collaborative and so on) and to the fact that these applications use mobile devices. Therefore, we must abstract all the common features (by analyzing the main features of Grid applications and constructing, for example,

Figure 1. Development process for secure Mobile Grid systems with SPEM 2.0

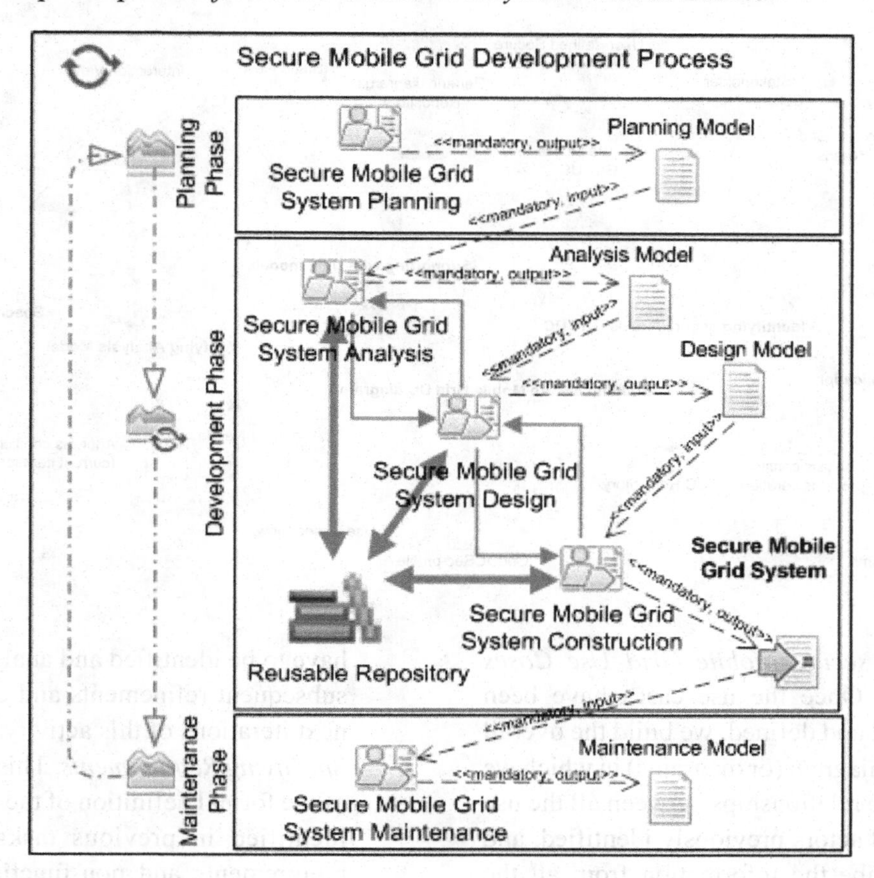

generic use case diagrams in which all these common features are represented) and make them available for the process (through the repository) in order to be able to use the common elements in any activity and adapt them to our needs.

The analysis activity is composed of tasks which build use case diagrams and specifications to obtain the analysis model in which the requirements are defined. This activity produces internal artifacts which are the output of some tasks and the input of others. All these internal artifacts are included in the analysis model to be used in the following activities if this is necessary. Figure 2 shows a graphical representation of the analysis activity tasks using SPEM 2.0 diagrams.

In this subsection, we describe the analysis activity, enumerating and describing briefly what tasks are parts of this activity. This analysis activity is composed of six tasks (see Figure 2):

1. *Defining Use Cases of the application.* The purpose of this task is to define the functional use cases of the application identified from the stakeholder needs and study the interactions with the user without considering the specific aspects of Mobile Grid environments.
2. *Identifying secure Mobile Grid Use Cases.* In this task we study the security aspects of the application within the Mobile Grid context and identify the possible security use cases and misuse cases that can be reused from those defined in the repository, for the system in development.

Figure 2. Tasks and artifacts of the Secure Mobile Grid System Analysis activity

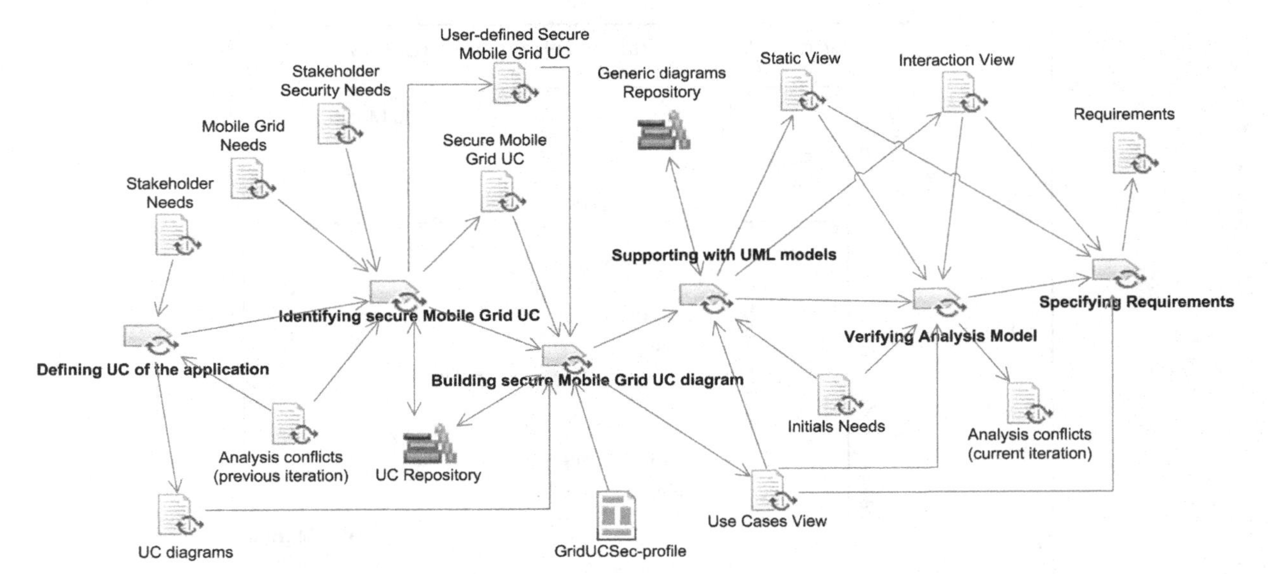

3. *Building secure Mobile Grid Use Cases diagram.* Once the use cases have been identified and defined, we build the overall use case diagram (or diagrams) in which we define the relationships between all the use cases and actors previously identified, and we describe the information from all the diagram's elements by following a new UML profile for Mobile Grid use cases (Rosado, Fernández-Medina, López, & Piattini, 2011a). We can also reuse and integrate some diagrams with common features of the repository which have been previously built for Mobile Grid environments.
4. *Supporting with UML models.* In this task we complete the analysis model with different UML models such as the sequence and collaboration diagrams according to use cases and scenarios, or class diagrams for an initial structural description of the system from the use cases diagrams built in previous tasks.
5. *Verifying Analysis Model.* The purpose of this task is to verify that the artifacts have been correctly generated and the possible conflicts or errors in the analysis model have to be identified and analyzed for their subsequent refinements and corrections in next iterations of this activity.
6. *Specifying Requirements.* This task consists of the formal definition of the requirements identified in previous tasks (functional requirements and non-functional requirements including security) in natural language (though a template of requirements specification will be defined in the future).

Once we have described the tasks of the analysis activity, we will explain the task 2, which is in charge of analyzing security requirements for the mobile grid system, and we apply the steps of this task in a case study. This task have been improved and updated with regard to the published work in (Rosado, Fernández-Medina et al., 2009b).

C. Task 2: Definition of secure Mobile Grid Use Cases

In this task, a study of the system security must be carried out before identifying the security use cases and misuse cases of the repository. First,

Figure 3. Task 2: Identifying secure Mobile Grid UC

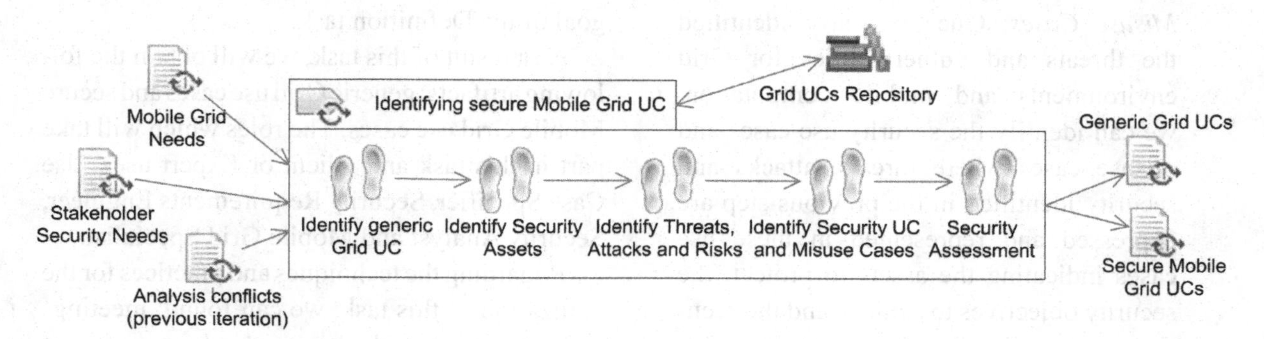

generic Grid use cases that are common to many Grid applications are identified of the repository because will take part in the application analysis. Secondly, assets that we wish to protect should be identified; thirdly, the possible threats and attacks to these assets should be defined and the risk associated with these threats should be studied. The security use cases and misuse cases should then be defined, thus obtaining certain elements of the reusable repository such as the misuse cases for the system and the security use cases that mitigate them. Finally, a security assessment should be carried out. Some of the security use cases and misuse cases identified for the application are therefore stored in the repository and can be reused for this specific application since they are part of the secure Mobile Grid UC output artifact.

During this task, it is possible to discover new use cases which are suitable for incorporation into the repository, or we may wish to modify or update existing use cases in the repository. The repository is an input and output artifact from which we can obtain different elements and add or create new ones. Also, we have to consider possible conflicts between Grid use cases, security use cases and misuse cases and solve them in this iteration.

A set of steps will serve as a guide for defining and specifying security requirements for mobile grid systems. Figure 3 shows the steps of this task using SPEM 2.0 diagrams.

- *Step 2.1. Identify generic Grid UC*: Once we have defined the use cases of the application in the task 1, we have to identify which are the generic Grid use cases that are related to the use cases of the application. To define the Grid use cases we will use the GridUCSec-profile defined as a model of the process (Rosado, Fernández-Medina, López et al., 2011a; Rosado, Fernández-Medina et al., 2010b) and using the repository where a large set of Grid use cases are defined.
- *Step 2.2: Identify Security Assets*: The security assets for a grid with mobile devices depend on the characteristics and type of system to be built. The CPU-intensive applications will consider resources as main assets while data-intensive applications will consider data as main assets to protect.
- *Step 2.3: Identify Threats, Attacks and Risks*. The threats analysis is the process of identifying, as many risks that can affect the assets as possible. A well-done threat analysis performed by experienced people would likely identify most known risks, providing a level of confidence in the system that will allow the business to proceed. In previous section the most important threats and attacks for these environments have been defined.

- *Step 2.4: Identify Security Use Cases and Misuse Cases*: Once we have identified the threats and vulnerabilities for Grid environments and mobile computation, we can identify the security use cases and misuse cases where threats, attacks and security identified in the previous step are expressed and represented in these use cases indicating the assets to protect, the security objectives to achieve and the security requirements that the system must fulfill through of our UML profile (Rosado, Fernández-Medina, López et al., 2011a; Rosado, Fernández-Medina et al., 2010b).
- *Step 2.5: Security Assessment*: It is necessary to assess whether the threats are relevant according to the security level specified by the security objectives. Then, we have to estimate the security risks based on the relevant threats, their likelihood and their potential negative impacts, in other words, we have to estimate the impact (what may happen) and risk (what will probably happen) which the assets in the system are exposed to. We have to interpret the meaning of impact and risk.

Therefore, the aim of this activity is identify security use cases and misuse cases correctly defined where all security requirements of our system are represented and identified.

We shall now provide a detailed description of this task that we have considered in our process using the SPEM 2.0 textual notation. We define the roles, steps, work products and guidance, which will be characterized according to the discipline that they belong to. According to SPEM, the task 2 is described by using the structure shown in Figure 4. Each task specifies WorkProductUse as both input and output respectively, the roles that perform or participate in this RoleUse task, and the collection of Steps defined for a Task Definition which represents all the work that should be carried out to achieve the overall development goal of the Definition task.

As a result of this task, we will obtain the following artifacts: generic Grid use cases and secure Mobile Grid use cases. The roles which will take part in this task are: Client or Expert user, Use Case Specifier, Security Requirements Engineer, Security Analyst and Mobile Grid Specialist.

Regarding the techniques and practices for the realization of this task, we can found: meetings and interviews with the involved, security use cases and misuse cases and cost/effort-benefit and analysis risks.

CASE STUDY

Our development process will be validated with a business application in the Media domain (see Figure 5) attempting to solve existing problems in this domain. The process will help us to build a Mobile Grid application, which will allow journalists and photographers (actors of media domain) to make their work available to a trusted network of peers the same instant it is produced, either from desktop or mobile devices.

With the explosion of ultra portable photo/video capture media (i.e. based on mobile phones, PDAs or solid state camcorders) everyone can capture reasonably good quality audiovisual material while on the move. We want to build a system that will cater for the reporter who is on the move with lightweight equipment and wishes to capture and transmit news content. This user needs to safely and quickly upload the media to a secure server to make it easier for others to access, and to avoid situations where his device's battery dies or another malfunction destroys or makes his media unavailable.

In the media domain, both the distributions of content, and the need for rapid access to this content, are apparent. News is inherently distributed everywhere and its value falls geometrically with time. These two reasons make the need for

Figure 4. Detailed description of the Task 2 using SPEM 2.0

Activity {kind = **Phase**}: **Secure Mobile Grid System Development**
Process: **Secure Mobile Grid Development**

Activity {kind = **Iteration**}: First **Secure Mobile Grid System Analysis**

TaskUse: **2. Identifying secure Mobile Grid UC**
ProcessPerformer {kind: primary}
RoleUse: **Client** {kind: in}
RoleUse: **UC specifier** {kind: in}
RoleUse: **Security Requirements engineer** {kind: in}
RoleUse: **Security Analyst** {kind: in}
RoleUse: **Mobile Grid specialist** {kind: in}
WorkDefinitionParameter {kind: in}
WorkProductUse: **Stakeholder security needs**
WorkProductUse: **Mobile Grid needs**
WorkProductUse: **Analysis conflicts**
WorkProductUse: **Repository of secure Mobile Grid Use Cases**
WorkDefinitionParameter {kind: out}
WorkProductUse: **Secure Mobile Grid UCs** {state: **initial draft**}
WorkProductUse: **Generic Grid UCs** {state: **initial draft**}
WorkProductUse: **Repository of secure Mobile Grid Use Cases** {state: **reviewed**}
Steps
Step: **Identify generic Grid UC** for the application
Step: **Identify security Assets** of the application in a Mobile Grid environment
Step: **Identify Threats, Attacks and Risk** of the application in a Mobile Grid environment
Step: **Identify the Security UCs and Misuse cases** from the repository
Step: **Security Assessment**
Guidance
Guidance {kind: **Checklist**}: **Catalogue of security assets to protect.**
Guidance {kind: **Checklist**}: **Catalogue of possible threats in the system.**
Guidance {kind: **Practice**}: **Well-defined misuse cases and security use cases for Mobile Grid environments.**
Guidance {kind: **Practice**}: **Cost/effort-benefit vs risk analysis**
Guidance {kind: **Practice**}: **Security use cases**
Guidance {kind: **Practice**}: **Misuse cases**
Guidance {kind: **Practice**}: **Meetings**
Guidance {kind: **Practice**}: **Interviews**

Grid technology evident in both scenarios which represent, however, a plethora of relevant business cases which share these two common characteristics: the need for fast access to distributed content.

Following the process of analysis defined in the definition of mobile grid security use cases activity aforementioned, we will identify and analyze security requirements involved in this case study helping of security repository and mobile grid security uses cases. For all possible use cases defined for this application, we are only going to consider three use cases (due to space constraints), defined in Table 1, which we are going to work with in the following tasks.

Once we have identified some of functional use cases of the application, now, we must identify all the use cases and security use cases for the Grid system that are related to the functional use cases of the application. These use cases for the Grid system include Grid use cases, security use cases, Grid security use cases, misuse cases and mobile use cases together with Grid actors and Misactors, all of them defined with the GridUCSec-profile.

Figure 5. Mobile Grid Computing system for Media application

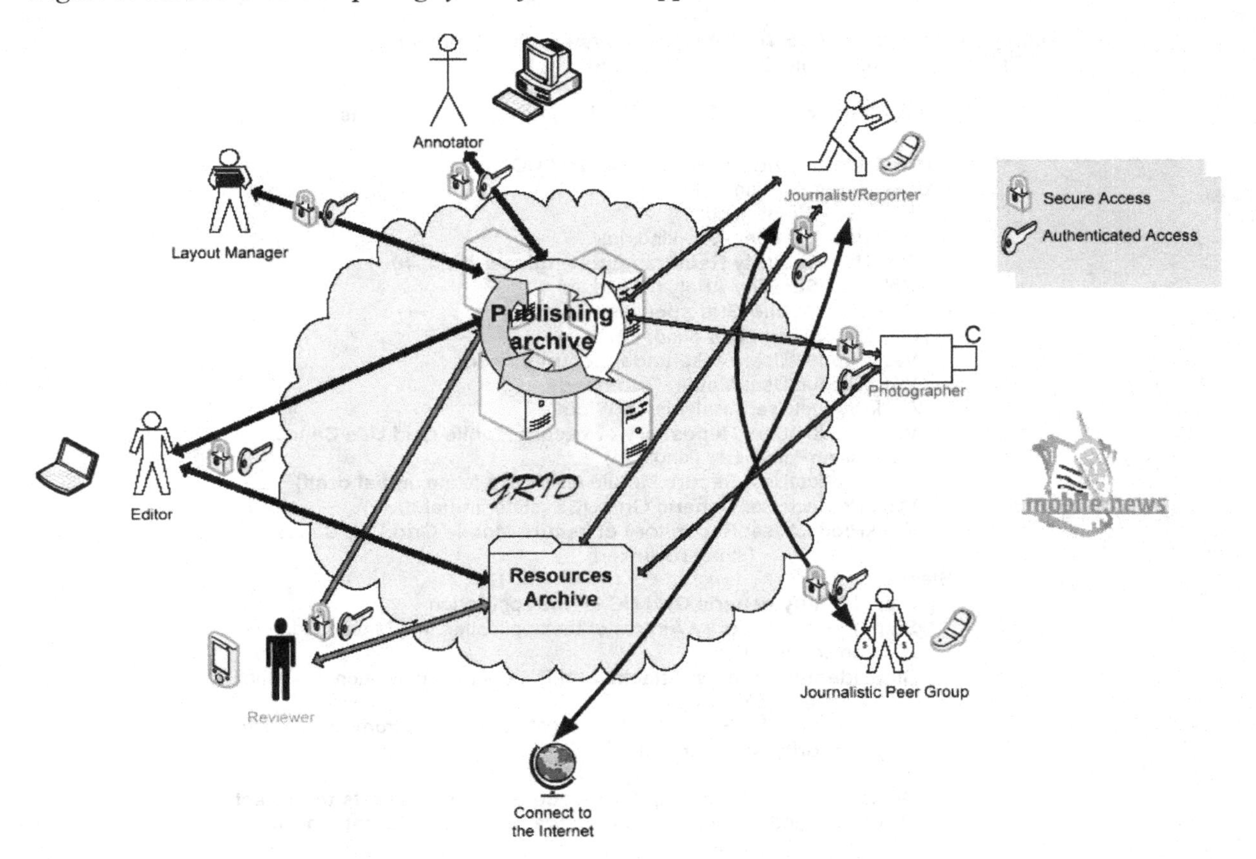

We use the reusable artifacts of the repository where many of these use cases for Grid systems and diagrams that can be easily used in this application and that help us obtain use cases, actors and associations that are necessary in this application are defined.

To identify the use cases and security use cases of the Grid system, we have to follow the steps defined in this task of the SecMobGrid process. Next we apply each one of these steps for this application.

Step 2.1: Identify Generic Grid UC for the application

We must act on the repository of Grid use cases to identify the generic Grid use cases that are needed to extract and that are related to the use cases defined in the previous task. In the repository we have a set of generic use cases which have a common behaviour for any Grid systems and have been identified in other executions of the process and that can be used in this application. We select some of these generic Grid use cases that have relation with the functional use cases identified previously and which are defined in Table 2.

Step 2.2: Identify Security Assets

On mobile Grid environments we can identify a set of assets that we must protect for obtaining a secure grid system, which are the following: User and system data (stored, transmitted); Identity information; Credentials (private keys, passwords); Accounting; CPU-/Storage-/Mobile devices-/Network-resources; General system.

Table 1. Use cases

Use Case Name	Add/edit Mobile user
Goals/Description	Provide authentication mechanisms
Scenario example	All users must be subscribed in the Grid
Description	- A new user fills in or edits an electronic subscription form with his/her profile information - The Grid administrator adds a new user to the system by approving the form or approves the user profile change
Use Case Name	**Search for news**
Goals/Description	A journalist can search for news material through the system interface in: 1. public sources 2. his organisation's historical archive 3. trusted commercial portals according to the subscriptions paid-for.
Scenario example	The journalist familiarizes himself with the topic
Description	- A user formulates a search query - The user selects sources to search from a list - The user submits the query
Use Case Name	**Get query results**
Goals/Description	Receive query results from available repositories
Scenario example	The Journalist receives a list with the results of the search query
Description	- The system returns results based on the metadata description of the stored material. - Results can be sorted according to the journalist's needs, such as thematic groups. - Visualization of results is based on the end user device capabilities (low resolution video for mobile devices)

In this first iteration of our case study, we define the most important assets related to the use cases aforementioned that we must protect and that are the reference for the identification of threats, attacks and security use cases. These assets are:

- *Personal information* about the journalist or editors: name, age, address, subscriptions, salaries. All this personal information is stored in the system and must be protected from unauthorized access.
- *Media information* used: photos, articles, recordings, videos, intellectual property rights. This information is of a professional nature and will be exchanged between Grid users and stored in different localizations of the Grid system for an easy and quick acess.
- *Exchange information*: messages, queries, transactions. The data transmitted between Grid elements (users, resources, server, etc.) which contain sensitive information that have to be protected from external disclosure or alteration.

Step 2.3: Identify Threats, Attacks and Risks

The set of threats and attacks that can occur in a Mobile Grid system is similar to that produced in a distributed system by adding those occurring in the mobile environment with wireless network and limited resources.

Examples of threats are unauthorized disclosure of information, attacks to the content of a message through wireless links, denial-of-service attacks, network authentication related attacks, physical node attacks, alteration of information, and so on. In Table 3 we can see the threats considered for the assets identified in Mobile Grid environments.

Table 2. Generic Grid Use Cases defined in the repository

Grid UC Name	**User Register**
Goals/Description	Register a user in the Grid before the user can send jobs or access to the Grid.
Scenario example	A new user fills in a form with information (username, role, domain, resource, credential type, etc.) and the form is stored in the Grid.
Description	- A user gives information to register in the Grid system - The Grid system processes this information and stores it in the Grid - The user obtains the Grid system a username and password to log in.
Grid UC Name	**Request of query**
Goals/Description	Make a query to the Grid
Scenario example	A user wants to obtain information about a topic (pictures, news, videos, etc.) and s/he requests the Grid with this query and waits for the results.
Description	- A query is received in the Grid - The Grid processes the query and sent it to appropriate target - The target executes the query and returns results
Grid UC Name	**Data Retrieve**
Goals/Description	Retrieve data requested
Scenario example	The Grid retrieves data of the resources indicated by the request
Description	- A request of retrieval of data has been authorized - The request is processed and the task is sent to the resource where data is stored - The resource returns requested data
Grid UC Name	**Send results**
Goals/Description	The results obtained are sent to the mobile device which initiated the request.
Scenario example	The results of a query are appropriately formatted to be shown on the screen of the mobile device.
Description	- The result of a query o request is obtained in the Grid when the task or subtasks have finished. - The Grid studies the sender to know the resource display, memory, cpu, etc. and to send the results in the right format

In this first iteration, we can identify several possible types of threats to Information:

- *Unauthorized access* to Grid system. In this scenario, the user wants to login the system, so that we must ensure authorized access.
- *Unauthorized disclosure* and *alteration of information*. The user can send information to the system or receive from the system, so that we must protect the information both transmitted and stored. Also we must protect the personal information that is transported through credentials.
- *Masquerade*. An attacker masquerades as a certain user, access the Grid and sends requests and obtains data from the Grid with the stolen credentials of a legal user. Such an attack could permit the disclosure or modification of information, the execution of unauthorized transactions, etc.

Step 2.4: Identify the Security Use Cases and Misuse Cases

Once we have defined the most significant threats and major assets to be protected in this first iteration, we start with the identification, definition and incorporation of security use cases and misuse cases for the application.

In the repository, the main security use cases for Mobile Grid environments, and misuse cases that capture the behaviour of the main threats identified in these environments are defined. We

Table 3. Assets and threats

Assets	Threats
User and system data (stored, transmitted)	- Unauthorized access (stored data) - Eavesdropping (transmitted data) - Unauthorized publishing - Manipulation - Erroneous data
Identity information	- Eavesdropping - Manipulation
Credentials (private keys, passwords)	- Theft / Spoofing (masquerade as a certain user, illegal use of software) - Publishing
Accounting	- Manipulation of log entries, CPU/memory usage, number and size of processes - Acquire information about competitor's work
CPU-/Storage-/Mobile devices-/Network-resources	- Misuse (e.g. Spambot) - Denial of Service
General System	- Security holes / exploits - Malicious / compromised resources - Backdoors, viruses, worms, Trojan horses

can identify those security use cases and misuse cases that fit in with the attacks and threats for this application identified in the previous step.

In this first iteration, the misuse cases that we have found in the repository and that fit in with the threats identified for this application are: Alteration info, Disclosure info, Unauthorized access and Masquerade.

In the repository, these misuse cases are defined in a generic way, therefore, we have to adapt them to this case study with the specific elements (actors, messages, assets, etc.) of this application. Table 4 shows the definition of these misuse cases.

With these misuse cases, we can identify security use cases that mitigate them observing the information offered by the repository for security use cases and the diagrams defined where we can see the relationships of mitigation between security use cases and misuse cases. In case that the required use cases are not in the repository we can define them and specify relationships as it is convenient.

We find in the repository the security use cases (including Grid security use cases and Grid actors) that are related to the misuse cases identified. These security use cases are: Authenticate, Authorize access, Ensure Confidentiality and Ensure Integrity.

Some security use cases have different instances depending on the use case path defined (Firesmith, 2003) so that we have to define some of them relating to the assets and misuse cases identified in this first iteration. For example, "Ensure Integrity" security use case has three instances, one related to message integrity from Grid to user, other related to message from user to Grid, and other related to data stored in the Grid. All these paths are important to be taken into account in the application, but here we only show one of them for simplicity making the same analysis for the rest of paths of these security use cases.

Table 5 shows the instances of the security use cases selected in this first iteration and which are defined in the Grid use cases repository. These security use cases selected are related to misuse cases identified previously mitigating the threats and attacks defined in such misuse cases.

Table 4. Misuse Cases for the case study

Misuse Case	**Alteration of information (MC1)**
Attack	Attack on the content of a message (integrity).
Summary	The external attacker type gains access to the message exchanged between the journalist and the Grid system, and modifies the part of the message that contains the media information with the intention of changing its meaning by modifying some aspects of the information like authors, dates, or secrecy information.
Preconditions	
1) The external attacker has physical access to the message.	
2) The external attacker has a clear knowledge of where the secrecy information is located within the message.	
Interactions	
1 User Interactions	The journalist sends a query message for obtaining media information
2 Misuser Interactions	The external attacker intercepts it and identifies the part of the message to modify the media information and he/she forwards it to the media Grid.
3 System Interactions	The Media Grid receives the corrupted message and processes it incorrectly due to its altered semantic content. That is, it establishes that the journalist wishes as new media information that media information which has been modified by the attacker
Postconditions	
1) The Media Grid will remain in a state of error with regard to the original intentions of the journalist.	
2) In the register of the system in which the media Grid was executed, the request received with an altered semantic content will be reflected.	
Misuse Case	**Disclosure of information (MC2)**
Attack	Attack on the confidentiality of a message from Grid system to user
Summary	The external attacker type gains access to the message exchanged between the journalist and the Grid system, and reads a specific piece of information.
Preconditions	
1) The external attacker has physical access to the message.	
Interactions	
1 User Interactions	The journalist sends a query message for obtaining media information
2 System Interactions	The Grid system receives the query message and processes it. The Grid system returns the media information related to the query to the journalist
3 Misuser Interactions	The external attacker intercepts it and reads the part of the message that contains the media information and he/she forwards it to the journalist
4 User Interactions	The journalist wishes as new media information that media information which has been intercepted by the attacker.
Postconditions	
1) The Grid system will remain in a normal state and the journalist continues without realizing the interception of information by the attacker	
Misuse Case	**Unauthorized access (MC3)**
Attack	Attack on the access rights and privileges to the Grid system.
Summary	The external attacker type gains access to the Grid system.
Preconditions	
1) The external attacker has physical access to the system and access messages.	
Interactions	
1 Misuser Interactions	The unauthorized user wants to login the system with the username/password or presenting a certificate.
2 System Interactions	The Grid system receives the access request and it allows the access to the Grid.
3 Misuser Interactions	The attacker sends queries to the Grid to obtain sensitive information or for storing harmful data for the system.
4 System Interactions	The Grid system receives the queries processes them and executes them.

continued on following page

Table 4. continued

Misuse Case	Alteration of information (MC1)
Postconditions	
1) The Grid system must not allow the access to unauthorized users	

Misuse Case	Masquerade (MC4)	
Attack	Attack on authorized user	
Summary	The external attacker type pretends to be an authorized user of a system in order to gain access to it or to gain greater privileges than those it is authorized for.	
Preconditions		
1) The external attacker has physical access to the system and the messages exchanged between the user and the Grid.		
Interactions		
1 User Interactions	The journalist sends a request to the Grid to execute certain task.	
2 Misuser Interactions	The attacker intercepts the request and obtains privileges information and authorized information of the user (credentials, roles, rights, etc.)	
3 Misuser Interactions	The attacker sends requests to the Grid presenting authorized credentials of certain authorized user.	
4 System Interactions	The Grid system receives these requests of the authorized attacker and executes the harmful actions.	
Postconditions		
1)	The Grid system must check the identity of the user who sends requests.	
2)	The Grid system must check the privileges and certificates presented by the user and the authenticity of the certificates.	

Table 5. Security use cases for the case study

Security Use Case		Ensure Integrity (SUC1)
Use Case Path		System Message Integrity
Security Threat		A misuser corrupts a message from the system to a user.
Preconditions		
1)	The misuser has the means to intercept a message from the system to a user.	
2)	The misuser has the means to modify an intercepted messag	
3)	The misuser has the means to forward the modified message to the user.	
Interactions		
1	System Interactions	The system sends a message to a user.
	System Actions	The system ensures that modifications to the message will be obvious to the user
2	Misuser Interactions	The misuser intercepts and modifies the system's message and forwards it to the user.
3	User Interactions	The user receives the corrupted message.
	System Actions	The system will recognize that the message was corrupted.
4	System Interactions	The system will notify the user that the message was corrupted
Postconditions		None

continued on following page

Table 5. continued

Security Use Case		**Ensure Confidentiality (SUC2)**
Use Case Path		User Message Integrity
Security Threat		A misuser accesses a private message from the user to the system
Preconditions		
1)	The misuser has the means to intercept a message from the user to the system	
2)	The system has requested private information from the user.	
Interactions		
1	Interactions	The user sends a private message to the system.
2	System Actions	The system makes the private message illegible while in transit.
3	Misuser Interactions	The misuser intercepts the user's private message.
Postconditions		The misuser cannot read the user's private message
Security Use Case		**Authenticate (SUC3)**
Use Case Path		Attempted Spoofing using Valid User Identity.
Security Threat		The application authenticates a misuser as if the misuser were actually a valid user.
Preconditions		
1)	The misuser has a valid means of user identification.	
2)	The misuser has an invalid means of user authentication.	
Interactions		
1	System Interactions	The system shall request the misuser's means of identification and authentication.
2	Misuser Interactions	The misuser provides a valid means of user identity but an invalid means of user authentication
3	System Actions	1) The system shall misidentify the misuser as a valid user. 2) The system shall fail to authenticate the misuser.
4	Misuser Interactions	The system shall reject the misuser by cancelling the transaction
Postconditions		
1)	The system shall not have allowed the misuser to steal the user's means of authentication.	
2)	The system shall not have authenticated the misuser.	
3)	The system shall not have authorized the misuser to perform any transaction that requires authentication.	
4)	The system shall record the access control failure.	
Security Use Case		**Authorize Access (SUC4)**
Use Case Path		Attempted Spoofing using Social Engineering
Security Threat		The misuser gains access to an unauthorized resource.
Preconditions		
1)	The misuser has a valid means of user identification enabling the impersonation of a valid user that is authorized to use a protected resource.	
2)	The misuser does not have an associated valid means of user authentication.	
3)	The misuser has basic knowledge of the organization including the ability to contact the contact center.	
Interactions		
1	Misuser Interactions	The misuser contacts the contact center.

continued on following page

Table 5. continued

2	Contact center Interactions	A user support agent shall request the misuser's identity and authentication.
3	Misuser Interactions	1) The misuser provides the valid user identity. 2) The misuser states that he or she has a temporary inability to authenticate himself or herself. 3) The misuser states that he or she has an urgent need to access a protected resource requiring authentication and authorization.
4	Contact center Interactions	The user support agent shall request one or more alternate forms of authentication. The user support agent shall check the appropriate procedures for the proper action.
	Contact center Actions	The user support agent shall request one or more alternate forms of authentication. The user support agent shall check the appropriate procedures for the proper action.
5	Misuser Interactions	The misuser fails to provide a valid alternate form of authentication.
6	Contact center Interactions	The user support agent shall refuse authentication and authorization to the requested resource.
Alternative Paths		The misuser can quit at any point.
Postconditions		
1)	The system shall not have authenticated the misuser.	
2)	The system shall not have authorized the misuser to access the protected resource.	
3)	The system shall record the access control failure.	

Task 2.5: Assessment of Security

Finally, it is necessary to assess whether the threats are relevant according to the security level specified by the security objectives. Therefore we must estimate the security risks based on the relevant threats, their likelihood and their potential negative impacts, in other words, we have to estimate the impact (what may happen) and risk (what will probably happen) to which the assets in the system are exposed. We must therefore interpret the meaning of impact and risk. In Table 6 we define the impact and risk for the threats identified previously. We are going to evaluate risk and impact with five possible values: Very Low, Low, Medium, High and Very High. The likelihood of a threat could be: Very Frequent (daily event), Frequent (monthly event), Normal (once a year), Rare (once in several years).

As we can see in the previous table, all threats have to be dealt with because they cause a high or very high value of risk in the worst case, therefore, misuse cases that represent these threats must be studied and analyzed in this first iteration and will take part of the Grid use cases diagram that we will build in the next task. For example, for alteration and disclosure of information we can see that if the information is sensitive (personal data, bank data), these treats represent a high risk for our system and we must ensure that attacks (modifying or altering information) do not attain their objectives. In this case we must strongly protect the information stored and transmitted between user and system. This assessment must be present in the next activities and it must take into account when we design the security service oriented architecture.

FUTURE RESEARCH DIRECTIONS

The main future lines of research open are detailed below:

- Define the process with a tool that supports the SPEM notation, such as EPF (Eclipse

Table 6. Assessment of impact and risk

Threat	Unauthorized access to Grid system	
Impact	MEDIUM if the authorization privileges are very limited (i.e. only reading).	VERY HIGH if the opposite is the case
Attack	Unauthorized access	
Probability	Normal	Normal
Risk	HIGH	VERY HIGH
Threat	**Unauthorized alteration of information**	
Impact	LOW if there is no personal information modified	HIGH if the opposite is the case
Attack	Modification of information	
Probability	Frequent	Frequent
Risk	LOW	HIGH
Threat	**Unauthorized disclosure of information**	
Impact	LOW when the disclosed information is not sensitive or important	HIGH if the opposite is the case
Attack	Interception of information	
Probability	Frequent	Very Frequent
Risk	LOW	HIGH
Threat	**Masquerade as a certain user**	
Impact	LOW when the exchanged information with the fooled entity is not sensitive or important	HIGH if the opposite is the case
Attack	Masquerade	
Probability	Frequent	Normal
Risk	MEDIUM	VERY HIGH

Process Framework), and enables its automated integration with the processes of other methodologies based on UML as UP, OPEN, OpenUP, etc.

- Concrete and refine the generic tasks of the used development processes that have been incorporated into our process.
- Refine and improve the parameters and tagged values of the GridUCSec-profile for capturing the most important aspects and features of Mobile Grid systems to take them into account in the design and construction activities of the process.
- Improve the reference security architecture for that the security aspects considered in the analysis activity through the GridUCSec-profile can easily be incorporated as parameters into the interfaces of the security architecture, into the definition of policies of the system or into the decisions of implementation.
- Study and incorporate security patterns into the design activity to facilitate and ensure the correct incorporation of architectural elements that define already proven security solutions and help us construct the security architecture specific for mobile Grid systems.
- Define templates for the specification of security requirements based on IEEE std. 1233, 12207.1, 830 standards, SIREN, etc. that impose a format and a specific method for the definition and extraction of information for functional and non-functional requirements, especially those of security, identified in the analysis activity and that

must be completed and managed in the rest of activities of the process.

- Carry out new case studies for a continuous improvement of the process in other environments and dominions apart from the one developed here.
- Extend the applicability of the process and adapt its tasks and artifacts in order to develop secure systems oriented to Cloud Computing.
- Extend the GridUCSec-profile to define not only stereotypes for use cases but also stereotypes for other kind of UML models such as the models of interaction, deployment, collaboration, etc., that can be used in the different activities of the process.
- Implement all security services and interfaces of the reference security architecture using the most advanced and used programming languages such as Java, .Net o C#.

CONCLUSION

The interest in incorporating mobile devices into Grid systems has arisen with two main purposes. The first one is to enrich users of these devices while the other is that of enriching the Grid's own infrastructure. Both benefit from this fact since, on the one hand, the Grid offers its services to mobile users to complete their work in a fast and simple way and, on the other hand, the mobile devices offer their limited resources, but millions of them, in any place and at any time, endorsed by the fast advance in the yield and capacity that is being carried out in mobile technology.

In many cases, constrained wireless networks are made up of devices that are physically constrained and therefore have little room for memory, batteries, and auxiliary chips. Security over the mobile platform is more critical due to the open nature of wireless networks. In addition, security is more difficult to implement into a mobile platform due to the limitations of resources in these devices.

Due to this difficulty when we want to incorporate mobile devices into a grid system and due to the fact that we must take into account security aspects throughout the life cycle, it is necessary to provide a systematic process to developers for building this kind of system considering grid characteristics, mobile computing and security aspects throughout the development process. This process must always be flexible, scalable and dynamic, so that it adapts itself to the ever-changing necessities of mobile Grid systems.

In this paper we have presented a process for designing and building a secure mobile grid system based on an iterative, incremental and reusable process. This process is composed of several stages and activities and in each one of them the stakeholders carry out their tasks. An important activity of the process is the security requirements analysis which we have proposed with a set of tasks to obtain security requirements for mobile grid systems based in security use cases. Considering a case study for media domain, we have applied the analysis activity for analyzing security requirements in this real application using techniques of uses cases, misuse cases, security use cases and risk assessment where we obtain a specification of security requirements of our system analyzed on several refinements.

Applying this set of tasks we have been able to incorporate security requirements into our analysis and into our system. The application of this case study has allowed us to improve and refine some activities, tasks and artifacts of the process.

ACKNOWLEDGMENT

This research is part of the following projects: MARISMA (HITO-2010-28), SISTEMAS (PII2I09-0150-3135) and SEGMENT (HITO-09-138) financed by the "Viceconsejería de Ciencia y Tecnología de la Junta de Comunidades de Castilla-La Mancha" (Spain) and FEDER, and MEDUSAS (IDI-20090557), BUSINESS (PET2008-0136), PEGASO/MAGO (TIN2009-13718-C02-01) and ORIGIN (IDI-2010043(1-5) financed by the "Ministerio de Ciencia e Innovación (CDTI)" (Spain). Special acknowledgment to GREDIA (FP6-IST-034363) funded by European Commission.

REFERENCES

Basin, D., & Doser, J. (2002). SecureUML: A UML-based modeling language for model-driven security. Paper presented at the 5th International Conference on the Unified Modeling Language. *Lecture Notes in Computer Science 2460.*

Basin, D., Doser, J., & Lodderstedt, T. (2003). *Model driven security for process-oriented systems*. Paper presented at the ACM Symposium on Access Control Models and Technologies, Como, Italy.

Bellavista, P., & Corradi, A. (2006). *The handbook of mobile middleware*. Auerbach Publications. doi:10.1201/9781420013153

Bhanwar, S., & Bawa, S. (2008). *Securing a Grid*. Paper presented at the World Academy of Science, Engineering and Technology.

Bradford, P. G., Grizzell, B. M., Jay, G. T., & Jenkins, J. T. (2007). Cap. 4. Pragmatic security for constrained wireless networks. In Xaio, Y. (Ed.), *Security in distributed, Grid, mobile, and pervasive computing* (p. 440). Tuscaloosa, USA: The University of Alabama.

Bresciani, P., Giorgini, P., Giunchiglia, F., Mylopoulos, J., & Perin, A. (2004). TROPOS: An agent-oriented software development methodology. *Journal of Autonomous Agents and Multi-Agent Systems*, *8*(3), 203–236. doi:10.1023/B:AGNT.0000018806.20944.ef

Castro, J., Kolp, M., & Mylopoulos, J. (2001). *A requirements-driven development methodology*. Paper presented at the 13th Int. Conf. on Advanced Information Systems Engineering, CAiSE'01.

Dail, H., Sievert, O., Berman, F., & Casanova, H. YarKhan, A., Vadhiyar, S., et al. (2004). Scheduling in the Grid application development software project. *In Grid resource management: State of the art and future trends* (pp. 73-98).

Enterprise Grid Alliance Security Working Group. (2005). *Enterprise Grid security requirements*, version 1.0.

Fernández-Medina, E., Jurjens, J., Trujillo, J., & Jajodia, S. (2009). Special issue: Model-driven development for secure Information Systems. *Information and Software Technology*, *51*(5), 809–814. doi:10.1016/j.infsof.2008.05.010

Fernández-Medina, E., & Piattini, M. (2005). Designing secure databases. *Information and Software Technology*, *47*(7), 463–477. doi:10.1016/j.infsof.2004.09.013

Firesmith, D. G. (2003). Security use cases. *Journal of Object Technology*, 53-64.

Flechais, I., Sasse, M. A., & Hailes, S. M. V. (2003). *Bringing security home: A process for developing secure and usable systems*. Paper presented at the New Security Paradigms Workshop (NSPW'03), Ascona, Switzerland.

Foster, I., & Kesselman, C. (2004). *The Grid2: Blueprint for a future computing infrastructure* (2nd ed.). San Francisco, CA: Morgan Kaufmann Publishers.

Foster, I., Kesselman, C., Tsudik, G., & Tuecke, S. (1998). *A security architecture for computational Grids*. Paper presented at the 5th ACM Conference on Computer and Communications Security, San Francisco, USA.

Gartner. (2007). *Gartner says worldwide PDA shipments top 17.7 Million in 2006*. Gartner Press Release. Retrieved from http://www.gartner.com/it/page.jsp?id=500898

Gartner. (2009). *Gartner says worldwide mobile phone sales declined 8.6 per cent and smartphones grew 12.7 per cent in first quarter of 2009*. Gartner Press Release. Retrieved from http://www.gartner.com/it/page.jsp?id=985912

Georg, G., Ray, I., Anastasakis, K., Bordbar, B., Toahchoodee, M., & Houmb, S. H. (2009). An aspect-oriented methodology for designing secure applications. *Information and Software Technology*, *51*(5), 846–864. doi:10.1016/j.infsof.2008.05.004

Giorgini, P., Mouratidis, H., & Zannone, N. (2007). Modelling security and trust with secure tropos. In Giorgini, H. M. P. (Ed.), *Integrating security and software engineering: Advances and future visions* (pp. 160–189). Hershey, PA: Idea Group Publishing.

Graham, D. (2006). *Introduction to the CLASP process*. Retrieved from https://buildsecurityin.us-cert.gov/daisy/bsi/articles/best-practices/requirements/548.html

Guan, T., Zaluska, E., & Roure, D. D. (2005). *A Grid service infrastructure for mobile devices*. Paper presented at the First International Conference on Semantics, Knowledge, and Grid (SKG 2005), Beijing, China.

Herveg, J., Crazzolara, F., Middleton, S. E., Marvin, D. J., & Poullet, Y. (2004). *GEMSS: Privacy and security for a medical Grid*. Paper presented at the HealthGRID 2004, Clermont-Ferrand, France.

Humphrey, M., Thompson, M. R., & Jackson, K. R. (2005). *Security for Grids*. Lawrence Berkeley National Laboratory. (Paper LBNL-54853).

Jacobson, I., Booch, G., & Rumbaugh, J. (1999). *The unified software development process*. Addison-Wesley Professional.

Jameel, H., Kalim, U., Sajjad, A., Lee, S., & Jeon, T. (2005). *Mobile-to-Grid middleware: Bridging the gap between mobile and Grid environments*. Paper presented at the European Grid Conference EGC 2005, Amsterdam, The Netherlands.

Jana, D., Chaudhuri, A., & Bhaumik, N. B. (2009). Privacy and anonymity protection in computational Grid services. *International Journal of Computer Science and Applications*, *6*(1), 98–107.

Jurjens, J. (2001). *Towards development of secure systems using UMLsec*. Paper presented at the Fundamental Approaches to Software Engineering (FASE/ETAPS).

Jurjens, J. (2002). *UMLsec: Extending UML for secure systems development*. Paper presented at the 5th International Conference on the Unified Modeling Language (UML), Dresden, Germany.

Jürjens, J. (2005). *Secure systems development with UML*. Springer.

Jürjens, J., Schreck, J., & Bartmann, P. (2008). *Model-based security analysis for mobile communications*. Paper presented at the International Conference on Software Engineering, Leipzig, Germany.

Kolonay, R., & Sobolewski, M. (2004). *Grid interactive service-oriented programming environment*. Paper presented at the Concurrent Engineering: The Worldwide Engineering Grid, Tsinghua, China.

Kruchten, P. (2000). *The rational unified process: An introduction* (2nd ed.). Addison-Wesley.

Kumar, A., & Qureshi, S. R. (2008, March 29). *Integration of mobile computing with Grid computing: A middleware architecture*. Paper presented at the 2nd National Conference on Challenges & Opportunities in Information Technology (COIT-2008), Mandi Gobindgarh, India.

Kwok-Yan, L., Xi-Bin, Z., Siu-Leung, C., Gu, M., & Jia-Guang, S. (2004). Enhancing Grid security infrastructure to support mobile computing nodes. *Lecture Notes in Computer Science*, *2908*, 42–54. doi:10.1007/978-3-540-24591-9_4

Litke, A., Skoutas, D., & Varvarigou, T. (2004). *Mobile Grid computing: Changes and challenges of resource management in a mobile Grid environment*. Paper presented at the 5th International Conference on Practical Aspects of Knowledge Management (PAKM 2004).

Mouratidis, H. (2004). *A security oriented approach in the development of multiagent systems: Applied to the management of the health and social are needs of older people in England*. University of Sheffield.

Mouratidis, H., & Giorgini, P. (2006). *Integrating security and software engineering: Advances and future vision*. Hershey, PA: IGI Global.

Nagaratnam, N., Janson, P., J. Dayka, Nadalin, A., Siebenlist, F., Welch, V., et al. (2003). *The security architecture for open Grid services*.

Open Grid Forum. (2006). *The open Grid services architecture*, version 1.5 o.

Open Group. (2009). *TOGAF™ version 9 - the open group architecture framework*. Retrieved from http://www.opengroup.org/architecture/togaf9-doc/arch/

Phan, T., Huang, L., Ruiz, N., & Bagrodia, R. (2005). Integrating mobile wireless devices into the computational Grid. In Ilyas, M., & Mahgoub, I. (Eds.), *Mobile computing handbook*. Auerbach Publications.

Pitzmann, A., & Köhntopp, M. (2001). *Anonymity, unobservability, and pseudonymity — a proposal for terminology. Designing Privacy Enhancing Technologies* (pp. 1–9). LNCS.

Popp, G., Jürjens, J., Wimmel, G., & Breu, R. (2003). *Security-critical system development with extended use cases*. Paper presented at the Tenth Asia-Pacific Software Engineering Conference (APSEC'03).

Rosado, D. G., Fernández-Medina, E., & López, J. (2009a). *Applying a UML extension to build use cases diagrams in a secure mobile Grid application*. Paper presented at the 5th International Workshop on Foundations and Practices of UML, in conjunction with the 28th International Conference on Conceptual Modelling, ER 2009, Gramado, Brasil.

Rosado, D. G., Fernández-Medina, E., & López, J. (2009b). Obtaining security requirements for a mobile Grid system. *International Journal of Grid and High Performance Computing*, *1*(3), 1–17. doi:10.4018/jghpc.2009070101

Rosado, D. G., Fernández-Medina, E., & López, J. (2009c). *Reusable security use cases for mobile Grid environments*. Paper presented at the Workshop on Software Engineering for Secure Systems, in conjunction with the 31st International Conference on Software Engineering, Vancouver, Canada.

Rosado, D. G., Fernández-Medina, E., & López, J. (2011a). Towards an UML extension of reusable secure use cases for mobile Grid systems. *IEICE Transactions on Information and Systems*, *94-D*(2), 243–254.

Rosado, D. G., Fernández-Medina, E., & López, J. (2011b). Security services architecture for secure mobile Grid systems. *Journal of Systems Architecture. Special Issue on Security and Dependability Assurance of Software Architectures*, *57*(3), 240–258.

Rosado, D. G., Fernández-Medina, E., López, J., & Piattini, M. (2008). *PSecGCM: Process for the development of secure Grid computing based systems with mobile devices*. Paper presented at the International Conference on Availability, Reliability and Security (ARES 2008), Barcelona, Spain.

Rosado, D. G., Fernández-Medina, E., López, J., & Piattini, M. (2010a). Analysis of secure mobile Grid systems: A systematic approach. *Information and Software Technology, 52*, 517–536. doi:10.1016/j.infsof.2010.01.002

Rosado, D. G., Fernández-Medina, E., López, J., & Piattini, M. (2010b). Developing a secure mobile Grid system through a UML extension. *Journal of Universal Computer Science, 16*(17), 2333–2352.

Rosado, D. G., Fernández-Medina, E., López, J., & Piattini, M. (2011). (in press). Systematic design of secure mobile Grid systems. *Journal of Network and Computer Applications*. doi:10.1016/j.jnca.2011.01.001

Sajjad, A., Jameel, H., Kalim, U., Han, S. M., Lee, Y.-K., & Lee, S. (2005). *AutoMAGI - an autonomic middleware for enabling mobile access to Grid infrastructure*. Paper presented at the Joint International Conference on Autonomic and Autonomous Systems and International Conference on Networking and Services - (icas-icns'05).

Steel, C., Nagappan, R., & Lai, R. (2005). Chapter 8-the alchemy of security design methodology, patterns, and reality checks. In *Core security patterns: Best practices and strategies for J2EE™, Web services, and identity management* (pp. 10-88). Prentice Hall PTR/Sun Micros.

Talukder, A., & Yavagal, R. (2006). Security issues in mobile computing. In *Mobile computing*. McGraw-Hill Professional.

Trujillo, J., Soler, E., Fernández-Medina, E., & Piattini, M. (2009). An engineering process for developing secure data warehouses. *Information and Software Technology, 51*(6), 1033–1051. doi:10.1016/j.infsof.2008.12.003

Vivas, J. L., López, J., & Montenegro, J. A. (2007). Grid security architecture: Requirements, fundamentals, standards, and models. In Xiao, Y. (Ed.), *Security in distributed, Grid, mobile, and pervasive computing* (p. 440). Tuscaloosa, USA.

Welch, V., Siebenlist, F., Foster, I., Bresnahan, J., Czajkowski, K., Gawor, J., et al. (2003). *Security for Grid services*. Paper presented at the 12th IEEE International Symposium on High Performance Distributed Computing (HPDC-12 '03).

Chapter 13
Trusted Data Management for Grid-Based Medical Applications

Guido J. van 't Noordende
University of Amsterdam, The Netherlands

Silvia D. Olabarriaga
Academic Medical Center - Amsterdam, The Netherlands

Matthijs R. Koot
University of Amsterdam, The Netherlands

Cees T.A.M. de Laat
University of Amsterdam, The Netherlands

ABSTRACT

Existing Grid technology has been foremost designed with performance and scalability in mind. When using Grid infrastructure for medical applications, privacy and security considerations become paramount. Privacy aspects require a re-thinking of the design and implementation of common Grid middleware components. This chapter describes a novel security framework for handling privacy sensitive information on the Grid, and describes the privacy and security considerations which impacted its design.

INTRODUCTION

Most current Grid middleware is designed primarily for high-performance and high-throughput computing and data storage (LHC, n.d.; Foster, Kesselman, & Tuecke, 2001). Initially, Grid infrastructure aimed mostly at the Physics community, but recently many other domains, such as Biology, Pharmaceutics, and Medical research have shown increasing interest in using Grids for their applications. Grid middleware, including gLite (gLite, n.d.) and the Globus Toolkit (Globus, n.d.), hides many aspects such as data distribution and replication from users of the system. As a result, users are often unaware that jobs and data are transferred through multiple Grid components in different administrative domains implicitly. This makes it hard for users to understand the

DOI: 10.4018/978-1-60960-603-9.ch013

Figure 1. A use-case for medical imaging research showing grid resources in different administrative domains, with an emphasis on data and job flow

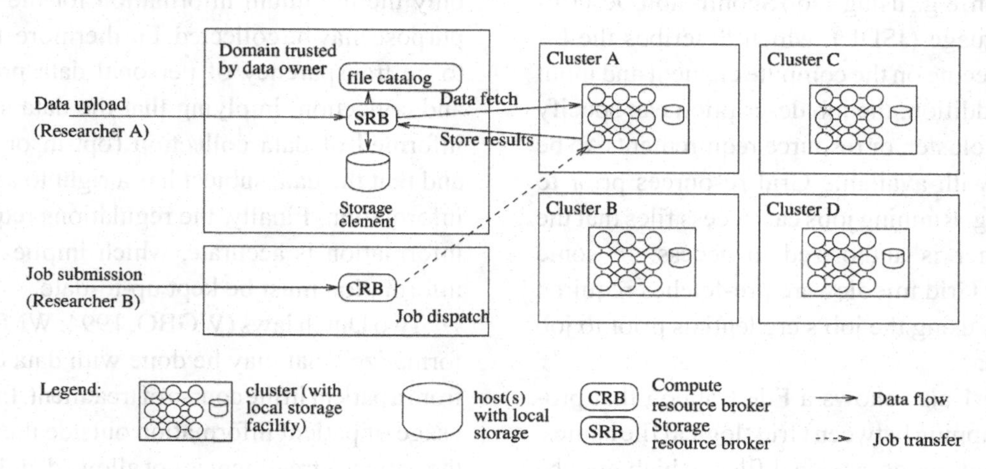

security implications of using Grid middleware, in particular when using it for applications that use privacy sensitive information.

Medical applications have very strict requirements on data handling and storage due to privacy concerns and regulations. Therefore, Grid middleware intended for usage in the medical domain should support policies that define where particular data may be stored, in what form, and what jobs from which users may access this data from what hosts or administrative domains.

This paper presents a new framework for managing privacy-sensitive data on the Grid, that allows for explicit data-owner control over data access and distribution related aspects. It makes a clear distinction between data storage components, access control, job authentication aspects, and auditing mechanisms for data related operations.

This paper is organized as follows: first we describe a use-case for medical research, based on our own experience (Olabarriaga, Nederveen, Snel & Belleman, 2006). Next, we analyze legal requirements with regard to medical data and technical aspects that are relevant when using Grid infrastructure to manage privacy-sensitive data. Finally, we describe a framework that allows data owners to express fine-grained data distribution and access control policies to allow for secure handling of medical data on the Grid. We conclude with an overview of some usability aspects.

USAGE SCENARIO

Figure 1 shows a typical Grid infrastructure deployment for medical research. A Grid storage system in one trusted administrative domain is used for storing medical research data. Although data is often replicated across different domains to enhance availability and reliability, we assume here that all storage facilities reside in only one administrative domain trusted by the data owner. Different incarnations of storage infrastructure exist, e.g., SDSC SRB and dCache (dCache, n.d.). In this paper, we refer to the storage infrastructure as a Storage Resource Broker (SRB) in a general way, without referring to a particular implementation.

First, Researcher A (data owner) uploads the data to an SRB he or she trusts, e.g., using gridFTP. Researcher B can now submit a job on the Grid through a Compute Resource Broker (CRB) which can reside in any administrative domain. The CRB transparently selects a cluster, typically based on load, where the job is scheduled for execution.

The user controls job submission via some job description, e.g., using a Job Submission Description Language (JSDL), which describes the binary to execute on the compute element and input files. In addition, the job description can specify a specific cluster, or resource requirements, to be matched with available Grid resources prior to scheduling. Running jobs can access files that the job's owner is authorized to access. In some cases, the Grid middleware pre-fetches required input files using the job's credentials prior to job execution.

Figure 1 also shows a File Catalog that provides a mapping between Grid 'logical file names' and the underlying physical files, which may be replicated on different storage systems on the Grid. Additionally, an SRB may also maintain a metadata service (not shown). Since metadata and file names may contain privacy sensitive information, both services should be managed by a trusted domain.

LEGAL REQUIREMENTS

The European Union (EU) has produced legislation on handling personal information and privacy (EC, 1995). This section focusses on EU and selected Dutch regulations. Countries outside the EU have adopted or are adopting legal measures to allow exchange of personal data with the EU countries (e.g., U.S. Safe Harbor Framework). For more information about other countries see (Fischer-Huebner, 2001; EC; Herveg, 2006; U.S. Congress, 1996).

EU regulations can be seen as leading guidelines for handling personal data (Fischer-Huebner, 2001). The data protection regulations can be summarized as follows. First, there must be a necessity for data collection and processing. Related to that, for each data collection, there has to be a clear purpose binding which specifies what is done with the information. Usage of data beyond this specified purpose is not allowed. In addition, a minimality principle exists, which states that only the minimum information for the required purpose may be collected. Furthermore, there has to be transparency of personal data processing and collection, implying that the data subject is informed of data collection (opt-in or opt-out) and that the data subject has a right to access the information. Finally, the regulations require that information is accurate, which implies that the information must be kept up-to-date.

Two Dutch laws (WGBO, 1994; WMO, 1998) formalize what may be done with data collected from a patient in the course of treatment. In general, usage of patient information outside the scope of the patient's treatment is not allowed, unless there is considerable public interest or similar necessity to do so. Medical scientific research is often considered such an exception (Herveg, 2006).

If a patient explicitly consents with usage of his data for medical research, that data is purpose-bound to a specific medical research activity. The data may not be disclosed beyond this activity. The physician or medical researcher who determines the purpose and means of processing is legally responsible for ensuring an appropriate level of security to protect data.

The restrictions described above only apply to personal data. In some situations, the data can be de-personalized to circumvent these restrictions, e.g., as done in (Kalra et al., 2005; Montagnat et al., 2007; Erberich et al., 2007). However, complete de-identification is hard to get right, and re-identification is often possible (Sweeney, 2002; Malin, 2002). For this reason, de-identified information should be considered confidential, and appropriate distribution and access control mechanisms are required.

BASIC GRID SECURITY INFRASTRUCTURE

The Grid Security Infrastructure (GSI) (Foster, Kesselman, Tsudik and Tuecke, 1998) is the de-facto standard for user and host authentication on the Grid. GSI is used by most mature Grid middleware implementations. Shortcomings of this infrastructure are described later in this paper; here we introduce the basic GSI infrastructure.

GSI essentially comprises a Public Key Infrastructure (PKI) that is used to sign user identity and host certificates. Users can create limited-lifetime Proxy certificates which allow them to send credentials with their jobs for authentication, without the risk of compromising the user's private key. Proxy certificates are used for all transactions by a job, such as gridFTP transactions. We here assume that all authorization decisions with regard to data are based on GSI user authentication by means of Proxy certificates. Other approaches (such as role-based or attribute-based authorization, as proposed in (Alfieri et al., 2004) are possible, but not required for our framework. Many Grid infrastructures manage access control to resources and storage based on virtual organization (VO) membership information. However, VO-based authorization is often too course-grained for protecting medical information: there may be many users (e.g., researchers) in a VO, which may not all be equally trusted to access particular data. Therefore, we assume authorization based on user identities in this paper.

PROBLEM ANALYSIS

Grids are, by nature, distributed across multiple administrative domains, only a few of which may be trusted by a specific data owner. Grid middleware, and thus jobs, typically run on an operating system (OS), such as Linux, that allows administrators to access all information on the system. A job or data owner does not have control over the hardware or software that runs on some remote system. Besides OS and middleware vulnerabilities, these systems might also not be well protected against physical attacks, such as stealing hard disks. Such aspects should be part of a risk assessment when decisions are made on which sites are trusted to store or access particular information.

Given legal constraints, trust decisions will and should be conservative. For example, unencrypted data, file names, and other sensitive metadata should only be stored in trusted domains, e.g., in the hospital. This aspect is even more prevalent in systems where jobs on remote machines can access medical data. Current OSs such as Linux provide little assurance that information stored on the system cannot be leaked to external parties (van 't Noordende, Balogh, Hofman, Brazier and Tanenbaum, 2007).

Even if files are removed after the job exits (e.g., temporarily created files), the contents could be readable by administrators or possibly attackers while the job executes. Furthermore, disks may contain left-over information from a job's previous execution, which is readable by an attacker who gains physical access to a storage device, if the system is not properly configured (NIST). As another example, it is possible to encrypt swap space in a safe way, but this is an option that has to be explicitly enabled in the OS. For these reasons, it is important for a data owner to identify critical aspects of the administration and configuration of a remote host, before shipping data to (a job running on) that host.

Another problem is that a data owner cannot control nor know the trajectory that a job took before it was scheduled on a host, since this is implicit and hidden in current Grid middleware. Therefore, even if the host from which a job accesses data is trusted by the data owner, there is a risk that the job was manipulated on some earlier host.

Current middleware does not provide a way to securely bind jobs to Proxy certificates: a certificate or private key bundled with a program can easily be extracted and coupled to another program which pretends to be the original program. In Grids, this issue is exacerbated by the fact that a job may traverse several middleware processes (e.g., a CRB) in different domains before it is scheduled at some host. Each of these hosts or domains may be malicious, and the administrator or an attacker that gains access to one of these hosts may replace the original job with another program that leaks information to an external party. Alternative authentication schemes (e.g., username/password-based) do not improve this situation.

For this paper, we assume that the implementation of a job is trusted when this job's owner is trusted. In particular, we assume that medical researchers are aware of confidentiality aspects regarding medical data and treat this data as confidential information – and as a result use only trusted programs to make use of this data. In the proposed framework, jobs can only access data from hosts that are trusted by the data owner, and we assume that a job submitted by a trusted user will not leak information to unauthorized parties. A mechanism is presented later in this paper that allow users to seal jobs in such a way that tampering with these jobs is not possible.

Note that mechanisms exist that limit the capabilities of a possibly untrusted program to export information to arbitrary external parties, e.g., using the jailing system described in (van 't Noordende, Balogh, Hofman, Brazier, and Tanenbaum, 2007). Such solutions can be considered as additional measures to increase security, but are outside the scope of this paper.

For this paper, we assume that jobs do not ship potentially privacy-sensitive (output) data back to the possibly untrusted CRB through which the they entered the system. Instead, jobs should be programmed to encrypt output data with the job owner's public key before returning to their CRB, or they should store any potentially sensitive (output) data only on secure storage, preferably the system that contained the input data.

Summarizing, a number of implementation issues should be solved before we can be sure that privacy-sensitive information cannot be accessed by unauthorized parties. First, a secure binding between jobs and Proxy certificates must be provided. Second, a data owner should be able to express in a policy which administrative domains he or she trusts to handle privacy sensitive information in a safe way, based on a risk assessment. Third, a data owner should be able to express policies with regard to a remote system's configuration details which are relevant to privacy and security and the way in which data is handled.

THE TSRB FRAMEWORK

We propose a framework for secure handling of privacy sensitive information on Grids that allows for controlling data access and distribution aspects. The components and interactions of the framework are presented in Figure 2.

The framework is centered around a secure storage infrastructure called Trusted Storage Resource Broker (TSRB). There may be many TSRBs on the Grid, possibly managed by different administrative domains in different VOs. The TSRB is coined ``trusted'', because (1) it is deployed in an administrative domain trusted by the data owner, and (2) it is trusted to enforce data-owner specified access control policies. The TSRB controls access to data items or collections by combining User-based Access Control Lists (User ACLs) and Host-ACLs. Host ACLs contain required host properties that must be met by a remote host before the data can be accessed by a job on this host.

Required host properties are described by the data owner in a Remote Host Property List (RHPL). Each host has a Host Property List (HPL) that contains host configuration details. The HPL contents are matched with the data's RHPL at connection time. The HPL is maintained by the remote host (Cluster A in Figure 2), and is signed by the host's administrator. The TSRB also maintains for each data collection or item a Host ACL containing a list of administrative domains

Figure 2. The TSRB framework: files, file names and metadata are managed by a Trusted SRB. Dotted lines depict microcontract establishment and auditing, solid lines depict data flow and job transfers

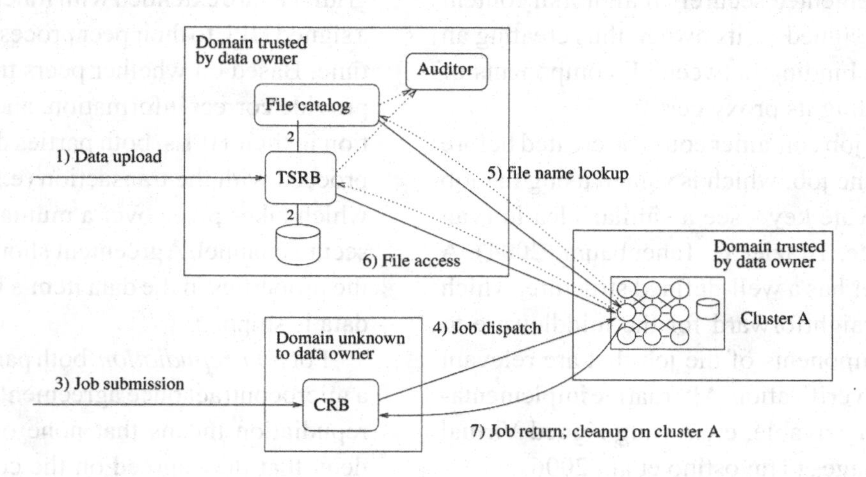

or hosts, who are trusted by the data owner both for confidentiality (of the administrators) and for providing correct information in their HPL.

The main actions are illustrated in Figure 2. A user uploads data to the TSRB, e.g., using gridFTP (step 1). The data is stored in a storage system maintained in the TSRB domain. Metadata can be stored in a separate service managed by the TSRB, e.g., a File Catalog in case of storing files (step 2). A job is submitted through a CRB (step 3), about which the data owner has no information. Eventually, the CRB submits the job to a cluster (step 4) that must be trusted by the data owner before the job can access data.

As part of the protocol before data access is authorized, user (job) and host authentication takes place, and the data's RHPL and the remote host's HPL are compared (details are given later). If RHPL and HPL match, a microcontract is established, which is a statement containing agreed-upon host properties and signed by both the TSRB and the remote host. Microcontracts are established for all authorization decisions, including, e.g., resolving file names in a File Catalog (step 5), and accessing the data item itself (step 6).

Only after the TSRB receives a microcontract, are the data shipped to the job or middleware acting on the job's behalf. In step 7 a job returns to its CRB where it can be collected by its owner. Subject to agreement in the microcontract, Cluster A ensures that no data from the job's execution remains on the host.

Auditing is important to allow data owners to track which jobs applied which operations on their data, on behalf of which users, and from which hosts. All established microcontracts are shipped to an auditor process (see Figure 2), which can be used by data owners to trace the transactions. Auditing can help establish trust (e.g., using reputation-based mechanisms), and enables tracking of potential sources of information leakage.

CONCEPTS AND INTERACTIONS

Job Authentication

A solution to provide a secure binding between jobs and Proxy certificates is to combine job integrity verification with a trust-based mechanism. Only if a data owner trusts a remote system to verify the integrity of incoming jobs properly, can he or she assume the the job-Proxy certificate binding to be valid, and can Proxy certificate-based authentication be trusted. Job integrity verification can be implemented securely if all initial content of the job is signed by its owner, thus creating an unforgeable binding between all components of a job, including its proxy certificate.

A secure job container could be created before submitting the job, which is signed using the job owner's private key - see a similar idea in (van 't Noordende, Brazier & Tanenbaum, 2004). A job container has a well-defined structure, which makes it straightforward for the middleware to find the components of the job that are relevant for integrity verification. Alternative implementations are conceivable, e.g., using signed Virtual Machine images (Travostino et al., 2006).

Host Property Lists

For risk assessment and policy enforcement, hosts should announce security relevant properties of their operating system, its configuration, and the used middleware, including properties regarding job integrity verification, in their Host Property List (HPL). The host administrator has the responsibility to fill in the HPL correctly. As a concrete example, the HPL could report on whether the operating system was configured to use encrypted swap space, on whether the middleware is capable of job integrity verification, and provides jobs with a private file system that is removed after the job exits.

HPLs allow for run-time assessment on whether a host adheres to the requirements for secure data handling as imposed by a data owner. This assessment takes place at the time that a connection is made to the TSRB. Because HPL matching takes place at connection time, no external trusted repository of HPLs is required for security.

Microcontracts

Microcontracts state the obligations that the site holds with regard to a transaction. Our framework requires that all Grid middleware components that are concerned with data transfer aspects (e.g., gridFTP) are extended with functionality to report a signed HPL to their peer processes at connection time. Based on whether peers trust each other to provide correct information, and on the information in their HPLs, both parties decide whether to proceed with the transaction (e.g., data transfer), which takes place over a mutually authenticated secure channel. Agreement should be reached on the properties in the data item's RHPL *before* any data is shipped.

For *non-repudiation,* both parties must co-sign a microcontract once agreement is reached. Non-repudiation means that none of the parties can deny that they agreed on the contract's content. To allow for auditing the exact operations on a particular data item, the microcontract has to be bound to each individual transaction, by including e.g., a hash over the data and the operation in the microcontract.

Trusted Storage Resource Broker

The TSRB is the key component for managing all privacy sensitive data in our framework. The TSRB is the central reference monitor and access point for data stored through this TSRB. In particular, the TSRB enforces the access control policies outlined in this paper. For clarity of exposition, we assume that the TSRB is a non-distributed service running in a single domain. The TSRB (and by implication, domain) is determined as trusted by a data owner prior to storing data on it.

Although we refer to the TSRB as a resource broker here, the TSRB is effectively an abstraction for a secure storage system. In case where the TSRB uses distributed facilities (e.g., untrusted storage elements managed by different domains), the TSRB can implement broker functionality. In this case, the TSRB should make sure that it stores only encrypted data on untrusted storage, using cryptographic filenames. Example storage systems that are implemented as a broker for encrypted data are described in (Montagnat et al., 2007; Xu, 2005).

Naming and Metadata Services

The TSRB can offer metadata services for managing and querying metadata about the stored data. Metadata is useful to search for data items of interest in large data collections. File names can be seen as metadata specific to file systems.

Naming or metadata services must be integrated into the TSRB, since access to file names and other sensitive metadata should be carefully protected. For example, careless encoding of file names could enable attackers to identify patient or hospital information from a file name and re-identify a patient. Naming or metadata services may be private to a VO, or part of some hierarchical naming service. In either case, file name lookup requests are subject to data-owner specified access control policies as outlined in this paper.

Access Control Lists

Access control in our system is enforced on the basis of ACLs. ACLs can be associated with individual data items or with a grouping (set) of data items. In case of files, grouping may be facilitated by e.g., associating ACLs with directory names. Unauthorized users should not even be able to find out if a given data item exists.

The User ACL contains a list of principals (job owners) that are allowed to access a (set of) data item(s), together with these principals' access rights on that data. The Host ACL specifies from what hosts or domains authorized jobs may access particular data, and with what access rights. Access rights from the User and Host ACLs are combined such that only the minimum set of rights for this data is granted to a job of a given user running on a given host.

The trusted domains or hosts in the Host ACL are determined by the data owner, e.g., based on whether he or she trusts the administrator of a particular administrative domain. Host ACLs are expressed as GSI host/domain name patterns, which match with the common name field of the x509 GSI host certificate, e.g., *.sara.nl, or host1.amc.nl. Specific patterns override wildcarded patterns. Also associated with data items or sets of data is a Remote Host Property List (RHPL). Before evaluating a remote host's HPL, it is checked that this host is in the Host ACL; only then is the HPL information considered trusted.

We chose to separately store an RHPL with each (set of) data items, in addition to the basic User and Host ACLs, because of the dynamic nature of Grid systems. Different domains may contain many machines or clusters, each of which with different configuration and job or data handling properties, which may even change over time. Connection-time RHPL / HPL matching allows the system to evaluate these properties at runtime, without relying on a (trusted) central repository of these properties.

Job Submission Procedure

At job submission time, a host must be selected from which the job's input data is accessible. Since CRBs are generally not trusted[1], client-side software should be used which contacts the TSRB before job submission. A file naming convention combined with a naming service (e.g., DNS) allows the client job submission program to locate the TSRB where the data is stored.

Client-side software can authenticate directly to the TSRB using the job owner's identity key. If authorized, it can fetch the relevant access control and HPL information, using which a job description is created. To allow for selection of suitable hosts by the CRB, HPLs could be published in a (global) information system. Note that because of run-time (R)HPL evaluation, the information system does not need to be completely consistent or trusted. This is important for scalability, as keeping a possibly global information system fully up-to-date may be infeasible.

Auditing

Auditing is important to allow for tracing all operations on a particular data item. For convenience and scalability, we use a trusted auditor process per TSRB, managed by the TSRB. Copies of the co-signed microcontracts of all transactions are sent to and strored by the auditor. This allows the data owners to trace all transactions that involve a particular data item in a way that ensures non-repudiability.

PUTTING IT ALL TOGETHER

Authorization of a data access requires that the connecting job's owner is on the User ACL, that the host on which the connecting job runs is on the Host ACL, and that the properties in the RHPL match the properties in the connecting host's HPL. Authorization of a data request consists of the following steps, assuming GSI host/Proxy certificate based authentication.

- At connection time, the connecting process (either a job or middleware, in case of data pre-fetching) authenticates with the TSRB using the job's Proxy certificate, resulting in an authenticated and encrypted SSL/TLS channel.
- The information from the Proxy certificate is matched against the User ACL to see if access is allowed. If not, an error is returned that does not indicate whether the data exists or not.
- The TSRB and the connecting process engage in a protocol for matching RHPL and HPL properties. If the connecting process is the middleware (e.g., during data pre-fetch), it can directly sign the microcontract. If the connecting process is a job, it has to request its local middleware (using a runtime interface) to match the RHPL of the TSRB with the host's HPL, and to have it sign a microcontract on its behalf if these properties match. The microcontract includes the (hash over the public key of the) Proxy certificate of the job to which it was issued.
- The signature over the microcontract (shipped together with the GSI host certificate that was used for signing) is compared with the Host ACL, to see if the HPL information is trusted and if access is allowed from this host.

The above mechanisms suffice to establish the required combination of Host ACL and User ACL based authorization, together with obtaining a microcontract signed by the connecting host before the data is shipped. If all provided information matches the data owner's requirements, the data is shipped to the requesting job or middleware, and the microcontract is logged in the auditor process.

USABILITY

Determining an appropriate Host ACL and HPL specification may be difficult for non-technical data owners. However, system administrators who support users may define template (R)HPLs with basic properties that hosts must adhere to when running jobs that access sensitive information. Such templates may be provided with the client-side software used for data uploading, and may be adapted by data owners and/or local system administrators at the time of use. Similarly, local (VO) administrators may help by composing default lists of trusted domains for particular data types or groups of users. Such measures allow secure usage of the system by researchers without burdening them with too many details. Dynamic adaptation of RHPLs for long-term storage of data is an open issue that needs to be addressed.

CURRENT STATUS AND FUTURE WORK

We have implemented a proof-of-concept implementation of the TSRB framework based on a gridFTP server from the Globus toolkit. We extended the gsi-FTP server with an authenticated key-exchange protocol to authenticate the client and establish a secure connection for data transfer; FTP commands were modified to include TSRB concepts such as HPL exchange and microcontracts. The resulting system's performance is as well as can be expected from a protocol that uses encryption to protect data transferred from server to client. Performance results are described in a separate report (Coca, 2011).

One of the more difficult issues to address when using our system, is how to decide whether a given system setup is secure. We have experimented with HPLs to describe various Linux systems. To determine a system's security, we used information obtained from the Common Vulnerability and Exposures (CVE) vulnerability database (http://cve.mitre.org), to locate potentially vulnerable packages on the system. A vulnerability score (Scarfone and Mell, 2009) is associated with each entry in the CVE database, which indicates the potential impact of a vulnerability on security of the system. However, Grid systems generally have different characteristics than desktop systems, for which the scoring method was devised.

Grid clusters are typically batch systems, and worker nodes within a cluster are usually not directly exposed to the Internet. Rather, the most important threats may originate from *within* the cluster, for example from malicious jobs that run concurrently with a job in the same cluster, or from jobs that compromised a machine some time earlier. We are currently studying whether the CVE-based vulnerability scoring can be adapted to Grid-specific characteristics. We are also studying ways to facilitate dynamic evaluation of HPL-based policies, such that users or administrators do not have to be overly burdened by (manually) updating policies or analyzing vulnerability reports to assess a system's security.

RELATED WORK

Montagnat et al. (2007) describe a Medical Data Manager (MDM) for DICOM images and associated metadata in a secure way. MDM is deployed inside hospitals, and provides read-only access to automatically de-identified DICOM images to grid jobs outside the hospital's domain. Data is encrypted before it becomes accessible to Grid jobs, so jobs must first acquire a key from a key store before they can access the data. However, MDM does not constrain from which hosts jobs may access the data or keys. MDM's reliance on automatic de-identification of DICOM headers may prove a vulnerability, e.g., in case of images which contain facial features of a patient as part of the binary data.

Globus MEDICUS (Erberich, Silverstein, Chervenak, Schuler, Nelson, & Kesselman, 2007) is an approach for sharing medical information (metadata and files) through Grid infrastructure. Encryption can be used to store information securely on untrusted storage elements in the Grid. One of the weak points of the system is that it does not clearly describe where the different components reside physically, i.e., what the trust model is. For example, metadata is stored in a meta catalog service which may be operated outside the hospital domain. In addition, the system depends on GSI for authentication, which makes the lack of a clear trust model even more worrisome.

Blancquer et al. (2009) describe an approach for managing encrypted medical data, building upon Hydra (Xu, 2005) and the ideas presented in Montagnat et al. (2007). The contribution of this approach is that key management and authorization are integrated with common Grid management concepts such as Virtual Organizations. However, like MDM and Hydra, the approach chosen by Blancquer et al. does not deal with the problem that the machine where the data is decrypted (by the job) may be compromised.

None of the related work considers trust in the hosts or clusters from which data are accessed, nor with the properties of the software running on these hosts.

DISCUSSION

We presented a trust-based security framework for Grid middleware that allows for enforcement of access control and data export policies for privacy-sensitive data. The framework proposes a Trusted SRB to manage data and enforce fine-grained access control policies on behalf of data owners. Access control policies combine user-based access control and trusted hosts lists with a runtime evaluation of properties of remote hosts from which jobs request data access. Microcontracts allow for establishing data handling agreements, and an auditing mechanism based on microcontracts allows for tracing all operations on the data.

The focus of this paper is on usage scenarios where Grid-based storage and data sharing is required. Our framework emphasizes data-owner specified user and host (property) based access control policies, to ensure that privacy sensitive information is only made accessible to authorized jobs running on hosts trusted by the data owner. This way, we can ensure that the data owner's requirements for secure data handling are met. More generally, we believe that the basic concepts presented in this paper, such as remote host property list evaluation, microcontracts, and auditing, can be of value for any distributed system or Grid middleware component in which precise control is required over where data or code may be distributed, and under what constraints.

ACKNOWLEDGMENT

We thank Oscar Koeroo, Dennis van Dok, and David Groep (NIKHEF) for valuable insight in the gLite-based VL-e infrastructure. Keith Cover (VU Medical Center) provided valuable information on privacy aspects of his job farming application. Berry Hoekstra and Niels Monen worked on a student project on HPLs and vulnerability scoring. Razvan Coca (UvA) is thanked for recent contributions to implementing the framework described in this paper. This work has been carried out as part of the Dutch research project Virtual Laboratory for e-Science (VL-e).

REFERENCES

Alfieri, R., Cecchini, R., Ciaschini, V., dell'Agnello, L., Frohner, A., Gianoli, A., et al. Spataro, F. (2004). Voms, an authorization system for virtual organizations. *European Across Grids Conference, LNCS 2970*, (pp. 33-40). Springer, 2004.

Blancquer, I., Hernández, V., Segrelles, D., & Torres, E. (2009). Enhancing privacy and authorization control scalability in the Grid through ontologies. *IEEE Transactions on Information Technology in Biomedicine*, *13*(1), 16–24. doi:10.1109/TITB.2008.2003369

Coca, R. (2011). Security enhancements of GridFTP:Description and Measurements. *Technical Report UVA-SNE-2011-01*, University of Amsterdam.

Dcache. (n.d.). *Dcache storage system*. Retrieved from http://www.dcache.org/

E.C. (1995). Directive 95/46/EC. *European commission data protection regulations overview page*. Retrieved from http://ec.europa.eu/justice_home/fsj/privacy/

Erberich, S., Silverstein, J. C., Chervenak, A., Schuler, R., Nelson, M. D., & Kesselman, C. (2007). Globus medicus - federation of dicom medical imaging devices into healthcare grids. *Studies in Health Technology and Informatics*, *126*, 269–278.

Fischer-Huebner, S. (2001). *IT-security and privacy: Design and use of privacy-enhancing security mechanisms*. New York, NY: Springer-Verlag.

Foster, I., Kesselman, C., Tsudik, G., & Tuecke, S. (1998). A security architecture for computational grids. *Proc. 5th ACM Conf. on Computer and Communication Security*, (pp. 83-92).

Foster, I., Kesselman, C., & Tuecke, S. (2001). The anatomy of the grid: Enabling scalable virtual organizations. *Int'l J. Supercomputer Applications, 15*(3).

Glite. (n.d.). *Glite middleware*. Retrieved from http://glite.web.cern.ch/glite

Globus. (n.d.). *Globus alliance toolkit homepage*. Retrieved from http://www.globus.org/toolkit/

Herveg, J. (2006). *The ban on processing medical data in European law: Consent and alternative solutions to legitimate processing of medical data in healthgrid. Proc. Healthgrid* (*Vol. 120*, pp. 107–116). Amsterdam, The Netherlands: IOS Press.

JSDL. (n.d.). *Job submission description language (jsdl) specification, v.1.0*. Retrieved from http://www.gridforum.org/documents/GFD.56.pdf

Kalra, D., Singleton, P., Ingram, D., Milan, J., MacKay, J., Detmer, D., & Rector, A. (2005). Security and confidentiality approach for the clinical e-science framework (clef). *Methods of Information in Medicine*, *44*(2), 193–197.

LHC. (n.d.). *LHC computing grid project*. Retrieved from http://lcg.web.cern.ch/LCG

Malin, B. (2002). *Compromising privacy with trail re-identification: The Reidit algorithms*. (CMU Technical Report, CMU-CALD-02-108), Pittsburgh.

Montagnat, J., Frohner, A., Jouvenot, D., Pera, C., Kunszt, P., & Koblitz, B. (2007). A secure grid medical data manager interfaced to the glite middleware. *Journal of Grid Computing*, *6*(1).

NIST. (2007). *Special publication 800-88: Guidelines for media sanitization by the national institute of standards and technology*. Retrieved from http://csrc.nist.gov/publications/nistpubs/#sp800-88

Olabarriaga, S. D., Nederveen, A. J., Snel, J. G., & Belleman, R. G. (2006). *Towards a virtual laboratory for FMRI data management and analysis. Proc. HealthGrid 2006* (*Vol. 120*, pp. 43–54). Amsterdam, The Netherlands: IOS Press.

Scarfone, K., & Mell, P. (2009) An analysis of CVSS version 2 vulnerability scoring. *Proceedings of the 3rd. Int'l Symposium on Empirical Software Engineering and Measurement (ESEM'09),* (pp. 516-525).

Sweeney, L. (2002). K-anonymity: A model for protecting privacy. *International Journal of Uncertainty. Fuzziness and Knowledge-Based Systems*, *10*(5), 557–570. doi:10.1142/S0218488502001648

Travostino, F., Daspit, P., Gommans, L., Jog, C., de Laat, C. T. A. M., & Mambretti, J. (2006). *S*eamless live migration of virtual machines over the man/wan. *Future Generation Computer Systems*, *22*(8), 901–907. doi:10.1016/j.future.2006.03.007

U.S. Congress (1996). *Health insurance portability and accountability act,* 1996.

U.S. Safe Harbor Framework. (n.d.). Retrieved from http://www.export.gov/safeharbor/

Van 't Noordende, G., Balogh, A., Hofman, R., Brazier, F. M. T., & Tanenbaum, A. S. (2007). *A secure jailing system for confining untrusted applications*. 2nd Int'l Conf. on Security and Cryptography (SECRYPT), (pp. 414-423). Barcelona, Spain.

Van 't Noordende, G. J., Brazier, F. M. T., & Tanenbaum, A. S. (2004). *Security in a mobile agent system*. 1st IEEE Symp. on Multi-Agent Security and Survivability, Philadelphia.

WGBO. (1994). *Dutch ministry of health, welfare and sport – WGBO*. Retrieved from http://www.hulpgids.nl/wetten/wgbo.htm

WMO. (1998). *Dutch ministry of health, welfare and sport - WMO.* Retrieved from http://www.healthlaw.nl/wmo.html.

Xu, L. (2005). *Hydra: A platform for survivable and secure data storage systems*. ACM StorageSS.

ENDNOTE

1 Note that if any (untrusted) CRB could query the TSRB directly for the locations from which data is available, the result can reveal whether a given data file exists or not. Such information may be considered sensitive in itself, as outlined earlier.

Section 4
Applications

Chapter 14
Large-Scale Co-Phylogenetic Analysis on the Grid

Heinz Stockinger
Swiss Institute of Bioinformatics, Switzerland

Alexander F. Auch
University of Tübingen, Germany

Markus Göker
University of Tübingen, Germany

Jan Meier-Kolthoff
University of Tübingen, Germany

Alexandros Stamatakis
Ludwig-Maximilians-University Munich, Germany

ABSTRACT

Phylogenetic data analysis represents an extremely compute-intensive area of Bioinformatics and thus requires high-performance technologies. Another compute- and memory-intensive problem is that of host-parasite co-phylogenetic analysis: given two phylogenetic trees, one for the hosts (e.g., mammals) and one for their respective parasites (e.g., lice) the question arises whether host and parasite trees are more similar to each other than expected by chance alone. CopyCat is an easy-to-use tool that allows biologists to conduct such co-phylogenetic studies within an elaborate statistical framework based on the highly optimized sequential and parallel AxParafit program. We have developed enhanced versions of these tools that efficiently exploit a Grid environment and therefore facilitate large-scale data analyses. Furthermore, we developed a freely accessible client tool that provides co-phylogenetic analysis capabilities. Since the computational bulk of the problem is embarrassingly parallel, it fits well to a computational Grid and reduces the response time of large scale analyses.

INTRODUCTION

The generation of novel insights in many scientific domains such as biology, physics, or chemistry increasingly relies on compute-intensive applications that require high-performance or large-scale, distributed high-throughput computing technology and infrastructure. In the discipline

DOI: 10.4018/978-1-60960-603-9.ch014

of bioinformatics, biological insight is typically generated via data analysis pipelines that use a plethora of distinct and highly specialized tools. Most commonly, bioinformaticians and biologists collaborate to analyze data extracted from large databases containing DNA and/or protein data in order to study, e.g., the function of living beings, the effect and influence of diseases and defects, or their evolutionary history. Early "classic" bioinformatics tools, such as CLUSTALW (Thompson et al., 1994) or BLAST (Altschul et al., 1997) that have been ported to Grid computing environments deal with biological sequence search, analysis, and comparison. Typically, these programs are embarrassingly parallel and therefore represent ideal candidate applications for Grid computing environments (Stockinger et al., 2006).

The study of the genome represents a way to obtain new insight and extract novel knowledge about living beings. In particular, stand-alone phylogenetic analyses have many important applications in biological and medical research. Applications range from predicting the development of emerging infectious diseases (Salzberg et al., 2007), over the study of Papillomavirus evolution that is associated with cervical cancer (Gottschling et al., 2007), to the determination of the common origin of Caribbean frogs (Heinicke et al., 2007).

Recent years have witnessed significant progress in the field of stand-alone phylogeny reconstruction algorithms, which represent an NP-complete optimization problem (Chor and Tuller, 2005), with the release of programs such as TNT (Goloboff, 1999), RAxML (Stamatakis, 2006), MrBayes (Ronquist and Huelsenbeck, 2003) or GARLI (Zwickl, 2006). Because of the continuous explosive accumulation and availability of molecular sequence data coupled with advances in phylogeny reconstruction methods, it has now become feasible to reconstruct and fully analyze large phylogenetic trees comprising hundreds or even thousands of sequences (organisms). However, current meta-analysis methods for phylogenetic trees such as programs that conduct co-phylogenetic tests can currently not handle such large datasets.

To alleviate this bottleneck in the meta-analysis pipeline, we recently parallelized, and released the highly optimized co-phylogenetic analysis program AxParafit (Axelerated Parafit - Stamatakis et al., 2007) that implements an elaborate statistical test of congruence between host and parasite trees (Legendre et al., 2002). AxParafit is a typical stand-alone Linux/Unix command line program. AxParafit has been integrated and can be invoked via a user-friendly graphical interface for co-phylogenetic analyses called CopyCat (Meier-Kolthoff et al., 2007). In this article, we present an enhanced version of this tool suite (henceforth denoted as CopyCat(AxParafit)) for co-phylogenetic analyses, that is packaged into a client tool which makes use of a world-wide Grid environment and thereby allows for large-scale data analysis. In the current version, the underlying Grid middleware is gLite (Laure et al., 2006) that is coupled with an efficient submission and execution model called Run Time Sensitive (RTS) scheduling and execution (Stockinger et al., 2006).

The remainder of this article is organized as follows: initially, we provide a brief introduction to the field of phylogenetic inference, co-phylogenetic analyses, and related software packages in Section 2. Next, we discuss the implementation and architecture of our new approach for efficient adaptation of the CopyCat(AxParafit) tool-suite to a Grid environment. Finally, we provide detailed performance results on the EGEE (Enabling Grids for E-SciencE, http://www.eu-egee.org) Grid infrastructure (where the gLite middleware is deployed in production mode) and demonstrate the performance as well as scalability of our proposed bioinformatics tool.

BACKGROUND

Phylogenetic (evolutionary) trees are used to represent the evolutionary history of a set of s

currently living organisms, roughly comparable to a genealogical tree of species rather than individual organisms. Phylogenetic trees or simply phylogenies are typically unrooted binary trees. The *s* organisms, which are represented by their DNA or AA (Amino Acid/Protein) sequences that are used as input data for the computation, are located at the leave nodes (tips) of the tree while the inner nodes of the topology represent common extinct ancestors. There exist various methods and models to reconstruct such trees which differ in their computational complexity and also in the accuracy of the final results, i.e., there exists a "classic" trade-off between speed and accuracy. As already mentioned in the introduction, phylogenetic analysis has many important applications in medical and biological research. In Figure 1, we provide a simple example for the phylogenetic tree of monkeys.

In the context of this article, however, we will not address stand-alone phylogenetic inference, but consider the problem of co-phylogenetic analysis. Given two phylogenetic trees that represent the evolutionary histories of hosts and their respective parasites, the "classic" example being mammals and lice, and given the extant associations between the former and the latter, we want to determine whether the parasite phylogeny is more similar to the phylogeny of the respective hosts than expected by chance alone. The main interpretation of such a congruence between the trees is that parasites have been associated with respect to their evolutionary history and mostly speciated in parallel (co-speciated) with their hosts (Page, 2002). Given a parasite tree with *n* organisms and a host tree with *m* organisms (sequences), their associations can be represented as a *n* times *m* binary matrix, that contains information of the type: does parasite *x* *(x=1...n)* occur or live on host *y* *(y=1...m)*? In addition to the question of global congruence, one may also be interested in whether individual associations significantly increase the agreement between the phylogenies. Such associations can be interpreted as being caused mainly by co-speciation.

As previously mentioned, recent advances in stand-alone phylogenetic inference methods in combination with the increasing availability of

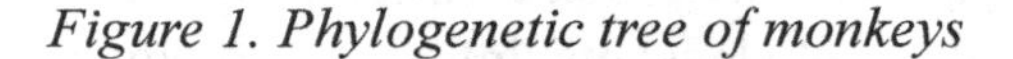

Figure 1. Phylogenetic tree of monkeys

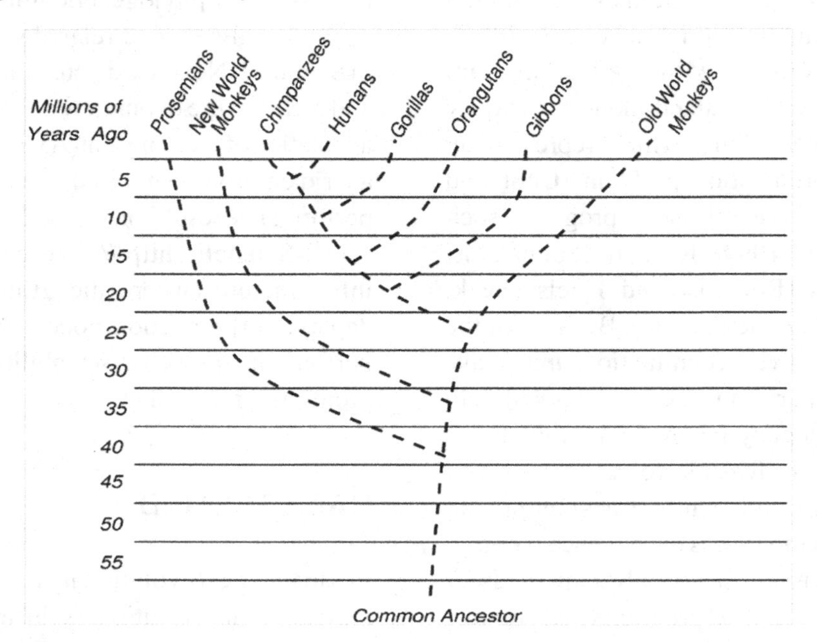

appropriate sequence data, allow for large-scale phylogenetic analyses with several hundred or thousand sequences (Stamatakis, 2006). Thus, large-scale co-phylogenetic studies have, in principle, become feasible. However, most common co-phylogenetic tools or methods such as BPA, Component, TreeMap, TreeFitter (cf. review in Charleston, 2006) or Tarzan (Merkle, 2006) are not able to handle datasets with a large number of organisms or have not been tested in this regard with respect to their statistical properties and scalability. Faster methods based on topological distances between trees, like, e.g., I_{cong} (de Vienne, 2007) are even limited to the analysis of bijective associations only. In this context bijectivity means that each parasite can only be associated to one single host, and vice versa. Therefore, there is a performance and scalability gap between tools for phylogenetic analysis and meta-analysis. The capability to analyze large datasets is important to infer "deep co-phylogenetic" relationships which can otherwise not be assessed (Meier-Kolthoff et al., 2007; Stamatakis et al., 2007). Deep relationships are relationships that determine the extant associations between parasite and host organisms at a high taxonomic level, such as, e.g., families and orders.

Parafit (Legendre, 2002) and the analogous highly optimized AxParafit (Stamatakis et al., 2007) program implement a statistical test to assess hypotheses of global congruence between trees as well as the impact of individual associations. This test is based on the permutation of the entries in the association matrix. The null hypothesis is that the global similarity between the trees, or the respective impact of an individual local association on the similarity, is not larger than expected by pure chance. Extensive simulations have shown that the Parafit test is statistically well-behaved and yields acceptable error rates. The method has been successfully applied in a number of biological studies (Hansen et al., 2003; Ricklefs et al., 2004; Meinilä et al., 2004).

In addition, the type-II statistical error of Parafit decreases with the size of the dataset (see Legendre, 2002), i.e., this approach scales well on large phylogenies of hosts and parasites in terms of accuracy. The AxParafit program is a highly optimized version of Parafit which yields exactly the same results. The sequential version of AxParafit is up to 67 times faster than the original Parafit implementation, while the speedup increases with increasing input size, caused by higher cache efficiency. The speedup of AxParafit has been achieved via low-level optimizations in C, re-design of the algorithm, omission of redundant code, reduction of memory footprint, and integration of highly optimized BLAS (Basic Linear Algebra Subroutines, http://www.netlib.org/blas/) routines.

Earlier work describes these optimizations together with a respective performance study. Moreover, the program was used to conduct the largest co-phylogenetic analysis on real-world data to date. The underlying data were smut fungi and their respective host plants (Stamatakis et al., 2007). Smut fungi are parasitic mushrooms that cause plant diseases. For economically important hosts, such as barley and other cereals, smut fungi can for instance cause considerable yield losses (Thomas and Menzies, 1997).

Workflow of a Co-Phylogenetic Analysis with CopyCat and AxParafit

In this section, we provide an outline of the work-flow for a full co-phylogenetic analysis using CopyCat(AxParafit). The input for a co-phylogenetic analysis with CopyCat(AxParafit) are the host and parasite phylogenies, that might have branch lengths, depending on which method/model was used to calculate the trees. The aforementioned associations are represented as a plain text file containing a list of sequence (organism) name pairs of hosts and parasites, i.e., an adjacency list. This input data representation is henceforth also referred to as list of host-parasite associations.

Initially, these files are parsed and transformed into the appropriate file format by CopyCat. In a first step, a principal coordinate analysis is conducted on the respective tree-based distance matrices induced by the host and parasite trees. This analysis is carried out by the AxPcoords (Axelerated Principal Coordinates) program (Stamatakis et al., 2007), which is an optimized version of the analogous DistPCoA program (Legendre and Anderson, 1998). The output of AxPcoords for the host and parasite trees is then parsed and appropriately prepared for the AxParafit analysis which takes the two principal coordinates matrices and the binary matrix with the associations as input. The output of this computation is a list of probabilities for the individual null hypotheses that a certain association does not improve the fit between host and associate phylogenies. In addition, a probability for the global null hypothesis of the absence of congruence between host and parasite trees is computed. Upon termination of AxParafit the output files are read by the CopyCat tool and presented in a human-readable format. It is important to note that the computations with AxParafit represent the by far largest part (over 95%) of the computational effort required to conduct such a co-phylogenetic analysis. Therefore, the AxPcoords and CopyCat parts of the workflow can be handled sequentially and executed locally. We will, thus, mainly focus on the parallel and gridified versions of AxParafit in the next sections. The basic workflow is outlined in Figure 2 (at the end of the article).

Parallel AxParafit

The most compute-intensive operation (95% of execution time) conducted by AxParafit to compute the statistics is a dense matrix-matrix multiplication of double precision floating point numbers. This is the rationale for integration of highly optimized BLAS routines. In the remainder of this article, we thus always refer to the BLAS-based version of AxParafit.

Initially, the program will compute the statistics for the global congruence of the complete list of host-parasite associations. This part of the computation is significantly less expensive than the individual tests for each host-parasite association, which take *nz* times longer, where *nz* is the number of non-zero entries in the binary association matrix, i.e., number of entries in the original host-parasite association list. For large datasets that require parallel and distributed computing resources as well as a sufficient amount of memory typically $nz >> 1$. The statistics computed during the global test of congruence are required as input data for the individual tests of host-parasite associations, hence there is a sequential dependency: global test $\rightarrow$ *nz* local tests. Thus, in the MPI-based parallel implementation we only parallelized these *nz* local tests which can be computed independently of each other via a straight-forward master-worker scheme. The master simply distributes the *nz* individual host-parasite association tests to the worker processes.

The potential bottleneck induced by the sequential part of the computations can be alleviated by using, e.g., the respective shared-memory implementations of BLAS. With respect to a gridification, this sequential dependency actually has advantages. Since the inference time as well as memory footprint of the global test of congruence are nearly identical (same type of operation, identical matrix sizes, permuted input data) to each of the individual *nz* tests, the information on run-times and memory requirements collected during the global tests can be used for scheduling decisions, as well as to determine an optimal level of granularity and to assess respective resource requirements.

FIT FOR THE GRID

In the following section, we describe how CopyCat(AxParafit) has been adapted and modified for use in a Grid environment. The overall

Figure 2. Detailed work- and dataflow for co-phylogenetic analysis on the Grid

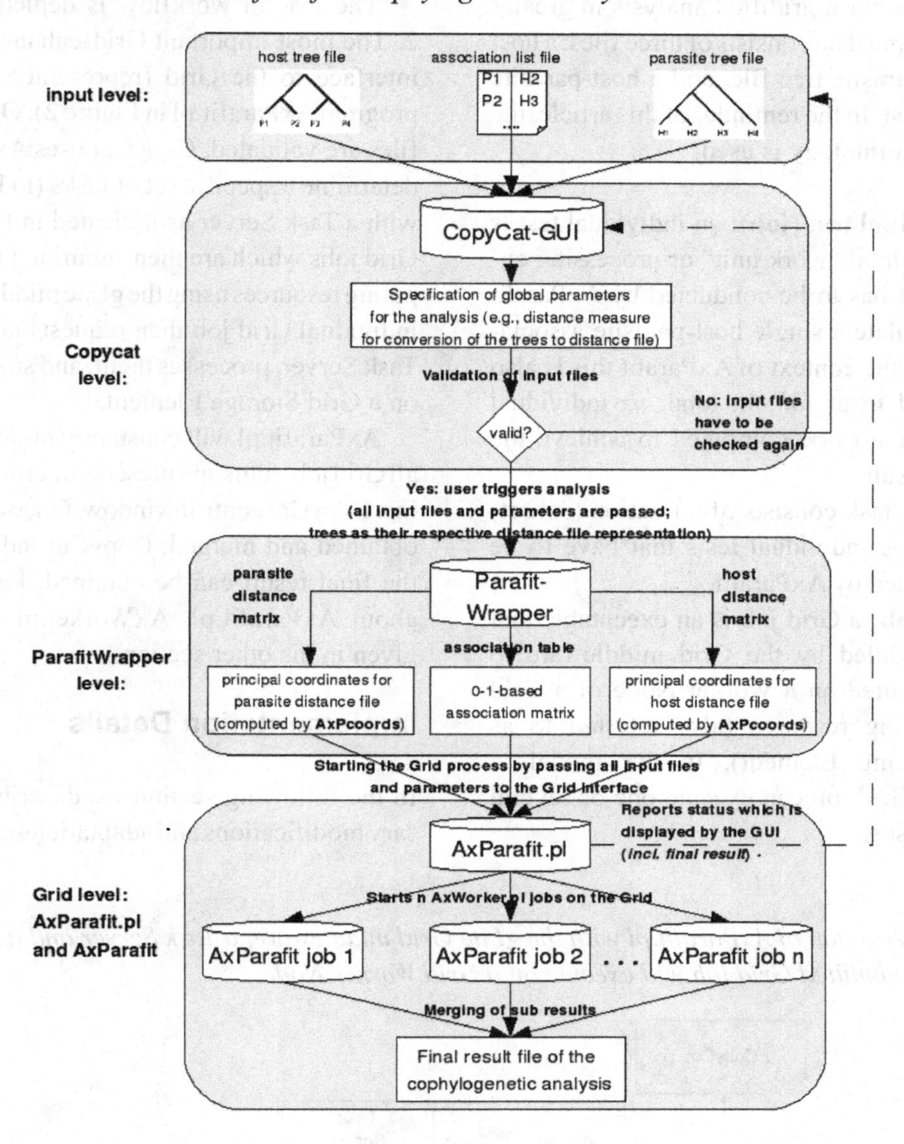

architecture of the client tool will be explained as well as the integration with an existing middleware toolkit.

An important design goal of the Grid-based system for co-phylogenetic analyses was to re-use the current graphical user interface of CopyCat such that the deployment of Grid resources is hidden from the end-user. One fundamental difference between the standard and Grid-enabled versions of CopyCat (AxParafit) is that specific Grid credentials are required (an X.509 user certificate) since Grid jobs can only be submitted by authenticated and authorized users.

Overall Architecture

The basic workflow of a co-phylogenetic study using CopyCat and the AxPcoords/AxParafit programs has already been outlined in the above section. Here, we will describe the architecture and workflow for a gridified analysis in greater detail. The input data consists of three files: a host tree file, a parasite tree file, and a host-parasite association list. In the reminder of this article, the following terminology is used:

- **Individual test (job):** an individual test is the minimal "work unit" or processing entity that has to be conducted by AxParafit to calculate a single host-parasite association. In the context of AxParafit this is also referred to as *job*. In total, *nz* individual tests have to be computed to achieve the final result.
- **Task:** a task consists of a fraction (subset) of the *nz* individual tests that have to be conducted by AxParafit.
- **Grid job:** a Grid job is an executable that is scheduled by the Grid middleware to be executed on a Worker Node of a Grid computing resource (also referred to as Computing Element). In our model, a single Grid job can execute one or several such tasks.

The overall workflow is depicted in Figure 2. The most important Grid-enhancement is the interface to the Grid (represented by the Perl program AxParafit.pl in Figure 2). Once the input files are validated, CopyCat uses AxParafit.pl to determine a specific set of tasks (to be registered with a Task Server as indicated in Figure 3) and Grid jobs which are then submitted to Grid computing resources using the gLite middleware. Each individual Grid job then requests tasks from the Task Server, processes them, and stores the result on a Grid Storage Element.

AxParafit.pl will constantly monitor the overall Grid job status and presents intermediate results in a CopyCat control window. Once all results are obtained and merged, CopyCat indicates where the final result can be obtained. Further details about AxParafit.pl, AxWorker.pl etc. will be given in the other section.

Implementation Details

In the following section we describe the necessary modifications and adaptations of the existing

Figure 3. Interaction of AxParafit.pl with the gLite Grid middleware, a Task Server and a Storage Element. Each submitted Grid job will execute on a Grid Worker Node

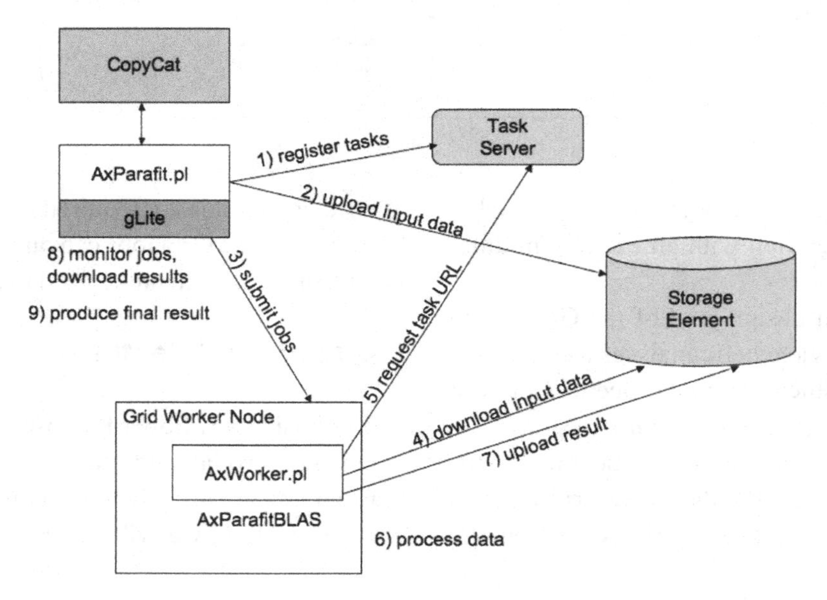

CopyCat and AxParafit tools as well as additional components that were necessary to implement the system outlined in Section 3.1.

CopyCat

Previous versions of CopyCat already provided straight-forward GUI-based functionality for the preparation and analysis of co-phylogenetic datasets. The CopyCat GUI is implemented in Java using the Standard Widget Toolkit (SWT). Upon startup, the user can load the host and parasite trees (represented in the standard Newick tree format: http://evolution.genetics.washington.edu/phylip/newicktree.html), together with a host-parasite association list in a simple plain-text format that contains one host-parasite association per line.

When starting an analysis, the user can now utilize a new Grid interface that connects CopyCat to the gridified program AxParafit. Instead of directly calling the AxParafit executable, the interface invokes a Perl script (AxParafit.pl) which hides the Grid-related parts from the user and CopyCat. By delegating the invocation process to a script, dependencies between the user front-end and the Grid software are minimized. Thus, future modifications like the development of a Web interface for job submissions (see Conclusion) or the usage of a different middleware system are possible.

The AxParafit.pl script entirely manages the execution of AxParafit on the Grid and provides status updates to the standard output stream at the same time. As CopyCat is listening to the output stream of the external programs it invokes, it also receives the status updates generated by the aforementioned Perl-script and writes them to the CopyCat log-message window, thus keeping the user informed about the progress of Grid jobs. Upon termination of the script, the output of the Grid jobs (individual tests of host-parasite associations), as well as the global significance test results, are read by CopyCat. The results can then be displayed and further analyzed via the CopyCat evaluation window.

Within the context of an automated Grid-driven simultaneous analysis of several distinct datasets (and other potential script-based applications, based on CopyCat), the program has been extended by a command-line interface. As a side-effect, this enables CopyCat users to speed-up certain analyses by simply executing a specific command-line call with a defined set of parameters (please refer to the CopyCat manual for detailed information on the command-line options).

Application-Side Modifications of AxParafit

As outlined in other section, the parallel MPI implementation of AxParafit uses a simple master-worker scheme. In order to devise a distributed version of AxParafit we modified the code as follows: initially, we appropriately modified the global test of congruence in AxParafit to write an additional file called "gridData.RUN-ID" where RUN-ID is the output file name appendix for a specific analysis that is passed to AxParafit via a command line parameter (for details see the AxParafit Manual at http://icwww.epfl.ch/~stamatak/). This file contains the necessary data to make scheduling decisions for the distributed computation of the *n* individual tests of host-parasite associations, i.e., the number of jobs *nz,* e.g. Jobs=2000, and the approximate execution time per job in seconds, e.g., ComputeTime=10 . This data can then be used to determine the level of granularity for individual Grid tasks since in the current example the scheduling overhead induced by distributing 2,000 jobs of 10 seconds each, along with the comparatively large input datasets on the Grid, would be immense. We have, thus, extended the implementation of the individual host-parasite association tests in AxParafit by two additional command line parameters -l (lower limit) and -u (upper limit). These parameters allow for computation of several host-parasite associations in one single program run. The lower and upper limits just refer to the order of the *nz* non-zero entries in the binary association matrix. Thus, in the present example, we can schedule larger, in terms of execution times, Grid jobs by only distributing two Grid jobs with -l 0 -u 1000 and -l 1000 -u 2000 that would require approximately 10,000 seconds of execution time each, i.e., Grid job 0 would compute statistics for the first 1,000 host-parasite associations and Grid job 1 for the remaining 1,000 associations. The result files of these distributed Grid jobs only need to be recovered and concatenated in the order of the associations they computed, and the respective result file can then be read and visualized by CopyCat.

Grid-Side Adaptation

Parafit.pl provides the actual link between CopyCat and the gridified version of AxParafit. First, it reads the file "gridData.RUN-ID" to determine the number of tasks to be created (registered) for execution on the Grid (Step 1 in Figure 3). As outlined in Section 3.2.1, the basic idea consists of combining appropriate fractions (subsets) of the *nz* individual tests into a single *task*, i.e., a set of individual tests $k < nz$ are executed by a Grid job. In order to make efficient use of the Grid and to reduce scheduling overhead, a task contains a minimum of k individual tests, such that the respective job requires at least 30 seconds on an average CPU. After the number of tasks has been determined, a certain number of Grid jobs (approximately nz/k) needs to be submitted (Step 3 in Figure 3) which then ask for tasks to be executed, i.e., issue work requests. An individual Grid job can request and execute several tasks, as long as the Task Server can provide more work (Steps 5 and 6 in Figure 3). The protocol used for the Task Server is HTTP which allows for fast communication between the client and the server. For additional background and fault tolerance features of this processing model with a Task Server please refer to Stockinger et al. (2006).

Before Grid jobs can be submitted, AxParafit.pl creates the Grid job specification, i.e. the job description file to decide which files (data and/or executables) to send to Grid computing resources. A typical job description file looks as follows:(See Box 1).

The wrapper code (identified as "Executable" in the JDL file above) is AxWorker.pl using the command line arguments specified by "Arguments". Once AxParafit.pl is running on a Grid Worker Node, it is responsible for requesting tasks from a Task Server and executing AxParafitBLAS. The two programs (AxWorker.pl and AxParafit-BLAS) are transferred to the Grid Worker Node as specified in the InputSandbox in the example above, i.e. gLite provides the means to transfer data from the client machine to the actual computing resource.

In parallel to the execution of Grid jobs, the script AxParafit.pl monitors the status and is responsible for providing and assembling the final result (Steps 8 and 9 in Figure 3).

In particular, when tasks have been processed successfully, they are downloaded from the Storage Element and transferred to the client. Note that an alternative implementation option is to transfer the output of individual tasks via the gLite middleware (using the OutputSandbox). However, because of performance and reliability considerations, it has turned out to be more efficient to store files at an external Storage Element and retrieve them from there: one reason is that the actual job output can only be retrieved if gLite indicates that a job has been finished. However, because of update latencies in the Grid-wide information and motoring system, jobs might have finished already several dozens of seconds or even a few minutes ago while the job status is still indicated as pending or running.

As a final remark: since the gLite services can only be accessed by authorized users, the execution of the AxParafit.pl script requires the usage of a valid X.509 proxy certificate.

Box 1.

```
Executable = "AxWorker.pl";
Arguments = "-j ax-May1319-41-28 -p 100 -1 2048 -2 2048 -3 2025 -4 2031 \
-A gsiftp://example.org/dpm/home/biomed/heinz/selection_2048.mat-ax-
May1319-41-28 \
-B gsiftp://example.org/dpm/home/biomed/heinz/selection_2048_P.pco-ax-
May1319-41-28 \
-C gsiftp://example.org/dpm/home/biomed/heinz/selection_2048_H.tra-ax-
May1319-41-28 -i 1";
Stdoutput = "output.txt";
InputSandbox = {"/home/stockinger/AxWorker.pl", "/home/stockinger/AxParafitB-
LAS"};
OutputSandbox = {"output.txt"}
```

EXPERIMENTAL RESULTS

The main goal of the gridified version of CopyCat(AxParafit) is to accelerate and facilitate large-scale analyses. We present two experiments with large computational demands and study their performance on the Grid. The performance improvement is outlined with respect to running the application sequentially on a single machine. Moreover, we conduct a performance comparison between a dedicated compute cluster and the Grid.

Test Environment

The Grid platform that is supported by our application is gLite 3. Tests are conducted using gLite on the EGEE production infrastructure. In particular, we use the Virtual Organization (VO) that is dedicated to biomedical applications: "biomed". Members of this VO have access to about 50 Computing Elements (acting as front-ends to computing clusters), each having between 2 and a few hundred processing cores. The exact number of processing cores available to a single user at a given time cannot be easily obtained since it depends on the current system load as well as the general availability of a Computing Element at a certain point in time. Currently, gLite does not support resource reservation nor job priorities, which means that experimental results can not be fully reproduced. However, once one is correctly registered with the Virtual Organization, one can use it any time of the day.

On the client side, we used gLite on GNU/Linux on an AMD Opteron machine (2 GB RAM, 2.2 GHz CPU) located in Lyon, France – previous tests (in particular with the installation of CopyCat and the Grid interface have been conducted on a machine located in Lausanne, Switzerland). The gLite components used are the workload management system (for job submission and status monitoring) as well as data management clients for file transfer. Additionally, we deployed and used a Task Server that is located in Lausanne, using resources provided by the Vital-IT group of the Swiss Institute of Bioinformatics. In the second experiment, we used a dedicated compute cluster with 128 CPUs. In contrast to the Grid, the cluster had to be reserved in advance.

Experiment with Real-world Data

In the first experiment we are interested in the raw performance (response time) of AxParafit.pl, i.e., how long does it take to fully process a set of tasks on the Grid. In this experiment, we do not include CopyCat but directly invoke AxParafit.pl as shown in Box 2.

The parameters -1, -2, -3 and -4 specify the number of rows and columns in the association matrix as well as the number of rows and columns in the parasite and host matrices; -p represents the number of permutations conducted by the statistical test; -A, -B, and -C are used to read the plain-text input files; -n specifies a run ID that is appended to all output files (for details on the AxParafit program parameters please refer to the AxParafit manual at http://icwww.epfl.ch/~stamatak/). The dataset we used is the aforementioned (Section 2) dataset for the study of smut-fungi, that was used to demonstrate performance of the stand-alone AxParafit code by Stamatakis et al. (2007). As already mentioned, this dataset represents the largest *real-world* co-phylogenetic study conducted to date. While the sequential execution time for this dataset still appears to be acceptable, such studies were previously not feasible with Parafit which is between 1-2 orders of magnitude slower than AxParafit. Since the host-parasite association list contains *nz*=2,362 entries, 2,362 individual tests need to be performed. The execution of AxParafit to compute global congruence of the trees returned an estimated run time of 3 seconds per job, i.e., an overall expected run time of almost two hours (2,362 x 3 seconds). The main goal of the first test is therefore to minimize the expected response time. We also executed the full test, as specified above, on a single machine and observed that the estimated run time of about 2 hours (7,000 seconds) is almost identical to the measured run time (7,200 seconds). Therefore, we deduce that the run time prediction mechanism is sufficiently accurate for our application. In our experiments, we varied the number of tasks (in the range between 60 and 162) as well as the number of parallel Grid jobs (in the range between 24 to 124) to experimentally determine the minimal response time. However, because of varying response times of the Grid (i.e. the various Computing Elements and their job queues etc.) it was not possible to determine an optimal number of Grid jobs and tasks. Finally, in the experiment we used 124 Grid jobs and 150 tasks which have been proposed by the work distribution algorithm outlined in aforementioned section. The overall response time to produce the final output was 11 minutes and 15 seconds (cf. Figure 4). Consequently, we observe a clear runtime improvement with respect to a single, sequential run. Note that the AxParafit.pl program had to be adapted to allow for parallel downloads of the individual results: originally, results were downloaded sequentially, which increased the overall response time by several minutes. By overlapping communication with computation, this problem was resolved.

Experiment with Synthetic Data

In another experiment, we used a larger (*synthetic*) test dataset that had been extracted from a larger empirical dataset to test scalability of AxParafit and compared the runtime of the Grid with the infiniband cluster at the Technical University of Munich equipped with 128 AMD Opteron 2.4 GHz CPUs. In the association list, there were *nz*=2,048 non-zero entries (equivalent to 2,048 tasks) and we used 100 permutations. The expected runtime of a single task was 568 seconds, i.e., about 10 minutes. As a result, the expected sequential response time to finish all 2,048 tasks is about 13.4 days. We used the wrapper as shown in Box 3.

Box 2.

```
AxParafit.pl -p 10 -1 413 -2 1400 -3 1390 -4 411 \
      -A smuts010907.mat -B smuts010907_P.pco -C smuts010907_H.tra -n RUN_1
```

Figure 4. Comparison of smut-fungi dataset on a single CPU and on a Grid using 124 Grid jobs and 150 tasks. Note that there is a rather high redundancy in Grid jobs and not all 124 jobs really participate in the overall calculation because of start-up latencies. In fact, a few Grid jobs (AxWorker.pl) started, requested tasks and found out that there were no more tasks available and gracefully finished

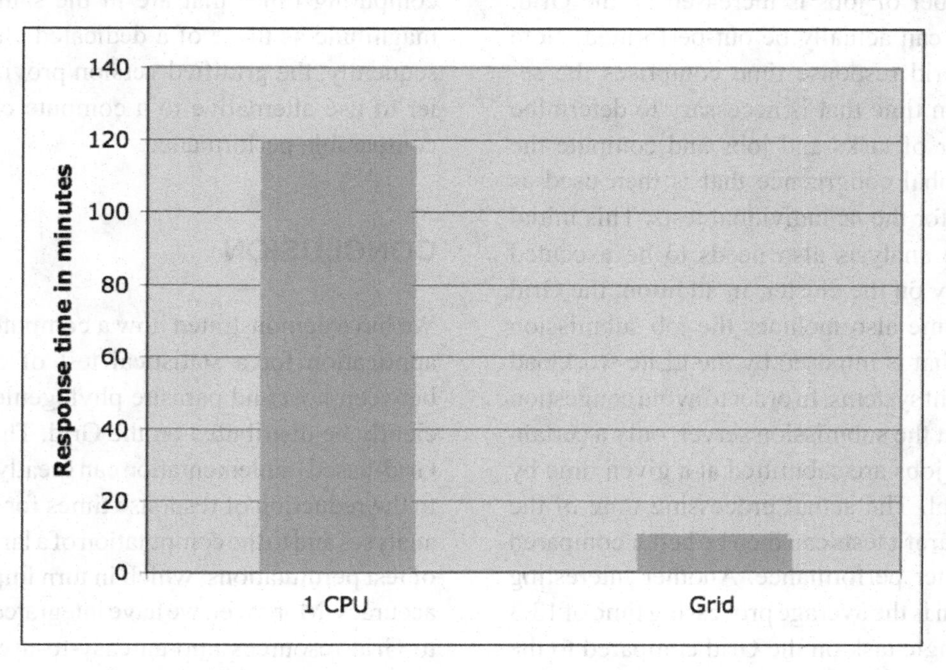

Note that the input files are bigger than in the previous experiment: they cannot be directly submitted with the Grid job but they are uploaded to a Storage Element and then dynamically downloaded by Grid jobs when needed.

A direct performance comparison between the cluster and a Grid is not feasible since the cluster we used had several favorable features that a multi-institutional Grid does not have: a shared file system between all processing nodes which minimizes the data transfer time; homogeneous hardware infrastructure; pre-defined number of CPUs that are available which does not require an automatic task assignment, no overhead for job submission etc. However, the cluster needed to be reserved in advance (larger slots can only be obtained overnight) which means that it was only available at a specific time, whereas Grid resources are available on demand at any time. Intuitively, one expects a cluster to provide a better response time to a large size application but it has a considerable "reservation latency", a fact that should not be underestimated.

The final performance results of the experiments are depicted in Figure 5. For the Grid execution, we used between 90 and 175 parallel jobs

Box 3.

```
AxParafit.pl -p 100 -1 2048 -2 2048 -3 2025 -4 2031 -A selection_2048.mat \
 -B selection_2048_P.pco -C selection_2048_H.tr a -n RUN_2
```

(the number varied during the overall execution time). Given the number of parallel jobs used in the Grid, the cluster performed better. However, if the number of jobs is increased on the Grid, the cluster can actually be out-performed. Note that, the Grid response time comprises the sequential run time that is necessary to determine the number of tasks and jobs and compute the test for global congruence that is then used as input data for the *nz* individual tests. This initial part of the analysis also needs to be executed sequentially on the cluster. In addition, the Grid response time also includes the job submission overhead that is imposed by the gLite workload management systems. In order to avoid congestion problems at the submission server, only a certain number of jobs are submitted at a given time by AxParafit.pl. The actual processing time of the 2,048 AxParafit tests can then be better compared to the cluster performance. Another interesting observation is the average processing time of 13.3 min per single task on the Grid compared to the local execution time of 11 min on the Grid client machines. This indicates that distinct Computing Elements have CPUs with rather different CPU speeds and latencies.

Overall, our Grid-based approach requires computing times that are in the same order of magnitude as those of a dedicated cluster. Consequently, the gridified version provides an easier to use alternative to a compute cluster with comparable performance.

CONCLUSION

We have demonstrated how a compute-intensive application for a statistical test of congruence between host and parasite phylogenies can efficiently be distributed on the Grid. The proposed Grid-based implementation can greatly contribute to the reduction of response times for large-scale analyses and to the computation of a larger number of test permutations, which in turn improve upon accuracy. Moreover, we have integrated the access to Grid resources into an easy-to-use Graphical User Interface (CopyCat) which entirely hides the technical details related to the exploitation of Grid

Figure 5. Performance comparison of a 128 CPU cluster with a Grid using between 90 and 175 parallel jobs. Note that the number of Grid job varied and was never constant

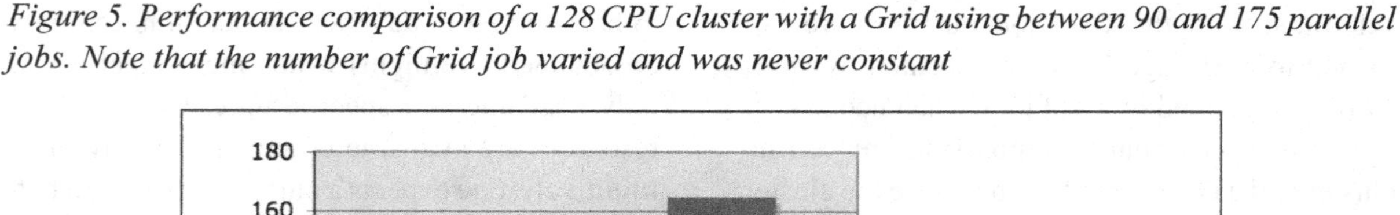

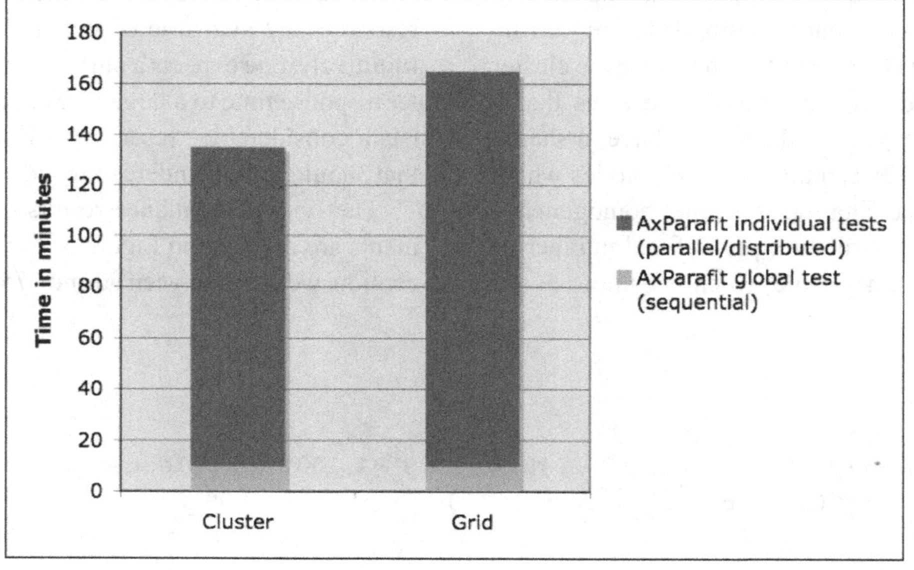

resources from the user. Note that in particular for non-expert users, easy accessibility and usability of HPC resources represents a major criterion for the selection of software and systems. We thus believe that the proposed architecture will greatly facilitate access to HPC resources for real-world biological studies on host-parasite evolution. Nonetheless, the requirement to obtain access and accreditation to use Grid resources (valid X.509 proxy certificate) will possibly hinder a large amount of potential users to exploit these new possibilities offered by the Grid. Based on previous experience with the development of the freely accessible RAxML Web servers for phylogenetic reconstruction (Stamatakis et al., 2008, over 8,000 job submissions in the first 8 months of operation) that are however scheduling jobs to dedicated clusters instead of the Grid, we believe that a freely accessible Web server for this Grid-enabled system for co-phylogenetic analyses can contribute to the generation of biological insights, by further simplifying the access to HPC resources. Thus, future work will concentrate on the development of such a Web server, as well as the integration with the aforementioned RAxML servers such as to provide a comprehensive phylogenetic and co-phylogenetic analysis pipeline.

ACKNOWLEDGMENT

This work was funded in part by the EU project EMBRACE Grid which is funded by the European Commission within its FP6 Program, under the thematic area "Life sciences, genomics and biotechnology for health", contract number LUNG-CT-2004-512092. The Exelixis lab (AS) is funded under the auspices of the Emmy-Noether program by the German Science Foundation (DFG).

REFERENCES

Altschul, S. F., Madden, T. L., & Schaffer, A. A. (1997). Gapped BLAST and PSI-BLAST: a new generation of protein database search programs. *Nucleic Acids Research, 25*(17), 3389–3402. doi:10.1093/nar/25.17.3389

Charleston, M. A., & Perkins, L. (2006). Traversing the tangle: Algorithms and applications for co-phylogenetic studies. *Journal of Biomedical Informatics, 39*, 62–71. doi:10.1016/j.jbi.2005.08.006

Chor, B., & Tuller, T. (2005). Maximum likelihood of evolutionary trees: hardness and approximation. *Bioinformatics (Oxford, England), 21*(1), 97–106. doi:10.1093/bioinformatics/bti1027

de Vienne, D. M., Giraud, T., & Martin, O. C. (2007). A congruence index for testing topological similarity between trees. *Bioinformatics (Oxford, England), 23*(23), 3119–3124. doi:10.1093/bioinformatics/btm500

Goloboff, P. (1999). Analyzing Large Data Sets in Reasonable Times: Solutions for Composite Optima. *Cladistics, 15*(4), 415–428. doi:10.1111/j.1096-0031.1999.tb00278.x

Gottschling, M., Stamatakis, A., & Nindl, I. (2007). Multiple Evolutionary Mechanisms Drive Papillomavirus Diversification. *Molecular Biology and Evolution, 24*(5), 1242–1258. doi:10.1093/molbev/msm039

Hansen, H., Bachmann, L., & Bakke, T. A. (2003). Mitochondrial DNA variation of *Gyrodactylus* spp. *Monogenea, Gyrodactylidae* populations infecting Atlantic salmon, grayling, and rainbow trout in Norway and Sweden. *International Journal for Parasitology, 33*(13), 1471–1478. doi:10.1016/S0020-7519(03)00200-5

Heinicke, M. P., Duellman, W. E., & Hedges, S. B. (2007). From the Cover: Major Caribbean and Central American frog faunas originated by ancient oceanic dispersal. *Proceedings of the National Academy of Sci*

Laure, E., Fisher, S., & Frohner, A. (2006). Programming the Grid with gLite. *Computational Methods in Science and Technology*, *12*(1), 33–45.

Legendre, P., & Anderson, M. J. (1998). DistPCOA program description, source code, executables, and documentation: http://www.bio.umontreal.ca/Casgrain/en/labo/distpcoa.html

Legendre, P., Desdevises, Y., & Bazin, E. (2002). A Statistical Test for Host-Parasite Coevolution. *Systematic Biology*, *51*(2), 217–234. doi:10.1080/10635150252899734

Meier-Kolthoff, J. P., Auch, A. F., Huson, D. H., & Göker, M. (2007). COPYCAT: Co-phylogenetic Analysis tool. *Bioinformatics (Oxford, England)*, *23*(7), 898–900. doi:10.1093/bioinformatics/btm027

Meinilä, M., Kuusela, J., Zietara, M. S., & Lumme, J. (2004). Initial steps of speciation by geographic isolation and host switch in salmonid pathogen *Gyrodactylus salaris (Monogenea: Gyrodactylidae). International Journal for Parasitology*, *34*(4), 515–526. doi:10.1016/j.ijpara.2003.12.002

Merkle, D., & Middendorf, M. (2005). Reconstruction of the cophylogenetic history of related phylogenetic trees with divergence timing information. *Theory in Biosciences*, *123*(4), 277–299. doi:10.1016/j.thbio.2005.01.003

Ricklefs, R. E., Fallon, S. M., & Birmingham, E. (2004). Evolutionary relationships, cospeciation, and host switching in avian malaria parasites. *Systematic Biology*, *53*(1), 111–119. doi:10.1080/10635150490264987

Ronquist, F., & Huelsenbeck, J. (2003). MrBayes 3: Bayesian phylogenetic inference under mixed models. *Bioinformatics (Oxford, England)*, *19*(12), 1572–1574. doi:10.1093/bioinformatics/btg180

Salzberg, S. L., Kingsford, C., & Cattoli, G. (2007). Genome analysis linking recent European and African influenza (H5N1) viruses. *Emerging Infectious Diseases*, *13*(5), 713–718.

Stamatakis, A. (2006). RAxML-VI-HPC: maximum likelihood-based phylogenetic analyses with thousands of taxa and mixed models. *Bioinformatics (Oxford, England)*, *22*(21), 2688–2690. doi:10.1093/bioinformatics/btl446

Stamatakis, A., Auch, A. F., Meier-Kolthoff, J., & Göker, M. (2007). AxPcoords & parallel AxParafit: statistical co-phylogenetic analyses on thousands of taxa. *BMC Bioinformatics*, *8*, 405. doi:10.1186/1471-2105-8-405

Stamatakis, A., Hoover, P., & Rougemont, J. (2008). (in press). A Rapid Bootstrapping Algorithm for the RAxML Web Servers. *Systematic Biology*. doi:10.1080/10635150802429642

Stockinger, H., Pagni, M., Cerutti, L., & Falquet, L. (2006). Grid Approach to Embarrassingly Parallel CPU-Intensive Bioinformatics Problems. *2nd IEEE International Conference on e-Science and Grid Computing (e-Science 2006)*, IEEE Computer Society Press, Amsterdam, The Netherlands.

Thomas, P. L., & Menzies, J. G. (1997). Cereal smuts in Manitoba and Saskatchewan, 1989-95. *Canadian Journal of Plant Pathology*, *19*(2), 161–165. doi:10.1080/07060669709500546

Thompson, J. D., Higgins, D. G., & Gibson, T. J. (1994). CLUSTAL W: improving the sensitivity of progressive multiple sequence alignment through sequence weighting, position-specific gap penalties and weight matrix choice. *Nucleic Acids Research*, *22*(22), 4673–4680. doi:10.1093/nar/22.22.4673

Zwickl, D. (2006). Genetic algorithm approaches for the phylogenetic analysis of large biological sequence datasets under the maximum likelihood criterion. *PhD Thesis*, The University of Texas at Austin.

This work was previously published in International Journal of Grid and High Performance Computing (IJGHPC), Volume 1, Issue 1, edited by Emmanuel Udoh & Ching-Hsien Hsu, pp. 39-54, copyright 2009 by IGI Publishing (an imprint of IGI Global).

Chapter 15
Persistence and Communication State Transfer in an Asynchronous Pipe Mechanism

Philip Chan
Monash University, Australia

David Abramson
Monash University, Australia

ABSTRACT

Wide-area distributed systems offer new opportunities for executing large-scale scientific applications. On these systems, communication mechanisms have to deal with dynamic resource availability and the potential for resource and network failures. Connectivity losses can affect the execution of workflow applications, which require reliable data transport between components. We present the design and implementation of π-channels, an asynchronous and fault-tolerant pipe mechanism suitable for coupling workflow components. Fault-tolerant communication is made possible by persistence, through adaptive caching of pipe segments while providing direct data streaming. We present the distributed algorithm for implementing: (a) caching of pipe data segments; (b) asynchronous read operation; and (c) communication state transfer to handle dynamic process joins and leaves.

INTRODUCTION

Heterogeneous distributed systems are the emergent infrastructures for scientific computing. From peer-to-peer, volunteer computing systems to the more structured ensembles of scientific instruments, data repositories, clusters and supercomputers such as computational grids (Foster and Kesselman, 1999), these systems are heterogeneous and dynamic in availability. Furthermore, the wide-area links that interconnect these resources are prone to transient or permanent failures. These dynamic characteristics introduce

DOI: 10.4018/978-1-60960-603-9.ch015

unique challenges for executing large-scale scientific applications.

This research is motivated by the need to support fault-tolerant communication within scientific workflows. A workflow consists of multiple processing stages, where intermediate data generated in one stage are processed in subsequent stages. A workflow component can be a device or an application, which is often modified to enable communication. Thus, a scientific workflow is a computational/data-processing pipeline; with data being captured, processed and manipulated as it pass through various stages (Figure 1). Currently, the data transfers between component applications are realised by: (a) file transfers (e.g. GridFTP); (b) remote procedure calls (e.g. RPC-V, GridRPC, OmniRPC); and (c) custom mechanisms (e.g. Web Services).

For coupling workflow components, we propose the π-channel, an asynchronous and persistent pipe mechanism. It is part of the π-Spaces/π-channels programming model which features:

1. Simplified application coupling using string channel names through π-Spaces. A π-Space is a name space for π-channels, enabling dynamic binding of channel endpoints between processes.
2. π-channel data are adaptively cached to achieve persistence. This allows π-channels to be created and written to, even in the absence of the reader. Persistence also makes π-channels accessible even after the writer has terminated.
3. Asynchronous receives are made possible through a communication thread; thus, an application is able to accept pipe segments even when it is busy in computation.

Figure 1. A simple four-stage workflow application. Arrows indicate data flow between component applications. Application B is an n-process parallel application.

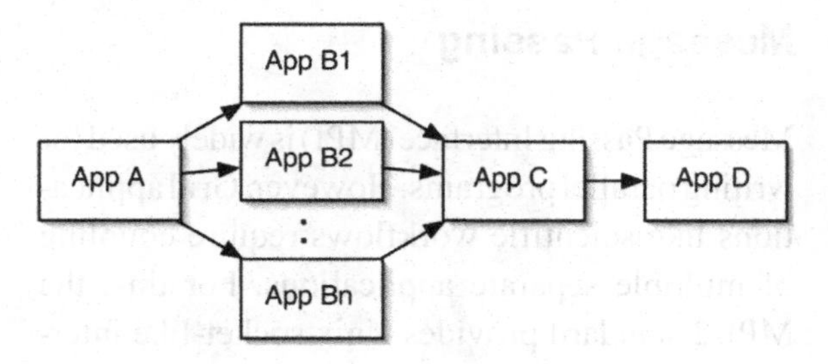

This article focuses on how π-channel persistence relates to fault-tolerant communication in scientific workflows. The extended API and semantics for π-Space/π-channels are presented. We describe the design and implementation of π-channels, including the server that implements this model along with the underlying distributed algorithm.

This article is organised as follows: We review related work in the next Section § 2. Then, we present the π-Spaces/π-channels programming model in § 3, including its application programming interface, semantics, and how fault-tolerance is achieved for workflows. In § 4, we discuss in detail its design and implementation, describing the distributed algorithm. Experimental results are presented in § 5, followed by the conclusions.

RELATED WORK

We briefly review the major models for communication on distributed environments highlighting their differences from π-Spaces/π-channels.

Pipe/Channel Models

The pipe/channel is a well-known IPC mechanism and appears in many forms: Unix pipes, named pipes, and TCP sockets (Stevens, 1998). Sockets with TCP, while used in network programming, are too low-level for scientific application programming. In particular, since communication endpoints are identified using IP/host addresses and port numbers, it is tedious to use in a dynamic, failure-prone environment. In the event of a link failure, TCP primitives will generate a "broken pipe" exception, which require explicit handling.

The channel abstraction has its early beginnings in Kahn's Process Networks (Kahn, 1974) and Communicating Sequential Processes (Hoare, 1985). This abstraction is the basis of current process calculi. Many coordination languages use the channel model, e.g. MoCha (Guillen-Scholten and Arbab, 2005), POLYLITH (Purtilo, 1994), Programmer's Playground (Goldman et al., 1995), Conic (Magee et al., 1989), and Netfiles (Chan and Abramson, 2001). Channels are provided in Grid programming environments, e.g. Ibis (van Nieuwpoort et al., 2005) and Vishwa (Reddy et al., 2006). Vishwa applications communicate through a pipe mechanism called DP (Johnson and Ram, 2001). However, these systems do not support fault-tolerance in the communication.

Communication persistence was previously explored by at least the following: (a) persistent pipes for transactions (Hsu and Silberschatz, 1991); (b) persistent connections over TCP (Zhang and Dao, 1995); and more recently, (c) the persistent streaming protocol (Hua et al., 2004) and (d) NapletSocket (Zhong and Xu, 2004). Unlike π-Spaces/π-channels, these lack a logical name space for communication endpoint coupling. For example, IP addresses and ports are necessary for configuring endpoints in (Hsu and Silberschatz, 1991) and (Zhang and Dao, 1995). Moreover, π-Spaces/π-channels support asynchronous communication.

Message-oriented middleware (MoM), e.g. IBM's Websphere MQ (IBM Websphere MQ, 2008) and the Microsoft Message Queueing System (MSMQ) (Microsoft, 2008), present a suite of asynchronous communication services suitable for general transaction processing. The message transfer times are in the order of minutes instead of seconds or milliseconds (Tanenbaum and Steen, 2007), reducing their applicability for high-performance scientific applications. Furthermore, queue management requires tedious setup and configuration, while the π-Spaces/π-channels model is designed for efficient pipe creation/retrieval, including dynamic binding of channel endpoints.

π-Channels in Context with Netfiles and GriddLeS

Our earlier work on Netfiles (Chan and Abramson, 2001, 2008) investigated file I/O as metaphor for interprocess communication. The idea was inspired from the Nimrod project (Abramson et al., 1995, 1997). Nimrod is a middleware for the executing large-scale parametric models (or sweeps) over distributed systems. These parametric sweep applications can be built without modifying the existing programs. For many such applications, the component programs communicate by through data files, with file transfers performed by the runtime.

In GriddLeS (Abramson and Kommineni, 2004), the file I/O metaphor is extended for wide-area environments like Grids, implemented over Web Services. The GriddLeS runtime provides an I/O multiplexer, which transparently performs file transfers and buffered remote I/O operations to couple applications that read/write files. This enables Grid workflows to be composed without rewriting any program code, a feature useful when existing legacy codes are executed over computational grids.

The π-channel abstraction extends Netfiles and GriddLeS with persistence and efficient asynchronous operations. Furthermore, GriddLeS offer static associations of names to process locations, while π-Spaces/π-channels provide dynamic π-channels binding.

Message Passing

Message Passing Interface (MPI) is widely used for writing parallel programs. However, Grid applications like scientific workflows require coupling of multiple separate applications. For this, the MPI-2 standard provides Unix socket-like interface for accepting and establishing connections between two MPI applications, enabling communication with MPI_Send() and MPI_Recv(). Although fault-tolerance may be incorporated into this mechanism, current projects (Fagg and Dongarra, 2004; Batchu et al., 2004; Bouteiller et al., 2006; Gropp and Lusk, 2004) are focused on fault-tolerant IPC within an application.

In the MPI model, each process is identified by an integer rank. Elegant and simple, this model works very well on SPMD applications where the number of processes is known and fixed. Phoenix (Taura et al., 2003), for example, modifies the process naming scheme so that processes may join and leave the computation without the need to re-assign ranks. When coupling multiple applications in a workflow, it is useful to have a user-intuitive convention to identify communication endpoints (Chan and Abramson, 2007, p. 6).

Generative Communication Models

Linda (Carriero and Gelernter, 1989) is a generative communication model that features decoupled communication. In Linda, processes communicate by posting and retrieving ordered sequences of values called tuples onto a logical shared space called tuplespace. Its elegance has inspired many systems such as Sun's Javaspaces (Freeman et al., 1999), and IBM's T-spaces (Wyckoff et al., 1998). The Linda tuplespace model encourages decoupled communication along two dimensions. First, tuples are posted and retrieved anonymously from tuplespace, achieving space decoupling. Second, since tuples are persistent, temporal decoupling is possible, allowing non-concurrent processes to communicate.

Workflows require efficient data transfers, which is challenging to achieve in Linda. Extensions have thus been proposed, e.g. WCL (Rowstron, 1998) supports bulk transport of tuples. Taskspaces (Sterck et al., 2003, 2005) provide direct communication using the tuples to identify communication endpoints (IP/port) and TCP to connect processes. However, it lacks support for communication fault-tolerance.

Π-CHANNELS: A PERSISTENT PIPE MECHANISM

A π-channel is an enhanced unidirectional (MRSW) pipe that has a unique user-specified string name and a FIFO sequence of arbitrary-length typed data segments, each treated as indivisible units. Fault-tolerant communication is achieved by persistence, enabling π-channels to be created and written to at any time, independent of the sink/reader, thus encouraging temporal decoupling.

During π-channel creation, if the matching reader is known, a direct connection (if possible) is used to efficiently transfer pipe segments. Due to persistence (Chan and Abramson, 2007), delivery of the π-channel to the π-Space continues (Figure 2a). This enables π-channel writes to proceed even if the link and/or reader have failed during communication. At this point, the pipe segments will be written to the π-Space (Figure 2b). When the reader resumes, a communication state transfer re-establishes the connections between the reader, the π-Space, and the writer.

π-Channels: Programming Model and Semantics

Table 1 summarises the API for π-channels. This is inspired from file-based I/O and connection-oriented socket communication. The key difference is that a thread is employed to receive data asynchronously. The read/write operations resemble the standard Unix I/O operations, with specifications of the segment data type, for heterogeneous communication.

Figure 2. Dual π-channel behaviours: (a) when reader-to-writer link is available; and (b) during a link failure, writer continues transmitting data to the space, delivery to the reader resumes upon link restoration.

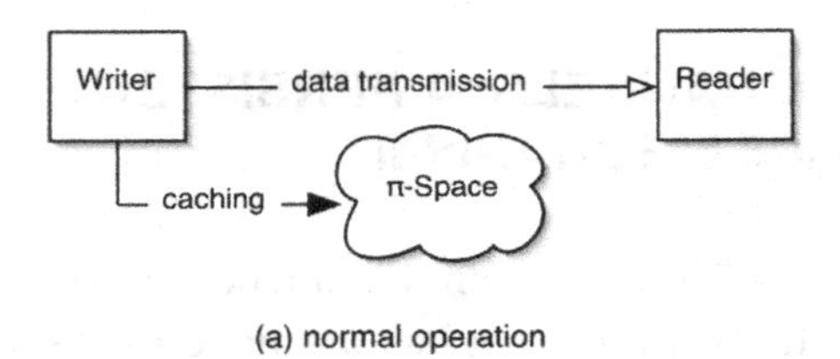

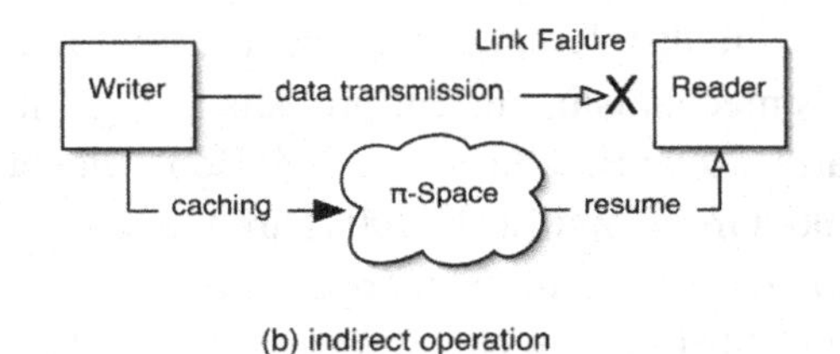

THE Π-CHANNEL API

The π-channels programming interface consists of the following:

1. pi_attach(str n, int s): attaches to a π-channel with name n on space s, does not block even if π-channel is absent. It initiates asynchronous receive, allowing segments to be buffered locally even before the pi_read() is issued. This primitive returns a descriptor representing the "read" end.
2. int pi_create(str n, int s, int mode): creates a π-channel with name n on space s. The mode argument is used to specify if this is a CREATE – for new π-channels; or APPEND – to resume writing. On success, it returns a descriptor representing the "write" end of the created π-channel.
3. int pi_read(int d, ptr b, int len, dtype_t[1]t): reads a segment of len elements of type t into buffer b from the descriptor d, blocks if no segments are available. It returns the number of elements successfully read or -1 if end of the pipe is reached.
4. int pi_write(int d, ptr b, int len, dtype_t t): writes a segment of len elements of type t from buffer b into the descriptor d.
5. int pi_close(int d), int pi_detach(int d): closes the "write" and "read" end of the π-channel d, respectively.
6. int pi_seek(int d, int seg_id): moves the logical pipe pointer for d to start reading from segment seg_id, with 0 as the first segment. This only works at the read end. On success, it returns 0.
7. pi_tell(int d): returns the segment ID of the logical pipe pointer of the π-channel.

Table 1. A brief overview of the key π-channels primitives

π-channel Primitive	Brief Description
pi_create()	Creates a new π-channel.
pi_attach()	Attach and retrieve a given π-channel.
pi_write()	Writes a pipe segment.
pi_read()	Reads a pipe segment.
pi_seek()	Seek to a new read segment position.
pi_tell()	Return the segment ID of upcoming segment.
pi_close()	Closes a π-channel that is opened for writing.
pi_detach()	Detach from reading a π-channel.
pi_unlink()	Marks a closed π-channel for deletion.

8. pi_unlink(str n): marks the π-channel n for deletion, returns -1 for open π-channels.

With persistence, a π-channel behaves as both: (a) an archival file – writes can proceed without readers; and (b) an online pipe – when the reader/s and the writer are concurrent. This duality makes pi_seek() possible, when such an operation would be meaningless on conventional pipes. All π-channels are immutable, so pi_seek() is disabled at the "write" end.

SEMANTICS OF Π-CHANNEL OPERATIONS

Figure 3 shows a state-transition diagram for π-channels. The top three states show the life-cycle of a π-channel as it is created, written-to, closed, and deleted. The remaining states show the transitions when there is a concurrent reader. In particular, the Read/Write applies when the π-channel has one active writer and at least one active reader. At the Reading state, the π-channel behaves like a stored file object, with data segments retrieved from π-Space.

This simple semantics facilitates a straightforward failure recovery mechanism. The key is to enable the application to resume π-channel operations upon recovery, without affecting its correctness. During writer recovery, pi_create() and pi_write() operations are "redone." If segments are already cached, the sequence of pi_write() operations are replayed, without changing π-channel segments that are already on π-Space. Eventually, a new segment is written marking to start of normal operation. During reader recovery, all pi_attach() and pi_read() are performed on cached π-channels, with pipe segments delivered from the cache. Processes are assumed to be piecewise deterministic, thus able to repeat the results it generated in a previous failed execution.

π-Spaces: Abstraction of Shared Space for π-Channels

Process coordination is through one or more π-Spaces. Similar to Linda tuplespace, a π-Space is a shared space abstraction for π-channels, which

Figure 3. The states of a π-channel

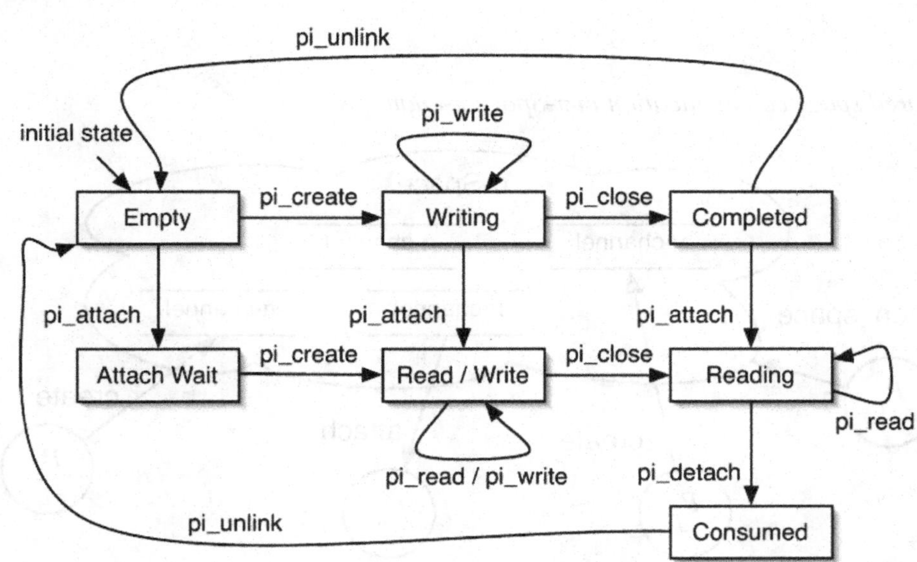

are explicitly posted and retrieved by processes (Figure 4). A π-Space is a logical name space for π-channels, providing dynamic binding of channel endpoints.

Persistence fits elegantly with the model of shared spaces. When a π-channel is created, a copy is automatically posted on the π-Space. During a pi_attach(), the π-Space is accessed to retrieve the named π-channel. Within a single π-Space, a name is bound to at most one π-channel. To reuse an existing name, the π-channel has to be marked for deletion. The programming model includes operations to create, access, and close π-Spaces, enabling the use of multiple spaces within a single application.

Support for Fault-Tolerance and Application Migration

During a workflow execution, the following events may occur:

1. The source application (writer) leaves the workflow. Unless the downstream components do not require any further data from this writer, the entire workflow may be stalled. This also occurs when all outstanding data segments has been consumed.
2. The sink application (reader) leaves the workflow. The writer continues streaming to the cache (π-Space), ignoring the loss of the reader. When the reader recovers, it can resume reading from the π-Space.
3. The link between applications is severed. Assuming that the π-Space is implemented as a reliable service; this is considered as a combination of source failure – from the perspective of the sink; and sink failure – from the perspective of the source.

An application may leave a workflow voluntarily or involuntarily. A voluntary departure occurs when it migrates to another resource. Communication state transfer is employed to re-establish connections with migrated applications. An involuntary departure may be due to machine failure and/or application crash. We assume the fail-stop failure model, i.e., the process crashes and performs no further communication. A connection loss is treated as a component failure and is detected when a communication operation cannot be completed.

Figure 4. Shared space communication in π-Space/π-channels

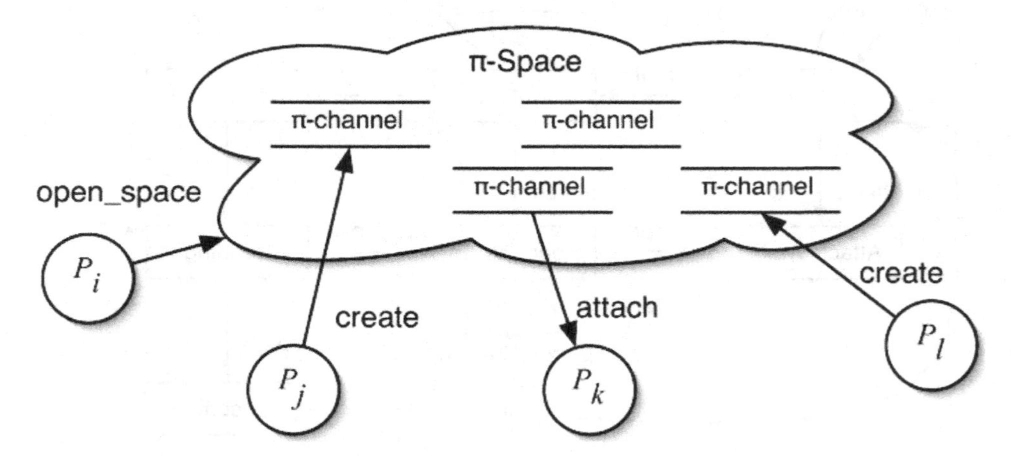

DESIGN AND IMPLEMENTATION

The π-Spaces/π-channels model is implemented as an API with a runtime system, and a multi-threaded server that provides the dynamic lookup and storage of π-channels. A π-Server manages a single π-Space, which encourages deploying multiple servers to improve the distribution of load. Furthermore, a π-Server may execute at the cluster head node, serving as a communication gateway to processes on remote clusters.

Basic Components and General Functional Description

Figure 5 (right) presents the π-Server components. The server maintains a thread pool for incoming and outgoing transmissions, providing a non-archival pipe storage service. The thread count is configurable to support different application loads. The in-bound threads handle incoming data streams from sources, while out-bound threads forward/push π-channel data to sinks. Data streams are transported using a TCP protocol, while look-ups are implemented using a lightweight UDP protocol with retransmission capability.

Figure 5 (left) shows the client-side components. Each π-Spaces/π-channels application is capable of accepting incoming transmissions asynchronously. During a pi_attach(), the reader informs the server of its identity. This allows data streams to be forwarded to the reader while it is busy performing computations. A single event-driven in-bound thread stores incoming segments into a buffer (by the data store component). When buffer capacity is reached, data segments are stored into a local file, identified by the pipe_id. During a pi_read(), this buffer/file is accessed to retrieve the requested segment.

General Description of the Distributed Algorithm

Table 2 presents the distributed algorithm for the π-Spaces/π-channels. We adopt the Python convention to indicate block structure (i.e., the statement alignment determines a block). The notation **(#line_no)** is used in-text when referring to the algorithm.

The pair (pipe_id, space_id) represents a system-wide π-channel identifier, and pipe_id is unique within each π-Space. Since the algorithm assumes a single π-Space, we remove reference to the space_id.

Each participating application maintains the following variables:

Figure 5. Client-side components and design of the π-server

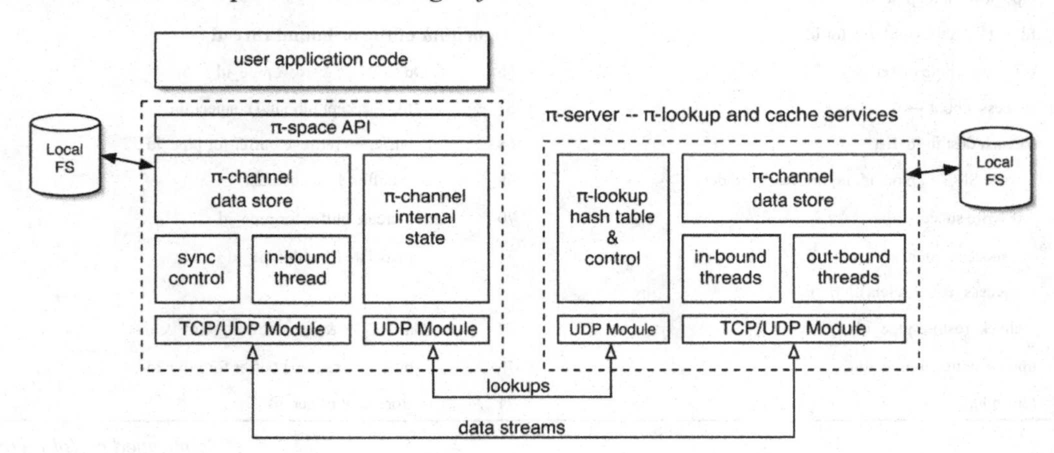

Table 2. Distributed algorithm for π-spaces/π-channels

Client API

```
1    pi_attach (n):
2      send 〈 get, n, my_id 〉 to π-Server
3      recv 〈 get, &pipe_id 〉 from π-Server
4      id ← free entry on local id_tab
5      if buffer for pipe_id does not exist:
6        create buffer for pipe_id
7      associate pipe_id & buffer with id
8      return id

9    pi_create (n):
10     send 〈 put, n, my_id 〉 to π-Server
11     recv 〈 get, &pipe_id, &dest_list 〉 from π-Server
12     foreach dest in dest_list:
13       fd_list [dest] ← connect to dest
14     id ← free entry on local id_tab
15     associate fd_list & pipe_id with id
16     return id

17   pi_read (id, m, len):
18     off ← compute offset
19     buffer ← retrieve buffer for id
20     block until seg [ off, len ] in buffer || eoc
21     if segment was found:
22       get seg [ off, len ] from buffer
23       store segment into m
24       return len
25     return –1 /*end of channel */

26   pi_write (id, m, len):
27     pipe_id ← map pipe_id from id
28     fd_list ← retrieve fd_list for id
29     off ← compute offset
30     success_count ← 0
31     foreach dest in fd_list:
32       write SEG 〈 pipe_id, m, off, len 〉 to dest
33       if write successful:
34         success_count++
35     if success_count < len(fd_list):
36       check_restore(pipe_id)
37     update status of pipe_id
38     return len

39   check_restore (pipe_id):
40     query π-Server for status of migrated reader
41     if migrated reader found:
42       fd ← connect to migrated reader
43       update fd_list with new fd
44       remove reader from hold_list
```

Algorithm of the π-server

```
45   On 〈 get, &n, &src 〉 :
46     entry ← find n on π-table
47     if entry does not exist:
48     entry ← create n on π-table
         entry.reader ← src
50     entry.has_reader ← true
51     pipe_id ← entry.pipe_id
52     send 〈 get, pipe_id 〉 to src
53     if entry.is_cached:
54       initiate forwarding to src

55   On 〈 put, &n, &src 〉 :
56     entry ← find n on π-table
57     if entry does not exist:
58       entry ← create n on π-table
59     dest_list ← ()
60     if entry.has_reader:
61       append entry.reader to dest_list
62     append my_id to dest_list
63     pipe_id ← entry.pipe_id
64     entry.is_cached ← true
65     send 〈 put, pipe_id, dest_list 〉 to src
```

Algorithm of the in-bound thread

```
66   On CON 〈 &src, &pipe_id 〉 :
67     fd ← accept inbound connection
68     buffer ← retrieve buffer for pipe_id
69     if buffer does not exist:
70       create buffer for pipe_id
71     associate fd with pipe_id

72   On SEG 〈 &pipe_id, &m, &off, &len 〉 :
73     buffer ← retrieve buffer for pipe_id
74     store m at offset off
```

continued on following page

Table 2. continued

Algorithm of the out-bound thread	
75	if forwarding π-channel pipe_id to dest:
76	fd ← open connection to dest
77	foreach segment ∈ local buffer:
78	send segment to fd

1. my_id – unique ID, implemented as an IP/port pair. The port number is dynamically generated during application startup.
2. id_tab – a local table **(#4, #14)** associating the pipe_id with open file descriptors fd_list, channel read-write pointers, and other local state information. This table corresponds to the π-channel internal state component in Figure 5.

During a pi_create(), a put request (#10) is sent to the π-Server, which creates (#58–#65) an entry for this π-channel on a hash table and returns a unique pipe_id. It replies (#58, #61, #62) with a list (possibly empty) of destinations. If the reader's identity is known, the reader's address appears first, followed by the server's address. The pi_create() establishes (#12, #13) a connection with the destinations and associates the pipe with the file descriptors. It returns as descriptor (#14, #16) the position of the π-channel on id_tab.

The pi_attach() sends a get request to the server (#2), which replies (#51, #52) with the unique pipe_id for the π-channel, even if non-existent. The server creates an entry for this π-channel, storing the reader's address for use in channel creation.

A pi_read() does not read directly from the open connection with the source. Instead, incoming data segments are handled by the in-bound thread (#66–#74), which listens and accepts TCP operations on behalf of the application. The π-Server manages a thread pool for the same purpose of enabling asynchronous read operations. When the in-bound thread accepts a π-channel, it allocates a buffer (#70) for pipe segments. Each segment (#72–#74) contains type information, length, offset, and pipe_id. The received segments are stored in a shared buffer, so that pi_read() can retrieve (#19, #22) them. The out-bound thread pushes π-channels to sinks (#54, #75–#78). During reader recovery, these threads send missed segments to the reader.

We only outline the migration mechanism (Table 3), showing when application state is saved and restored after migration. The idea (#79–#86) is to attempt a graceful connection shutdown before migrating. Since pipes are cached, undelivered data segments can be retrieved from the π-Server. The hold_list (#89, #44) identifies the migrating processes.

Communication state migration, similar to (Chanchio & Sun, 2004), performs a connection hand-over with the migrated reader (#39–#43). In Figure 6, the migrated peer re-establishes connection with the writer so that: (1) Seg 2 is retrieved from the π-Server; and at the same time (2) Seg 3 is streamed from the writer.

EXPERIMENTAL RESULTS

Two aspects of the implementation are evaluated. First, we measure the rate in which π-channel lookup operations are handled by the π-Server under two scenarios: (a) π-Server and clients are on one cluster; and (b) clients perform lookups over a wide-area network. Second, we measure the throughput when communication takes place between two applications over our WAN testbed. This test shows how asynchronous read operations improve the bandwidth utilisation. Table 4 lists the resources we used. VPAC (Victorian Partnership for Advanced Computing) is an HPC consortium

Table 3. The communication migration protocol

79	if I am migrating:
80	disable sending acks for heartbeats
81	migrating ← true
82	foreach open π-channel:
83	save offset into checkpoint
84	flush and close all connections
85	perform local state checkpoint
86	send checkpoint to the new location
87	if a peer is migrating:
88	/* reject connections from this list */
89	add peer_addr to hold_list

of universities in Victoria, Australia. Our wide-area testbed uses both Monash and VPAC resources.

π-Server Lookup Performance

We evaluate the request-handling rate of the π-Server, with up to 32 clients concurrently generating lookup requests. Table 5 presents results conducted over mahar, measuring the execution time of all clients when looking up 25 366 unique but randomly generated π-channel names. Each client performed 40 000 lookups, without channel read/write operation. Clients were assigned on execute nodes, with the π-Server on the head node. At least 12 runs were performed for each test case, using only the timings from the middle 10 runs.

On Table 6, we present the timings for lookups on a WAN between Monash and VPAC. Clients ran on mahar compute nodes with the π-Server running on wexstan's head node, using the same parameters as in the LAN tests. These results indicate that the bottleneck for grid applications will most likely be the high latencies between the π-Server and the clients.

Figure 6. Time diagrams showing concurrent reading of a π-channel from π-Server and writer. In (a), the migrated reader resumes reading from cache. In (b), it also resumes connection with the writer.

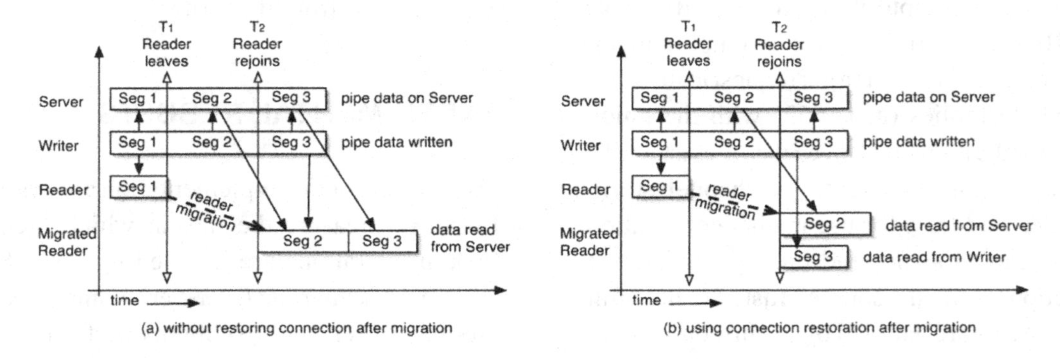

Table 4. Participating systems in our experiments

Name	Location	Processor	OS	#CPUs
mahar.infotech.monash.edu.au	Monash	Intel P4	Linux 2.4.27-3	50
edda.vpac.org	VPAC	IBM Power5	SLES 9 Linux	80
wexstan.vpac.org	VPAC	AMD Opteron	Red Hat Linux	246
tango.vpac.org	VPAC	AMD Opteron	CentOS 5 Linux	760

π-Channel Throughput on a Multi-Cluster Testbed

We evaluate and compare the data transfer throughput using π-channels under three scenarios: (a) matched create/attach operations; (b) create first, then attach; and (c) asynchronous read operations. Two processes are executed, using a pair of π-channels for communication.

For scenario one, processes use paired pi_attach() and pi_create() operations, i.e., when one process is writing, the other is reading. For the second scenario, each process executes all π-channel writes first, followed by reads. The π-Server caches most of the pipe segments during the write phase. Once the processes perform the pi_attach(), the pipe segments are retrieved from the π-Server rather than the writer.

In the third scenario, processes initiate a non-blocking pi_attach() on an inbound channel first, before a pi_create(), followed by the write and then read operations. This notifies the π-Server of a pending request for a channel, providing writers with the destination addresses and encouraging direct streaming. The performance results show a reduced dependence on the π-Server to store pipe segments.

Figure 7 shows the segment send rates while Figure 8 presents the measured bandwidth. The π-Server executed on edda, one application on tango and another on wexstan. The best performance is achieved under asynchronous operation (scenario three). The detailed results are presented on Table 7. The tests were conducted with at least 12 trials per case. Of these, the mean is computed using ten results, discarding the highest and lowest values. The standard deviation given is for the throughput, i.e., the mean message send rates for each of the segment sizes. Note the absence of observable performance differences for Scenario 1 and 2. This means that matching pi_attach() with pi_create() operations does not lead to any improvement in data transfer rates. Thus, π-Space/π-channels applications may be written without using an odd-even rule to match reads and writes.

In Scenario 3, the use of pi_attach() notifies the π-Server of a pending request to retrieve a

Table 5. UDP request-reply performance within a cluster

Number of Clients	2	4	8	16	32
Number of Requests Served	80 000	160 000	320 000	640 000	1 280 000
Mean Execution Times (s)	21.99	25.51	38.72	44.65	96.58
Standard Deviation	0.822	0.813	0.741	0.430	0.170
Request Rate (per second)	3 642.6	6 277.6	8 267.1	14 335	13 390
Standard Deviation	138.89	200.85	162.38	137.58	23.82

Table 6. UDP request-reply performance on a WAN

Number of Clients	2	4	8	16	32
Number of Requests Served	10 000	20 000	40 000	80 000	160 000
Mean Execution Times (s)	10.58	11.02	11.52	12.34	24.74
Standard Deviation	0.518	0.340	0.285	0.122	0.268
Request Rate (per second)	946.4	1 815.6	3 473.8	6 480.8	6 467.6
Standard Deviation	46.32	56.53	86.53	64.07	70.39

π-channel. This makes it possible for the writer of that π-channel to transmit the channel segments directly to the reader, showing a substantial improvement in the data transfer performance over that of the first two scenarios. Furthermore, the overlap of sends and receives results in better utilisation of the available bandwidth.

CONCLUSION

We have presented π-Spaces/π-channels, a communication mechanism for scientific workflows on dynamic environments, where resources may fail and network links may be disrupted. The key feature of π-channels is persistence, enabling communication despite process failures or departures. This article presents its design and implementation. In particular, we describe the distributed algorithm showing how persistence

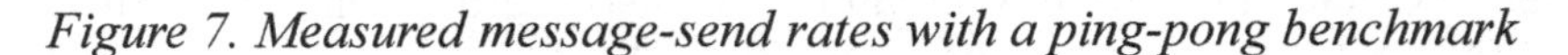

Figure 7. Measured message-send rates with a ping-pong benchmark

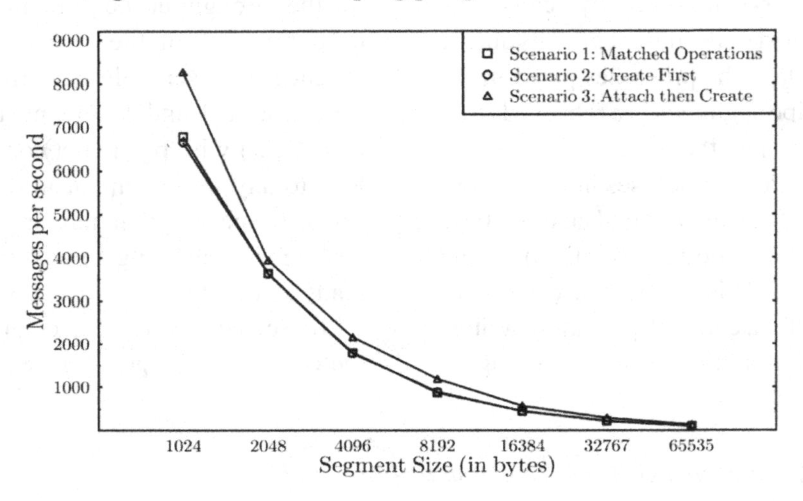

Figure 8. Measured bandwidth with a ping-pong π-Space/π-channels application. The horizontal bar shows the measured bandwidth using iperf[2]

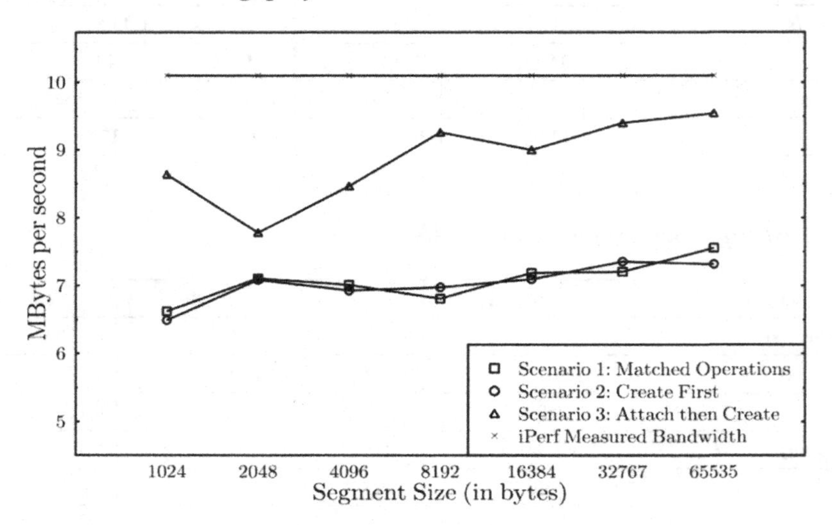

Table 7. Results from the bandwidth and throughput measurements

Size of Segment (bytes)	1024	2048	4096	8192	16384	32767	65535
Scenario 1							
Throughput	6781.586	3638.966	1794.872	871.818	460.053	230.603	120.902
Std. Dev.	186.270	25.994	37.553	7.248	7.471	3.927	1.467
Bandwidth (MB/s)	6.623	7.107	7.011	6.811	7.188	7.206	7.556
Scenario 2							
Throughput	6646.947	3627.420	1773.944	892.972	454.091	235.335	117.091
Std. Dev.	140.815	26.239	29.443	7.959	7.517	2.163	1.622
Bandwidth (MB/s)	6.491	7.085	6.929	6.976	7.095	7.354	7.318
Scenario 3							
Throughput	8841.626	3983.438	2166.548	1185.448	575.914	300.792	152.712
Std. Dev.	249.939	15.483	19.652	14.235	8.900	4.880	1.309
Bandwidth (MB/s)	8.634	7.780	8.463	9.261	8.999	9.399	9.544

is achieved with the caching mechanism and the asynchronous operation using the in-bound thread. A communication state transfer mechanism is employed, which re-establishes connections with migrated components to improve the data stream transfer time. Experimental results show that the caching mechanism is able to buffer channel data segments and when asynchronous operation is employed, throughput is substantially increased. With asynchronous operation, sending and receiving of messages can be overlapped, resulting in improved message sending rates than normal non-asynchronous operation.

ACKNOWLEDGMENT

We thank the Victorian Partnership for Advanced Computing (VPAC) for the use of their facilities and for their continuing support for this project.

REFERENCES

Abramson, D., Foster, I., Giddy, J., Lewis, A., Sosic, R., Sutherst, R., & White, N. (1997). Nimrod Computational Workbench: A Case Study in Desktop Metacomputing. In *Australian Computer Science Conference (ACSC 97)*. Macquarie University, Sydney.

Abramson, D., & Kommineni, J. (2004). A Flexible IO Scheme for Grid Workflows. In *Proc. of the 18th International Parallel and Distributed Processing Symposium*. Krakow, Poland.

Abramson, D., Sosic, R., Giddy, J., & Hall, B. (1995). Nimrod: A Tool for Performing Parameterised Simulations using Distributed Workstations. In *Proc. of the 4th IEEE Symposium on High Performance Distributed Computing*. Virginia. IEEE Press.

Batchu, R., Dandass, Y. S., Skjellum, A., & Beddhu, M. (2004). MPI/FT: A Model-Based Approach to Low-Overhead Fault Tolerant Message-Passing Middleware. *Cluster Computing*, 7(4), 303–315. doi:10.1023/B:CLUS.0000039491.64560.8a

Bouteiller, A., Hérault, T., Krawezik, G., Lemarinier, P., & Cappello, F. (2006). MPICH-V Project: A Multiprotocol Automatic Fault-Tolerant MPI. *International Journal of High Performance Computing Applications*, *20*(3), 319–333. doi:10.1177/1094342006067469

Carriero, N., & Gelernter, D. (1989). How to Write Parallel Programs: A Guide to the Perplexed. *ACM Computing Surveys*, *21*(3), 323–357. doi:10.1145/72551.72553

Chan, P., & Abramson, D. (2001). NetFiles: A Novel Approach to Parallel Programming of Master/Worker Applications. In *Proc. of the 5th International Conference and Exhibition on High-Performance Computing in the Asia-Pacific Region (HPCAsia 2001)*, Queensland, Australia.

Chan, P., & Abramson, D. (2007). π-spaces: Support for Decoupled Communication in Wide-Area Parallel Applications. In *Proc. of the Sixth International Conference on Grid and Cooperative Computing*, (pp. 3–10). Urumchi, Xinjiang, China: IEEE.

Chan, P., & Abramson, D. (2008). Netfiles: An Enhanced Stream-based Communication Mechanism. In J. Labarta, K. Joe, & T. Sato (Eds.), *High-Performance Computing, Revised Selected Papers. Sixth International Symposium, ISHPC 2005 and First International Workshop on Advanced Low Power Systems, ALPS 2006, 4759 of Lecture Notes in Computer Science*, (pp. 254–261). Springer-Verlag.

Chanchio, K., & Sun, X.-H. (2004). Communication State Transfer for Mobility of Concurrent Heterogeneous Computing. *IEEE Transactions on Computers*, *53*(10), 1260–1273. doi:10.1109/TC.2004.73

Fagg, G. E., & Dongarra, J. (2004). Building and Using a Fault-Tolerant MPI Implementation. *International Journal of High Performance Computing Applications*, *18*(3), 353–361. doi:10.1177/1094342004046052

Foster, I., & Kesselman, C. (1999). Computational Grids. In *The Grid: Blueprint for a New Computing Infrastructure*, (pp. 15–51).

Freeman, E., Hupfer, S., & Arnold, K. (1999). *JavaSpaces Principles, Patterns, and Practice*. Addison-Wesley.

Goldman, K. J., Swaminathan, B., McCartney, T. P., Anderson, M. D., & Sethuraman, R. (1995). The Programmers' Playground: I/O Abstraction for User-Configurable Distributed Applications. *IEEE Transactions on Software Engineering*, *21*(9), 735–746. doi:10.1109/32.464547

Gropp, W., & Lusk, E. (2004). Fault Tolerance in Message Passing Interface Programs. *International Journal of High Performance Computing Applications*, *18*(3), 363–372. doi:10.1177/1094342004046045

Guillen-Scholten, J., & Arbab, F. (2005). Coordinated Anonymous Peer-to-Peer Connections with MoCha. In N. Guelfi, G. Reggio, & A. Romanovsky, (Eds.), *Scientific Engineering of Distributed Java Applications, Revised Selected Papers. 4th International Workshop, FIDJI 2004, 3409* of *Lecture Notes in Computer Science*, (pp. 68–77). Springer-Verlag.

Hoare, C. (1985). *Communicating Sequential Processes*. Prentice Hall.

Hsu, M., & Silberschatz, A. (1991). Unilateral Commit: A New Paradigm for Reliable Distributed Transaction Processing. In *Proc. of the 7th International Conference on Data Engineering*, (pp. 286–293). IEEE Computer Society.

Hua, K. A., Jiang, N., Peng, R., & Tantaoui, M. A. (2004). PSP: A Persistent Streaming Protocol for Transactional Communications. In *ICCCAS 2004: Proc. of the 2004 International Conference on Communications, Circuits and Systems, 1*, 529–533. IEEE Computer Society.

IBM Websphere MQ. (2008). *The IBM Websphere MQ Family.* [online]. URL: http://www.ibm.com/software/websphere. (March, 2008).

Johnson, B. K., & Ram, D. J. (2001). DP: A Paradigm for Anonymous Remote Computation and Communication for Cluster Computing. *IEEE Transactions on Parallel and Distributed Systems, 12*(10), 1052–1065. doi:10.1109/71.963417

Kahn, G. (1974). The Semantics of Simple Language for Parallel Programming. In *Proc. of the 1974 IFIP Congress*, (pp. 471–475).

Magee, J., Kramer, J., & Sloman, M. (1989). Constructing Distributed Systems in Conic. *IEEE Transactions on Software Engineering, 15*(6), 663–675. doi:10.1109/32.24720

Microsoft. (2008). *Microsoft Message Queueing.* [online]. URL: http://www.microsoft.com/windowsserver2003/technologies/msmq/default.mspx (March, 2008).

Purtilo, J. M. (1994). The POLYLITH Software Bus. *ACM Transactions on Programming Languages and Systems, 16*(1), 151–174. doi:10.1145/174625.174629

Reddy, M. V., Srinivas, A. V., Gopinath, T., & Janakiram, D. (2006). Vishwa: A Reconfigurable P2P Middleware for Grid Computations. In *Proc. of the 2006 International Conference on Parallel Processing (ICPP 2006)*. IEEE Press.

Rowstron, A. (1998). WCL: A Co-ordination Language for Geographically Distributed Agents. *World Wide Web (Bussum), 1*(3), 167–179. doi:10.1023/A:1019263731139

Sterck, H. D., Markel, R. S., & Knight, R. (2005). A Lightweight, Scalable Grid Computing Framework for Parallel Bioinformatics Applications. In *HPCS'05: Proc. of the 19th International Symposium on High Performance Computing Systems and Applications*. IEEE Press.

Sterck, H. D., Markel, R. S., Pohl, T., & Rüede, U. (2003). A Lightweight Java Taskspaces Framework for Scientific Computing on Computational Grids. In *SAC2003: Proc. of the ACM Symposium on Applied Computing*, (pp. 1024–1030). New York, NY, USA: ACM Press.

Stevens, W. R. (1998). *Unix Network Programming: Networking APIs: Sockets and XTI, 1* (2nd ed.). Prentice-Hall PTR.

Tanenbaum, A. S., & Steen, M. V. (2007). *Distributed Systems: Principles and Paradigms*. Pearson Prentice Hall, 2 edition.

Taura, K., Kaneda, K., Endo, T., & Yonezawa, A. (2003). Phoenix: A Parallel Programming Model for Accommodating Dynamically Joining/Leaving Resources. In *PPoPP '03: Proc. of the Ninth ACM SIGPLAN Symposium on Principles and Practice of Parallel Programming*, (pp. 216–229), New York, NY, USA: ACM Press.

van Nieuwpoort, R. V., Maassen, J., Wrzesinska, G., Hofman, R. F. H., Jacobs, C. J. H., Kielmann, T., & Bal, H. E. (2005). Ibis: a Flexible and Efficient Java-based Grid Programming Environment. *Concurrency and Computation, 17*(7–8), 1079–1107. doi:10.1002/cpe.860

Wyckoff, P., McLaughry, S. W., Lehman, T. J., & Ford, D. A. (1998). T Spaces. *IBM Systems Journal, 37*(3), 454–474. doi:10.1147/sj.373.0454

Zhang, Y., & Dao, S. (1995). A 'Persistent Connection' Model for Mobile and Distributed Systems. In *ICCCN '95: Proc. of the 4th International Conference on Computer Communications*, (pp. 300–307). IEEE Computer Society.

Zhong, X., & Xu, C.-Z. (2004). A Reliable Connection Migration Mechanism for Synchronous Transient Communication in Mobile Codes. In *Proc. of the 2004 International Conference on Parallel Processing*. IEEE Press.

ENDNOTES

1. The dtype_t covers various data types, e.g. PI_INT, PI_FLOAT, etc.
2. http://sourceforge.net/projects/iperf

This work was previously published in International Journal of Grid and High Performance Computing (IJGHPC), Volume 1, Issue 3, edited by Emmanuel Udoh & Ching-Hsien Hsu, pp. 18-36, copyright 2009 by IGI Publishing (an imprint of IGI Global).

Chapter 16
Self-Configuration and Administration of Wireless Grids

Ashish Agarwal
Carnegie Mellon University, USA

Amar Gupta
University of Arizona, USA

ABSTRACT

A Wireless Grid is an augmentation of a wired grid that facilitates the exchange of information and the interaction between heterogeneous wireless devices. While similar to the wired grid in terms of its distributed nature, the requirement for standards and protocols, and the need for adequate Quality of Service; a Wireless Grid has to deal with the added complexities of the limited power of the mobile devices, the limited bandwidth, and the increased dynamic nature of the interactions involved. This complexity becomes important in designing the services for mobile computing. A grid topology and naming service is proposed which can allow self-configuration and self-administration of various possible wireless grid layouts.

INTRODUCTION

Foster (2002) offers a checklist for recognizing a "grid". A grid allows

- Coordination of resources that are not subject to centralized control;
- Use of standard, open, general-purpose protocols and interfaces; and
- Delivery of nontrivial qualities of service.

The emergence of the Wireless Grid meets all these criteria and is fueled by technological advances in grid computing and wireless technology. The ultimate vision of the grid is that of an adaptive network offering secure, inexpensive, and coordinated real-time access to dynamic, heterogeneous resources, potentially traversing geographic boundaries but still able to maintain the desirable characteristics of a simple distributed system, such as stability, transparency, scalability

DOI: 10.4018/978-1-60960-603-9.ch016

and flexibility. The technologies originally developed for use in a wired environment are now being augmented to operate in wireless situations. The development of the wireless technologies such as 802.11, GPRS, and 3G has extended the reach of wireless services to all the individuals. With the ubiquity and indispensability of wireless technologies established, these technologies are now making inroads into grids.

A wireless grid has to face added complexity due to the limited power of the mobile devices, the limited bandwidth, and the increased dynamic nature of the interactions involved. This added complexity has to be considered while designing service oriented architecture for mobile devices (Oliveira et al, 2006). This article highlights the key characteristics of the wireless grids and suggests various possible grid layouts. A grid topology and a naming protocol have been proposed to address the self-configuration and self-administration requirements of these grid layouts. This article is organized as follows. Section 2 describes the key characteristics of the wireless grids. Section 3 describes various possible grid layouts. Section 4 mentions the technical challenges associated with these layouts. Section 5 introduces a grid topology and a naming protocol to address the self configuration and self administration challenges. Section 6 concludes the article.

KEY CHARACTERISTICS

The development of the wireless grid technologies is governed by three driving forces:

- **New User Interaction Modalities and Form Factors***:* Traditional applications that can exist on the Wired Grid need to expand their scope by extending the interactions to mobile devices through adapting the user interface to small screens, small keyboards, and other I/O modalities such as speech. The mobile access interface needs to address the issue of connectivity of mobile devices.
- **Limited Computing Resources:** Wireless applications need to share the resources and to provide access to additional computational resources to mitigate the constraints imposed by limited storage, computational capability, and power of mobile devices.
- **Additional New Supporting Infrastructure Elements:** New applications, especially ones involving dynamic and unforeseen events, need to be addressed through the rapid provisioning of major amounts of computational and communications bandwidths. For example, the occurrence of an urban catastrophe could trigger a dynamic adaptive wireless network to alert people to organize remedial actions in a coordinated fashion, and to provide better control of available resources and personnel.

Grid Resources

A Wireless Grid must provide a virtual pool of computational and communications resources to consumers at attractive prices. Various grid resources are described below:

- **Computing Power:** Wireless devices possess limited computation power. Wireless grids can overcome this limitation by distributing the computational tasks across multiple power-constrained devices. But this raises the need for establishing appropriate collaborative processes between these geographically distributed tasks.
- **Storage Capacity:** Wireless devices possess limited storage capability. Grids can overcome this limitation by distributing the data storage over multiple devices. Data can be recombined into a single entity and then made available to the users. However,

this creates the need to enable data access and update to occur simultaneously and to avoid contention through the application of advanced synchronization techniques.

- **Communications Bandwidth:** Wireless grids can harness the power of wireless technology to allow remote access. At the same time, the grid infrastructure should be robust enough to ensure high Quality of Service (QoS).
- **Multiplicity of Applications:** Wireless Grids should allow the users ubiquitous access to a wide variety of applications. However, one needs to overcome the need to install these applications on separate mobile devices.

GRID LAYOUT

Drawing upon the paradigm of the wired grids (Gentzsch, 2001; Ong, 2003 and Tiang, 2003), various layouts of the wireless grids are possible. The classification schemes can be based on the architecture or on the function of the grids.

Classification by Architecture

One way to characterize the architecture of the wireless grid is by the degree of heterogeneity of the actual devices and the level of control exercised by those who own and administer the devices (figure 1). It can vary from a simple network of homogeneous devices bound by a single set of policies and rules to a complex network of heterogeneous devices spread across multiple organizational, political and geographical boundaries, as categorized below:

- **Local Cluster or Homogeneous Wireless Grid:** This simplest form involves a local collection of identical or similar wireless devices that share the same hardware architecture and the same operating systems. Because of the homogeneity of the end systems, the integration of these devices into the wireless grid, as well as the consequent sharing of resources, becomes a much easier task. Today, this type of organization is more likely to be found in a single division of an organization where one single administrative body exercises control over all the devices. An example

Figure 1. A simplified depiction of the 3-tier wireless Grid architecture (adapted from Ong, 2003)

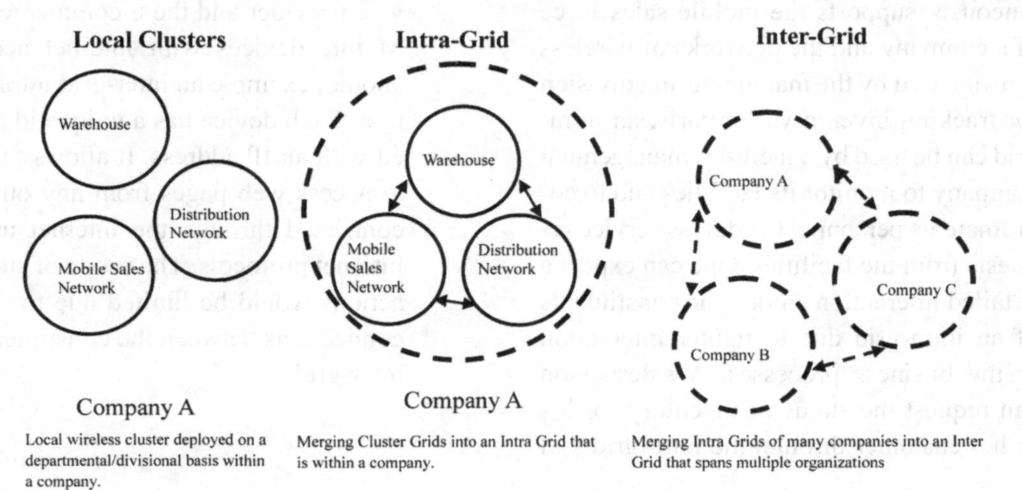

would be a network of mobile handheld devices for coordinating medical personnel in the hospital. A local cluster can be used to coordinate field personnel engaged in collaborative tasks such as construction, mining, or repair services. It can also be used in a remote sensor network for monitoring crops or seismic activities. It remains to be seen whether market forces will result in convergence of hardware (virtual or real) and software and the emergence of a dominant design, which can exploit resource-sharing strategies that are more intimately bound to the device.

- **Wireless Intra-Grids:** An intra-grid encompasses wireless devices that belong to multiple divisions or communities within an actual organization (AO). The divisions may be located in different geographies and maybe governed by a separate set of policies, but there exists a level of trust and oversight so that "ground truth" may be known with respect to identity and characteristics. AOs are the point where resolution can occur between the virtual presence of a wireless entity and its actual name and location. AOs also tend to be persistent in time, and become the point of composition among other AOs. An example of an intra-grid would be a wireless grid that simultaneously supports the mobile sales force of a company and the networks of wireless sensors used by the manufacturing division for tracking inventory. Similarly, an intra-grid can be used by a facilities management company to monitor its facilities and to coordinate its personnel to address service requests from the facilities. One can expect a detailed interaction among the constituents of an intra-grid due to tighter interaction in the business processes. A salesperson can request the status of inventory for his or her customer through the intra-grid that connects the sales network to the inventory tracking system.
- **Inter-Grid:** An inter-grid encompasses multiple AOs and transcends greater amounts of geographical, organizational, and other types of differences, such as ones related to intellectual property rights and national laws. Multiple AOs may come together to form Virtual Organizations (VOs) where they can collaborate and share resources such as information, knowledge, and even market access to exploit fast-changing market opportunities. The relationship can be long or short term (Ong, 2003). Resource management and policy integration (security, authentication and data management tasks) attain greater complexity due to the scalability requirements. To move beyond mere ad hoc composition of AOs, a (potentially) universal composition of declarative policies must be proposed and accepted. An example of an inter-grid interaction would be a scenario involving an American tourist visiting Japan and trying to conduct a local e-commerce transaction using his/her cell phone. The transaction would involve a handshake between the traveler's cell phone service provider, traveler's credit card company, the Japanese wireless service provider and the e-commerce vendor. Mobile devices with internet access are another example an inter-grid implementation. Each device has a unique id associated with an IP address. It allows the device to access web pages from any other node connected through the internet using the internet protocols. The scope of such interactions would be limited due to the loose connections between the constituents of an inter-grid.

Classification by Usage Pattern

Wireless grids can be classified by usage patterns as summarized in Table 1.

- **Computational Grid:** In a computational grid, the need for creating the wireless grid is driven primarily by the need to borrow computational resources from others. This arises, in part, because of the power constraints on mobile devices, which in turn limits their computational capability. The computational grid may be cooperative or parasitic (Barabasi et al, 2001). In a cooperative setup, inputs from multiple nodes are needed to analyze a particular scenario. For example, sensor network deployed in the battlefield would present the enemy's position. Similarly, a wireless sensor network will be used to monitor conditions for predicting natural calamities like earthquakes or volcanoes. In a parasitic setup, the nodes would rely on each other to manage the power constraints. Any remote setup, will allow for this possibility due to lack of other power resources. Some kind of redundancy would be built in such a setup.
- **Data Grid:** In this case, the need for creating the wireless grid is dictated primarily by the need to provide shared and secure access to distributed data. Since data can be presented in various contexts on various systems, reconciling the underlying semantics continues to challenge evolving technology. One example involves an urgent search for donors with a rare blood type. A hospital would issue a query to the medical history databases in the region through its mobile network. The mobile service providers will notify potential donors through the alert messages transmitted to their respective mobile devices, and the resulting responses would be processed and reconciled. Internet serves as a massive data grid where the information resides on multiple servers and such information can be accessed using portals and search engines or by directly request to a particular IP address.
- **Utility Grid:** Here the motivation for the wireless grid is derived from the need to provide ubiquitous access to specialized pieces of software and hardware. Users can request resources when needed (on-demand) and only be charged for the amount being used. This model can subsume both Computational and Data grids. For example, users might tap Wireless Utility grids for information such as the traffic conditions and routing, and for making instantaneous transactions related to commercial products and services.

TECHNICAL CHALLENGES

Among the many challenges wireless grids face, these grids must overcome the following set of initial technical challenges:

- **Dynamic Configurability:** Wireless grids are characterized by changing topology

Table 1. Wireless Grid usage patterns

Grid Type	Possible Architecture	Mainly Provides
Computational	Cluster, Intra, Inter	Computational Power
Data	Cluster, Intra, Inter	Data Access and Storage
Utility	Intra, Inter	On-demand Access various of Resources

due to the mobile nature of the grid components. Grids should provide self-configuring and self-administering capability to allow these dynamic changes for all possible grid layouts.

- **Routing Plasticity:** Efficient routing protocols are required to address the power limitation of the end devices along with the consideration for stable wireless connectivity, route optimization and efficient use of the limited bandwidth.
- **Discovery Semantics and Protocols:** Service description protocols are needed to describe the services provided by various components of the wireless grid. Once the services are published, a discovery protocol is needed to map the mobile resources to the services.
- **Security:** Because of the inherent nature of the wireless connection, the diversity of the link quality, the potential unreliability of the end-devices, the power constraints of the mobile device, and the enforcement of security and privacy policies all present major challenges in the wireless grid environment. Effective security requires adequate computational power to execute the security algorithms in acceptable times. In addition, sufficient radio power is required to achieve an effective signal-to-noise ratio (in the face of encrypted signaling streams) and to close the link. This suggests a careful husbanding of access points and the hand-over to ensure that the minimum possible power is required from each of the wireless devices.
- **Policy Management:** Grid architecture designers need to address policies that govern the usage, privileges, access to resources, sharing level agreements, quality of service, and the composability and the automated resolution of contradictory policies among organizations; as well as other technical issues mentioned above.

SELF-CONFIGURING AND SELF-ADMINISTERING DYNAMIC ADDRESS SERVICES ACROSS VIRTUAL ORGANIZATIONS

To flourish, grids must exist for the benefit of the members and users. To add tangible value, infrastructures that support wireless grids must address the issue of dynamic updates to the grid to account for network node failure, and the entry or exit of nodes. Previous work on Self-Configuring and Self-Administering Domain Name Service (DNS) has led to a reliable, intelligent and distributed lightweight protocol for automatically adapting to the changes in the networks (Huck et. al., 2002); this protocol can be modified and extended for use in the wireless grid environment.

Grid Topology

Several researchers have evaluated the topology and configuration of mobile networks (Nesargi and Prakash, 2002; Vaidya, 2002; Mohsin and Prakash, 2002; Weniger and Zitterbart, 2004). However, these ad hoc systems are standalone in nature. We believe that the commercial grids will possess some access to the wired Internet infrastructure and thereby follow a hybrid model (figure 2). It will consist of Mobile Ad-hoc Networks (MANET) type systems with multiple-hop paths between mobile nodes and access points to the wired network. An application of this hybrid setup has been the Mesh Networks (Bruno et al. 2005). Data will need to flow across the grid using a combination of Mobile IP (Perkins, 2002) or the new Mobile IPv6 and Ad-Hoc routing protocols such as Dynamic Source Routing Protocol (DSR) (Hu, Perrig and Johnson, 2005) and Ad hoc On-Demand Distance Vector Routing (AODV) (Perkins and Royer, 1999; Papadimitratos and Haas, 2005).At a high level, one needs to support the critical role of the management and composition of subnets and arbitrary collections of wireless members. There must be a Root Station (RS) present in some form as well as a Base Station (BS). The RS maintains cognizance over a set of wireless devices and serves as the final mapping of logical to physical devices. The BS manages and enforces policy within and among groups. A grid layout can include a root station for a community or an actual organization (AO) of wireless nodes (figure 3). A root station will maintain up-to-date information about its own network and the associated nodes as well as serve as the gateway to the wired network. Multiple organizations may come together to form a virtual organization (VO). An AO can belong to multiple VOs. A base station (BS) can be envisaged for a VO. A BS will maintain information about networks for various organizations and the associated root stations. For a homogeneous grid, the same server can perform both the RS and BS functions. In case of an inter-grid, which can span multiple virtual organizations, several BSs are needed to coordinate to maintain the inter-grid information. Redundancy can be maintained by having secondary servers to perform the RS and BS functions. Both RS and BS should not be resource-constrained devices. Instead, the RS and the BS could be a simple PC, workstation, or server equipped with an appropriate interface to communicate with the edge nodes such as sensor nodes or other mobile nodes.

Figure 2. A hybrid wireless network

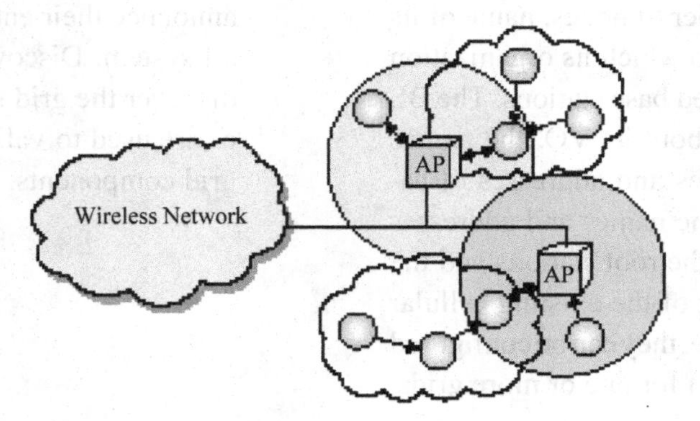

Self-Configuration and Administration of Wireless Grid

As previously stated, wireless grids possess a unique dynamic quality that is not found readily in the wired grids. Therefore, technologies that support self-configuration and self-administration are critical to the continued growth of the wireless grid paradigm. Wireless grids should allow:

- Configuration of addresses for the grid components: nodes, RS and BS
- Name- to- address resolution for the grid components
- Maintenance of the state information for the grid

The address for the nodes can be obtained in several different ways. It is possible that the address may not be an IP address in case the device is a sensor with no IP stack. We envision that an IP incapable node could use, as its own address, either the MAC address of the system chip or a unique serial number provided at the time of manufacturing the device. A name, unique to the AO domain, can be assigned to the device through an automatic handshake process between the device and the RS. RS and BS are connected to the wired infrastructure and can obtain IP addresses using the DHCP protocol (Droms, 1997).

Figure 3. Wireless Grid spanning multiple virtual organizations

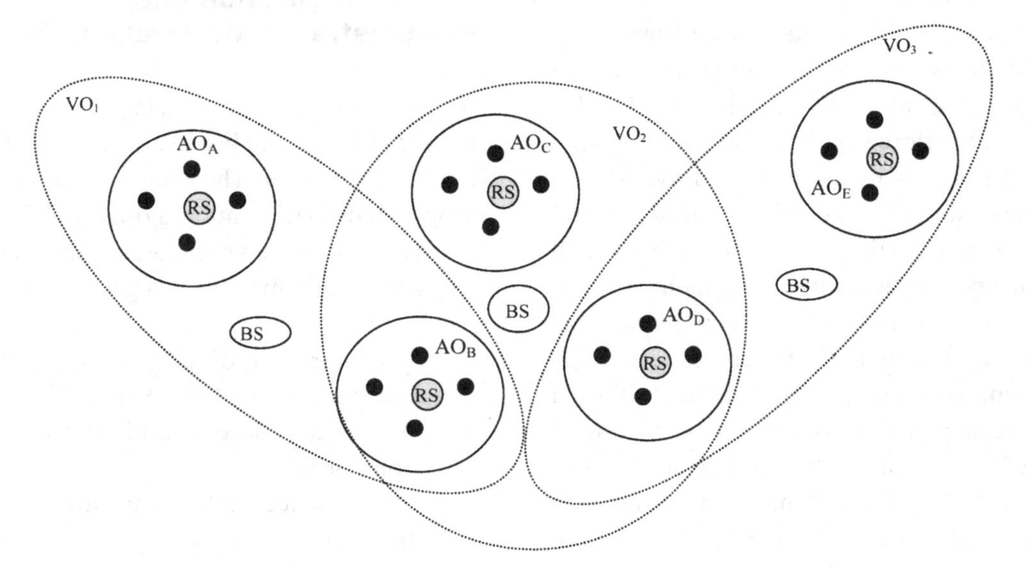

RS and BS provide the naming service for resource discovery across the AOs and VOs. The notion of grid service (Foster et. al, 2001) can be extended to the wireless grids. In such a scenario, the RS can provide a naming service for resource discovery based on service description (Winoto et. al., 1999; Zhu et al, 2003, Sharmin et al., 2006)) at the node level. Resource discovery can be extended to a virtual organization where a BS can provide a naming service for resource discovery within various actual organizations. Multiple BSs can coordinate to provide service discovery across multiple VOs.

Each node maintains information about itself and the AO it belongs to. The RS maintains information about its AO such as the name and address pairs for its nodes, number of nodes, name of its AO, names of the VOs to which its organization belongs and the associated base stations. The BS maintains information about its VO, the names of associated AOs, names and addresses of associated RSs, and also the names and addresses of other BSs. Note that the root stations and the base stations can be part of the existing cellular and internet infrastructure; they can be configured to handle communication for one or more grids. In such a case, the grid owners pay a fee to the internet and cellular service providers to handle their specific communication requests.

- **Messages:** Messages are used for communication between the grid components and are a mechanism for resource discovery. Figure 4 shows the structure of a message. It consists of a three-field header followed by a payload section. The header fields are explained in Table 2. The payload holds the data from the message specific to each Opcode. Table 3 lists the possible opcode values.
- **Message Behavior:** Enter and leave messages are used by the grid components to announce their entry or exit from the overall system. Discover messages are used to discover the grid resources. Hello messages are used to validate the existence of the grid components.

GRID OPERATION

Node Management

Node Entry or Exit

Mobile nodes register <address, name> tuple with the Root Station (RS) as they enter the network under the RS coverage. Node sends an enter_node message. If the node cannot directly establish connection with the RS, it uses multiple hops to pass on the registration information. This can happen in a setup where the wired node (RS) is out of operation or when reach and wireless signals are weak in strength. When the RS receives the request, it sends enter_node response to the node and adds the information related to the node. The response includes the information about RS and AO. For example, in an emergency situation, appropriate personnel such as police or fire workers may arrive or leave the site. A local cluster can be formed to handle both voice and data communications. Entry and exist messages can help to maintain the status of the emergency workers, to efficiently distribute critical data they may possess, and to better coordinate the activities.

Node Discovery

A chain-of-responsibility pattern (Gamma et. al., 1995) is used for node discovery. A node N_o sends a discover_node request to RS_o seeking connection to a different node N_a. The request contains the name of the requested node. RS_o looks up its AO_o information to locate the node and sends a discover_node response to node N_o with the address information of N_a. If node N_a does not exist in the AO_o then RS_o sends the discover_node request to the BS_o. This request includes the RS_o

Figure 4. Message format

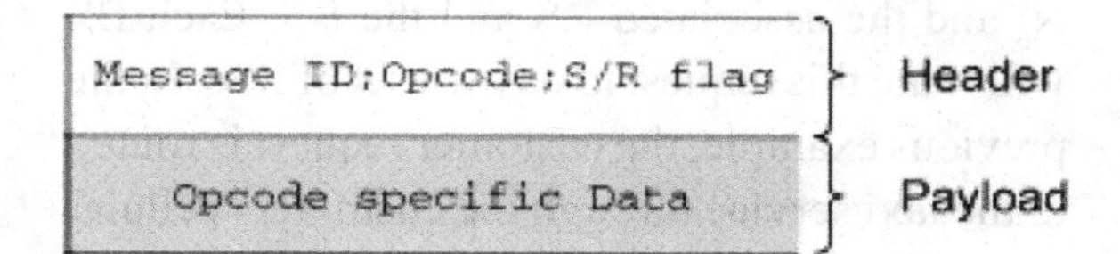

Table 2. Message header fields

Header Field	Description
Message Id	The unique message id for the message
Opcode	The operation code for the message.
S/R flag	Send/Response Flag. A flag indicating whether the message is a send request or response to a send request

Table 3. Opcode values

Opcode Values	Brief Description
Enter_node	Informs members of the entry of the node
Leave_node	Informs member of the exit of the node
Enter_RS	Informs members of the entry of the RS
Leave_RS	Informs member of the exit of the RS
Enter_BS	Informs members of the entry of the BS
Leave_BS	Informs member of the exit of the BS
Discover_node	Used to discover node
Hello	Used to verify if the members exist

information. BS_o in turn broadcasts the request to all the RSs associated with its VO_o. If RS_a locates the node in its AO_a then it notifies the BS_o about the availability of the node. BS_o in turn sends the discover_node response to the requesting RS_o with the address of N_a., which is then forwarded to N_o. For example, a taxi dispatch service may be designed to respond to a customer request by broadcasting messages to all the taxis. These can be routed through root stations associated with different localities. An empty taxicab nearest to the customer location picks up the message and sends a confirmation back. The same mechanism can work if the nodes belong to different virtual organizations. In this case, the request will be routed to all other base stations by the BS_o when it fails to hear back from the RSs in its VO. The broadcast request to the BS will include information about the requested node Na, requesting node N_o and the associated RS and the BS. Each BS will route this request to its own set of RSs. In the previous example, the customer request is routed to the taxi service through the customers cellular provider which connects both the customer and the taxi service.

RS Management

Business partners can engage in a dynamic relationship to form a virtual organization (Walton and Whicker, 1996). This can lead to ad hoc creation of a VO, dynamic changes to the VO and the need for resource discovery across several AOs within a VO. For example, during a disaster event, several agencies can come together for purposes of disaster management. In such a situation, one can envisage a VO being formed between several agencies to facilitate communication. The rules of engagement are pre-determined. A central agency may coordinate activities of several agencies. Depending on the requirement, new agencies can be called upon to deal with the situation. Once their work is done, these agencies leave. An appropriate RS management protocol can ensure that the VO structure is transparent at all times and that the process of exchange of is efficient.

VO Creation

Several AOs will come together to form a VO. An assumption is that a BS will be available to create a VO with a unique name and address. Each RS will send an enter_RS message to the BS with the information about AO such as AO name, RS name and RS address. In its response, BS will send the VO information such as the VO name, BS name and BS address. BS will maintain a list of all AOs and the associated RS names and addresses. RS Entry or Exit

It is possible that a new AO can join a VO, or an existing AO can leave a VO. Any AO can be associated with multiple VOs at the same time. In such a scenario, we need to provide a capability to dynamically configure the RS. Entry mechanism and registration will be the same as the VO creation. In case an AO is leaving the VO, the RS will broadcast exit_RS message to the associated base stations and delete information about the VO. On receiving the message, the BS will delete the RS and AO information from its record.

BS Management

In dynamic markets, two or more virtual organizations can come together to conduct business. This will lead to dynamic associations between the VOs and the need for resource discovery across several VOs. In the example for disaster management, it is possible that the concerned agencies are grouped under different VOs which in turn coordinate with each other. So a virtual organization can handle the relief work for people affected. This can include coordinating food supplies, shelter, and medicine for the victims through different agencies. Another virtual organization can deal with the reconstruction work that involves activities like assessing the magnitude of damage and managing the process of repairs. Coordinating activities across multiple

VOs will require the protocol to enable dynamic access to multiples AOs with each VO.

BS Entry or Exit

It is possible that two or more BS discover one another. In that case, they will send broadcast messages describing their VO. Each BS will receive an acknowledgement in response and the information about other BS. Through such interactions each BS will be able to generate a list of existing VOs and the names and addresses of the associated BSs. A BS will broadcast its entry or exit. Remaining BSs will update their lists accordingly.

Multiple RS and BS

In the description so far, we have assumed that there is only one RS for each AO and only one BS for each VO. However, depending on the size of the network and the distances between the components, there could be several RSs per AO and several BSs per VO to facilitate address assignment and resource discovery.

Nodes

Within an AO, the nodes will register with the nearest RS. Each RS will maintain information about all other RS within an AO. It is possible that registration request for a node is sent to more than one RS. In this case, the first RS to receive the information will send an enter_node response to all other RSs within the AO in order to avoid duplicate registration. For the node discovery, the RS will first check with the local RSs before forwarding the discover request across the VO.

RS

When a new RS is added to the grid, it will send enter_RS message to all the existing RSs within an AO, as well as to the associated VOs and BSs. In response, the RSs will send their information to the new RS. BS will update its list of RSs and send the VO information back to the RS.

BS

In case of multiple BSs within a VO, the entry of a new BS will be broadcasted to all the existing BSs. In response, the BSs will send their information to the new BS. This will include information about their AOs. For the node discovery, a BS will first check with the BSs within the VO before forwarding the request across multiple VOs.

Addressing Transient Nature of Wireless Grid

Wireless networks are characterized by weak transmission signals and message losses. Power constrained nodes may suddenly crash. These types of events can create inconsistencies in the information maintained by the grid components

Node Failure: In order to detect the node failure, the RS can periodically send hello requests to the registered nodes. In case of no response from the node, the RS will send the hello requests to the specific node. After a threshold number of requests, the RS assumes that the node has failed and deletes the node information.

Message Losses: Message losses can manifest themselves in the same fashion as the node failure. The message initiator, i.e., node, RS or BS, will make multiple attempts to elicit a response from others. One of the retries will succeed in obtaining the response. There may be cases where the messages are lost only for a set of recipients. A RS or a BS can lookup its organization information and send messages to only the set of the recipients that did not respond to the previous attempts. A leaving node or a RS may not wait for a confirmation from all the recipients. Existing members in the network can periodically send hello messages to confirm their individual presence. When a RS or BS does not receive a response to the hello messages from certain members, they make an assumption that the members are no longer part of the network.

Other Considerations

Redundancy: We have assumed that there is only one RS per AO and only one BS per VO. However, depending on the size of the network and the distances between the components there can be several RSs per AO and several BSs per VO to facilitate address assignment and resource discovery. This will also be important in order to increase the throughput capacity of the network (Liu et. al., 2003). The concept of electing a new leader when the group DNS server leaves the group (Huck et. al., 2002) could be extended to the network of RS or BS nodes that communicate, share and manage hand-offs across boundaries. In the case where one RS or BS leaves the group, a pre-configured secondary BS can take over the concerned responsibility automatically.

Security Issues: Throughout our discussions we have assumed that nodes or the stations do not operate in a malicious. A rogue node or a station can manipulate the configuration of the network. By such actions, the rogue node can corner a number of addresses, making them unavailable for other nodes that may wish to join the AO. Subsequently, the rogue node can also respond on behalf of the phantom nodes making it difficult to clean up their addresses. If IP addresses are in short supply, such an action can prevent some bona-fide nodes from joining the AO. Also, the rogue node can significantly overload the system by generating several requests within a short time. It is also possible for a malicious node to generate exit messages for nodes that are still part of the network.

Many approaches assume the existence of a Security Association (SA) between the end hosts, which choose to employ a secure communication scheme and, consequently, need to authenticate each other (Papadimitratos and Haas, 2002). This SA could have been established via a secure key exchange (Asokan and Ginzboorg, 2000), or through initial distribution of credentials.

The attacks mentioned above can be thwarted by the use of digital certificates that the nodes may have obtained a priori from some trusted Authentication Servers (ASs). Using such certificates and knowledge of the AS public key, the grid nodes and stations can authenticate each other and sign their messages even when the AS is not reachable. Further work is needed to evaluate all possible security mechanisms.

Policy Management: Since the end-devices or nodes can be power constrained, one cannot assume that the devices are capable of running complex protocols such as Lightweight Directory Access Protocol (LDAP) or Common Open Policy Service (COPS). The technical aspects of policy management, such as privileges and access to resources, can be potentially handled through the root stations and the base stations. The RS should be capable of not only handling the resource intensive protocols but also maintaining the latest information on the nodes in the network and their capabilities. RS could maintain the policy database that could be populated manually or through a messaging mechanism between the nodes and the RS. When a node leaves the local grid, the policies relevant to the node are discarded. Similarly, when a new node enters the local grid, it can configure its policies on the RS through lightweight messaging. Alternatively, the policies could be pre-configured on the RS based on a classification of the resources into one of several classes, i.e., low power resource class, highly secure class, etc. This means that the devices, when they register must also communicate their capabilities.

Similar to the RS, a base station (BS) for centralized control can be envisaged for the enterprise or the virtual organization with intra-grid architecture. For an inter-grid, two or more BSs need to interact in order to conform to end-to-end Quality of Service guarantees while traversing across multiple enterprises.

CONCLUSION

In the real world, a grid environment is usually heterogeneous at least for the different computing speeds at different participating sites. The heterogeneity presents a challenge for effectively arranging load sharing activities in a computational grid. This article develops adaptive processor allocation policies based on the moldable property of parallel jobs for heterogeneous computational grids. The proposed policies can be used when a parallel job, during the scheduling activities, cannot fit in any single site in the grid. The proposed policies require users to provide estimations of job execution times upon job submission. The policies are evaluated through a series of simulations using real workload traces. The results indicate that the adaptive processor allocation policies can further improve the system performance of a heterogeneous computational grid significantly when parallel jobs have the moldable property. The effects of inexact runtime estimations on system performance are also investigated. The results indicate that the proposed adaptive processor allocation policies are effective as well as stable under different system configurations and can tolerate a wide range of estimation errors.

REFERENCES

Adjie-Winoto, W., Schwartz, E., Balakrishnan, H., & Lilley, J. (1999). The Design and Implementation of an Intentional Naming System. *Proc. 17th ACM SOSP*, Kiawah Island, SC, Dec.

Asokan, N., & Ginzboorg, P. (2000, November). Key Agreement in Ad Hoc Networks. *Computer Communications*, *23*(17), 1627–1637. doi:10.1016/S0140-3664(00)00249-8

Barabasi, A.-L., Freeh, V. W., Jeong, H., & Brockman, J. B. (2001). Parasitic Computing. *Nature*, 412.

Bruno, R., Conti, M., & Gregori, E. (2005). Mesh Networks: Commodity Multihop Ad Hoc Networks. *IEEE Communications Magazine*, *43*(3), 123–131. doi:10.1109/MCOM.2005.1404606

Droms, R. (1997). Dynamic Host Configuration Protocol. *IETF RFC 2131*.

Foster, I. (2002). *What is the Grid? A Three Point Checklist*. Argonne National Laboratory, http://www- fp.mcs.anl.gov/~foster/Articles/WhatIs-TheGrid.pdf.

Foster, I., Kesselman, C., & Tuecke, S. (2001) The Anatomy of the Grid: Enabling Scalable Virtual Organizations. *International J. Supercomputer Applications, 15*(3).

Gamma, E., Helm, R., Johnson, R., & Vlissides, J. (1995). *Design Patterns*. Reading, MA: Addison-Wesley.

Gentzsch, W. (2001). *Grid Computing: A New Technology for the Advanced Web*. White Paper, Sun Microsystems, Inc., Palo Alto, CA.

Huck, P., Butler, M., Gupta, A., & Feng, M. (2002). A Self-Configuring and Self-Administering Name System with Dynamic Address Assignment. *ACM Transactions on Internet Technology*, *2*(1), 14–46. doi:10.1145/503334.503336

Liu, B., Liu, Z., & Towsley, D. (2003). On the Capacity of Hybrid Wireless Networks. *Proc. of IEEE Infocom*.

Mohsin, M., & Prakash, R. (2002). IP Address Assignment in a Mobile Ad Hoc Network. *IEEE Military Communications Conference (MILCOM 2002), 2*(10), 856-861.

Nesargi, S., & Prakash, R. (2002). MANETconf: Configuration of Hosts in a Mobile Ad Hoc Network. *Proceedings of INFOCOM'02,* (pp. 1059-1068.L).

Oliveira, L., Sales, L., Loureiro, E., Almeida, H., & Perkusuch, A. (2006). Filling the gap between mobile and service-oriented computing: issues for evolving mobile computing towards wired infrastructures and vice versa. *International Journal of Web and Grid Services*, *2*(4), 355–378. doi:10.1504/IJWGS.2006.011710

Ong, S. H. (2003). *Grid Computing: Business Policy and Implications*. Master's Thesis, MIT, Cambridge, MA.

Papadimitratos, P., & Haas, Z. J. (2005). Secure Routing for Mobile Ad Hoc Networks. *Advances in Wired and Wireless Communication, IEEE/Sarnoff Symposium*, (pp. 168-171).

Perkins, C., & Royer, E. (1999). Ad Hoc On-Demand Distance Vector Routing. *In 2nd IEEE Workshop on Selected Areas in Communication, 2,* 90–100.H.

Perkins, C. E. (2002). Mobile IP. *Communications Magazine, IEEE*, *40*(5), 66–82. doi:10.1109/MCOM.2002.1006976

Sharmin, M., Ahmed, S., & Ahamed, S. I. (2006). *An Adaptive Lightweight Trust Reliant Secure Resource Discovery for Pervasive Computing Environments*. Proceedings of the Fourth Annual IEEE International Conference on Pervasive Computing and Communications, March, 258-263.

Tiang, H. (2003). *Grid Computing as an Integrating Force in Virtual Enterprises*. Master's Thesis, MIT, Cambridge, MA.

Vaidya, N. H. (2002). *Weak Duplicate Address Detection in Mobile Ad Hoc Networks.* MIBIHOC2002, June.

Walton, J., & Whicker, L. (1996) Virtual Enterprise: Myth and Reality. *Journal of Control*, (pp. 22-25).

Weniger, K., & Zitterbart, M. (2004). Mobile ad hoc networks – current approaches and future directions. *Network, IEEE*, *18*(4), 6–11. doi:10.1109/MNET.2004.1316754

Zhu, F., Mutka, M., & Mi, L. (2003). *Splendor: A secure, private, and location-aware service discovery protocol supporting mobile services* (pp. 235–242). Pervasive Computing and Communications.

This work was previously published in International Journal of Grid and High Performance Computing (IJGHPC), Volume 1, Issue 3, edited by Emmanuel Udoh & Ching-Hsien Hsu, pp. 37-51, copyright 2009 by IGI Publishing (an imprint of IGI Global).

Chapter 17
Push-Based Prefetching in Remote Memory Sharing System

Rui Chu
National University of Defense Technology, China

Nong Xiao
National University of Defense Technology, China

Xicheng Lu
National University of Defense Technology, China

ABSTRACT

Remote memory sharing systems aim at the goal of improving overall performance using distributed computing nodes with surplus memory capacity. To exploit the memory resources connected by the high-speed network, the user nodes, which are short of memory, can obtain extra space provision. The performance of remote memory sharing is constrained with the expensive network communication cost. In order to hide the latency of remote memory access and improve the performance, we proposed the push-based prefetching to enable the memory providers to push the potential useful pages to the user nodes. For each provider, it employs sequential pattern mining techniques, which adapts to the characteristics of memory page access sequences, on locating useful memory pages for prefetching. We have verified the effectiveness of the proposed method through trace-driven simulations.

INTRODUCTION

The rapid developing of Internet has boosted the bloom of network computing technology. As typical systems, cluster computing, peer-to-peer computing, grid computing, as well as cloud computing, commonly focus on the goal of sharing various resources distributed in a certain network environment, and provide services for a large number of users. The resources to be shared in such systems include CPU cycles, storage, data, and, as particularly discussed in this work, the memory.

As one of the most important resources in computer architecture, memory plays a key role in the factors impacting the system performance. Especially for the memory-intensive applications that have large work sets, or the I/O-intensive appli-

DOI: 10.4018/978-1-60960-603-9.ch017

cations that massively access the disk, the memory capacity may dominate the overall performance. The ultimate reason is that there exist large gaps on performance and capacity between memory and disk (Patterson, 2004), thus the traditional computer systems have to supplement the memory capacity using the low-speed disk based virtual memory, or improve the disk performance using the limited memory based cache. Accordingly, an intermediate hierarchy between memory and disk is needed to relax such restrictions.

Remote memory sharing, which aggregates a large number of idles nodes in the network environment, and exploits their memory resources for fast storage, could meet the requirements of intermediate hierarchy with adequate performance and capacity (Feeley, *et al.*, 1995; Hines, Lewandowski, *et al.*, 2006; Newhall, *et al.*, 2008; Pakin, *et al.*, 2007). The memory-intensive applications can swap obsolete local memory pages to remote memory instead of local disk (Feeley, *et al.*, 1995), or the I/O-intensive applications can also benefit from the large data cache with better hit ratio (Vishwanath, *et al.*, 2008). Various remote memory sharing schemes were proposed in the past decades. Their difference mainly exists on the underlying network environments. The network memory or cooperative caching stands on a single cluster (Deshpande, *et al.*, 2010; Wang, *et al.*, 2007), while our previous work named RAM Grid devotes to the memory sharing in the high-speed wide-area network such as a campus network (Chu, *et al.*, 2006; Zhang, *et al.*, 2007), and the recently proposed RAM Cloud also tries to aggregate the memory resources in the data center (Ousterhout, *et al.*, 2010). Their common ground is to boost the system performance with shared remote memory.

In order to study the potential performance improvement of remote memory sharing system, we will use our previous work RAM Grid as an example, to compare the overheads of data access for an 8KB block over local disk, local network file system and remote memory resource across the campus network with average 2ms round-trip latency and 2MB bandwidth. From Table 1, we can observe that the remote memory access only reduces the overhead by 25%~30%, and the major overhead mainly comes from the network transmission cost (nearly 60%). Therefore, the performance of remote memory sharing can be obviously improved if we reduce or hide some of the transmission cost. Prefetching is such an approach to hide the cost of low speed media among different levels of storage devices (Shi, *et al.*, 2006; Vanderwiel, *et al.*, 2000; Yang, *et al.*, 2004). In this work, we will employ prefetching in remote memory sharing in order to reduce the overhead and improve the performance. Differing from traditional I/O devices, the remote nodes providing memory resources often have extra CPU cycles. Therefore, they can be exploited to decide the prefetching policy and parameters, thus releasing the user nodes, which are often dedicated to mass of computing tasks, from the process of prefetching. In contrast to traditional approaches, in which the prefetching data are decided by a rather simple algorithm in a user node, such a push-based prefetching scheme can be more effective.

To facilitate later description, we will classify the nodes in RAM Grid into different categories. The user node is the consumer of remote memory, while the corresponding memory provider is called the memory node. Before that, there also exist manager nodes which act as information

Table 1. Data access overhead in different ways

	remote memory	local disk	LAN file system
memory access	<0.01ms		
net latency	2ms		0.68ms
net transmit	4ms		0.06ms
disk latency		7.9ms	7.9ms
disk transmit		0.1ms	0.1ms
total	≈ 6ms	8ms	8.74ms

directories. In later sections, the system architecture and the prefetching design will be discussed among these nodes distributed in a high-speed wide-area network environment.

OVERVIEW

In traditional systems, an actual disk I/O operation only occurs when it misses the local file system cache in the operating system. Sarkar *et al.* mentioned that the cache must be large enough otherwise the costly disk accesses will dominate the system performance (Sarkar, *et al.*, 1996). The effect of RAM Grid, as well as other remote memory sharing systems, is that it provides abundant memory resources, which serves as an intermediate cache hierarchy between the local file system cache and local disk.

Another problem of the traditional file system cache comes from the mechanism of read ahead. The system often read several sequential blocks when accessing just the first block of the sequence. We can take the read ahead as a "blind" pull based prefetching; the shortcoming of such prefetching is two-fold. Firstly, the user node should decide the number of blocks that it needs to read ahead, which will unnecessarily take extra CPU cycles. Secondly, read ahead on sequential blocks without pattern analysis may have the risk of wasting disk or network bandwidth and memory buffers for the fact that not all of the applications will access sequential blocks, which is usually called "cache pollution". In this paper, we propose a push-based prefetching to solve the first problem, and a "smart" prefetching based on the pattern analysis instead of a "blind" one to address the second problem.

In order to study the operations of traditional file system cache, we collect disk access traces from a very busy running web server with about 2 million page views per day. The server configuration includes 2 Intel Pentium4 3.0GHz CPU with 2GB physical memory and 80GB SCSI hard disk, running Windows 2003 Enterprise Edition operating system and IIS 6.0. We collect the disk access trace using the DiskMon toolkit. Note that the web server is providing contents for real users, thus the disk I/O also comes from the real browsing activities. After record 2,380,370 disk accesses in 50 hours, including all of the hits and misses in the local cache, we can observe from the collected traces that many of the disk accesses have specific patterns, which results from the hyperlink relationships and the fact that the users often have their browsing habits. For a generic example, the access on sector 76120651 has 1,295 occurrences in our traces, and most of them are near to the access on sectors 76120707 and 76120735. Although they are not sequential numbers and there are often several outlying accesses between them, we can infer that 76120651...76120707...76120735 is a pattern. In most of cases, the access on sector 76120651 indicates that the access on 76120707 and 76120735 will come soon.

Therefore, we can design a prefetching algorithm based on pattern forecasting, which is executed by the memory nodes in RAM Grid. After a number of accesses fall into the remote caching provided by a memory node, it can forecast the most probable disk blocks referred by sequential accesses, and actively pushes these probable disk blocks to the user node. Such a push-based prefetching algorithm will make time overlapping in network communication and boost the system performance, as illustrated in Figure 1.

Compared with the traditional read ahead mechanism, the advantages of the push-based prefetching can be listed as follows. Firstly, the user nodes in RAM Grid are usually burdened with heavy workloads, while the memory nodes often have extra CPU cycles. Thus the latter fit for the forecast process of prefetching much better than the former, and a consumptive but precise prefetching algorithm can be employed. Secondly, besides the computational overhead, a prefetching algorithm may have considerable space consumption, and the memory nodes have

Figure 1. Time overlapping in network communication

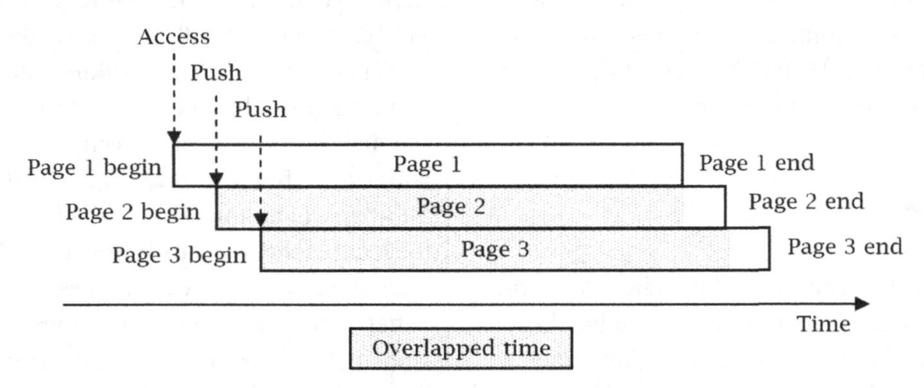

plenty resources to do that instead of the user nodes with limited available memory. Moreover, since the prefetching algorithm is executed by the memory node, the potential used data blocks can be pushed without a prefetching command comes from the user node, and the extra communication cost can also be prevented.

The basic operations of the remote caching are "put page" and "get page" upon the basic element "memory pages", which correspond to write or read operations on local disks upon disk blocks. In most cases, the "write" or "put page" can be overlapped by an asynchronous operation, thus, their access latency can be ignored. As a consequence, we only consider the "read" or "get page" operations and do not distinguish between them.

SYSTEM DESIGN

VanderWiel *et al.* concluded that a data prefetching mechanism should address three basic questions (Vanderwiel, *et al.*, 2000): 1) When is prefetching initiated, 2) where are prefetched data placed, and 3) what is prefetched? In this section, we will mainly discuss these questions.

Prefetching Buffer

For each prefetched disk block, memory pages must be allocated to hold it before the actual reading starts. If the free physical memory is not enough, the operating system has to evict some of the obsolete memory pages. However, the prefetched memory pages may not be used at all. In this case, the allocated memory pages are wasted. Therefore, we need to design a prefetching algorithm to maximize the possibility that a user node will use the prefetched memory pages pushed by a memory node.

We will firstly propose a system policy that a user node determines whether a prefetched memory page should be accepted. The policy is important because not all of the pushed memory pages should be accepted, otherwise the cache will be polluted, while the network bandwidth will also be wasted if the user node rejects too much pushed pages. In our scheme, a prefetching buffer is assigned for each user node. The prefetching buffer is a queue of free memory pages with the maximal size of k ($0 < k < F$), and F is the number of free memory pages in the user node. The system should maintain the prefetching buffer as follows:

- If the prefetched page can be found in the file system cache, just reject it; else accept it when $k > 0$;
- Else if the size of the current prefetching buffer is less than k, allocate memory for the accepted page and add the page to the tail of queue;

- Otherwise, the length of the queue equals k, it means that the size of the prefetching buffer has reached the maximum limit. Discard the oldest page at the head of the queue, and add the accepted page to the queue tail;
- When a page in prefetching buffer is actually accessed, it will be read into the file system cache, then remove it from the prefetching buffer.

The parameter k is a key factor for the prefetching buffer and it is related to the free physical memory of the system. If a user node lacks physical memory, its k should be set to a smaller value to minimize the memory waste, while k should be set larger to hold more prefetched pages when the free memory is larger. The relationship between k and free memory will be studied in the simulations.

Access Trace

In our prefetching policy, the memory node selects the memory pages to be pushed through the access patterns of a user node, which can be analyzed from the historical traces of the user. The memory node should also record current access traces for future analysis. Every "get page" operation from a user node contains the ID of the disk block corresponding to a desired memory page, it seems appropriate to record each disk block ID as historical trace data and analyze request patterns from it. Unfortunately, in most file systems, a file corresponds to a certain number of disk blocks; while their relationship may be changed at any moment. This means that a block may belong to different files in different time once the file was moved or deleted. Therefore, instead of disk block ID, we consider a file ID and an offset in the file for trace recording. When a user node gets a page from a memory node, it also sends the file ID (supported by the file system, such as the inode in Unix-like file systems) and offset within the file. The memory node will record and analyze both of the file ID and offset in the historical trace.

There is another problem in the access trace. When multiple applications are accessing the file system in parallel, the access pattern of one application can be interblended by other applications. For example, an application A may read block A_1, A_2, A_3, and application B will read B_1, B_2, B_3. While their access trace may be considered as A_1, A_2, A_3, B_1, B_2, B_3, or A_1, B_1, A_2, B_2, A_3, B_3,. We can also observe this occasion in the real web server traces described in last Section. The accesses on sectors 76120651, 76120707 and 76120735 can be taken as a pattern, but there are many outlying traces between every two of them. Therefore, the prefetching algorithm should recognize each of the patterns in a mixed access sequence, as explained later in the prefetching algorithm.

Trace Recording Process

When a memory node is recording the access trace of the user node, it maintains a sequence of file IDs and offsets which is ordered by their access timestamp. The sequence can be denoted as $S = \langle o_1, o_2, o_3, \ldots, o_n \rangle$, where each $o_i \left(i \in [1, n] \right)$ is a combination of file ID and offset.

Each sequence should be partitioned into some small ones when recording. If the difference of access time for two neighboring items o_i and o_{i+1} in a sequence is longer than a threshold t, we can split the sequence into two halves between the neighboring items. The rationale for splitting is that if the memory node pushes o_{i+1} when o_i is accessed, o_{i+1} will stay in the prefetching buffer for a long time and may possibly be discarded by the user node before it is actually accessed. In other words, prefetching for o_{i+1} is useless because its intended access time is too late. Therefore, we can partition a sequence into small sequences through a parameter t and find access patterns in each small sequence. We will call a partitioned small sequence a "trace item". Indeed, the selection of t can have an impact on the performance of prefetching. We will analyze this parameter in the simulations.

Besides of the memory nodes and user nodes, we also exploit the manager nodes, which often have idle CPU cycles and less churn, to collect and merge the trace items from each memory node and compose a trace library from the accumulated trace items. The manager nodes should also dispatch the trace library to new memory nodes for prefetching. There is a maximum number for the trace items in a trace library, denoted as M. It is related with the system performance. Indeed, the larger M means more user patterns and more accurate results in the prefetching algorithm, while the larger M also needs more memory space to hold the trace library and more transmission cost between manager nodes and the memory nodes. We will evaluate the impact of M later.

PREFETCHING ALGORITHM

The goal of the prefetching algorithm is to predict the most likely pages of future requests that start with the current request issued by the user node. The memory node needs to determine those pages based on the user patterns derived from the trace library and other necessary parameters, such as the maximum prefetching buffer length of the user node, k. In order to reduce the network communication cost, we want the predicted and prefetched memory pages to have the highest probability to be used by the user node. In fact, selecting proper pages is a data mining problem, which can be defined as follows.

Let $\{o_1, o_2, \ldots, o_n\}$ be a set of all possible items, where each item is a recorded access. A sequence S, which can be denoted as $S = \langle o_1, o_2, \ldots, o_n \rangle$, is an ordered list of items. The number of items in sequence S, denoted as *length*(S). A sequence $\alpha = \langle a_1, a_2, \ldots, a_n \rangle$ is contained in another sequence $\beta = \langle b_1, b_2, \ldots, b_m \rangle$, denoted as $\alpha \subseteq \beta$, *iff* there exist integers $1 \le i_1 < i_2 < \ldots < i_n \le m$, such that $a_1 = b_{i_1}$, $a_2 = b_{i_2}, \ldots, a_n = b_{i_n}$. A sequence can be appended to another sequence using a concatenation operator '+'. For example, $\alpha + \beta = \langle a_1, a_2, \ldots, a_n, b_1, b_2, \ldots, b_m \rangle$. The first occurring position of an item a in a sequence S is denoted as $first(S, a)$, thus $first(\beta, b_n) = n$ ($b_1 \ne b_n, b_2 \ne b_n, \ldots, b_{n-1} \ne b_n$). A trace library L is an ordered set consisting of multiple sequences, that is, $L = \langle S_1, S_2, \ldots, S_n \rangle$.

For a given sequence S, if there exists a trace library $L = \langle S_1, S_2, \ldots, S_n \rangle$, and $S \subseteq S_i \left(i \in [1, n]\right)$, then we say that L supports S, and the support of S in trace library L is the number of S_i in L which satisfies $S \subseteq S_i$. The problem of mining prefetching sequences can be described as follows.

For a given access o_c, search $S = \langle o_c, o_1, o_2, \ldots, o_n \rangle$ with the maximum support in the trace library, where $length(S) \ge 2$. S is called the prefetching list. The memory node obtains the prefetching list S and sequentially pushes pages in the list to the user node, when the latter performs a "get page" operation with an access o_c.

According to the definition of $\alpha \subseteq \beta$, the common items in α and β is not necessary to be consecutive. Supposing $S_1 = \langle A_1, B_1, A_2, B_2, A_3, B_3 \rangle$, $S_2 = \langle A_1, A_2, A_3 \rangle$, then we have $S_2 \subseteq S_1$. This definition solves the problem mentioned in last section.

In fact, the background of the prefetching algorithm is inspired by the traditional sequential pattern mining (Agrawal, *et al.*, 1995, 1996). Although the there exist similarities, the problem is quite different. Firstly, an item in sequential pattern mining can be composed by several numbers, while for prefetching it is just a single number that indicates the block identification. Moreover, the output of sequential pattern mining is the sequence whose support is higher than a threshold

H, in prefetching, however, the problem focuses on the maximum support among the sequences. Such differences make the prefetching algorithm much more efficient. Furthermore, for practical usage, we design some additional constraints for mining prefetching pages:

- $length(S) \leq k+1$, where S is the prefetching list and k is the maximum prefetching buffer size of the user node. Supposing we push more than k pages to the user node each time, some of the pushed pages may be discarded due to the limit of the prefetching buffer size.
- We can add a constraint in the definition of $\alpha \subseteq \beta$ that $i_{x+1} - i_x \leq d$, where $x \in [1, n-1]$ and d is a given integer threshold. Because the prefetched pages should not be far away from the accessed pages in a sequence of the trace library, otherwise its may be discarded before accessed.

Our prefetching algorithm named PrefixSpan-Prefetching (PSP) can be described as Algorithm 1.

The algorithm PSP is efficient since it has no recursion. The time cost of PSP primarily comes from the iteration from step 4 to step 26, the condition to stop this iteration is that $length(S)>k+1$ or it cannot find possible item P when spanning S in step 16. In other words, in the worst case this iteration should run k times for a prefetching list of length k. In each iteration, steps 5-17 need to scan the trace library, whose maximum size is M. Suppose that the maximum length of sequences in the trace library is m, in order to perform a binary search in each sequence, we build a sorted index in advance for the first occurrence of each item in the sequence. Therefore, for each item in the prefetching list, the time cost of PSP is $O(M\log_2 m)$ in the worst case, and the extra space cost from each sorted index is $O(m)$ in the worst case.

The correctness of PSP is discussed as follows.

Lemma. The support of a sequence $\alpha+\beta$ is always not greater than the support of α.

Proof. Assuming that $\{S_i\}$ is the set of sequences where $\alpha + \beta \subseteq S_i$, it is obvious that for each S_i, we have $\alpha \subseteq S_i$, which indicates the support of $\alpha+\beta$ is always not greater than α. ■

From the lemma, we have the corollary as follows:

Corollary. If the support of a sequence α is not the maximum, the support of a sequence $\alpha+\beta$ cannot be the maximum either.

Based on the lemma and the corollary, the following theorem proves the correctness of algorithm PrefixSpan-Prefetching:

Theorem. After algorithm PrefixSpan-Prefetching, the support of sequence S is the maximum for all possible S, where $2 \leq length(S) \leq k+1$.

Proof. Step 14 in PSP is only executed in the first iteration and it can get a sequence $S=<o_c,P>$, which has the maximum support; step 16 is executed after the first iteration, it only accepts item P if the support of $S+<P>$ is equal to that of S. Before step 19, if the support of any sequence T is less than that of S, then there is no sequence that has the maximum support, with the prefix of T (due to the corollary). Thus, T should not be spanned in step 19. Hence, the support of sequence S is always the maximum in each iteration. ■

Algorithm 1. PrefixSpan-Prefetching (PSP)

Input: Current access o_c, the trace library L, factor k and d.
Output: Prefetching list S, where $2 \le length(S) \le k+1$.

1 As an initial sequence, let $S = < o_c >$;
2 let $L' = L$. L' is a copy of L and will prevent the latter from any modifies during the algorithm;
3 let $last = \{o_c\}$. The set last will contain all possible postfix items in the prefetching list;
4 **while** $length(S) \le k+1$ **do**
5 **for** each $S_i \in L'$, **do**
6 **if** $\forall P(P \in last \Rightarrow P \in S_i)$ **then**
7 select $P_x \in last$ where $first(S_i, P_x)$ is the minimal;
8 supposing $first(S_i, P_x) = n$, trim the first n items of S_i;
9 **else**
10 delete S_i in L'
11 **end**
12 **end**
13 **if** $S = < o_c >$ **then**
14 scan L' once, find all possible items P, where $(P \in S_i) \wedge (S_i \in L') \wedge first(S_i, P) \le d$, and <P> has the maximum support in L';
15 **else**
16 scan L' once, find all possible items P, where $\forall S_i(S_i \in L' \Rightarrow \exists P(P \in S_i \wedge first(S_i, P) \le d))$;
17 **end**
18 let $last = \varphi$;
19 **for** each possible items P in steps 13-17, **do**
20 let $S = S + < P >$;
21 let $last = last \cup \{P\}$;
22 **end**
23 **if** $last = \varphi$ **then**
24 terminate the algorithm;
25 **end**
26 **end**

PERFORMANCE EVALUATION

Simulation Methodology

Our application scenario is composed of abundant PCs and several server stations loosely coupled in a high-speed wide-area network. Some PCs are idle and have free memory resources, whereas servers are usually busy for tasks with mass non-consequence data accesses (such as a web server or DBMS), whose local physical memory is intended to be utilized as much as possible. A typical example of this configuration is a campus or enterprise network with many heterogeneous computers. In order to simulate such scenario by the disk I/O traces of the very busy running web server that we have collected, we assume that servers in our scenario are all web servers with many users. The disk I/O traces in our simulation have already been mentioned before.

We have built a discrete event based simulator of the environment with 1000 different nodes. The simulation topology of 1000 nodes is generated using the ASWaxman model through the topology generator BRITE (Medina, *et al.*, 2001). We use the TopDown method in BRITE to generate a 2-level network topology, which includes 10 ASes and each AS has 100 router-level nodes respectively, the nodes placement follows the heavy-tailed distribution. The generated topology is a DAG, where vertices are simulation nodes and each edge is an overlay path between two vertices. The routing between any two vertices is the shortest path between them computed by *Dijkstra's Algorithm*.

We define parameters of hard disks and the remote memory to calculate local and remote I/O overheads. When performing a disk read with n successive blocks, the overhead is given by:

$$T_S + T_L + (n-1) \times T_W + n \times \frac{S_p}{B_d}$$

where T_S is the seeking time, T_L the latency time, T_W the waiting time between two successive readings, S_p the block size, and B_d the disk bandwidth. Typical values of these parameters are T_S =4.9 milliseconds, T_L =3.0milliseconds, T_W =0.2 milliseconds, S_p =4KB, and B_d =80MB/s.

For the remote memory, the read overhead for n successive block readings is given by:

$$T_U + T_{RTT} + n \times \frac{S_p}{B_N}$$

where T_U is the start-up time, T_{RTT} the round-trip time, and B_N the network bandwidth. In our simulation, T_U is set to 5 microseconds, T_{RTT} varies from 1 millisecond to 4 milliseconds following a uniform distribution, and B_N varies from 0.5MB/s to 3MB/s following a uniform distribution. These parameters are from the actual testing of our campus network.

Results

Simulation 1. The effect of proportion of user nodes

In this set of simulations, we test the effect of the proportion of user nodes on average overheads. The proportion of user nodes is set to around c $(0 \leq c \leq 1)$ in our simulation. Both overheads of RAM Grid without or with prefetching would change when the proportion of user nodes changes. As illustrated in Figure 2. When the proportion of user nodes is within the range from 20% to 30% the overhead changes rapidly. When the proportion falls out of this range, curves become flat. This is reasonable, since when the proportion of user nodes is less than 20%, most of them can obtain sufficient memory resources, and if it is more than 30%, the number of user nodes that can capture resources becomes smaller and curves thus change little with the increasing proportion of user nodes. Therefore, bounds 20% and 30% can be considered as critical proportions. In Figure 3, we compare three types of hit ratios in the proposed scheme: 1) the hit ratios of local and remote memory, which means the percentage of all accesses except the ones that do not hit any type of cache and cause the actual disk I/O operations; 2) the hit ratios of local buffer cache, meaning the percentage of all accesses which hit the local cache of file system, or hit the prefetching buffer in our scheme; 3) the hit ratios of prefetching buffer only, that is, the probability of hitting the prefetching buffer if the access does not hit the local cache of the file system. It is interesting that the hit ratios of remote and local memory decease rapidly with the increasing of user nodes proportion, however hit ratios of the local buffer cache and prefetching buffer are slightly increasing. We can infer that the increasing overhead in Figure 2 is mainly due to the insufficient memory in the entire environment. However, the performance of our prefetching algorithm would not decease in this case.

Simulation 2. The effect of prefetching buffer size

The prefetching buffer is one of key factors in our scheme. It shares the free memory capacity with the file system cache, and its maximum size is restricted. In this set of simulations, we let the upper bound of prefetching buffer to be from 1/2 to 1/128 of the free physical memory capacity. The overhead and hit ratio with prefetching are reported in Figure 4 and Figure 5, respectively. Similarly, there exists a critical proportion range for the prefetching buffer. When the proportion of the prefetching buffer is within the range from 1/4 to 1/16, the performance of the algorithm changes rapidly. The proportion of 1/3 is close to the optimal, since it does not take much local memory, and thus has the good performance of prefetching. We can also see from Figure 5 that the hit ratio of prefetching buffer decreases rapidly with the decreasing prefetching buffer, whereas that of the local cache and the prefetching buffer does not change very much. This lies on the fact that the free local memory is constant. The decreasing prefetching buffer causes the increase of the local cache, which reduces the effect of prefetching buffer.

Simulation 3. The effect of trace library size

The effectiveness of the PSP algorithm is related to the amount of user patterns contained in the trace library. We set a maximum trace library size and the old traces should be discarded. By default, the trace library contains at most 3000 traces. Indeed, it is near optimal value in Figure 6 and Figure 7. Obviously, when trace library becomes small, the performance of PSP drops rapidly. This is because there are not enough training data to get the right prefetching list. However, the larger size of the trace library can also decrease the performance of the algorithm. Both the overhead and hit ratio in the case of 8000 traces of the trace library are the worst in Figure 6 and Figure 7. This situation is true with not only the transmission overhead of large trace library,

Figure 2. Reading overhead with different user node proportion

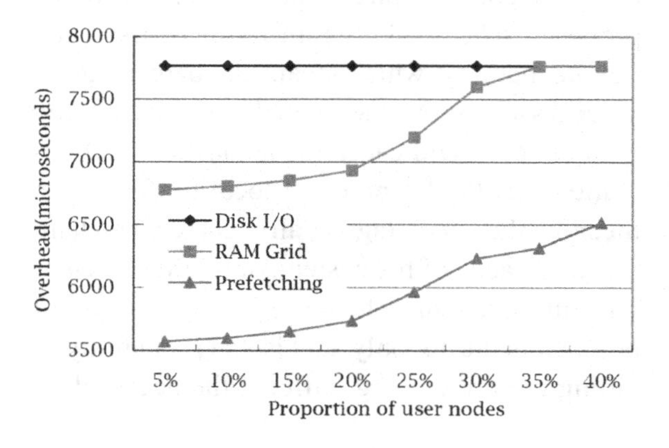

Figure 3. Hit ratio with different user node proportion

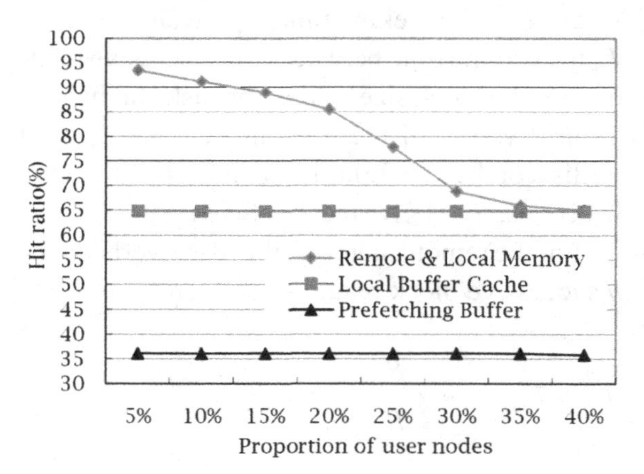

Figure 4. Reading overhead with different maximum prefetching buffer proportion

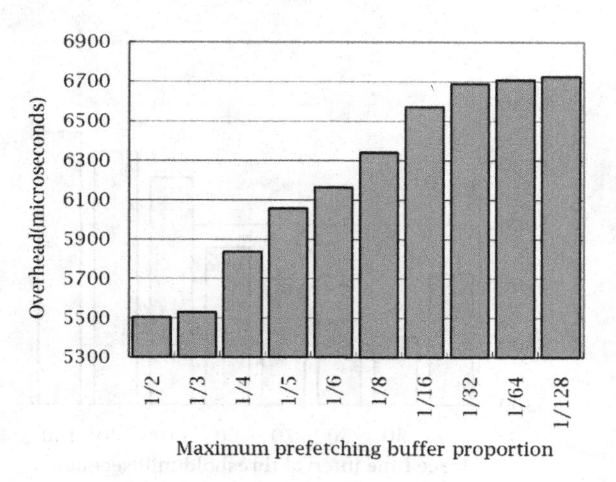

Figure 5. Hit ratio with different maximum prefetching buffer proportion

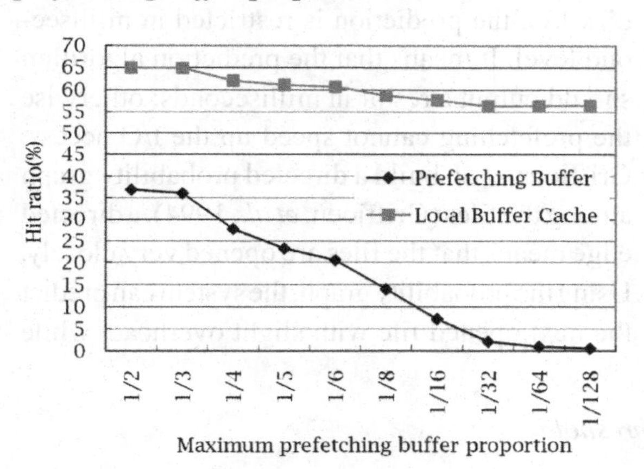

but also old traces that are kept in the large trace library for a long time, which would not help improve the accuracy of the algorithm. Therefore, the desired algorithm should either wash out the old traces in time, or reduce their impact, which are interesting topics for our future work.

Simulation 4. The effect of splitting threshold

According to the trace recording process, a sequence is split into small ones when the recording time interval of two neighboring items is greater than a threshold. In the last set of simulations, we vary this time interval threshold from 10ms to 150ms (the default value is 30ms), and illustrate the overhead and hit ratio with prefetching in Figure 8 and Figure 9, respectively. We observe that here also exists an optimal value for this threshold. In particular, short traces would miss some long sequences during the trace collecting process, while a large time interval threshold that may keep longer traces would take a long time to collect enough traces for prefetching, resulting in the slow increase of the trace library size and thus less accuracy of the algorithm.

RELATED WORK

The history of remote memory sharing system can be retrospect to 1990s. As an initial work, several memory sharing schemes, which are usually called network memory systems, have been proposed. We can category these systems into three major types based on the objectives, which are high-speed paging device (Feeley, *et al.*, 1995; Hines, Lewandowski, *et al.*, 2006; Hines, Wang, *et al.*, 2006; Markatos, *et al.*, 1996; Oleszkiewicz, *et al.*, 2004), data cache for local

Figure 6. Reading overhead with different maximum trace library size

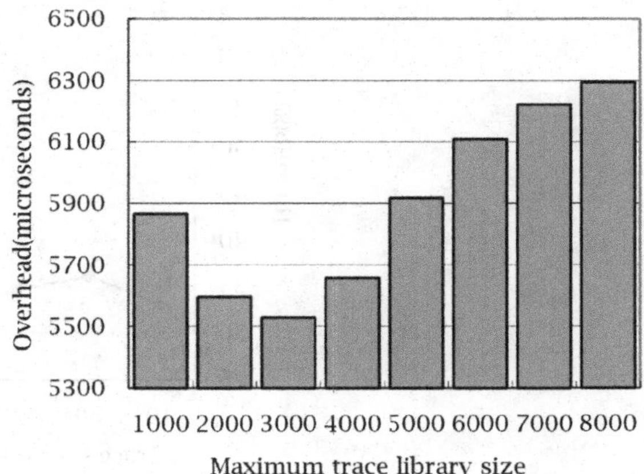

Figure 7. Hit ratio with different maximum trace library size

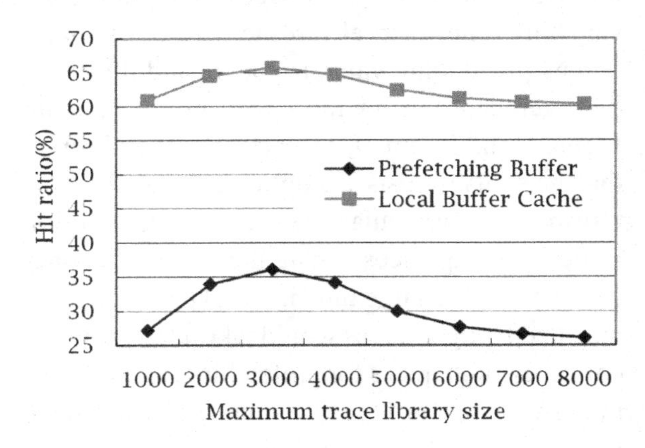

Figure 8. Reading overhead with different trace time interval threshold

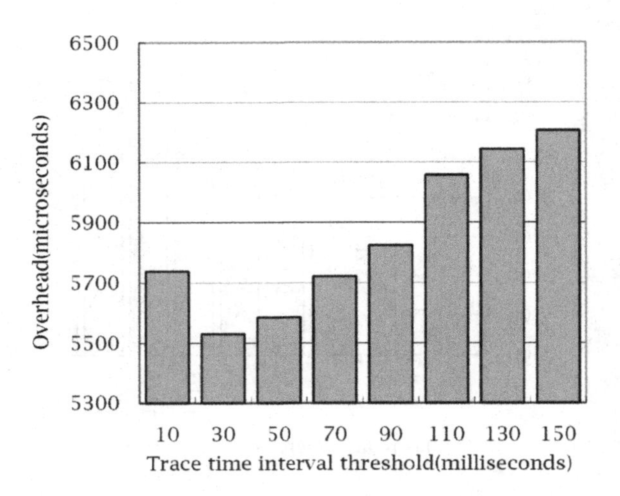

or networked file systems (Chang, *et al.*, 1999; Dahlin, *et al.*, 1994; Jiang, *et al.*, 2006; Sarkar, *et al.*, 1996; Voelker, *et al.*, 1998), or remote RAM disk respectively (Flouris, *et al.*, 1999). Unlike network memory schemes, RAM Grid tries to share the plentiful memory resources distributed in a wide area network (Chu, *et al.*, 2006). It aggregates resources in a large scale and avoids the inadequate idle memory resources problem within a single cluster, while it must also deal with the dynamic and heterogeneous resources effectively using a decentralized architecture.

The effect of prefetching mainly depends on the prediction of the data access. For magnetic disk I/O, the prediction is restricted in millisecond level. It means that the prediction algorithm should output a result in milliseconds; otherwise the prefetching cannot speed up the I/O access. Griffioen *et al.* build a directed probability graph among the files (Griffioen, *et al.*, 1994), a directed edge means that the files are opened very closely. Using the probability graph, the system can predict the next opened file with slight overhead, while

Figure 9. Hit ratio with different trace time interval threshold

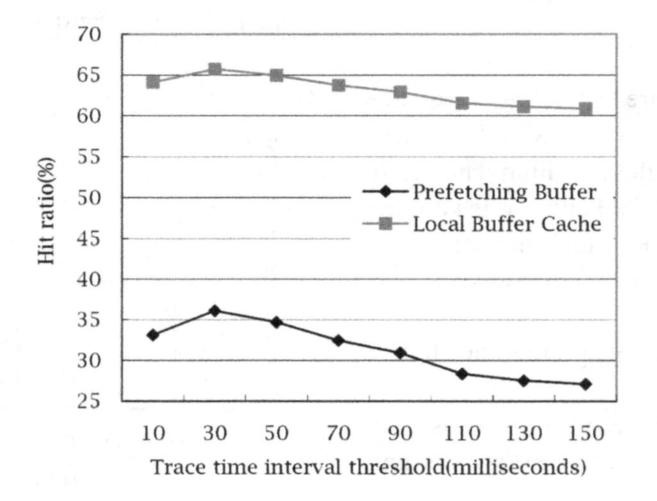

the prediction has a coarse granularity. Choi *et al.* present a prediction based on disk blocks(Choi, *et al.*, 2000). The algorithm classifies disk access into several predefined patterns, and predict current pattern when accessing disk blocks. The work of Gniady *et al.* predicts I/O access using the program counter (Gniady, *et al.*, 2004). The system maintains a hash table from the program counter to the access pattern and predict the pattern. Different with those previous works, our system predict the I/O access using a data mining method, which usually has finer granularity and better precision.

Our algorithm for collecting the trace and inferring the pattern of users is based on the problem of sequential pattern mining. Agrawal *et al.* first defined the problem of sequential patterns (Agrawal, *et al.*, 1995, 1996). However, their algorithms are not applicable to for very long sequences which are often the case in grid environments. Pei *et al.* proposed PrefixSpan algorithm (Pei, *et al.*, 2001), which improves upon Apriori and reduces the overhead. These algorithms are based on the general problem of sequential pattern mining in very large databases (Ayres, *et al.*, 2002); while the background of our algorithm is quite specific, some of the restrictions in traditional sequential pattern mining can be released and the algorithm is also more effective.

CONCLUSION

With the rapid development of the network technology, several remote memory sharing systems have been proposed to aggregate memory resources through definite network environment. Our previous work, RAM Grid, made use of the remote memory to boost the performance of memory-intensive and I/O-intensive applications. In this paper, in order to reduce the network communication cost of accessing the remote memory, based on a push strategy and inspired by traditional sequential patterns mining techniques, we propose a prefetching algorithm to push more pages to a user node. By mining the historical information, a memory node can push the required data to user nodes efficiently. We demonstrate the efficiency and effectiveness of the proposed prefetching scheme through comprehensive trace-driven simulations.

ACKNOWLEDGMENT

The work was supported by the National Natural Science Foundation of China under Grant No.61003076 and the National Basic Research Program of China (973) under Grant No.2011CB302600.

REFERENCES

Agrawal, R., & Srikant, R. (1995). *Mining sequential patterns.* Paper presented at the 17th International Conference on Data Engineering.

Agrawal, R., & Srikant, R. (1996). *Mining sequential patterns: Generalizations and performance improvements.* Paper presented at the 5th International Conference on Extending Database Technology: Advances in Database Technology.

Ayres, J., Flannick, J., Gehrke, J., & Yiu, T. (2002). *Sequential pattern mining using a bitmap representation.* Paper presented at the 8th International Conference on Knowledge Discovery and Data Mining Edmonton, Alberta, Canada.

Chang, E., & Garcia-Molina, H. (1999). *Medic: A memory and disk cache for multimedia clients.* Paper presented at the IEEE International Conference on Multimedia Computing and Systems, Florence, Italy.

Choi, J., Noh, S. H., Min, S. L., & Cho, Y. (2000). *Towards application/file-level characterization of block references: A case for fine-grained buffer management.* Paper presented at the ACM SIGMETRICS International Conference on Measurement and Modeling of Computer Systems Santa Clara, California, United States.

Chu, R., Xiao, N., Zhuang, Y., Liu, Y., & Lu, X. (2006). *A distributed paging RAM Grid system for wide-area memory sharing.* Paper presented at the 20th International Parallel and Distributed Processing Symposium, Rhodes Island, Greece.

Dahlin, M. D., Wang, R. Y., Anderson, T. E., & Patterson, D. A. (1994). *Cooperative caching: Using remote client memory to improve file system performance.* Paper presented at the 1st Symposium on Operating Systems Design and Implementation, Monterey, California.

Deshpande, U., Wang, B., Haque, S., Hines, M., & Gopalan, K. (2010). *MemX: Virtualization of cluster-wide memory.* Paper presented at the International Conference on Parallel Processing.

Feeley, M. J., Morgan, W. E., Pighin, F. H., Karlin, A. R., Levy, H. M., & Thekkath, C. A. (1995). *Implementing global memory management in a workstation cluster.* Paper presented at the Symposium on Operating Systems Principles, Copper Mountain Resort, Colorado.

Flouris, M. D., & Markatos, E. P. (1999). The network RamDisk: Using remote memory on heterogeneous NOWs. *Cluster Computing, 2*(4), 281–293. doi:10.1023/A:1019051330479

Gniady, C., Butt, A. R., & Hu, Y. C. (2004). *Program-counter-based pattern classification in buffer caching.* Paper presented at the 6th Symposium on Operating Systems Design and Implementation, San Francisco, CA.

Griffioen, J., & Appleton, R. (1994). *Reducing file system latency using a predictive approach.* Paper presented at the USENIX Summer Conference.

Hines, M., Lewandowski, M., Wang, J., & Gopalan, K. (2006). *Anemone: Transparently harnessing cluster-wide memory.* Paper presented at the International Symposium on Performance Evaluation of Computer and Telecommunication Systems, Calgary, Alberta, Canada.

Hines, M., Wang, J., & Gopalan, K. (2006). *Distributed Anemone: Transparent low-latency access to remote memory in commodity clusters.* Paper presented at the International Conference on High-Performance Computing, Bangalore, India.

Jiang, S., Petrini, F., Ding, X., & Zhang, X. (2006). *A locality-aware cooperative cache management protocol to improve network file system performance.* Paper presented at the 26th IEEE International Conference on Distributed Computing Systems, Lisbon, Portugal.

Markatos, E. P., & Dramitinos, G. (1996). *Implementation of a reliable remote memory pager.* Paper presented at the USENIX Annual Technical Conference, San Diego, CA.

Medina, A., Lakhina, A., Matta, I., & Byers, J. (2001). *BRITE: An approach to universal topology generation.* Paper presented at the International Workshop on Modeling, Analysis and Simulation of Computer and Telecommunications Systems, Cincinnati, Ohio.

Newhall, T., Amato, D., & Pshenichkin, A. (2008). *Reliable adaptable network RAM.* Paper presented at the International Conference on Cluster Computing.

Oleszkiewicz, J., Xiao, L., & Liu, Y. (2004). *Parallel network RAM: Effectively utilizing global cluster memory for large data-intensive parallel programs.* Paper presented at the International Conference on Parallel Processing, Montreal, Quebec, Canada.

Ousterhout, J., Agrawal, P., Erickson, D., Kozyrakis, C., Leverich, J., & Mazieres, D. (2010). The case for RAMClouds: Scalable high-performance storage entirely in DRAM. *ACM SIGOPS Operating Systems Review*, *43*(4), 92–105. doi:10.1145/1713254.1713276

Pakin, S., & Johnson, G. (2007). *Performance analysis of a user-level memory server.* Paper presented at the International Conference on Cluster Computing.

Patterson, D. A. (2004). Latency lags bandwith. *Communications of the ACM*, *47*(10), 71–75. doi:10.1145/1022594.1022596

Pei, J., Han, J., Mortazavi-Asl, B., & Pinto, H. (2001). *PrefixSpan: Mining sequential patterns efficiently by prefix-projected pattern growth.* Paper presented at the 17th International Conference on Data Engineering.

Sarkar, P., & Hartman, J. (1996). *Efficient cooperative caching using hints.* Paper presented at the Symposium on Operating Systems Design and Implementation, Seattle, WA.

Shi, X., Yang, Z., Peir, J.-K., Peng, L., Chen, Y.-K., Lee, V., et al. (2006). *Coterminous locality and coterminous group data prefetching on chip-multiprocessors.* Paper presented at the 20th International Parallel and Distributed Processing Symposium, Rhodes Island, Greece.

Vanderwiel, S. P., & Lilja, D. J. (2000). Data prefetch mechanisms. *ACM Computing Surveys*, *32*(2), 174–199. doi:10.1145/358923.358939

Vishwanath, V., Burns, R., Leigh, J., & Seablom, M. (2008). Accelerating tropical cyclone analysis using LambdaRAM, a distributed data cache over wide-area ultra-fast networks. *Future Generation Computer Systems*, *25*(2), 184–191. doi:10.1016/j.future.2008.07.005

Voelker, G. M., Anderson, E. J., Kimbrel, T., Feeley, M. J., Chase, J. S., Karlin, A. R., et al. (1998). *Implementing cooperative prefetching and caching in a globally-managed memory system.* Paper presented at the Joint International Conference on Measurement and Modeling of Computer Systems, Madison, Wisconsin, United States.

Wang, N., Liu, X., He, J., Han, J., Zhang, L., & Xu, Z. (2007). *Collaborative memory pool in cluster system.* Paper presented at the International Conference on Parallel Processing.

Yang, C.-L., Lebeck, A. R., Tseng, H.-W., & Lee, C.-H. (2004). Tolerating memory latency through push prefetching for pointer-intensive applications. *ACM Transactions on Architecture and Code Optimization*, *1*(4), 445–475. doi:10.1145/1044823.1044827

Zhang, Y., Li, D., Chu, R., Xiao, N., & Lu, X. (2007). *PIBUS: A network memory-based peer-to-peer IO buffering service*. Paper presented at the 6th International IFIP-TC6 Conference on Ad Hoc and Sensor Networks, Wireless Networks, Next Generation Internet.

Chapter 18
Distributed Dynamic Load Balancing in P2P Grid Systems

You-Fu Yu
National Taichung University, Taiwan, ROC

Po-Jung Huang
National Taichung University, Taiwan, ROC

Kuan-Chou Lai
National Taichung University, Taiwan, ROC

ABSTRACT

P2P Grids could solve large-scale scientific problems by using geographically distributed heterogeneous resources. However, a number of major technical obstacles must be overcome before this potential can be realized. One critical problem to improve the effective utilization of P2P Grids is the efficient load balancing. This chapter addresses the above-mentioned problem by using a distributed load balancing policy. In this chapter, we propose a P2P communication mechanism, which is built to deliver varied information across heterogeneous Grid systems. Basing on this P2P communication mechanism, we develop a load balancing policy for improving the utilization of distributed computing resources. We also develop a P2P resource monitoring system to capture the dynamic resource information for the decision making of load balancing. Moreover, experimental results show that the proposed load balancing policy indeed improves the utilization and achieves effective load balancing.

INTRODUCTION

Recently, grid computing is one of attractive architectures for high-performance computing. The grid computing system is an Internet-scale distributed computing system for sharing distributed resources across the traditional organization boundary. In grid systems, the most important issues include how to integrate the dynamically heterogeneous distributed resources, and how to improve the utilization of these integrated resources (Dandamudi, 1995). Although these various grid projects aim at sharing distributed resources from different virtual organizations (VOs), it is still difficult to share distributed

DOI: 10.4018/978-1-60960-603-9.ch018

resources due to the different goals in building different VOs.

The peer-to-peer (P2P) computing system is another Internet-scale computing model where computers share distributed resources via exchanges among the participating computers (Androutsellis-Theotokis et al., 2004; Li et al., 2006). The widespread deployment of P2P computing systems offers great potential for resource sharing. The P2P system has the similar objective of the grid system to coordinate large sets of distributed resources. Therefore, many projects attempt to integrate these two complementary technologies to form an ideal distributed computing system (Amoretti et al., 2005; Shan et al., 2002; Shudo et al., 2005)

In this chapter, we propose a P2P-based mechanism to form a P2P Grid platform for achieving load balancing of distributed computing resources. In general, the job submission in grid systems is carried out by a global resource broker to distribute load. Here, we propose a campus-to-campus Uni-P2P communication model to integrate the Taiwan UniGrid (Taiwan UniGrid, 2009) and the Taiwan TIGER system (Yang et al., 2005) by using a P2P communication mechanism which builds the communication pipes among sites in different grid systems. This campus-to-campus Uni-P2P communication model also supports a P2P resource monitoring system that captures the dynamic resource usage. In the P2P Grid platform, super peers are employed to manage grid sites. The concept of super peers, which exhibit more powerful computing ability, bandwidth and hardware capacity, is also considered in this Uni-P2P communication model to improve the efficiency of searching distributed resources. Moreover, we propose a dynamic distributed load balancing policy to improve the idle resource utilization in the P2P Grid platform.

The rest of this chapter is structured as follows: related works are discussed in section 2 followed by the discussion of the system architecture in sections 3. Experimental results are shown in section 4. Section 5 describes conclusions and future research directions.

RELATED WORKS

There are many middlewares (e.g., Globus Toolkit, Unicore, gLite, etc.) which have been developed for grid systems. Most of them focus on providing the core middleware services for supporting the development functionality of high-level applications. However, they usually depend on specialized servers to maintain the distributed resource information. On the other hand, P2P systems adopt decentralized resource discovery approaches and thus do not rely on any specialized servers to capture distributed resource information. In this section, we present the related works of grid information systems and load balancing policies.

Resource Monitoring Systems

There are resource monitoring software for capturing the resource information, such as Ganglia, Gstat (LCG), MDS, NWS and REMOS. Ganglia is a distributed resource monitoring system; it monitors system performance and system information such as CPU load, memory usage, hard disk usage, I/O load, and network bandwidth. Gstat is the resource monitoring tool developed by ASGC in order to support the members of EGEE in handling global grid resources. Gstat supports information such as the number of CPUs and their load, the number of waiting/running jobs, and the response time from GIIS. MDS (Monitor and Discovery System) is one of the Globus Toolkits; it supports information services and monitors/searches grid resources. NWS (Network Weather Services) is also a distributed resource monitoring system. It monitors the performance of networks and computing resources periodically, and then predicts future system performance by real time information. REMOS (REsource MOnitoring System) allows the application to capture the shared resource information in the distributed computing environment. However, the above resource monitoring systems do not support the P2P mechanism for sharing resource information among sites, and result in the system bottleneck in the hierarchical architecture. Therefore, we propose a dynamic, distributed resource monitoring systems in the Uni-P2P communication model for the P2P Grid platform to capture the dynamic distributed resource status.

Resource Broker

The load balancing mechanisms in grid systems can be classified into the global approach and the local approach. The global load balancing approach usually adopts the resource broker to distribute load. Resource brokers consider the usage information of grid resources, e.g., CPU load, hard disk usage, memory load, etc., to make decisions in order to achieve better system performance. gLite is the middleware developed by the E-Science project (Enabling Grids for E-Science and Industry in Europe). The global resource broker takes charge of distributing jobs to different VOs in the gLite middleware. These distributed jobs are sent to the job queues in each VO for execution. After job submission, the global resource broker cannot dynamically adjust the load in each job queue, i.e., the global resource broker does not support the function of dynamic job migration.

In previous studies (Hu et al., 2006; Xia et al., 2006; Xu et al., 2006), authors propose distributed load balance mechanisms for computational grids with the unstructured P2P architecture. In their systems, every computing node has a job queue to manage the job execution. These studies demonstrate that their model always converges to a steady load balancing state without complete knowledge about other nodes. However, they assume that the computational grid is a homogeneous unstructured P2P network where computing nodes in the grids are homogeneous. It is not practical under the assumption that different computing nodes have the same processing speed, memory size, and storage space. These studies also assume that the process of load balancing is relatively short, during which there are neither new tasks submitted nor old tasks finished. They claim that they could solve the problem of archiving perfect load balance in decentralized architecture. However, load balancing is a time-consuming process even when new jobs are submitted or submitted jobs are migrated. In this paper, our Uni-P2P communication model can support dynamic job migration to balance loads among different grid sites.

Process Migration

Process (or Job) migration is the action which transfers a process between two computing nodes. A process migration (Tanenbaum, 2007) involves data, stacks, register contents, and the state for the underlying operating system, such as parameters related to process, memory, and file management information. Process migration could improve the load balance (Eager et al., 1986; Eager et al., 1988; Hu et al., 2006, Iyengar et al., 2006) and the reliability of distributed computing systems. Recently, some migration technology has been raised by adopting the checkpoint/restart technology in the migration process. A previous study (Milojičić et al., 2000) mentions about many process migration algorithms in distributed computing systems. Eager copy is the simple and most common process migration algorithm. Many previous studies (e.g., Lazy copy and Pre-copy) (Richmond et al., 1997) also focus on how to enhance the effectiveness of process migration, such as the information state transfer, the transfer order, the process resumption, and the network traffic reduction.

Load Balance Policy

To fully exploit the P2P Grid computing system, load balancing is one of the key issues in achieving high performance. There are three goals of the load balancing policy shown in the following:

- To distribute the workload from high-loading sites to low-loading sites.
- To maximize the resource utilization.
- To minimize the job execution time.

According to the decision making approach, load balancing policies can be categorized into two types.

Static Policy

The static load balancing policy (Pan et al., 2007) makes the balance decision by the resource information before executing jobs. It is easy to implement the static load balancing policy, and the overhead of implementing the static policy is lower than that of implementing the dynamic policy. However, it is more difficult for the static policy to obtain the optimal performance due to that it can not adjust the decision at runtime. In the high variation system, the performance of the static policy is very poor.

Dynamic Policy

The dynamic load balancing policy (Chen et al., 2008; Duan et al., 2008) makes decisions of the resource allocations by the runtime information. Although the dynamic policy (e.g., JRT (Wu et al., 2008), Max-Min (Ali et al., 1999), Min-Min (Ali et al., 1999) and RESERV (Vincze et al., 2008)) brings better performance, it is difficult to be implemented due to that it needs to collect the dynamic information for making the optimal decision. In general, the dynamic policy has the better performance than that of the static policy. In addition, the dynamic policy can maximize the system performance in the high variation environment.

On the other hand, according to the management approach, load balance policies could also be categorized into two types.

Centralized Policy

The centralized load balancing policy adopts one computing node to be the resource manager and makes the load balance decision. The centralized resource manager manages global resource information by collecting information from all sites. The global resource information facilitates the resource manager to allocate resources. Therefore, the centralized policies manage resources easily and achieve better performance. However, centralized resource manager could be the bottleneck of the system; moreover, it may become the single point of failure.

Distributed Policy

The distributed load balancing policy allows every computing site in the distributed system to make load balance decisions. In addition, the computing site only needs to collect the information from its linked sites. Although the cost of obtaining and maintaining the dynamic system information is very high, the distributed policy still could make the decision successfully when one or more sites join or leave the system. Therefore, the stability of the distributed policy is better than that of the centralized policy. The distributed policy is usually used in distributed system. Shah et al. (2007) propose a decentralized load balancing algorithm which employs the job arrival rates and the job response for making load balancing decisions. Lei et al. (2007) make the load balancing decisions according to the CPU and memory status. Tang et al. (2008) improve the system stability through the resource-constrained load balancing controller. Liang et al. (2008) propose an adaptive load balancing algorithm which makes the workload of all nodes as evenly as possible. Subrata et al. (2008) propose a decentralized game-theoretic

Figure 1. P2P Grid system

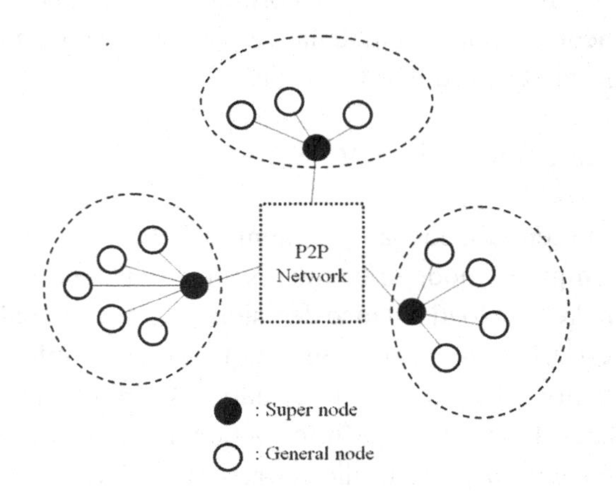

approach which not only provides the similar performance with centralized approach, but also remain the advantage of distributed approaches. Li et al. (2009) propose a hybrid strategy to balance the workload according to the average-based and immediate resource information. Moreover, the hybrid strategy improves the performance of sequential tasks.

This paper proposes a Uni-P2P communication model to connect grid sites and also proposes a P2P resource monitoring system to collect the resource information. A preliminary load balancing prototype (Huang et al., 2010) for P2P Grid systems is also implemented. It employs limited system information to achieve the load balancing and improve the resource utilization. In this chapter, we integrate the proposed load balancing prototype into the Uni-P2P communication model to make P2P Grid systems more efficient.

SYSTEM ARCHITECTURE

P2P Grid System

The P2P Grid computing system is a distributed computing systems based on the grid computing system, which employs the P2P approach to exchange information. In the P2P Grid system, each site consists of one super node and several general nodes. Super nodes exchange the site information with each other by adopting the P2P approach and manage resources and jobs in general nodes. General nodes are responsible both for job execution and for supplying the resource information of the general node to the super node. The P2P Grid system architecture is shown in Figure 1.

Figure 2. P2P Grid system architecture

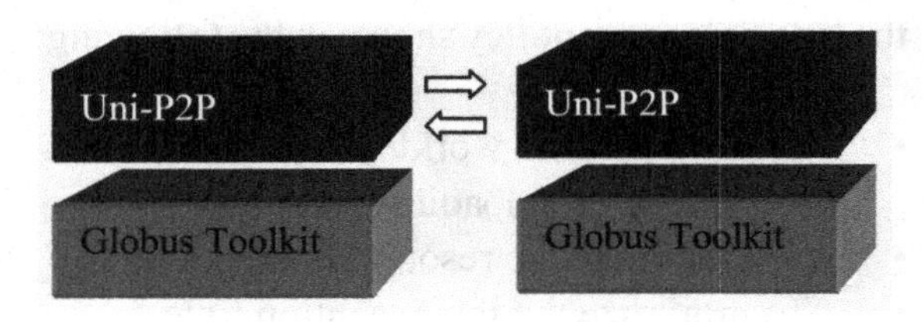

Uni-P2P Communication Model

The Uni-P2P communication model is developed by JXTA. The JXTA project (Gong et al., 2002) was proposed to enable P2P routing services which locate and communicate with peers.

In our system architecture, we build the Uni-P2P communication model on the Globus Toolkit as shown in Figure 2.

Uni-P2P communication model includes five modules, as shown in Figure 3: the configuration module, the information service module, the file transfer module, the load balance module, and the execution management module.

The functions of the configuration module include the basic parameters setup, P2P pipe startup, and the initialization of peers. This module is fundamental in the Uni-P2P communication model. The file transfer module supports the universal pipes among computing nodes to transfer job files, data files, command messages, and the job description files. The information service module includes three sub-modules: the resource discovery, the resource collection and the resource aggregation. The information service module manages the global resource information among

Figure 3. System architecture of the Uni-P2P communication model

Application
Portals | Collaboratories | High Energy Physics | . .
High Level Service
Information Service | Execution Management | File Transfer
Configure | Load Balancing | Uni-P2P communication model
Core Services
Peer Discovery Protocol | Peer Information Protocol | Pipe Binding Protocol
Peer Resolver Protocol | Peer Endpoint Protocol | Rendezvous Protocol
Fabric
Disk | Network | Operating Systems | Databases | . .

sites and the local resource information among computing nodes. The load balance module takes charge of load measurement, best cost site decision, job queue check and job description generation. The execution management module is responsible for job execution. Jobs waiting in the Condor queue can be handled in this module.

Because security is not a major concern in P2P systems, we omit the security consideration and instead focus on research issues about job migration in this study.

In this chapter, the computing nodes in each site are classified into super peers and general peers. The general peer that starts with JXTA RDV and Relay Service becomes a super peer, and then the super peer starts up the resource load balance module. The attributes and resource information of computing nodes for general peers are sent to the super peer. In our Uni-P2P communication model, there are six function modules to handle P2P communication, as shown in Figure 4.

The P2P communication mechanism first configures the basic setting before starting the Uni-P2P services. Then, the resource discovery module searches and records the peer information in the host table. The resource loading module collects the resource usage (e.g., CPU and memory load) of general peers and records the information in the resource table. The message receiving/sending module listens to the services at any

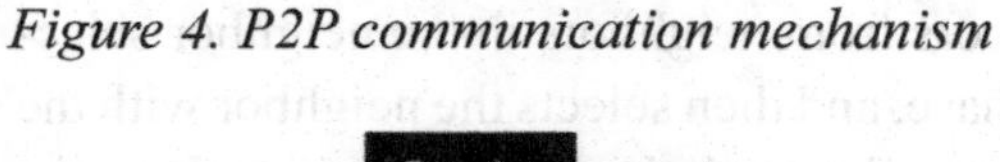

Figure 4. P2P communication mechanism

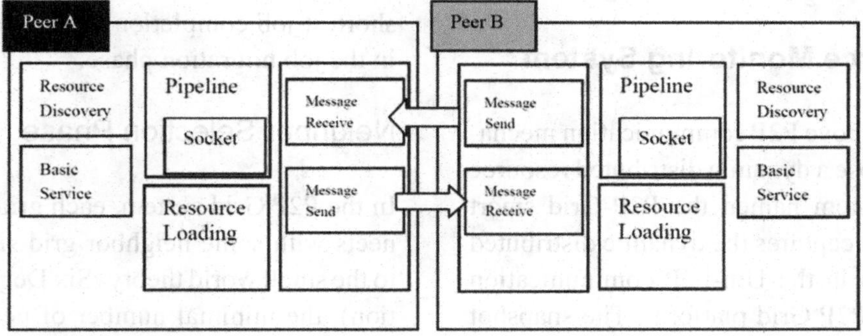

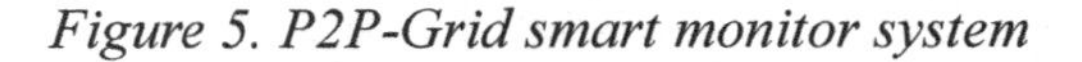

Figure 5. P2P-Grid smart monitor system

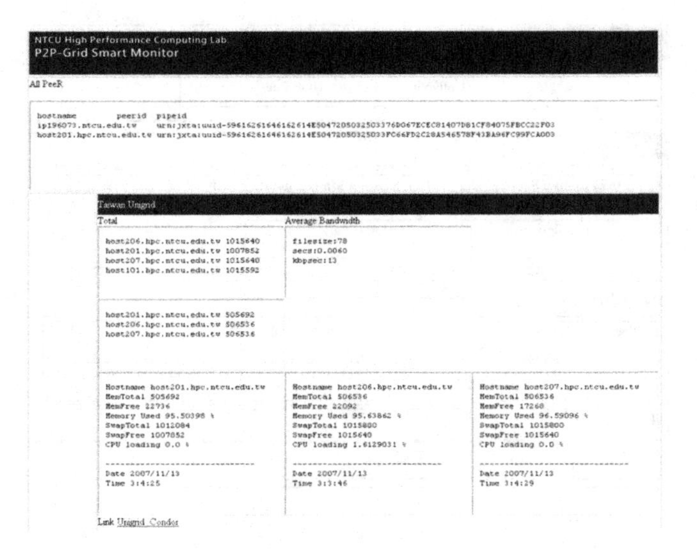

time for receiving and sending messages. The pipeline module establishes the pipeline between peers. The input-pipe and output-pipe are used in message passing or file transferring. After establishing the pipeline between peers, the P2P communication mechanism delivers messages, files, and tasks via the socket module.

In order to measure the resource load, the resource loading module is also built in the super peer. When the member peer sends a request to the super peer, the super peer will actively search for the appropriate resources, and the super peer selects the suitable resources according to the load balance policy. The pipeline and socket modules start the job migration. And then, the overloaded jobs would be migrated to other super peers which have enough (or suitable) computing resources.

P2P Resource Monitoring System

Basing on the above P2P communication mechanism, we propose a dynamic distributed resource monitoring system named the P2P-Grid smart monitor, which captures the dynamic distributed resource status in the Uni-P2P communication model for the P2P Grid platform. The snapshot of the P2P-Grid smart monitor system is shown in Figure 5.

Load Balance Policy

In this section, we present the proposed Self-Adaptive Load Balance (SALB) policy for P2P Grid systems. SALB is a distributed dynamic load balance policy by applying the sender-initiated strategy. The sender-initiated strategy means that when the node becomes overloaded, it starts to find out other nodes to migrate jobs. SALB consists two phases: the neighbor selection phase and the job migration phase. Therefore, when the grid site is overloaded, it picks several low-loading neighbors from neighbors in the neighbor selection phase, and then selects the neighbor with the shortest job completion time for migrating jobs in the job migration phase.

Neighbor Selection Phase

In the P2P Grid system, each grid site only connects with some neighbor grid sites. According to the small world theory (Six Degrees of Separation), the minimal number of neighbors of one site should be greater than six square root of the number of all the grid sites. We denote the number of all the grid sites by N, and the minimal number of neighbors should be greater than or equal to $\sqrt[6]{N}$. Therefore, each grid site maintains at least $\sqrt[6]{N}$ neighbors to ensure the network connectivity. In addition, in order to improve the load balancing, SALB estimates the remaining resources capability of each neighbor site and selects the neighbors with more remaining capabilities for migrating jobs. Suppose that a grid site has m kinds of resources, and each resource's remaining capability of site s is denoted by $RC_i(s)$, where $i = 1 \ldots m$. The weight of each resource is denoted by $W_i(s)$. Assume that $DRC(l, r)$ is the difference of remaining capability between the local site l and the remote site r. Then,

$$DRC(l,r) = \sum_{i=1}^{m} W_i(l) \frac{RC_i(l)}{RCi(r)} \tag{1}$$

When a gird site joins the P2P Grid system, it randomly selects $2\sqrt[6]{N}$ sites as the candidates of neighbors and picks the first $\sqrt[6]{N}$ sites with the smaller DRC as its neighbors. Thus, these neighbors have more remaining resources capability for load balancing. To avoid the out-of-date information, we also set a time interval t for each grid site to re-select neighbor sites periodicity. In addition, SALB estimates the relative loading (RL) and the absolute loading (AL) of grid sites to determine whether the local site is sendable and which neighbor sites are receivable. We define the utilization of resource i in site s by $U_i(s)$. Besides, the average utilization, the maximum utilization and the minimum utilization of resource i of site s are defined by $U\_AVG_i(s)$, $U\_MAX_i(s)$ and $U\_MIN_i(s)$ respectively. Therefore, the RL of site s is defined as

$$RL(s) = \sum_{i=i}^{m} W_i(s) \frac{U_i - U\_AVG_i(s)}{U\_MAX_i(s) - U\_MIN_i(s)} \tag{2}$$

while the AL of site s is defined as

$$AL(s) = \sum_{i=1}^{m} W_i(s) U_i(s) \tag{3}$$

Moreover, SALB predefines the high threshold of RL by RL_H, the low threshold of RL by RL_L, the high threshold of AL by AL_H, and the low threshold of AL by AL_L. The status of different combinations of RL and AL are described as follows: When RL is greater than or equal to RL_H, the status is set to be "High". When RL is less than RL_H and is greater than RL_L, the status is set to be "Moderate". When RL is less than or equal to RL_L, the status is set to be "Low". When AL is greater than or equal to AL_H, the status is set to be "High". When AL is less than AL_H and is greater than AL_L, the status is set to be "Moderate". When AL is less than or equal to AL_L, the status is set to be "Low". The statuses of different combinations of RL and AL are shown in table 1.

According to the statuses of RL and AL, we classify the status of local sites to be "Sendable"

Table 1. States of RL and AL

	RL status		*AL* status
$RL \geqq RL\_H$	High	$AL \geqq AL\_H$	High
$RL\_H > RL > RL\_L$	Moderate	$AL\_H > AL > AL\_L$	Moderate
$RL \leqq RL\_L$	Low	$AL \leqq AL\_L$	Low

Table 2. The status of grid sites

RL status	*AL* status	Local site's status	Neighbor site's status
High	High	Sendable	Unreceivable
High	Moderate	Sendable	Unreceivable
High	Low	Sendable	Unreceivable
Moderate	High	Sendable	Unreceivable
Moderate	Moderate	UnSendable	Receivable
Moderate	Low	UnSendable	Receivable
Low	High	Sendable	Unreceivable
Low	Moderate	UnSendable	Receivable
Low	' Low	UnSendable	Receivable

or "Unsendable", and the status of its neighbor sites to be "Receivable" or "Unreceivable". In order to further improve the load balancing performance, we refine the definition of the status set. As long as the status of one of *RL* or *AL* is "High", the grid site becomes "Sendable". Moreover, when the status of only one of *RL* or *AL* is "High", the grid site becomes "Unreceivable". When a grid site becomes "Sendable", it means that the grid site is overloaded. Therefore, the sendable site starts to pick out its neighbors which are "Receivable", and enables the job migration phase. The statuses of different combinations in gird sites are shown in table 2.

Job Migration Phase

In the job migration phase, the sendable site calculates the possible job turnaround time for each receivable neighbor site if this neighbor site is the candidate site. And then, the job is migrated to the neighbor site which has the minimal job turnaround time. The job turnaround time is the sum of the forecasted waiting time, the execution time and the migration time which are respectively denoted by T_w, T_e, and T_m. T_w is the forecasted waiting time of the migrated job *J* in the destination site *s*. SALB forecasts the total remaining execution time, T_r, of the running jobs and the total execution time, T_i, of the idle jobs in the destination site *s*. Then, assume that the number of CPU in the site *s* is N_{CPU}. Then, the forecasted waiting time is defined as

$$T_w = \frac{T_r + T_i}{N_{CPU}} \tag{4}$$

T_e is the executing time of job *J* in the site *s*. T_m is the migration time of job *J* which is migrated from the local site to the destination site. The migration involves the job's program code and some required data files. Thus,

$$T_m = \frac{\text{The size of program code file and data file}}{\text{The bandwidth between the local site and destination site}} \tag{5}$$

Denoting the job turnaround time by T_{jt}, then

$$T_{jt} = T_w + T_e + T_m. \tag{6}$$

Figure 6 shows the algorithm of SALB. Each grid site changes their neighbors periodically to improve the global load balancing. In addition, due to that SALB applies the sender-initiated strategy, only when the grid site becomes "Sendable", the load balancing policy is enabled. The sendable site picks out the neighbors which are "Receivable" according to the statuses of *RL* and *AL*. Then, the neighbor's job turnaround time (T_{jt}) is calculated and the neighbor which has the minimal job turn-

Figure 6. SALB algorithm

```
Algorithm SALB;
Input : local site l, total number of sites N, time interval t.
Output : load balancing status.
Begin
  every time interval t
     site l connects to 2⁶√N grid sites randomly.
     pick out the first ⁶√N gird sites as neighbors from the 2⁶√N grid sites according to the
   increasing SRC.
     when site l is "Sendable"
       calculate the T_jt of job J in each "Receivable" neighbor. // job J is the last job in site l
       pick out the "Receivable" neighbor with the minimal T_jt as the destination site r
        if site l's T_jt > site r's T_jt
           migrate job J to site r
        else
           retain job J
End
```

around time is chosen to be the destination site r. Finally, T_{jt} for each site is compared. If T_{jt} of site r is less than that of the site l, site l will migrate job J to site r, otherwise site l retains job J until the next time interval.

EXPERIMENTAL RESULTS

This section introduces the experimental environment and results of SALB. In our discussion of the experimental results, we particularly focus on the efficiency of load balancing.

Experimental Environment

In this experiment, we adopts JXTA with version 2.5.1, Java with version 1.6.0 and Condor with version 6.7.20 to implement SALB in Taiwan UniGrid for evaluating load balancing. In addition, we construct five grid sites, and each site consists of one super node and some general nodes. Super nodes are responsible for the communication with other neighbor sites, assigning jobs to their general nodes and executing SALB strategy. General nodes are responsible for executing the jobs assigned by the super nodes. Moreover, SALB can be extended to larger scale systems. Table 3 shows the specification of the experimental platform.

Table 3. System specification

Site	Hosts	Peer Types	CPU clock	Memory
1	Host201	Super node	Intel P-D 3.40GHz x 2	512M
1	Host204	General node	Intel P-D 3.40GHz x 2	512M
2	Host205	Super node	Intel P-4 3.40GHz x 2	512M
2	Host208	General node	Intel P-4 3.40GHz x 2	512M
3	Host206	Super node	Intel P-4 3.40GHz x 2	512M
3	Host207	General node	Intel P-4 3.40GHz x 2	512M
4	Host221	Super node	Intel P-4 3.40GHz	256M
4	Host223	General node	Intel P-4 3.40GHz	256M
5	Host222	Super node	Intel P-4 3.40GHz x 2	512M
5	Host224	General node	Intel P-4 3.40GHz x 2	512M

Table 4. Definitions of relate arguments

RL_H	20%
RL_L	-20%
AL_H	60%
AL_L	40%
W_{CPU}	60%
W_{Memory}	30%
$W_{Bandwith}$	10%
t	60 second

This experiment employs five benchmarks, including f77split, fd_predator_prey, fd1d_heat_explicit, satisfiability and linpack_bench. Because each benchmark consumes different computing resources, we employ these five benchmarks as five different jobs. All the jobs are firstly submitted to site 5. In addition, we compare the load balancing performance of SALB with those of FIFO (Fist In First Out) strategy and JRT strategy when the numbers of jobs are 50, 100, 150 and 200 respectively. We employ CPU, memory and bandwidth to represent the computing resources, and their weights are denoted as W_{CPU}, W_{Memory} and $W_{Bandwidth}$ respectively. Table 4 shows the definitions of related arguments.

Evaluation of SALB

Figure 7 shows the average execution time of each strategy with different numbers of jobs. We can observe that SALB spends a little more time than JRT. This is because SALB spends more time in picking out neighbors and calculating the job turnaround time. Therefore, the time SALB spends for load balancing is longer than those other strategies spend when the number of jobs is small. However, SALB performs better with the increasing number of jobs.

Figure 8 shows the average CPU utilization of each strategy with different numbers of jobs. It shows that SALB has the maximum average CPU utilization. This is because SALB picks the neighbors with the most remaining resources for migrating jobs. Therefore, the average utilization of computing resources could be improved.

Figure 7. Average execution time

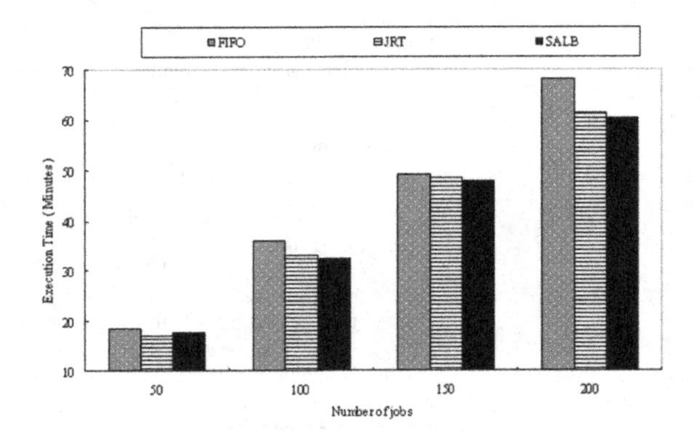

Figure 9, 10, 11 and 12 show the average CPU utilization of each strategy with different numbers of jobs. In these figures, we can observe that SALB is steadier than other strategies. This is because it is more possible for the grid sites with more remaining resources to be the candidates for migrating jobs. Therefore, the CPU utilization keeps steadily until finishing all the jobs.

CONCLUSIONS AND FUTURE RESEARCH DIRECTIONS

In this chapter, we propose a Uni-P2P communication model which supports the resource discovery,

Figure 8. Average CPU utilization

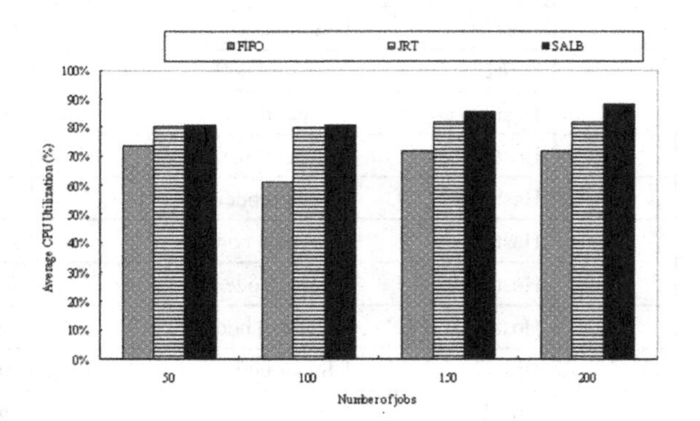

Figure 9. Average CPU utilizations of executing 50 jobs

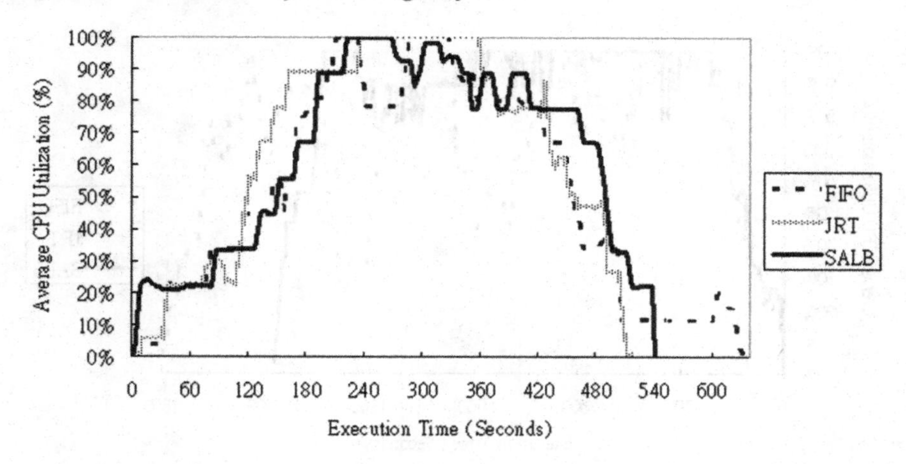

Figure 10. Average CPU utilizations of executing 100 jobs

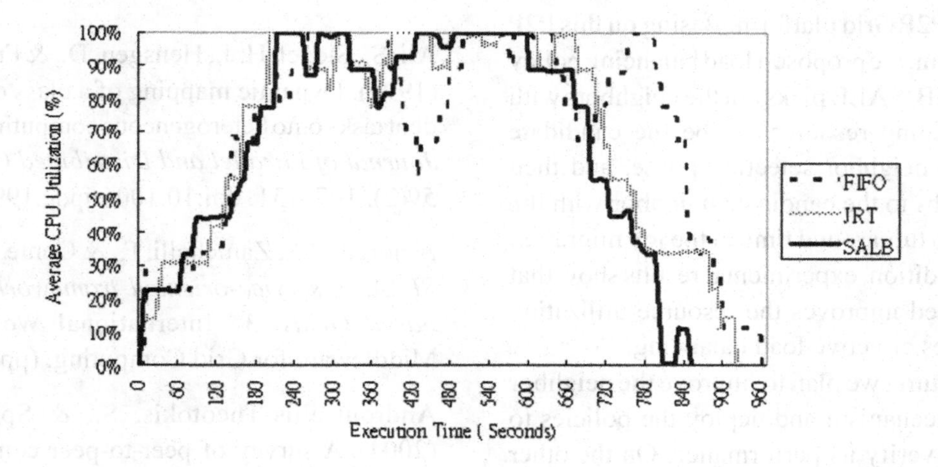

Figure 11. Average CPU utilizations of executing 150 jobs

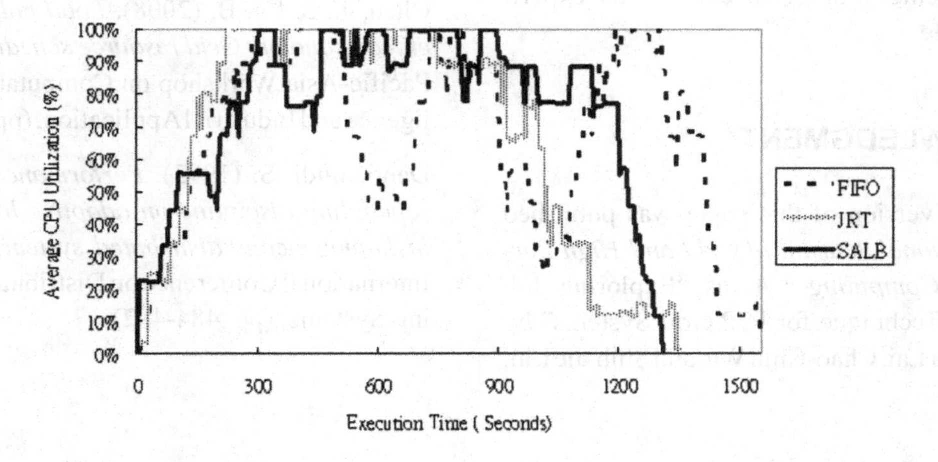

Figure 12. Average CPU utilizations of executing 200 jobs

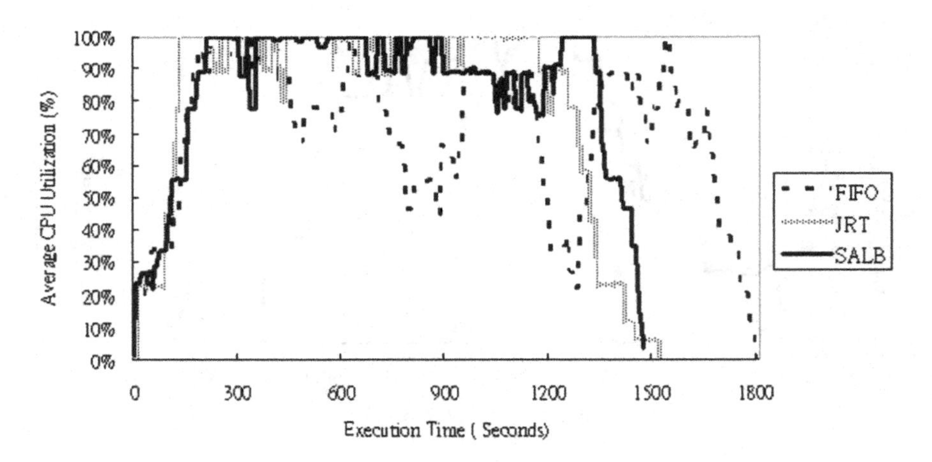

loading balance and job migration functions to establish a P2P Grid platform. Basing on this P2P Grid platform, we propose a load balancing policy named SALB. SALB picks out the neighbors with more remaining resources to be the candidate sites in the neighbor selection phase, and then migrates jobs to the candidate neighbor with the minimal job turnaround time in the job migration phase. In addition, experimental results show that SALB indeed improves the resource utilization and achieves effective load balancing.

In the future, we plan to improve the neighbor selection mechanism and deploy the policies to UniGrid to verify its performance. On the other hand, we will adopt the grid simulator, such as GridSim, as the experiment environment for our load balancing strategy to enlarge the experimental scale.

ACKNOWLEDGMENT

An earlier version of this paper was published in *International Journal of Grid and High Performance Computing* 1(4) as "Exploring Job Migration Technique for P2P Grid Systems" by Kuan-Chou Lai, Chao-Chin Wu, and Shih-Jie Lin.

REFERENCES

Ali, S., Siegel, H. J., Hensgen, D., & Freund, R. F. (1999). Dynamic mapping of a class of independent tasks onto heterogeneous computing systems. *Journal of Parallel and Distributed Computing*, *59*(2), 107–131. doi:10.1006/jpdc.1999.1581

Amoretti, M., Zanichelli, F., & Conte, G. (2005). *SP2A: A service-oriented framework for P2P-based Grids*. 3rd International Workshop on Middleware for Grid Computing, (pp. 1-6).

Androutsellis-Theotokis, S., & Spinellis, D. (2004). A survey of peer-to-peer content distribution technologies. *ACM Computing Surveys*, *36*(4), 335–371. doi:10.1145/1041680.1041681

Chen, J., & Lu, B. (2008). *Load balancing oriented economic Grid resource scheduling*. IEEE Pacific-Asia Workshop on Computational Intelligence and Industrial Application, (pp. 813-817).

Dandamudi, S. (1995). *Performance impact of scheduling discipline on adaptive load sharing in homogeneous distributed system*. 15th IEEE International Conference on Distributed Computing Systems, (pp. 484-492).

Duan, Z., & Gu, Z. (2008). *Dynamic load balancing in Web cache cluster.* 7th International Conference on Grid and Cooperative Computing, (pp. 147-150).

Eager, D. L., Lazowska, E. D., & Zahorjan, J. (1986). A comparison of receiver initiated and sender initiated adaptive load sharing. *Performance Evaluation*, *6*(1), 53–68. doi:10.1016/0166-5316(86)90008-8

Eager, D. L., Lazowska, E. D., & Zahorjan, J. (1988). *The limited performance benefits of migrating active processes for load sharing.* 1988 ACM SIGMETRICS Conference on Measurement and Modeling of Computer Systems, (pp. 63-72).

Gong, L., Oaks, S., & Traversat, B. (2002). *JXTA in a nutshell a desktop quick reference.* Sebastopol, CA: O'Reilly & Associates.

Hu, J., & Klefstad, R. (2006). *Decentralized load balancing on unstructured Peer-2-Peer computing Grids.* 5th IEEE International Symposium on Network Computing and Applications, (pp. 247-250).

Huang, P. J., Yu, Y. F., Chen, Q. J., Huang, T. L., Lai, K. C., & Li, K. C. (2010). A self-adaptive load balancing strategy for P2P grids. In C. H. Hsu, et al. (Eds.), *ICA3PP 2010, part II, LNCS 6082*, (pp. 348-357). Heidelberg/Berlin, Germany: Springer-Verlag.

Iyengar, M. S., & Singhalc, M. (2006). Effect of network latency on load sharing in distributed systems. *Journal of Parallel and Distributed Computing*, *66*(6), 839–853. doi:10.1016/j.jpdc.2005.09.005

Lei, S., Yuyan, S., & Lin, W. (2007). *Effect of scheduling discipline on CPU-MEM load sharing system.* 6th International Conference on Grid and Cooperative Computing, (pp. 242-249).

Li, J., & Vuong, S. (2006). *Grid resource discovery based on semantic P2P communities.* 2006 ACM Symposium on Applied Computing, (pp. 754-758).

Li, Y., Yang, Y., & Zhu, R. (2009). *A hybrid load balancing strategy of sequential tasks for computational Grids.* IEEE International Conference on Networking and Digital Society, (pp. 112-117).

Liang, G. (2008). *Adaptive load balancing algorithm over heterogeneous workstations.* 7th International Conference on Grid and Cooperative Computing, (pp. 169-174).

Milojičić, D. S., Douglis, F., Paindaveine, Y., Wheeler, R., & Zhou, S. (2000). Process migration. *ACM Computing Surveys*, *32*(3), 241–299. doi:10.1145/367701.367728

Pan, Y., Lu, W., Zhang, Y., & Chiu, K. (2007). *A static load-balancing scheme for parallel XML parsing on multicore CPUs.* 7th IEEE International Symposium on Cluster Computing and the Grid, (pp. 351-362).

Richmond, M., & Hitchens, M. (1997). A new process migration algorithm. *ACM SIGOPS Operating Systems Review*, *31*(1), 31–42. doi:10.1145/254784.254790

Shah, R., Veeravalli, B., & Misra, M. (2007). On the design of adaptive and decentralized load balancing algorithms with load estimation for computational Grid Environments. *IEEE Transactions on Parallel and Distributed Systems*, *18*(12), 1675–1686. doi:10.1109/TPDS.2007.1115

Shan, J., Chen, G., He, J., & Chen, X. (2002). *Grid society: A system view of Grid and P2P environment.* International Workshop on Grid and Cooperative Computing, (pp. 19-28).

Shudo, K., Tanaka, Y., & Sekiguchi, S. (2005). *P3: P2P-based middleware enabling transfer and aggregation of computational resources*. IEEE International Symposium on Cluster Computing and the Grid, (pp. 259- 266).

Subrata, R., Zomaya, A. Y., & Landfeldt, B. (2008). Game-theoretic approach for load balancing in computational Grids. *IEEE Transactions on Parallel and Distributed Systems*, *19*(1), 66–76. doi:10.1109/TPDS.2007.70710

Taiwan UniGrid. (n.d.). Retrieved October 13, 2009, from http://www.unigrid.org.tw/index.html.

Tanenbaum, A. S. (2007). *Modern operating systems* (3rd ed.). Prentice Hall.

Tang, Z., Birdwell, J. D., & Chiasson, J. (2008). Resource-constrained load balancing controller for a parallel database. *IEEE Transactions on Control Systems Technology*, *16*(4), 834–840. doi:10.1109/TCST.2007.916305

Vincze, G., Novák, Z., Pap, Z., & Vida, R. (2008). *RESERV: A distributed, load balanced Information System for Grid applications*. 8th IEEE International Symposium on Cluster Computing and the Grid, (pp. 596-601).

Wu, Y. J., Lin, S. J., Lai, K. C., Huang, K. C., & Wu, C. C. (2008). *Distributed dynamic load balancing strategies in P2P Grid systems*. 5th Workshop on Grid Technologies and Applications, (pp. 95-102).

Xia, Y., Chen, S., & Korgaonkar, V. (2006). *Load balancing with multiple hash functions in peer-to-peer networks*. IEEE 12th International Conference on Parallel and Distributed Systems, (pp. 411-420).

Xu, Z., & Bhuyan, L. (2006). *Effective load balancing in P2P systems*. 6th IEEE International Symposium on Cluster Computing and the Grid, (pp. 81-88).

Yang, C. T., Li, C. T., Chiang, W. C., & Shih, P. C. (2005). *Design and implementation of TIGER Grid: An integrated metropolitan-scale Grid environment*. 6th International Conference on Parallel and Distributed Computing Applications and Technologies, (pp. 518-520).

Chapter 19
An Ontology-Based P2P Network for Semantic Search

Tao Gu
University of Southern Denmark, Denmark

Daqing Zhang
Institut Telecom SudParis, France

Hung Keng Pung
National University of Singapore, Singapore

ABSTRACT

This article presents an ontology-based peer-to-peer network that facilitates efficient search for data in wide-area networks. Data with the same semantics are grouped together into one-dimensional semantic ring space in the upper-tier network. This is achieved by applying an ontology-based semantic clustering technique and dedicating part of node identifiers to correspond to their data semantics. In the lower-tier network, peers in each semantic cluster are organized as Chord identifier space. Thus, all the nodes in the same semantic cluster know which node is responsible for storing context data triples they are looking for, and context queries can be efficiently routed to those nodes. Through the simulation studies, the authors demonstrate the effectiveness of our proposed scheme.

INTRODUCTION

In recent years, the use of context information has attracted a lot of attention from researchers and industry participates in ubiquitous and pervasive computing. Users and applications are often interested in searching and utilizing widespread context information. Context information is characterized as an application's environments or situations (Dey et al., 2000). With the vast amount of context information spread over multiple context spaces and the increasing needs of cross-domain context-aware applications, how to provide an efficient context search mechanism is challenging in the context-aware research community.

One approach is to use a centralized search engine to store context data and resolve search requests. Although this approach can provide fast responses to a context query, it has limitations such as scalability, a single processing bottle-

DOI: 10.4018/978-1-60960-603-9.ch019

neck and a single point of failure. Peer-to-peer (P2P) approaches, on the other hand, have been proposed to overcome these obstacles and are gaining popularity in recent years. P2P systems such as Gnutella (Gnutella) and Freenet (Freenet) allow nodes to interconnect freely and have low maintenance overhead, making it easy to handle the dynamic changes of peers and their data. The past years have seen an increased focus on decentralized P2P systems (Han, et al., 2006, Li, et al., 2006, Liu, et al., 2004, Morselli, et al., 2005). However, a query has to be flooded to all the nodes in a network including the nodes that do not have relevant data. The fundamental problem that makes search in these systems difficult is that data are randomly distributed in the network with respect to their semantics. Given a search request, the system either has to search a large number of nodes or run a risk of missing relevant data. Other P2P systems such as Chord (Stoica, et al., 2001), CAN (Ratnasamy, et al., 2001), Pastry (Rowstron, et al., 2001) and Tapestry (Zhao, et al., 2004) typically implement distributed hash tables (DHTs) and use hashed keys to direct a search request to the specific nodes by leveraging a structured network. In these systems, a data object is associated with a key which can be produced by hashing the object name. A node is assigned with an identifier which shares the same space as the keys. Each node is responsible for storing a range of keys and corresponding objects. When a search request is issued from a node, the search message is routed through the network to the node responsible for the key. They can guarantee to complete search in a logarithmic number of steps. Over years, many applications have been developed, such as file sharing (LimeWire) and content distribution (Castro, et al., 2003).

In this article, we propose a two-tier semantic P2P network to search for context information in wide-area networks. The basic idea is to construct a two-level semantic P2P network based on metadata (i.e., context ontologies), which is essentially a semantic approach, to facilitate efficient search. In this system, context data are represented by a collection of RDF (RDF) triples. Peers with the same semantics are grouped together into a semantic cluster in the upper-tier network. All the semantic clusters are constructed as a one-dimensional semantic ring space. This is achieved by dedicating part of hashed node identifiers to correspond to their data semantics. Data semantic is extracted according to a set of schemas. Peers in each semantic cluster can be organized as a structured P2P network such as Chord identifier space in the lower-tier network. Thus, all the nodes in the same semantic cluster know which node is responsible for storing context data triples they are looking for, and context queries can be efficiently routed to those nodes.

The rest of the article is organized as follows. Section 2 presents the detail of the two-tier semantic P2P network. Section 3 evaluates the performance of our system using simulation and presents the results. Section 4 reviews related works, and finally Section 5 concludes our work.

THE TWO-TIER SEMANTIC P2P NETWORK

In this section, we first present an overview of the two-tier semantic P2P network, followed by a description of technical details. For ease of discussion, we use the terms node and peer interchangeably for the rest of the article.

OVERVIEW

In this network, a large number of nodes storing context data are grouped and self-organized into a two-tier semantic P2P network, in accordance with their semantics. A node can act as producer, consumer or both. Producers provide various context data for sharing whereas consumers obtain context data by submitting their context queries and receiving results. Each node maintains a local data repository which supports RDF-based query using RDQL (RDQL). Upon creation, each producer will first go through the ontology-based semantic mapping process to extract the semantics of its local data. It will then join a semantic cluster by applying the SHA1 hash function to the semantics of its main data. These semantic clusters logically form the upper-tier network in which each node builds its routing index based on the small world network model (Kleinberg, 2000). In the lower-tier network, nodes in each semantic cluster are organized as Chord for storing context data and routing context queries in a logarithmic number of hops. Upon receiving a context query, the node first pre-processes it to obtain the semantic cluster associated with the query, and then routes it to an appropriate semantic cluster. In the lower-tier, the node routes the query using its finger table. Nodes that receive the query do a local search, and return results.

ONTOLOGY-BASED SEMANTIC CLUSTERING

In this section, we describe how to use ontology-based metadata to extract the semantics of both RDF data and queries, and map them into appropriate semantic clusters. In our system, context data are described as RDF triples based on a set of context ontologies. We adopt a two-level hierarchy in the design of context ontologies. The upper ontology defines common concepts in a computing domain, e.g., context-aware computing, and it is shared by all peers. Each peer can define its own concepts in its lower ontologies. Different peers may store different sets of lower ontologies based on their application needs. The upper ontology can be extended with new concepts and properties upon the agreement among all the peers in the network.

To illustrate the semantic mapping process, we use an example of ontology as shown in Figure 1. All the leaf nodes in the upper ontology are used as semantic clusters, and denoted as set $E = \{Service, Application, Device, \ldots\}$. The mapping computation is done locally at each peer. For the mapping of RDF data, a peer needs to define a set of lower ontologies and store them locally. Upon joining the network, a peer first obtains the upper ontology and merges it with its local lower ontologies. Then it creates instances (i.e., RDF data) and adds them into the merged ontology to form its local knowledge base. A peer's local data may be mapped into one or more semantic clusters by extracting the subject, predicate and object of an RDF data triple. Let SCn_{sub}, SCn_{pred} and SCn_{obj} where $n = 1, 2, \ldots$ denote the semantic clusters extracted from the subject, predicate and object of a data triple respectively. Unknown subjects/objects (which are not defined in the merged ontology) or variables are mapped to E. If the predicate of a data triple is of type *ObjectProperty*, we obtain the semantic clusters using $(SC1_{pred} \bigcup SC2_{pred} \bigcup \ldots SCn_{pred}) \bigcap (SC1_{obj} \bigcup SC2_{obj} \bigcup \ldots SCn_{obj})$. If the predicate of a data triple is of type *DatatypeProperty*, we obtain the semantic clusters using $(SC1_{sub} \bigcup SC2_{sub} \bigcup \ldots SCn_{sub}) \bigcap (SC1_{pred} \bigcup SC2_{pred} \bigcup \ldots SCn_{pred})$. Examples 1 and 2 in Figure 2a show the RDF data triples about the location and light level in a bedroom provided by a producer peer. In Example 2, we first obtain the semantic clusters from both the subject and predicate, and then intersect their results to get the final semantic cluster – *IndoorSpace*.

A context query follows the same procedure to obtain its semantic cluster(s), but it needs all the sets of lower ontologies. In real applications, users may create duplicate properties in their lower ontologies which conflict with the ones in the upper ontology. For example, the upper ontology defines the *rdfs:range* of predicate *locatedIn* as *Location* whereas the lower ontology defines its *rdfs:range* as *IndoorSpace*. To resolve this issue, we create two merged ontologies, one for clustering peers and the other for clustering queries. If such a conflict occurs, we select the affected properties defined in the lower ontology

Figure 1. An example of ontology for illustration

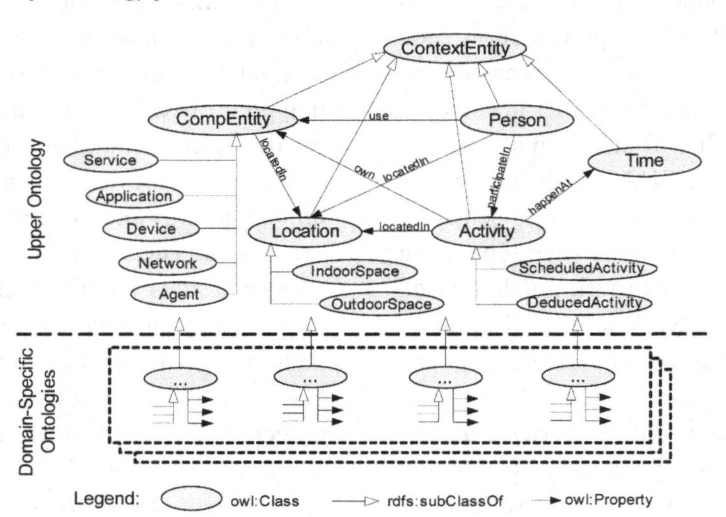

to generate the merged ontology for clustering peer and select the affected properties defined in the upper ontology to generate the merged ontology for clustering queries. With this scheme, a peer can extract the semantics of its data triples more precisely without losing generality for context queries. For example, predicate *locatedIn* may have the *rdfs:range* of *IndoorSpace* in the merged ontology for clustering peers (see Figure 2a) and have the *rdfs:range* of *Location* in the merged ontology for clustering queries (see Figure 2b). Data triple *<socam:John socam:locatedIn socam:Bedroom>* will be mapped to *IndoorSpace*; and query *<socam:John socam:locatedIn ?x>* will be mapped to both *IndoorSpace* and *OutdoorSpace* rather than only *IndoorSpace*. This is most likely the case of real life applications.

THE UPPER-TIER NETWORK

In this section, we describe the process of constructing the two-tier semantic P2P network. After obtaining the semantics from its local context data, a node needs to participate in the network. It will first join an appropriate semantic cluster in the upper-tier network, and then store its data triples and participate in the lower-tier network. As a node may obtain multiple semantics from its local data, we choose the semantic cluster corresponding to the largest set of data to place the node. We call this semantic cluster the major semantic cluster of this node. The remaining semantic clusters which a node's data corresponds to are called minor semantic clusters of this node.

A node is assigned with an ID upon joining the network. We use SHA1 hash function to generate nodes' identifier space. To incorporate semantic information associated with a node, we dedicate part of hashed node identifiers to correspond to the semantic cluster. More specifically, in a *k*-bits identifier space, we allocate *m*-bits for semantic cluster information and *n*-bits for its IP address, where $k = m + n$. An example of a node's ID generated by hashing its semantic cluster *Person* and its IP address "137.132.81.235" is given below.

$$\text{node id} = [\text{hash}_m(\text{“Person”})]\,[\text{hash}_n(\text{“137.132.81.235”})]$$

With this encoding scheme, we are able to construct the two-tier network and identify a node in the network, i.e., the first *m*-bits of a node's ID (called *semantic cluster ID* or *sid* in short) cor-

Figure 2. An example of semantic cluster mapping

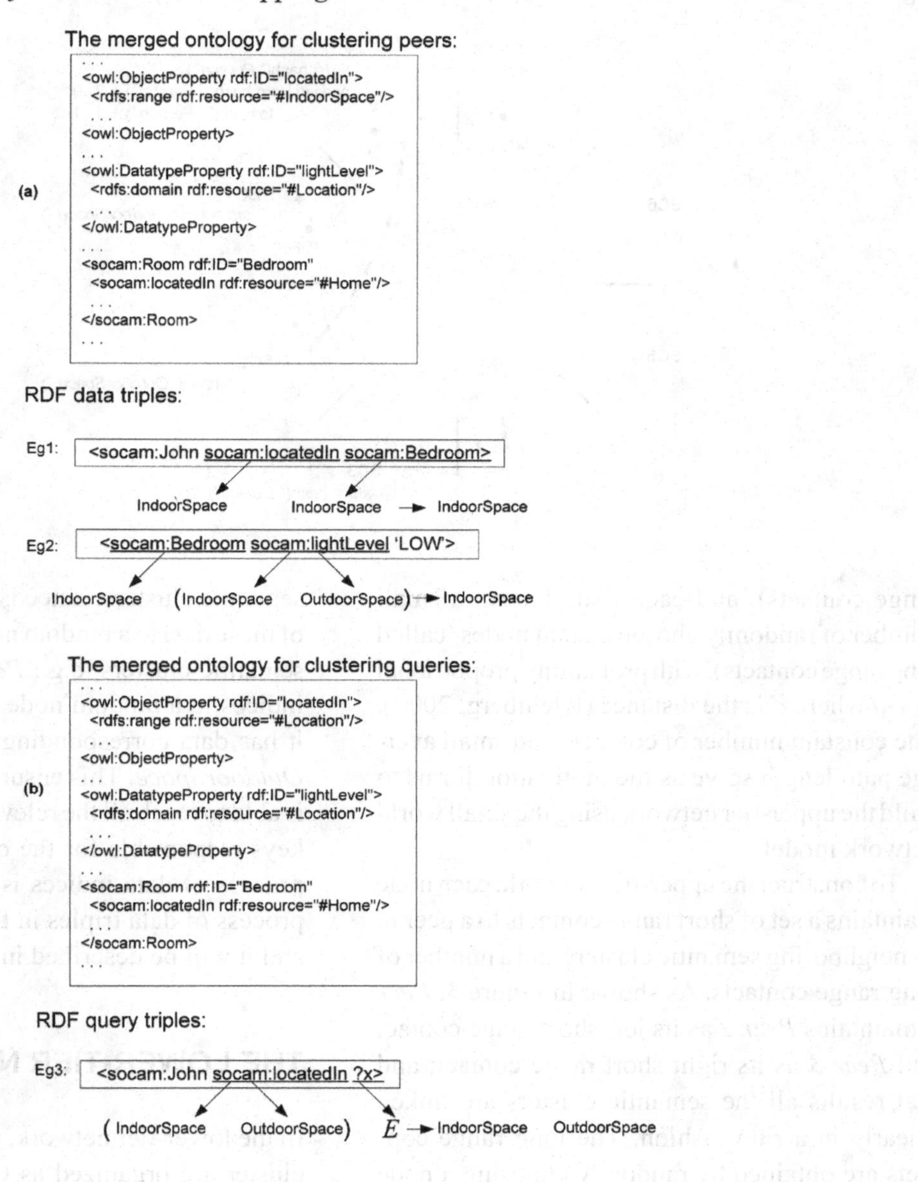

responds to the semantic cluster in the upper-tier and the last *n*-bits represents the node's ID in the lower-tier.

We follow the small world network model to construct the upper-tier network. The small network model is characterized as small average path length between two nodes in the network and large cluster coefficient defined as the probability that two neighbors of a node are neighbors themselves. Studies show that searches can be efficiently routed in small world networks when: Each node in the network knows its local neighbors (called short

Figure 3. The construction of the upper-tier network (note: the sign "+" represents appending)

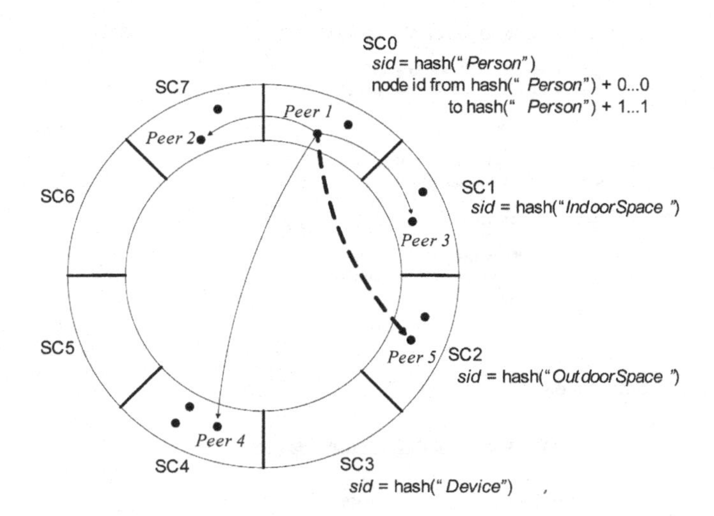

range contacts); and each node knows a small number of randomly chosen distant nodes (called long range contacts), with probability proportional to *1/d* where *d* is the distance (Kleinberg, 2000). The constant number of contacts and small average path length serve as the motivation for us to build the upper-tier network using the small world network model.

To construct the upper-tier network, each node maintains a set of short range contacts to a peer in its neighboring semantic clusters and a number of long range contacts. As shown in Figure 3, *Peer 1* maintains *Peer 2* as its left short range contact and *Peer 3* as its right short range contact; and that results all the semantic clusters are linked linearly in a ring fashion. The long range contacts are obtained by randomly choosing a node in the upper-tier based on a distribution function with its probability proportional to *1/d*, where *d* is the semantic distance (e.g., can be represented as Euclidean distance). The long range contacts aim at providing shortcuts to reach other semantic clusters quickly. Via short range and long range contacts, search in the upper-tier network can be guided greedily by comparing *sids* of the destination and the traversed nodes. In addition, if a peer has context data corresponding to its minor semantic clusters, it needs to register the indices of these data to a random node in each of its minor semantic clusters, e.g., *Peer 1* registers its data indices to a random node – *Peer 5* in *SC2* since it has data corresponding to semantic cluster – *OutdoorSpace*. This ensures that a context query is able to reach all the relevant nodes that store the keys responsible for the query. The registration process of data indices is similar to the storing process of data triples in the lower-tier network, and it will be described in the next section.

THE LOWER-TIER NETWORK

In the lower-tier network, peers in each semantic cluster are organized as Chord for storing data triples and routing context queries. This approach divides the one-dimensional Chord identifier space into multiple Chord identifier spaces. The number of neighbors maintained per node is logarithmic to the number of nodes in its semantic cluster. Hence, the maintenance cost can be reduced as compared to the original Chord.

A peer is organized into Chord based on the randomly chosen node identifier by applying the SHA1 hash function to its IP address. To facili-

Figure 4. An example of 3-bit Chord identifier space of 6 nodes (could hold up to 8 nodes) for the illustrating of storing data triples and query routing

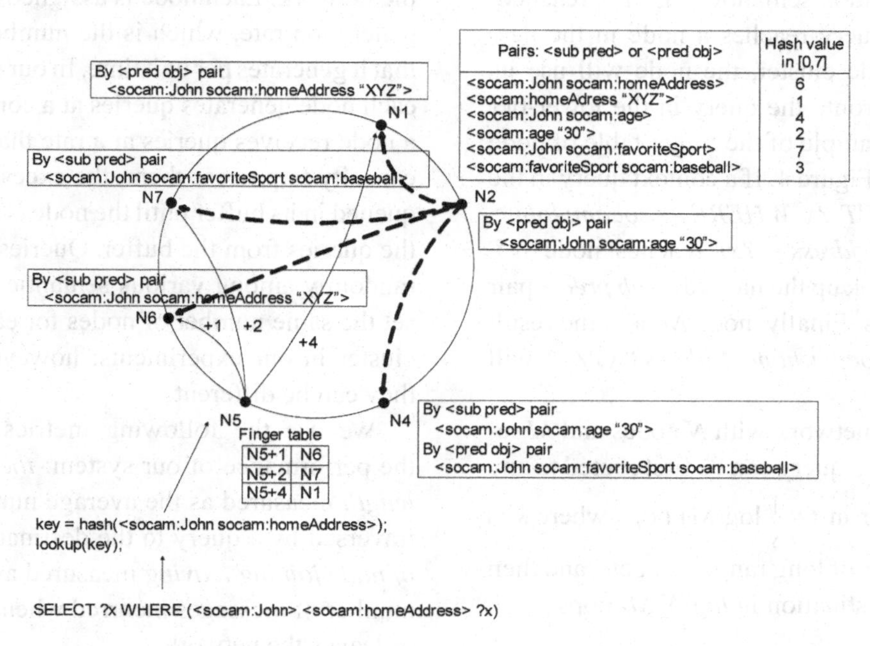

tate efficient context query, we build distributed indices for each data triple. Each data triple is in the form of subject, predicate, and object. Since the predicate of the triple is always given in a context query, we store each data triple two times in Chord. We apply the hash function to the *<sub pred>* and *<pred obj>* pairs to generate the keys for storing each data triple. Each data triple will be stored at the successor nodes of the hashed key values of *<sub pred>* and *<pred obj>* pairs. We define the *Store* procedure to perform the above storing process for each data triple. Figure 4 illustrates the process that node *N2* stores the following data triples in a 3-bit Chord identifier space of 6 nodes.

<socam:John socam:homeAddress "XYZ">
<socam:John socam:age "30">
<socam:John socam:favoriteSport
    socam:baseball>

To register the indices of data corresponding to the minor semantic cluster(s), a node first sends a *Register* message to a random node in each of its minor semantic clusters, and then it follows the same procedure as above to store the indices.

QUERY ROUTING

The query routing process involves two steps: inter-cluster routing and intra-cluster routing. A context query will be first forwarded to the appropriate semantic cluster and routed to destination peers in the lower-tier network. When a node receives a context query, the destination semantic cluster can be extracted from the query using the ontology-based semantic mapping technique (described in Section 2.2). First, we obtain the search key by hashing the destination semantic cluster. We then compare the search key with the most significant *m*-bits of its neighbors' identifiers, and forward the query to the closest neighboring node.

This forwarding process is recursively carried out until the destination semantic cluster is reached.

When the query reaches a node in the destination semantic cluster, the node will use its finger table to route the query in the lower-tier network. An example of the finger table of node *N5* is shown in Figure 4. If a context query in the form of *SELECT ?x WHERE (<socam:John> <socam:homeAddress> ?x)* reaches node *N5*, node *N5* will look up the hashed *<sub pred>* pair using its fingers. Finally, node *N6* and the result *<socam:John socam:homeAddress "XYZ">* will be returned.

For a given network with *N* nodes and *M* semantic clusters, a query can be first routed to any semantic cluster in $O(\frac{1}{s}\log^2 M)$ hops where *s* is the total number of long range contacts, and then routed to the destination in *log(N/M)* hops.

EVALUATION

We move on to evaluate our system using simulation and compare its performance to the original Chord. We first describe our simulation model and the performance metrics. Then we report the results from a range of simulation experiments. We also report the measurement results from the prototype system we developed.

Simulation Model and Metrics

We use the AS model to generate network topologies as previous studies (Saroiu, et al., 2002) have shown that P2P topologies follow both small world and power law properties. The simulation starts with having a pre-existing node in the network and then performing a series of join operations invoked by new coming nodes. A node joins its major semantic cluster based on its local data, and then stores its data triples and registers its data indices. After the network reaches a certain size, a mixture of node joining and leaving operations is invoked to simulate the dynamic characteristic of the network. Each node is assigned with a query generation rate, which is the number of queries that it generates per unit time. In our experiments, each node generates queries at a constant rate. If a node receives queries at a rate that exceeds its capacity to process them, the excess queries are queued in its buffer until the node is ready to read the queries from the buffer. Queries are selected randomly among various semantic clusters. We set the same number of nodes for each semantic cluster in our experiments; however, in reality they can be different.

We use the following metrics to measure the performance of our system: *the search path length* measured as the average number of hops traversed by a query to the destination; *the cost of node joining/leaving* measured as the average number of messages incurred when a node joins or leaves the network.

Simulation Results

First, we evaluate the efficiency of query routing in our system and compare it to Chord. We built the two-tier network by defining a number of semantic clusters in the upper-tier. In this experiment, we fix the number of semantic clusters to 16 and vary network size from 2^5 to 2^{13}. Hence, each semantic cluster in the lower-tier has a number of nodes ranged from 2 to 2^9. Figure 5 plots the average search path length of our system with 1 to 5 long range contacts on a logarithmic scale in comparison with Chord. The result shows that the two-tier network with 2 or more long range contacts has shorter search path as compared to Chord for a network size of 2^{13} nodes or less. It also shows that the search path length of the two-tier network is logarithmic to the number of nodes with a fixed number of semantic clusters.

In this experiment, we evaluate the impact of semantic clustering in our system. We fix the semantic cluster size to 8 (i.e., 8 nodes in each semantic cluster) and vary the number of seman-

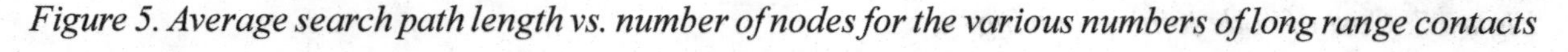

Figure 5. Average search path length vs. number of nodes for the various numbers of long range contacts

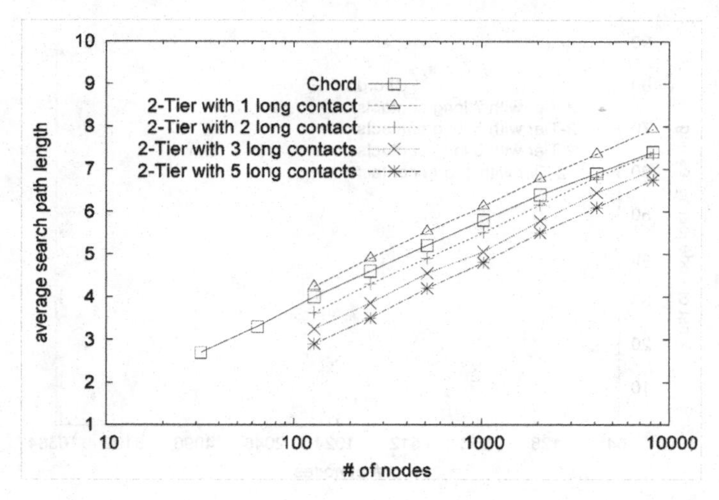

tic clusters in the upper-tier from 2^4 to 2^{11}. Since the number of nodes in each semantic cluster is fixed in this experiment, the average search path length in the lower-tier is a constant. Figure 6 plots search path length vs. number of semantic clusters in our system in the various settings of numbers of long range contacts. The result shows that increasing the number of long range contacts reduces search path length significantly. Figure 6 also reveals that search path length in the upper-tier matches the small world phenomenon.

We compare the cost of node joining and leaving between our system and Chord in this experiment. We vary network size from 2^5 to 2^{14}. In reality, the number of semantic clusters may increase when the network size increases. To simulate this behavior, we increase the number of semantic clusters with proportional to ccccccccc by making the number of semantic clusters equal

Figure 6. Average search path length vs. number of semantic clusters in the various settings of numbers of long range contacts

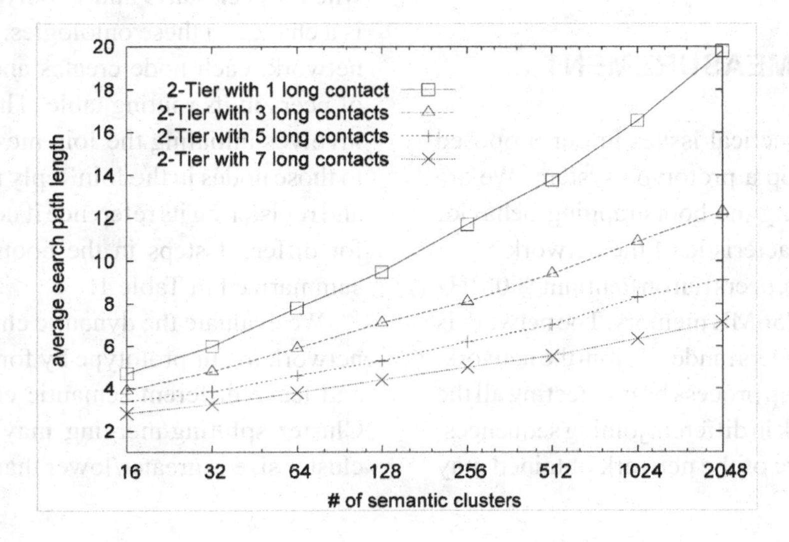

Figure 7. Cost of node joining/leaving

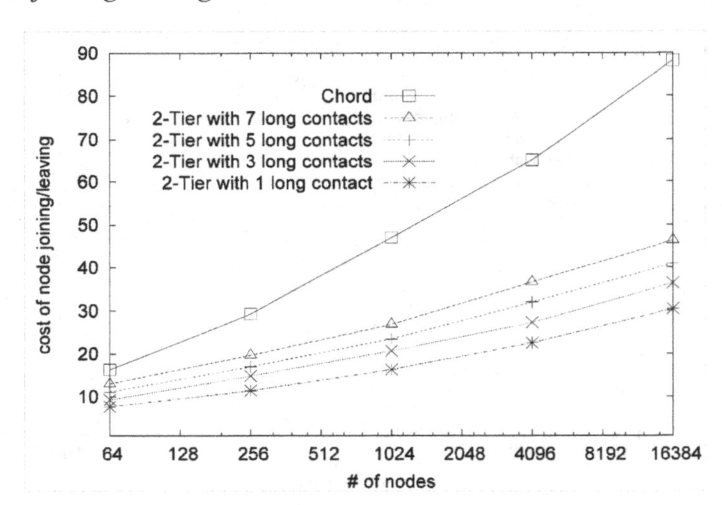

to the number of nodes in each semantic cluster. Figure 7 plots the average number of messages incurred when a node joins or leaves the network. The results show that our system reduces the cost of node joining/leaving significantly as compared to Chord whose update cost of node joining/leaving is $O(log^2N)$, where N is the total number of nodes in the network. This is also the effect of clustering, i.e., the number of nodes in a semantic cluster is much smaller than the number of nodes in the whole network. Hence, each node needs maintain a smaller size of finger table in our system as compared to Chord.

PROTOTYPE MEASUREMENT

Aim to explore practical issues in our proposed system, we develop a prototype system. We are interested in finding the bootstrapping behavior and dynamic characteristic of the network.

In the prototype, peers run on Pentium 800MHz desktop PCs with 256MB memory. The network is constructed when peers randomly join the network. We test the bootstrap process by connecting all the peers to the network in different joining sequences; hence, the structure of the network obtained may differ from one to another. When a peer starts, it first goes through the semantic clustering mapping process to identify which semantic cluster to join. The mapping process is done by iterating each of the RDF data triples and identifying its corresponding semantic cluster. Then the peer chooses the major semantic cluster to join. On average, the program initialization process takes about 4.26 seconds, and the mapping process for each RDF data triple takes about 0.251 ms. The initialization process involves reading and merging the ontology files stored locally and generating internal data structures for mapping. It is done only once when a peer starts and is only repeated if there is a change in these ontologies. Upon joining the network, each node creates and maintains a set of peers in its routing table. The joining process involves initiating the Join message, connecting to those nodes in the JoinReply message received and registering its reference if needed. The results for different steps in the bootstrap process are summarized in Table 1.

We evaluate the dynamic characteristic of the network in our prototype by forcing peers to join and leave different semantic clusters randomly. Cluster splitting/merging may occur when the cluster size is greater/lower than the default size.

Table 1. The results for the bootstrapping process

Processes	Average Time Taken
Program Initialization	4.26 s
Semantic Clustering Mapping	0.251 ms/RDF triple
Joining Process	2.56 s

For testing the dynamic characteristic of the network, we introduce a parameter: ***Time-to-Stability (TS)***. We define the steady state of a peer as the state in which a peer maintains live connections to the peers in its routing table. The steady state of a peer may collapse if one of the following events occurs:

- Its *short range contacts* or *long range contacts* leave the network or some of these peers change their major semantic clusters (due to their local data change).
- Its reference peer(s) leave the network or their major semantic clusters change.

Queries routing may be affected when peers are not in the steady state. The ***TS*** parameter is measured from the time when the steady state of a peer collapses until it reaches the steady state again. We measure the ***TS*** of the affected peers for different test cases and the results are summarized in Table 2 (note that no backup links are used in these cases).

In a highly dynamic network, peers leave and join frequently; this may result in high relapse rate. A high relapse rate may affect query routing in the network. To prevent this, we use a backup link for each type of connections. Once the steady state collapses, a peer can switch to the backup link immediately for the affected connection. With this backup scheme, we can minimize the disruption to query routing in the highly dynamic network where peers frequently leave and join.

Table 2. Results on TS

Test Cases (without backup links)	Average *TS*
Case 1: The short range contacts or long range contacts leaves the network or changes its major cluster or cluster splitting/merging occurs	271 ms per connection
Case 2: Reference hosting nodes leave/change	87 ms per reference

RELATED WORK

Centralized RDF repositories and lookup systems, such as RDFStore ([RDFStore) and Jena (Jena 2), have been implemented to support the storing and querying of RDF documents. These systems are simpler to design and reasonably fast for low to moderate number of triples. However, they have the common limitations of centralized approaches, such as single processing bottlenecks and single points of failure.

Schema-based P2P networks, such as Edutella (Nejdl, et al., 2003), are proposed to combine P2P computing and the Semantic Web. These systems build upon peers that use explicit schemas to describe their contents. They use super-peer based topologies, in which peers are organized in hypercubes to route queries. However, current schema-based P2P networks still have some shortcomings: queries have to be flooded to every node in the network, making the system difficult to scale. Crespo, et al. (2003) proposed the concept of Semantic Overlay Networks (SONs) in which peers are grouped by semantic relationships of documents they store. Each peer stores additional information about content classification and route queries to the appropriate SONs, increasing the chances that matching objects will be found quickly and reducing the search load. However, queries still need to be flooded in each overlay network resulting in redundant query messages in the network. Cai, et al. (2004) proposed a scalable and distributed RDF repository called RDFPeers based on a structured P2P system. RDFPeers organize into a multi-attribute addressable network (MAAN) (Cai, et al., 2003) which extends Chord to efficiently answer multi-attribute and range queries. When an RDF triple is inserted into the network, it will be stored three times by applying a globally-known hash function to its subject, predicate, and object. We take a similar approach to deploy Chord as the substrate for the lower-tier network, however, we store the *<sub pred>* and *<pred obj>* pairs for each data triple as the predicate is always known in a context query. Thus, the cost of inserting RDF triples into the network can be reduced. In addition, the identifier space of the lower-tier in our network is much smaller than the one in RDFPeers. Hence, the maintenance cost is lower as compared to RDFPeers since each peer maintains fewer neighbors. Tang, et al. (2003) applied classical Information Retrieval techniques to P2P systems and built a decentralized P2P information retrieval system called pSearch. The system makes use of a variant of CAN to build the semantic overlay and uses Latent Semantic Indexing (LSI) (Deerwester, et al., 1990) to map documents into term vectors in the space. Li, et al. (2004) built a semantic small world network in which peers are clustered based on term vectors computed using LSI. They proposed an adaptive space linearization technique for constructing link structures. While we take the semantic approach which is conceptually similar to (Tang, et al., 2003) and (Li, et al., 2004), we propose the use of schema-based metadata to extract data semantics. The formal design of ontologies minimizes the problems of synonyms and polysemy incurred by VSM, and incurs a lower overhead than LSI does. Kleinberg (Kleinberg, 2000) proposed the small world network model where every node maintains four links to each of its closest neighbors and one long distance link to a node chosen from a probability function. He has shown that a query can be routed to any node in $O(log^2 n)$ hops, where n is the total number of nodes in the network. We build the upper-tier network based on the small world network model. The small world model has many advantages, such as it is easy to construct and the number of state information that each node maintains is fixed and not proportional to the number of semantic clusters. In our earlier work (Gu, et al., 2005), we have proposed a semantic P2P network for context search by using a Gnutella-like network as the substrate. However, the flooding-based routing mechanism is not very efficient in terms of search path and scalability. This article proposes a more efficient and scalable semantic network based on a structured P2P network (i.e., Chord).

CONCLUSION

In this article, we present an ontology-based semantic P2P network for searching context information in wide-area networks. The preliminary results have shown that our system has good search efficiency and low cost of node joining and leaving, and our system can scale to a large number of peers. The use of our system is not limited to the context-aware computing domain; in fact, it applies to any P2P searching system where schemas are explicitly defined.

REFERENCES

Cai, M., & Frank, M. (2004). *RDFPeers: A Scalable Distributed RDF Repository based on A Structured Peer-to-Peer Network*. Paper presented at the Proceedings of the 13th International World Wide Web Conference, New York.

Cai, M., Frank, M., Chen, J., & Szekely, P. (2003). *MAAN: A Multi-attribute Addressable Network for Grid Information Services*. Paper presented at the Proceedings of the 4th International Workshop on Grid Computing.

Castro, M., Druschel, P., Kermarrec, A.-M., Nandi, A., Rowstron, A., & Singh, A. (2003). *Splitstream: High-bandwidth Content Distribution in a Cooperative Environment*. Paper presented at the Proceedings of the International Workshop on Peer-to-Peer Systems (IPTPS 2003).

Crespo, A., & Garcia-Molina, H. (2003). *Semantic Overlay Networks for P2P Systems*. Technical report, Stanford University.

Deerwester, S. C., Dumais, S. T., Landauer, T. K., Furnas, G. W., & Harshman, R. A. (1990). Indexing by Latent Semantic Analysis. *Journal of the American Society for Information Science American Society for Information Science*, *41*(6), 391–407. doi:10.1002/(SICI)1097-4571(199009)41:6<391::AID-ASI1>3.0.CO;2-9

Dey, A., & Abowd, G. (2000). *Towards a Better Understanding of Context and Context-Awareness*. Paper presented at the Proceedings of the Workshop on the What, Who, Where, When and How of Context-awareness at CHI 2000. Freenet. http://freenet.sourceforge.net.

Gnutella. http://gnutella.wego.com.

Gu, T., Tan, E., Pung, H. K., & Zhang, D. (2005). *A Peer-to-Peer Architecture for Context Lookup*. Paper presented at the Proceedings of the International Conference on Mobile and Ubiquitous Systems: Networking and Services (MobiQuitous 2005), San Diego, California.

Han, J., & Liu, Y. (2006). *Rumor Riding: Anonymizing Unstructured Peer-to-Peer Systems*. Paper presented at the Proceedings of IEEE ICNP, Santa Barbara, CA. Jena 2 - A Semantic Web Framework. http://www.hpl.hp.com/semweb/jena2.htm.

Kleinberg, J. (2000). *The Small-World Phenomenon: an Algorithm Perspective*. Paper presented at the Proceedings of the 32nd ACM Symposium on Theory of Computing. LimeWire. http://www.limewire.com/english/content/home.shtml.

Li, M., Lee, W.-C., & Sivasubramaniam, A. (2006). *DPTree: a Balanced Tree Based Indexing Framework for Peer-to-Peer Systems*. Paper presented at the Proceedings of IEEE ICNP, Santa Barbara, CA.

Li, M., Lee, W. C., Sivasubramaniam, A., & Lee, D. L. (2004). *A Small World Overlay Network for Semantic Based Search in P2P*. Paper presented at the Proceedings of the Second Workshop on Semantics in Peer-to-Peer and Grid Computing, in conjunction with the World Wide Web Conference.

Liu, Y., Liu, X., Xiao, L., Ni, L. M., & Zhang, X. (2004). *Location-aware Topology Matching in P2P Systems*. Paper presented at the Proceedings of IEEE INFOCOM, Hong Kong, China.

Morselli, R., Bhattacharjee, B., Srinivasan, A., & Marsh, M. A. (2005). *Efficient Lookup on Unstructured Topologies*. Paper presented at the Proceedings of ACM PODC, Las Vegas, NV, USA.

Nejdl, W., Wolpers, M., Siberski, W., Schmitz, C., Schlosser, M., Brunkhorst, I., & Lser, A. (2003). *Super-peer-based Routing and Clustering Strategies for RDF-based Peer-to-Peer Networks*. Paper presented at the Proceedings of the 12th World Wide Web Conference.

Ratnasamy, S., Francis, P., Handley, M., Karp, R., & Shenker, S. (2001). *A Scalable Content Addressable Network*. Paper presented at the Proceedings of ACM SIGCOMM.

RDF. http://www.w3.org/RDF. World Wide Web Consortium: Resource Description Framework. RDFStore. http://rdfstore.sourceforge.net.

RDQL. http://www.w3.org/Submission/2004/SUBM-RDQL-20040109/.

Rowstron, A., & Druschel, P. (2001). Pastry: Scalable. Distributed Object Location and Routing for Large-scale Peer-to-Peer Systems. *Lecture Notes in Computer Science*, *2218*, 161–172.

Saroiu, S., Gummadi, P., & Gribble, S. (2002). *A Measurement Study of Peer-to-Peer File Sharing Systems*. Paper presented at the Proceedings of Multimedia Computing and Networking.

Stoica, I., Morris, R., Karger, D., Kaashoek, F., & Balakrishnan, H. (2001). *Chord: A Scalable Peer-to-Peer Lookup Service for Internet Applications*. Paper presented at the Proceedings of ACM SIGCOMM.

Tang, C. Q., Xu, Z. C., & Dwarkadas, S. (2003). *Peer-to-Peer Information Retrieval Using Self-Organizing Semantic Overlay Networks*. Paper presented at the Proceedings of ACM SIGCOMM 2003, Karlsruhe, Germany.

Zhao, B. Y., Huang, L., Stribling, J., Rhea, S. C., Joseph, A. D., & Kubiatowicz, J. D. (2004). Tapestry: A Resilient Global-scale Overlay for Service Deployment. *IEEE Journal on Selected Areas in Communications*, *22*(1), 41–53. doi:10.1109/JSAC.2003.818784

This work was previously published in International Journal of Grid and High Performance Computing (IJGHPC), Volume 1, Issue 4, edited by Emmanuel Udoh & Ching-Hsien Hsu, pp. 26-39, copyright 2009 by IGI Publishing (an imprint of IGI Global).

Chapter 20
FH-MAC:
A Multi-Channel Hybrid MAC Protocol for Wireless Mesh Networks

Djamel Tandjaoui
Center of Research on Scientific and Technical Information, Algeria

Messaoud Doudou
University of Science and Technology Houari Boumediène, Algeria

Imed Romdhani
Napier University School of Computing, UK

ABSTRACT

In this article, the authors propose a new hybrid MAC protocol named H-MAC for wireless mesh networks. This protocol combines CSMA and TDMA schemes according to the contention level. In addition, it exploits channel diversity and provides a medium access control method that ensures the QoS requirements. Using ns-2 simulator, we have implemented and compared H-MAC with other MAC protocol used in Wireless Network. The results showed that H-MAC performs better compared to Z-MAC, IEEE 802.11 and LCM-MAC.

INTRODUCTION

Wireless mesh networks are an attractive field for several research labs, and they were the subject of many papers in the few last years. These intensive works try to solve different open issues which concern mainly the capacity of the wireless mesh network protocols, and especially MAC protocols capacity (Akyildiz, Wang and Wang, 2005).

MAC protocols for wireless networks suffer from many problems such as scalability; data throughput degrades significantly when increasing the number of nodes or hops in the network. Furthermore, many other MAC problems persist for example the interference effect and radio channel allocation strategies. These problems are caused by using advanced radio technologies such as directional antenna, omnidirectional antenna and multi-channel/multi-radio systems. Thus,

DOI: 10.4018/978-1-60960-603-9.ch020

all existing MAC protocols must be improved or reinvented.

Researchers have started revising the design of wireless networks MAC protocols, especially MAC protocols of ad hoc and sensors networks. The international standard groups are also working on the specification of new technologies for wireless mesh networks that includes IEEE 802.16, 802.11s, 802.15.5, and ZigBee. Several researches issues still exist and need to be solved. In particular, the interesting research problem related to the scalability issue of existing IEEE 802.11 networks. The most addressed solution intends to develop a hybrid MAC protocol that combines the strength of TDMA and CSMA while offsetting their (Akyildiz, Wang and Wang, 2005). In the wireless mesh network, it is important that the underlying MAC schemes could be able to provide high bandwidth by exploiting channel diversity and support QoS requirements. It must have the capacity of self-organizing, self-configuring, and self-healing.

In Wireless MAC protocols, using hybrid schemes outperform random-based and schedule-based schemes. In case of random-based schemes, throughput drops significantly when increasing traffic intensity, number of nodes, or hops in the network. In addition, random-based schemes cannot guarantee contention-free transmission. The one hop packet loss probability increase when the number of nodes trying to transmit simultaneously increase. This probability cumulates across multiple hops. Schedule-based schemes provide for contention-free transmission slots to each node. The schedule comprising of these transmission slots is based on the network traffic and topology. To derive and propagate the schedule, traffic and topology information needs to be collected, which involves network overhead. Thus, the frequent changes in the network conditions results in high overheads, and leading to poor performance of schedule-based schemes.

In this article, we study the problems which persist at wireless MAC layer in multi-hop wireless Network. In addition, we propose a new hybrid MAC scheme, called H-MAC (Hybrid MAC) for wireless mesh network that combines the strengths of TDMA and CSMA. H-MAC extends the hybrid multi-hops scheme defined in Z-MAC (Rhee, Warrier, Aia, and Min, 2005) to support channel diversity and QoS requirements for wireless mesh network. The main feature of H-MAC is its adaptability to the level of contention in the network. In fact, under low contention, H-MAC behaves like CSMA, and under high contention, it behaves like TDMA.

H-MAC uses two contention modes: Low Contention Level (LCL) and High Contention Level (HCL). It also implements two allocation algorithms. The first Receiver Based Channel Assignment Algorithm (RBCA) is used for channel allocation and the second Sender Based Slot Assignment Algorithm (SBSA) is used for slot allocation. We have evaluated the performances of our protocol by comparing it to other used MAC protocols. In this evaluation, we have used the ns-2 simulator and we have conducted several simulation scenarios. The obtained result showed that H-MAC performs better compared to Z-MAC, IEEE 802.11 and LCM-MAC.

This article is organized as follows. In the second section we describe the related works and discuss the different protocols proposed for wireless MAC. We present and detail H-MAC protocol in section 3. In section 4, we present our simulation and the obtained results. We conclude our work in section 5.

RELATED WORKS

We classify MAC solutions in three main classes. The first class is the hybrid protocols that combine CSMA and TDMA. The second class contains multi-channel MAC protocols, and the third class includes MAC protocols with QoS support. In the next sections, we will outline the strengths and weaknesses of these classes.

Figure 1. Throughput comparison between CSMA and TDMA

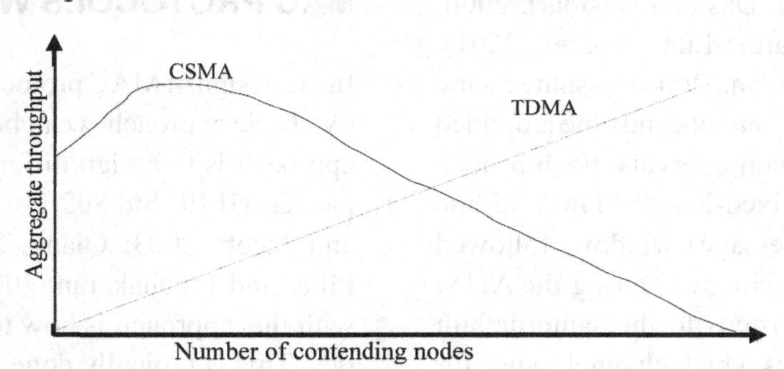

HYBRID MAC PROTOCOLS

Based on the access strategy used, MAC protocols can be sorted into three categories: random-access or contention-based, schedule based and hybrid.

A random-access scheme like CSMA works well with low contention and provides better throughput. However, the data throughput degrades significantly when increasing the number of contending nodes. A scheduled scheme like TDMA does not provide good throughput with low contention. But, the network throughput progresses proportionally according to the number of contending nodes (Krishna Rana, Hua Liu, Nyandoro and Jha, 2006; Chlamtac, Farago, Myers, Syrotiuk and Zaruba, 2000; Henderson, Kotz and Abyzov, 2004).

Some approaches combining the strength of random and schedule based schemes have been developed. In the schema described in (Koubias and Haralabidis, 1996), the default transmission is random-based. However, when detecting a collision, a round of token passing (contention-free) transmission mode is initiated. Thus, whenever collision probability increases, the scheme shifts to schedule-based contention-free transmission. PTDMA is a hybrid protocol presented by Emphremides and Mowafi (Ephremides and Mowafi, 1982). In this protocol the probability of collision is controlled by programming nodes to transmit with different probability. ADAPT (Myers, 2002) is another protocol that employs similar approach like PTDMA, but is much simpler.

Z-MAC (Rhee, Warrier, Aia and Min, 2005) is also an hybrid scheme based on the same approach as ADAPT. It has been optimized for multi-hop scenario and adapted to perform in sensor network. Z-MAC uses STDMA scheduling to reduce collision probability of CSMA based scheme (Gronkvist, 2004). Like ADAPT, by combining CSMA and TDMA, Z-MAC delivers a robust scheme which even in worst case, performs as well as CSMA scheme.

Bandwidth Aware Hybrid MAC (Krishna Rana, Hua Liu, Nyandoro and Jha, 2006) is another protocol similar to Z-MAC. It improves the hybrid schemes of ADAPT and Z-MAC by proposing an algorithm that allocates slots to the nodes in proportion to their bandwidth requirements.

MULTI-CHANNEL MAC PROTOCOLS

A large number of multi-channel MAC protocols and TDMA scheduling algorithms have been proposed in the literature (Kyasanur, Jungmin, Chereddi and Vaidya, 2006). Multi-channel MAC protocols have extended the DCF (Distributed Coordination Function) function of IEEE 802.11 protocol (IEEE 802.11 Working Group, 1997) and use certain type of control messages for frequency negotiation (So and Vaidya, 2004; Fitzek, Angelini, Mazzini and Zorzi, 2003; Li, Haas, Sheng and Chen, 2003; Jain, Das and Nasipuri, 2000; Tzamaloukas and Garcia-Luna-Aceves, 2001). MMAC (So and Vaidya, 2004) assumes time synchronization in the network and time is divided into fixed-length beacon intervals. Each beacon interval consists of a fixed-length ATIM (Ad-hoc Traffic Indication Message) window, followed by a communication window. During the ATIM window, each node listens to the same default channel and negotiates which channel to use for data communication. After the ATIM window, nodes that have successfully negotiated channels with their destinations send out data packets using 802.11 DCF for congestion avoidance (IEEE 802.11 Working Group, 1997).

Multi-channel MAC protocols in Wireless Sensor Networks (WSNs) are also studied (Zhou et al., 2006). Due to the limited size of the data packets used in WSNs, authors have proposed to use static frequency assignment to avoid the overhead of control packets for frequency negotiation. There are also many TDMA scheduling algorithms proposed for ad hoc networks (Chlamtac and Kutten, 1985; Chlamtac and Farago, 1994; Bao and Garcia-Luna-Aceves, 2001; Rajendran, Obraczka and Garcia-Luna-Aceves, 2003). These algorithms are mainly designed for sharing a single channel in the network and providing collision free access. For example, the TMMAC protocol presented in (Zhang, Zhou, Huang, Son and Stankovic, 2007) is one these algorithms that combines TDMA scheme and channel diversity to improve the network throughput. It is proved that TMMAC achieves 84% more aggregate throughput than MMAC (Zhang, Zhou, Huang, Son and Stankovic, 2007). MMSN (Zhou et al., 2006) is another MAC protocol that exploits channel diversity in sensors networks. MMSN omits exchanging RTS/CTS, because in WSN, the packet is very small, 30~50Bytes.

MAC PROTOCOLS WITH QOS

In the design of MAC protocols with QoS support, two basic approaches can be employed. The first approach is to assign different *priority levels* to packets (IEEE Std 802.11e, 2004; Ying, Ananda and Jacob, 2003; Qiang, Jacob, Radhakrishna Pillai and Prabhakaran, 2002). The major issue with this approach is how to assign these priorities. This is typically done by defining different intervals for both the random backoff period and AIFS (Arbitration Inter Frame Space) period, such as the EDCA (Enhanced Distributed Channel Access) function of IEEE 802.11e. In a single hop environment, EDCA offers better average delay and throughput than the usual DCF. The IEEE 802.11s working group plans to extend the 802.11e scheme for the multi-hop wireless mesh network (Conner, Kruys, Kim and Zuniga, 2006).

The second approach to support QoS is to *reserve resources* for a particular real-time traffic flow. For example, each node between particular source and destination nodes allocates some dedicated time slots for this flow before the actual transmission starts. This improves the end-to-end throughput. However, this reservation mechanism is much more complex than a priority mechanism. Typically, it adds signaling overhead to coordinate the nodes (all nodes between source and destination must agree in distributed manner on the reserved resources).

H-MAC PROTOCOL

In this section, we present our H-MAC protocol. This protocol extends the hybrid multi-hops schema defined in Z-MAC (Rhee, Warrier, Aia, and Min, 2005), which combines TDMA and CSMA according to the contention level. Compared to Z-MAC, H-MAC uses multi-channel hybrid schema which guarantees the QoS requirements for a multi-hop wireless mesh network.

The Network Model

In our protocol, we assume that each node is assigned a unique identifier. The network interface is equipped with a single half duplex radio transceiver. We also assume that the network card is capable to send either unicast or broadcast packets. The network topology is represented by an undirected graph $G = (V;E)$, where V is the set of nodes, and E is the set of links between nodes. The existence of a link $(u; v) \in E$ implies that $(v; u) \in E$, and that node u and v are within the transmission range of each other. In this case, u and v are called *one-hop neighbors* of each other. The set of one-hop neighbors of a node i is denoted by N_i^1. Two nodes are called *two-hop neighbors* of each other if they are not adjacent, but have at least one common one-hop neighbor. The neighbor information of node i refers to the union of the one-hop neighbors of i itself and the one-hop neighbors of i's one-hop neighbors, which is equal to:

This set contains the entire *one hop* and *two hops* neighbors of a node i.

PROTOCOL DESCRIPTION

H-MAC uses the two contention modes LCL and HCL similar to that of Z-MAC. It also implements two allocation algorithms. The first one is a Receiver Based Channel Assignment Algorithm (RBCA). In this algorithm, each node is assigned a unique channel in which it will receive all its packets. The second is the Sender Based Slot Assignment algorithm (SBSA) where each node is assigned a set of slots of which it will become the owner. These algorithms are an extension of NCR (Neighbor-aware Contention Resolution) algorithm defined in (Bao and Garcia-Luna-Aceves, 2003), which does not require any control message exchange. H-MAC uses a medium access function similar to the IEEE 802.11e EDCA techniques that support the QoS requirements (IEEE Std 802.11e, 2004).

H-MAC operates in two phases: initialization phase and communication phase. In the initialization phase, the following operations run in sequence: neighbor discovery, channel assignment, slot assignment, and finally global time synchronization. These operations run only once during the setup phase and does not run again until a significant change in the network topology (such as HELLO joining, or QUIT message) occurs. In the communication phase, each node performs channel negotiation and runs the LCL or HCL mode according to the contention level.

THE INITIALIZATION PHASE

a. Neighbor Discovery

At the initialization, each node broadcasts its ID. After that, it periodically broadcasts a ping message to its one-hop neighbors to build its one-hop neighbors list. A ping message contains the current list of its one-hop neighbors N_i^1. This message is sent at a random time in each second for 30 seconds. Through this process, each node gathers the information received from the pings from its one-hop neighbors which essentially constitutes its two-hop neighbor information (See Figure 2).

b. Channel Allocation Algorithm RBCA

The Receiver Based Channel Assignment (RBCA) is an implicit Consensus algorithm. Each node is assigned a unique channel in which it will receive all its packets. This algorithm uses pseudo-random generator similar to that used by the NCR algorithm (Bao and Garcia-Luna-Aceves, 2003). It solves a special election problem where an entity decide its leadership among a known set of contenders in any given contention context. Each node calculates a hash using its ID as a seed, and if its hash is

Figure 2. Neighbor discovery process

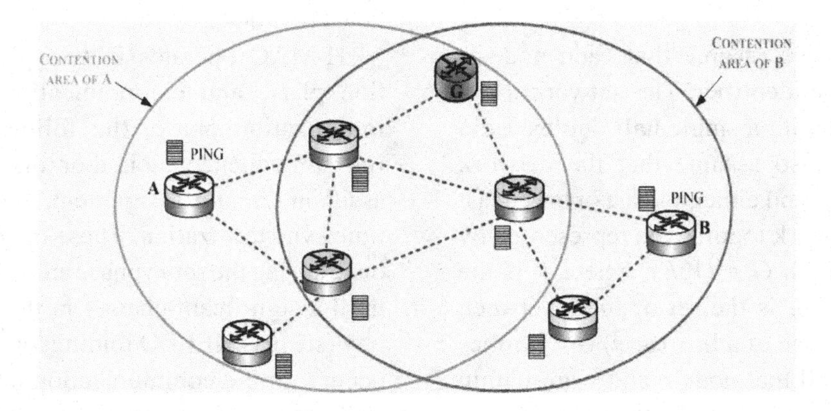

the biggest among its two-hop neighbors it wins the channel. Otherwise, it chooses the channel in which it has obtained its max hash. Then, it broadcasts this information to its two-hop neighbors. The RBSA algorithm has the following structure:

Let Hash(x) be a fast message digest generator. C_{max}: number of channels, V_2: two-hop neighbors, α a node, C_α: the channel number affected to α and '⊕' is the concatenation of two operands.

c. Slot Allocation Algorithm SBSA

The Sender Based Slot Assignment (SBSA) is also an implicit consensus algorithm. Each node is assigned a set of transmission slots of which it will become the owner. Thus, the node will have the highest priority to send during these slots. SBSA works in the same way as RBCA where a node determines for each channel its slot using the distributed election algorithm. We denote the set of contenders of an entity i by M_i, and thus its contention context by $t_i = (c_i, s_i)$, where c_i is the channel i and s_i is the slot i in channel i. To decide the leadership of an entity without incurring communication overhead among the contenders, we assign each node a priority that depends on the identifier of the node and the current contention context. Equation *(1)* provides a formula to derive the priority, denoted by H_i, for node i and contention context t_i

$$H_i = \text{Hash}(i \oplus t_i) \oplus i, \text{ where } t_i = (c_i \oplus s_i) \quad (1)$$

Where the function *Hash* is a fast message digest generator like MD4 or MD5 that returns a random integer in a predefined range, and the sign '⊕' is the concatenation of two operands. Note that, although the *Hash* function can generate the same number on different inputs, each number is unique because it is appended with the identifier of the node. The set of contexts is showed by the following matrix $\|T\|_{C*S}$.

A node α wins the slot $t_{ij} = (c_i \oplus s_j)$ if it has the highest hash value, i.e. the inequality presented below must be verified for a node α, and that the H_i are calculated using the equation (2):

$$\underset{i \in M_i \cup \{\alpha\}}{argmax}\, H_i = \alpha \quad (2)$$

argmax provides the argument of the maximum, that is to say, the value of the given argument for which the value of the given expression reaches its maximum value. The SBSA algorithm has the following structure:

Figure 3. H-MAC slot structure

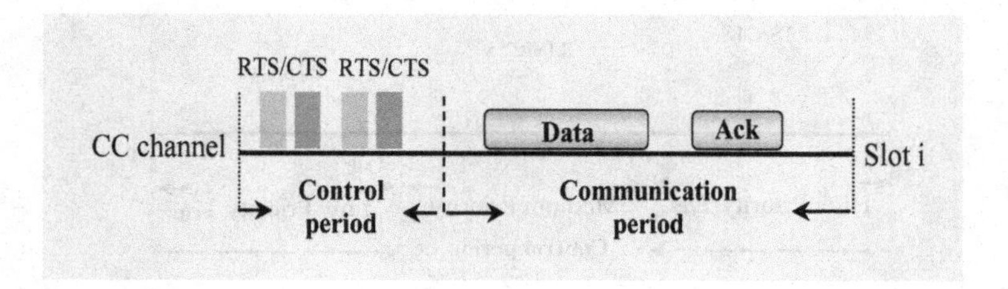

Let H be a pseudo-random hash function. S_{max}: number of slots, V_2: two-hop neighbors, α: a node and $List_{\alpha}$ is a list of slots.

```
SBSA algorithm (α, Listα);
 { Listα = Ø; j=0;
   repeat
   { i=1; found = false;
     repeat
     { for (k ∈ V2 ∪ {α}) Hk = H(k ⊕ Si ⊕ Cj) ⊕ k;
         if (∀k ∈ V2, Hα > Hk)
          then
           found = true; Listα = Listα ∪ Sij; break;
          else
             i++;
     } while (i< Smax);
     if (found == false)
       then i = arg max Hα ; Listα = Listα ∪ Sij;
         j++;
   } while (j< Cmax);
 }
Broadcast Listα to 2-hop neighbors.
```

THE COMMUNICATION PHASE

In H-MAC, a slot 0 of each local frame is reserved to broadcast packet transmission (access by CSMA). The channel negotiation is done in a dedicated *Control Channel* (CC); this channel can be used for transmission after the control period.

After the initialization phase, all nodes switch to the control channel CC at slot start, and they must be ready to run the transmission control. In H-MAC, a node can be in one of two modes: low contention level (LCL) or high contention level (HCL). A node is in HCL only when it receives an explicit contention notification (ECN) message from a two-hop neighbor within the last frame t_{ECN}. Otherwise, the node is in LCL. A slot is divided into:

- *Control period*: to negotiate the slot *i* on different channels using RTS/CTS with priority (QoS), and the first which succeed its CTS_{jn} (j: channel j, n: destination node) wins slot *Sij*.
- *Transmission period*: the winners and their destination nodes switch to the appropriate channel to exchange unicast packets (Figure 3).

a. The LCL mode

In LCL, any node can compete to transmit in any slot. The control phase is divided into 3 periods in this mode:

- *High priority* T_{HP}: it is reserved to owners or to high priority packets (real time traffic).
- *Medium priority* T_{MP}: it is reserved to one-hop neighbors or to medium priority packets (audio, video).

Figure 4. The structure of the control period

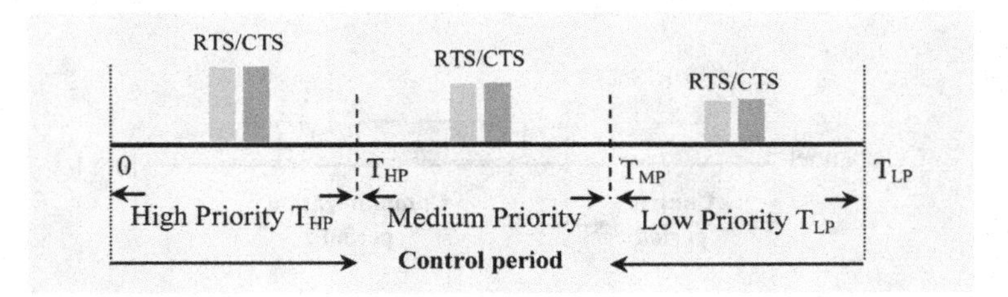

- *Low priority T_{LP}*: it is reserved to two-hop neighbors or to low priority packets (best-effort, background).

The transmission rule: according to Figure 4, as a node *i* acquires data to transmit, it checks whether:

- It is the owner of the current slot on its destination's channel or it has a high priority packet.
- It is the one-hop neighbor of the owner of the slot on its destination's channel or it has a medium priority packet.
- It is the two-hop neighbor of the owner of the slot on its destination's channel or it has a low priority packet.

b. The HCL Mode

In HCL, we have only the first and the second period. Consequently, a node can compete in the current slot if and only if:

- It is the owner of the slot on its destination's channel or it has a high priority packet.
- It is the one-hop neighbor of the owner of the slot on its destination's channel or it has a medium priority packet.

After the control phase, all nodes that have already succeed their negotiation switch to the channel of their destination nodes and start the data packet transmission for the rest of the slot.

c. The Priority Queues and QoS Support

H-MAC protocol uses the priority queue concept inspired from the IEEE 802.11e protocol to support the QoS requirements. Each node maintains 3 priority queues:

- *High priority queue*: contains real time packets (we can also integrate transient traffic i.e. not originated form the current node).
- *Medium priority queue*: contains audio and video packets.
- *Low priority queue*: contains best-effort and background packets.

d. Explicit Contention Notification (ECN)

ECN messages notify two-hop neighbors not to act as hidden terminals to the owner of each slot when contention is high. Each node makes a local decision to send an ECN message based on its local estimate of the contention level (Figure 5). The estimation is obtained by the noise level of the channel. ECN is similar to RTS/CTS in CSMA/CA. But the difference is that HCL uses topology information (i.e., slot information) to avoid two hop collision. The cost of ECN is also far less than RTS/CTS since it is triggered only when contention is high.

Figure 5. Explicit Contention Notification Scheme

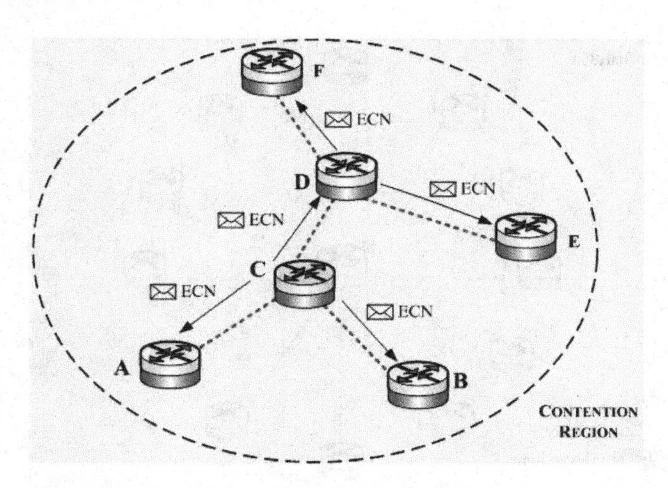

e. Local Time Synchronization

The protocol adopts the same synchronization technique used in Z-MAC. The advantage of such technique is that synchronization is required only among neighboring senders and also when they are under high contention. These points offer an excellent opportunity to optimize the overhead of clock synchronization because synchronization is required only locally among neighboring senders. In addition, the frequency of synchronization can be adjusted according to the transmission rates of senders so that senders with higher data rates transmit more frequent synchronization messages. In this scheme, receivers synchronize passively their clocks to the senders' clocks and do not have to send any synchronization messages.

PERFORMANCE EVALUATION

We have implemented H-MAC using the network simulator ns-2 (Fall and Vradhan, 1998) and compared its performance with the existing MAC protocols. In fact, we compared the performance of H-MAC with Z-MAC (Rhee, Warrier, Aia, and Min, 2005), MMAC (So and Vaidya, 2004), LCM-MAC (Maheshwari, Gupta and Samir, 2006), and 802.11 (IEEE 802.11 Working Group, 1997). The performance evaluation in our simulation is achieved through a set of tests which allows making comparison with other MAC protocols, and it takes the following aspects: The impact of the hybrid scheme and channel diversity on network throughput.

HYBRID SCHEME EVALUATION

In this simulation, we have chosen to make a comparison between H-MAC, Z-MAC, and 802.11 MAC protocol. We have measured and compared the effective channel utilization of H-MAC and Z-MAC. For this purpose, we have repeated the same simulation and used the default settings of Z-MAC as described in (Rhee, Warrier, Aia, and Min, 2005). We varied the backoff window sizes to see the impact of window sizes on channel utilization. We used three scenarios in our simulation: one-hop, two-hop and multi-hop scenarios.

One-hop scenario: in this scenario 21 nodes are placed equidistant from a receiver in a circle (Figure 6). Before each run, we ensured that all nodes are in a one-hop distance to each other so that there are no hidden terminals. This scenario is used to measure the achievable throughput of

Figure 6. One hop network scenario

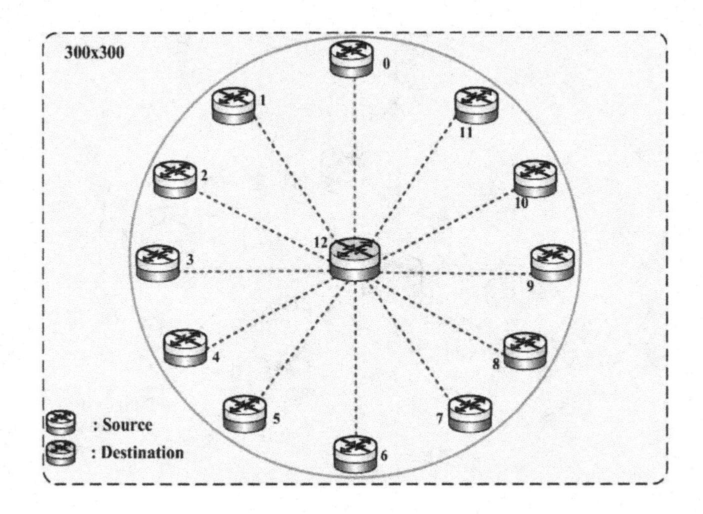

different MAC protocols for different levels of contention within a one-hop neighborhood. Since Z-MAC has the same test, we can compare its results to ours.

We fixed the frame size to 20 slots and varied the number of senders. HCL is disabled because the performance of HCL and LCL is the same when all nodes are in a one-hop distance to each other. Before running H-MAC, the channel allocation algorithm RBCA and the slot allocation algorithm SBSA are executed by each node in the network. In addition, H-MAC runs TPSN (Ganeriwal, Kumar and Srivastava, 2003) to synchronize the clocks of the senders.

The Figure 7 shows simulation results and the throughput comparison for one-hop scenario involving H-MAC and Z-MAC. The H-MAC protocol shows good performance, but with a mar-

Figure 7. Throughput comparison in a one hop scenario

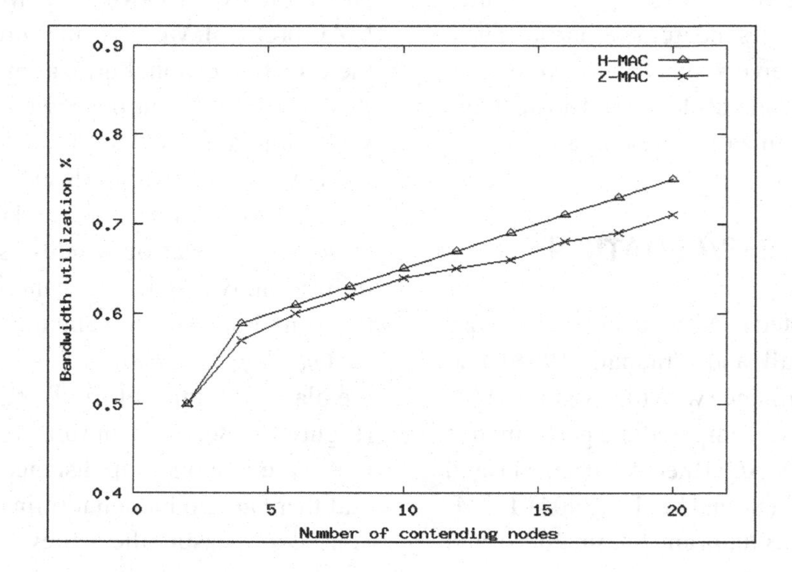

Figure 8. Tow-hop network scenario

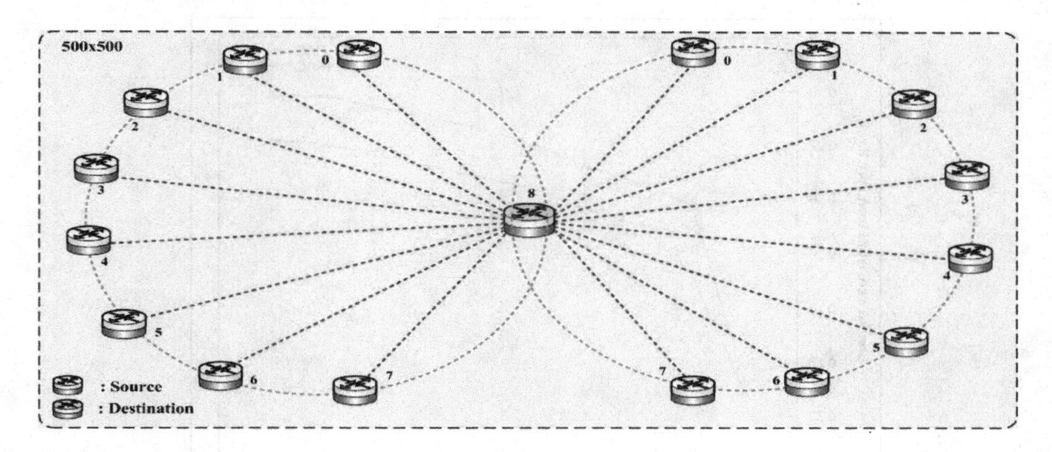

gin similar to that of Z-MAC. This performance similarity is explained by the fact that H-MAC uses the same medium access scheme as Z-MAC, and because all nodes are within one-hop distance from the destination, so the senders can be easily synchronized with each other.

Two-hop scenario: this scenario is used to measure the performance of the different protocols when hidden terminals are present. We organized nodes into two clusters as illustrated in Figure 8. The two clusters are placed approximately 5 meters apart. A receiver node (or routing node) is placed in the middle of the two clusters. We ensure that all senders find the receiver as a one-hop neighbor and all nodes are reachable by two hop communications. We also reduced the transmission power of senders to 1 dBm (1.3 mW) to control the number of hidden terminals.

In the tow-hop scenario, we measured the data throughput when hidden terminals are present. We varied the number of senders while fixing the number of neighbors. As in the one-hop benchmark, all senders have always data to send. Each additional sender is chosen from the alternating clusters.

For H-MAC tests, we set the frame size to 20 slots. In this test, we run H-MAC with the local clock synchronization protocol in which each sender sends one synchronization packet in every 100 packets transmitted. The data throughput reported by H-MAC includes the overhead of the clock synchronization and ECN.

The Figure 9 shows the two-hop tests results. With the ns-2 simulator, we verified that the two node clusters do not sense each other to maximize the number of hidden terminals. We noticed that despite using the RTS/CTS mechanism in H-MAC during the control period, H-MAC maintains the same good performance but with slightly degradation in channel utilization to 73%. Z-MAC has suffered from performance degradation that undergo until 68%. This performance degradation is caused by the presence of the hidden terminals, and by the overhead of ECN messages.

Multi-hop scenario: in this scenario, we created a network of 20 nodes, placed randomly in a 100*100m surface area. The maximum two-hop neighborhood size of all nodes is 19 and the maximum local frame size is set to 20 slots. We used fixed routing paths for all tests. The purpose behind this scenario is to measure the total network throughput in the multi-hop environment (See the Figure 10).

In the multi-hop scenario, each node has always data to send. All senders are transmitting at their full transmission power. The number of channels

Figure 9. The throughput comparison in a two-hop scenario

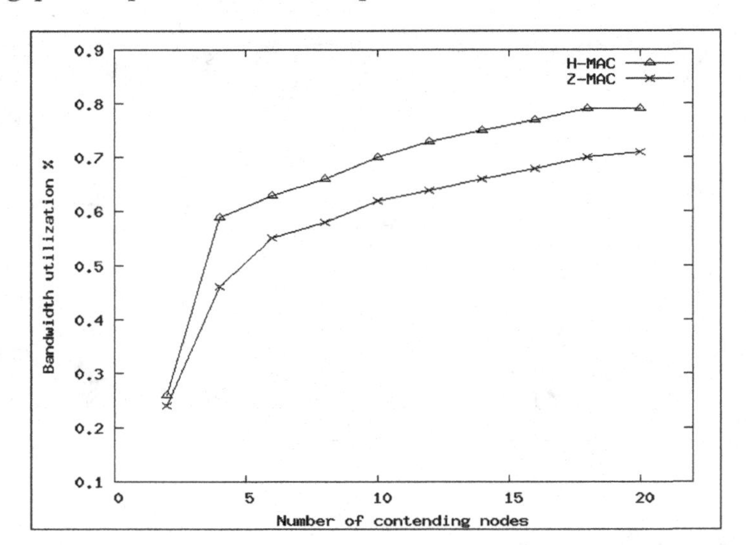

used by H-MAC is fixed to 3 channels, and the channel capacity is set to 1Mbps.

The Figure 11 shows the simulation results. We varied the number of contending node and we measured the aggregate data throughput. H-MAC obtains its highest performance in this simulation. With a number of sending nodes equal to 5, H-MAC achieves a data throughput of 2.282 Mbps than 1.251 Mbps achieved by Z-MAC. The throughput increases progressively with the number of sending nodes, and it can reach 3.431 Mbps with the number of sending nodes equal to 21. However, Z-MAC does not have any improvement in the data throughput, which stays stable

Figure 10. multi-hop network scenario

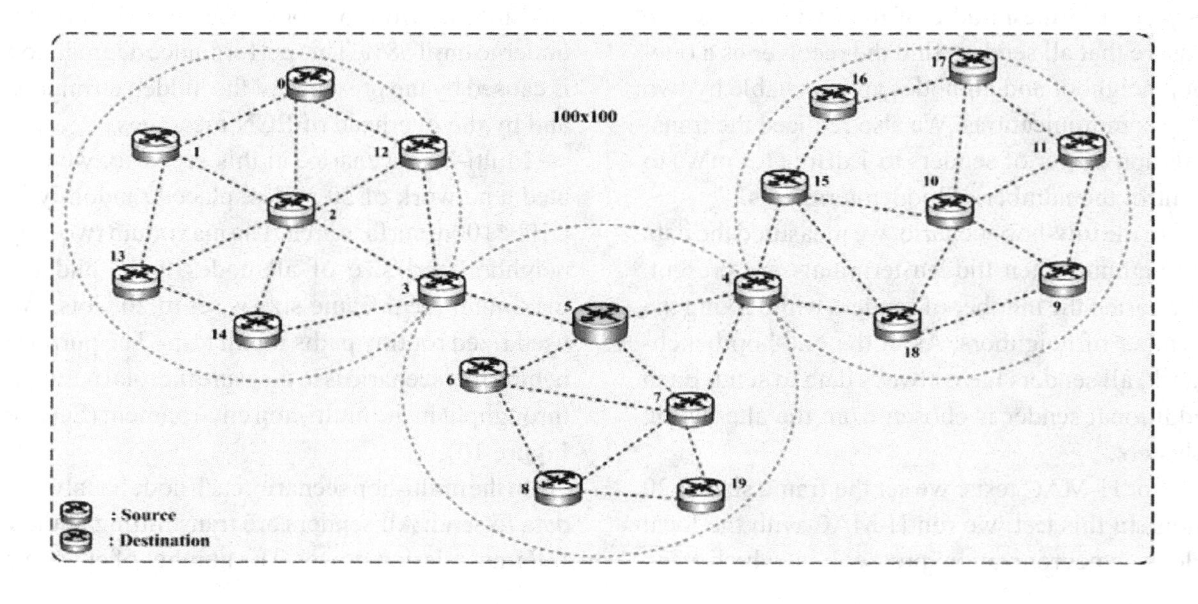

Figure 11. The throughput comparison in a multi-hop scenario

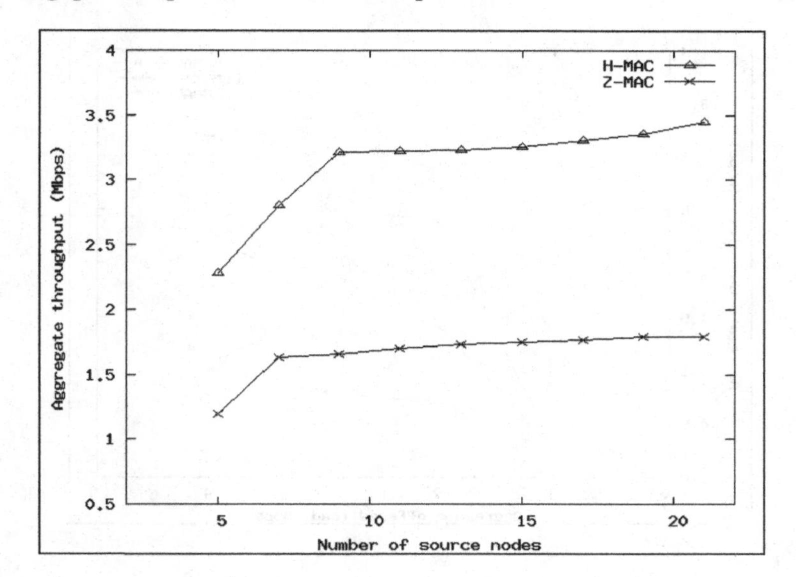

when increasing the number of sending nodes; and it goes no further than 1.59Mbps. This result explains the advantage of the utilization of channel diversity by H-MAC compared to Z-MAC which uses one single channel.

CHANNEL DIVERSITY EVALUATION

In this simulation, we evaluated H-MAC and compared it against two known multi-channel protocols, LCM-MAC and MMAC. The simulation scenario was performed with 100 nodes placed randomly in 500m × 500m area. All the radio parameters are being ns-2 defaults, and the nominal bit rate of each channel is set to 1 Mbps.

There are 50 CBR flows with randomly selected source-destination pairs. The shortest path routing is used. The data packet sizes are 1000 bytes. The data packet generation rate for each flow is varied to vary the load in the network and simulations are done for different number of channels. 6 and 13 channel results are presented in Figure 12 and Figure 13.

We have simulated three protocols H-MAC, LCM-MAC, and MMAC. For MMAC, the specified values in (So and Vaidya, 2004) of 80ms for data window and 20ms for the ATIM window are used. Note that it is fair to compare the three protocols H-MAC, LCM-MAC and MMAC together as they use one interface. LCM MAC performs better than (or similar to) MMAC at all times.

We noticed that, despite using time synchronization, MMAC's performance is not improved at low loads. This is due to the large data window size. At low loads senders run out of packets to send to the receivers present in their current channel. As they cannot change channel until the end of data window, this results in wastage of bandwidth. LCM-MAC also does not give proportional improvement with the increase in channels. Contrary to LCM-MAC and MMAC, H-MAC shows better performance in both simulations. By its dynamic adaptation to the contention level between CSMA and TDMA, H-MAC maintains its good performance, and thus the data throughput increases progressively with the increase in the number of used channels.

Figure 12. Throughput comparison in 500×500 scenario with 6 channels

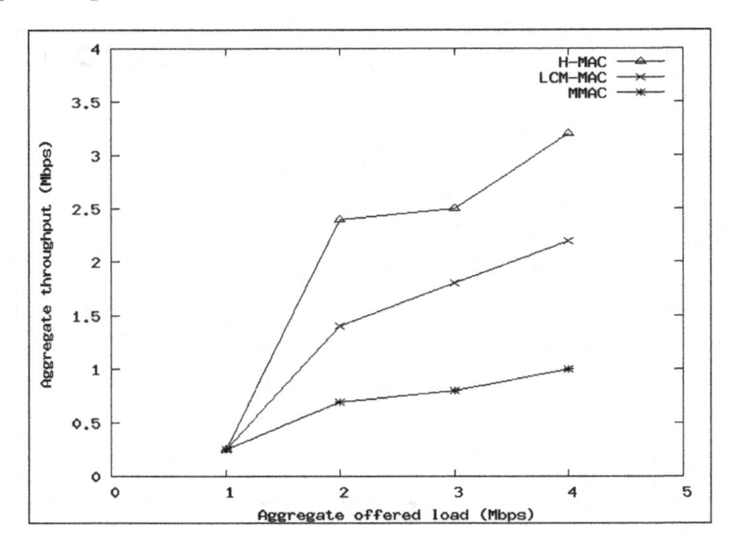

To demonstrate the performance benefit of using multiple channels in wireless networks, we plotted the average throughput of H-MAC and LCM-MAC, with varying number of channels (m) and compared them against single channel 802.11. Single channel 802.11 is only used for baseline comparison. The earlier mentioned scenario with 100 nodes in 500 × 500 m area is used for this plot.

In Figure 14 note that H-MAC's performance increases almost linearly with increase in number of channels. This demonstrates the efficiency of the H-MAC scheme. It does not face control channel bottleneck, nor does it face any control

Figure 13. Throughput comparison in 500×500 scenario with 13 channels

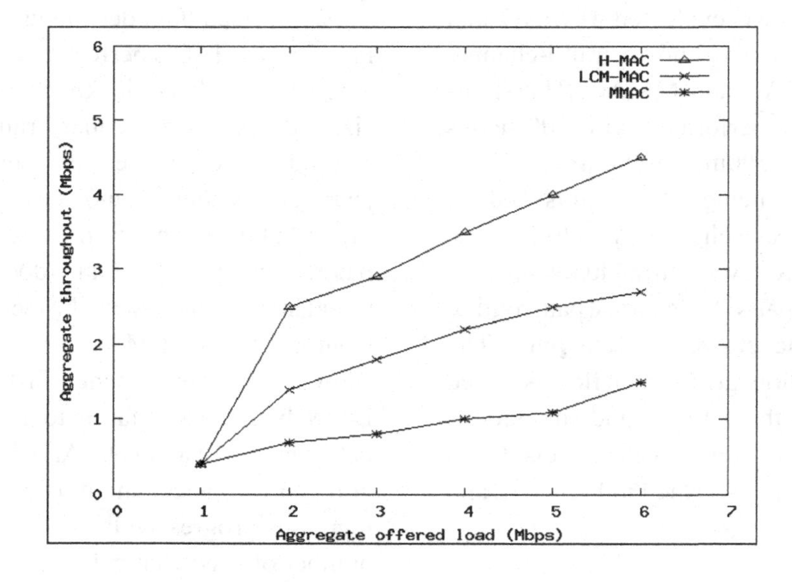

Figure 14. Throughput Comparison according to the number of used channels

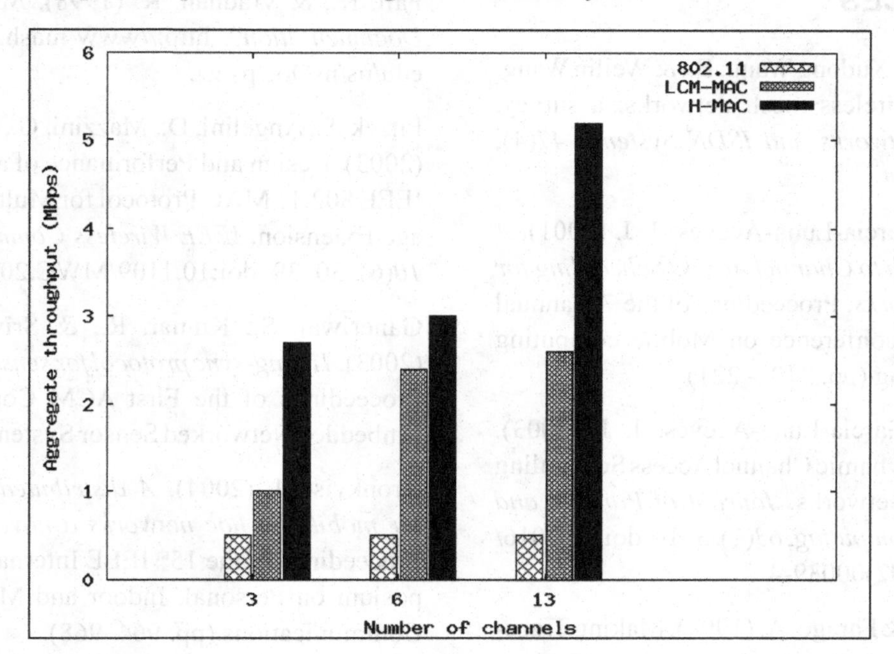

period inefficiencies as in LCM-MAC or MMAC. Also, note that H-MAC, in fact, provides k time the throughput relative to 802.11 while using k channel. This is because of using the hybrid scheme by H-MAC. LCM MAC also provides substantial improvement over 802.11, slightly less than k times for the 3 and 6 channel simulations. But, the throughput does not increase proportionately for 13 channels.

CONCLUSION

This article presents a new multi-channel MAC protocol, called H-MAC for the multi-hop wireless mesh networks. H-MAC can dynamically adjust the behavior of MAC between CSMA and TDMA depending on the level of contention in the network.

The observed simulation results show that our protocol provides much superior performance among all MAC protocols which use hybrid scheme and channel diversity with a single radio. H-MAC performs better than Z-MAC although their channel utilization rate is almost the same. In addition, the simulation results on channel diversity show that H-MAC provides a far superior performance compared to both LCM-MAC and MMAC.

Some of the issues not discussed in this article are the non-negligible channel switching delay and different data packet sizes as well as mechanisms for broadcasts in our protocols. Thus, we have not performed the simulation tests which allow to evaluate the QoS support and its impact on data throughput measurement. This is because of non availability of MAC protocols with QoS implementations during our simulation. In future work, we intend to study the above issues. We will implement and test the studied protocols in real wireless testbeds using different software-based MAC platforms.

REFERENCES

Akyildiz, I. F., Xudong Wang, B., & Weilin Wang, B. (2005). Wireless mesh networks: a survey. *Computer Networks and ISDN Systems*, *47*(4), 445–487.

Bao, L., & Garcia-Luna-Aceves, J. J. (2001). *A New Approach to Channel Access Scheduling for Ad Hoc Networks*. Proceedings of the 7th annual international conference on Mobile computing and networking (pp. 210 – 221).

Bao, L., & Garcia-Luna-Aceves, J. J. (2003). Distributed Dynamic Channel Access Scheduling for Ad Hoc Networks. *Journal of Parallel and Distributed Computing*, *63*(1), 3–14. doi:10.1016/S0743-7315(02)00039-4

Chlamtac, I., & Farago, A. (1994). Making Transmission Schedules Immune to Topology Changes in Multi-hop Packet Radio Networks. *IEEE/ACM Transactions on Networking*, *2*(1), 23 – 29.

Chlamtac, I., Farago, A., Myers, A., Syrotiuk, V., & Zaruba, G. (2000). *A performance comparison of hybrid and conventional mac protocols for wireless networks*. Proceedings of VTC 2000 (pp. 201–205).

Chlamtac, I., & Kutten, S. (1985). *A Spatial-Reuse TDMA/FDMA for Mobile Multi-hop Radio Networks*. Proceedings of IEEE INFOCOM (pp. 389-394).

Conner, W. S., Kruys, J., Kim, K. J., & Zuniga, J. C. (2006). *IEEE 802.11s Tutorial*. Overview of the Amendment for Wireless Local Area Mesh Networking. Intel Corp, Cisco Systems, TMicroelectronics, InterDigital Comm Corp.

Ephremides, A., & Mowafi, O. A. (1982). Analysis of hybrid access schemes for buffered users probabilistic time division. *IEEE Transactions on Software Engineering*, *SE-8*, 52–61. doi:10.1109/TSE.1982.234774

Fall, K., & Vradhan, K. (1998). *NS Notes and Documentation*". http://www-mash.cs.berkeley.edu/ns/nsDoc.ps.gz.

Fitzek, F., Angelini, D., Mazzini, G., & Zorzi, M. (2003). Design and Performance of an Enhanced IEEE 802.11 MAC Protocol for Multihop Coverage Extension. *IEEE Wireless Communications*, *10*(6), 30–39. doi:10.1109/MWC.2003.1265850

Ganeriwal, S., Kumar, R., & Srivastava, M. (2003). *Timing-sync protocol for sensor networks*. Proceedings of the First ACM Conference on Embedded Networked Sensor Systems (SenSys).

Gronkvist, J. (2004). *A distributed scheduling for mobile ad hoc networks a novel approach*. Proceedings of the 15th IEEE International Symposium on Personal, Indoor and Mobile Radio Communications (pp. 964–968).

Henderson, T., Kotz, D., & Abyzov, I. (2004). *The changing usage of a mature campus-wide wireless network*. Proceedings of the Tenth Annual International Conference on Mobile Computing and Networking (MobiCom) (pp. 187–201).

IEEE 802.11 Working Group (1997). *Wireless LAN Medium Access Control (MAC) and Physical Layer (PHY) Specifications*.

Jain, N., Das, S. R., & Nasipuri, A. (2000). *A Multichannel CSMA MAC Protocol with Receiver-Based Channel Selection for Multihop Wireless Networks*. Proceedings of the 10th IEEE International Conference on Computer Communications and Networks (pp. 432-439).

Koubias, S. A., & Haralabidis, H. C. (1996). Mition: A mac-layer hybrid protocol for multichannel real-time lans. *Proceedings of the Third IEEE International Conference on Electronics, Circuits, and Systems* (pp. 327 – 330).

Krishna Rana, Y., Hua Liu, B., Nyandoro, A., & Jha, S. (2006). Bandwidth Aware Slot Allocation in Hybrid MAC. *Proceedings of 31st IEEE Conference on Local Computer Networks* (pp. 89 – 96).

Kyasanur, P., Jungmin, C., Chereddi, S., & Vaidya, N. H. (2006). Multichannel mesh networks: challenges and protocols. *IEEE Wireless Communication, 13*(2), 30–36. doi:10.1109/MWC.2006.1632478

Li, J., Haas, Z. J., Sheng, M., & Chen, Y. (2003). *Performance Evaluation of Modified IEEE 802.11 MAC for Multi-Channel Multi-Hop Ad Hoc Network*. Proceedings of the 17th International Conference on Advanced Information Networking and Applications. (pp. 312–317).

Maheshwari, R., Gupta, H., & Samir, R. (2006). *Multichannel MAC Protocols for Wireless Networks*. Proceedings of the 3rd IEEE Communication Society on Sensor and Ad Hoc Communications Networks (pp. 393-401).

Myers, A. D. (2002). Hybrid MAC Protocols For Mobile Ad Hoc Networks. *PhD thesis, Computer Science, University of Texas at Dallas*.

Qiang, Q., Jacob, L., Radhakrishna Pillai, R., & Prabhakaran, B. (2002). MAC Protocol Enhancements for QoS Guarantee and Fairness over the IEEE 802.11 Wireless LAN. *Proceeding of the Conference on Computer Communication Network (ICCNC)*.

Rajendran, V., Obraczka, K., & Garcia-Luna-Aceves, J. J. (2003). Energy-Efficient, Collision-Free Medium Access Control for Wireless Sensor Networks. *Proceedings of the First ACM Conference on Embedded Networked Sensor Systems (SenSys)*.

Rhee, I., Warrier, A., Aia, M., & Min, J. (2005). ZMAC: a Hybrid MAC for Wireless Sensor Networks. *Proceedings of the 3rd international conference on Embedded networked sensor systems* (pp. 90-101).

So, J., & Vaidya, N. (2004). Multi-Channel MAC for Ad Hoc Networks: Handling Multi-Channel Hidden Terminals Using A Single Transceiver. *Proceedings of the 5th ACM international symposium on Mobile ad hoc networking and computing* (pp. 222 – 233).

IEEE Std 802.11e. (2004). *Medium Access Control (MAC) Enhancements for Quality of Service (QoS)*. IEEE Draft for Wireless Medium Access Control (MAC) and Physical Layer (PHY) Specifications, / Draft 11.0.

Tzamaloukas, A., & Garcia-Luna-Aceves, J. J. (2001). A Receiver-Initiated Collision-Avoidance Protocol for Multi-Channel Networks. *Proceedings of the 20th IEEE INFOCOM* (pp. 189-198).

Ying, Z., Ananda, A. L., & Jacob, L. (2003). A QoS Enabled MAC Protocol for Multi-Hop Ad Hoc Wireless Networks. *Proceeding of IEEE International Conference on Performance, Computing, and Communications* (IPCCC).

Zhang, J., Zhou, G., Huang, C., Son, S. H., & Stankovic, J. A. (2007). TMMAC: An Energy Efficient Multi- Channel MAC Protocol for Ad Hoc Networks. *Proceedings of IEEE International Conference on Communications* (pp. 24-28).

Zhou, G., Huang, C., Yan, T., He, T., Stankovic, J. A., & Abdelzaher, T. (2006). MMSN: Multi-Frequency Media Access Control for Wireless Sensor Networks. *Proceedings of the 25th IEEE INFOCOM* (pp. 1-13).

Zhou, G., Huang, C., Yan, T., He, T., Stankovic, J. A., & Abdelzaher, T. (2006). MMSN: Multi-Frequency Media Access Control for Wireless Sensor Networks. *Proceedings of the 25th IEEE INFOCOM* (pp. 1-13).

This work was previously published in International Journal of Grid and High Performance Computing (IJGHPC), Volume 1, Issue 4, edited by Emmanuel Udoh & Ching-Hsien Hsu, pp. 40-56, copyright 2009 by IGI Publishing (an imprint of IGI Global).

Chapter 21
A Decentralized Directory Service for Peer-to-Peer-Based Telephony

Fabian Stäber
Siemens Corporate Technology, Germany

Gerald Kunzmann
Technische Universität München, Germany[1]

Jörg P. Müller
Clausthal University of Technology, Germany

ABSTRACT

IP telephony has long been one of the most widely used applications of the peer-to-peer paradigm. Hardware phones with built-in peer-to-peer stacks are used to enable IP telephony in closed networks at large company sites, while the wide adoption of smart phones provides the infrastructure for software applications enabling ubiquitous Internet-scale IP-telephony.

Decentralized peer-to-peer systems fit well as the underlying infrastructure for IP-telephony, as they provide the scalability for a large number of participants, and are able to handle the limited storage and bandwidth capabilities on the clients. We studied a commercial peer-to-peer-based decentralized communication platform supporting video communication, voice communication, instant messaging, et cetera. One of the requirements of the communication platform is the implementation of a user directory, allowing users to search for other participants. In this chapter, we present the Extended Prefix Hash Tree algorithm that enables the implementation of a user directory on top of the peer-to-peer communication platform in a fully decentralized way. We evaluate the performance of the algorithm with a real-world phone book. The results can be transferred to other scenarios where support for range queries is needed in combination with the decentralization, self-organization, and resilience of an underlying peer-to-peer infrastructure.

DOI: 10.4018/978-1-60960-603-9.ch021

INTRODUCTION

Structured peer-to-peer overlay protocols such as Chord (Stoica et al 2001) are increasingly used as part of robust and scalable decentralized infrastructures for communication platforms. For instance, users connect to an overlay network to publish their current IP address and port number using a unique user identifier as the keyword. In order to establish a communication channel to a user, the user's identifier must be looked up in order to learn the TCP/IP connection data. Registration and lookup of addresses are realized using Distributed Hashtables (DHT).

However, users in such applications do not always know the unique identifier of the person to be contacted. Therefore, it must be possible to look up the identifier in a phone-book-like user directory. When looking up an identifier, the user might not know all data necessary to start an exact query. For example, the user might know the last name of the person to be searched, but not its first name or address. Moreover, people often are not willing to fill out all data fields, e.g. the address of the person to be called. Therefore, the phone book is required to support range queries, like queries for all people with a certain last name.

A challenge arises from the non-uniform distribution of people's names. Figure 1 shows the frequency of last names in the city of Munich, Germany. Last names are Pareto-distributed, or Zipf-distributed, i.e., there are a few last names that are very common, while most last names are very rare.

In this article, we propose the use of Extended Prefix Hash Trees (EPHTs) as a scalable indexing infrastructure to support range queries on top of Distributed Hash Tables. The EPHT is evaluated by using real-world phone book data; experiments show that our approach enables efficient distributed phone book applications in a reliable way, without the need for centralized index servers. A comparison with related work shows that this has not been possible using techniques introduced before.

Figure 1. Frequency of last names in Munich, Germany

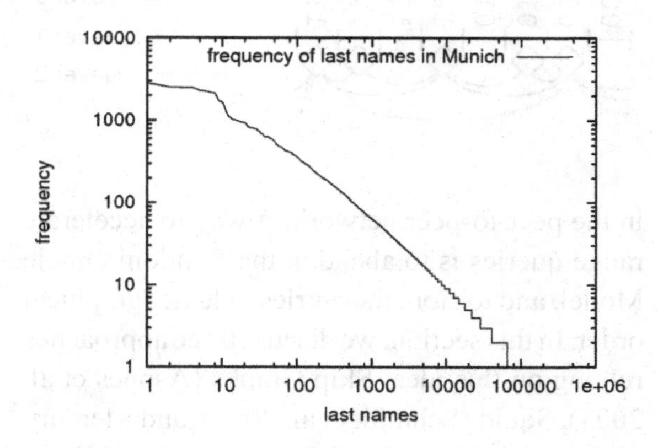

In the following section, we review related work and highlight the problems with current approaches. Then, we present the EPHT algorithm, and compare it with the original Prefix Hash Tree (PHT) algorithm. Then, we evaluate its performance by running a series of experiments. Finally, we summarize our results and show our conclusions.

RELATED WORK

When entries are stored in a Distributed Hash Table, the location of an entry is defined by the hash value of its identifier. A common way to achieve a uniform distribution of the entries among the peers in the DHT is to require the hash function used to calculate the hash value to operate in the Random Oracle Model (Bellare et al, 1993), i.e. even if two identifiers differ only in a single Bit, the hash values of these identifiers are two independent uniformly distributed random variables.

While this hash function allows for good balancing of the data load in a DHT, it makes range queries very costly. Iterating among a range of identifiers that are lexicographically next to each other means addressing nodes in a random order

Figure 2. Skip graph

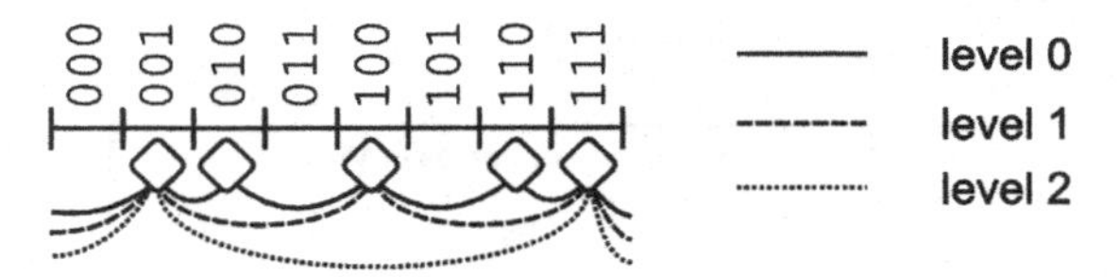

in the peer-to-peer network. A way to accelerate range queries is to abandon the Random Oracle Model, and to store the entries in lexicographical order. In this section, we discuss three approaches relying on this idea: Skip Graphs (Aspnes et al, 2003), Squid (Schmidt et al, 2004), and Mercury (Bharambe et al, 2004). We point out the difficulties arising with these approaches in scenarios like a distributed phone book.

A comparison between EPHTs and the original PHT algorithm (Rambhadran et al, 2004) is presented after we introduced the EPHT.

Skip Graphs

Figure 2 shows a linear three-Bit identifier space. The peers, as indicated by diamonds, are randomly distributed among the identifiers. Each peer is responsible for the identifiers in the range between itself and its predecessor or successor.

As shown in Figure 2, Skip Graphs introduce several levels of linked lists for traversing the peers. The higher the level of the list, the more peers are skipped, accelerating routing to specific ranges. By maintaining several independent lists on each level in parallel, Skip Graphs provide balancing of the traffic load and resilience to node failure.

However, the problem with Skip Graphs is that the entries' identifiers are not distributed uniformly among the linked list, while the peers are randomly distributed. Entries for a last name starting with 'S' are very common in the German phone book, while last names starting with 'Y' are very uncommon. Therefore, the peer being responsible for a common entry becomes a hot spot in terms of network traffic and data load.

Figure 3. Squid

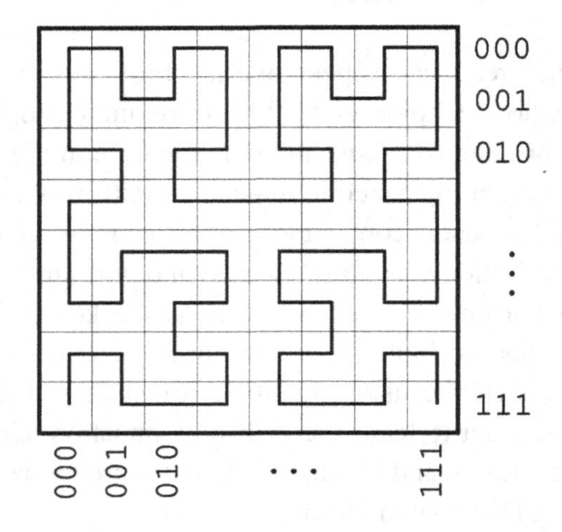

Squid

Squid (Schmidt et al, 2004) is an approach for combining several keywords when determining the position of an entry in the Distributed Hash Table. Squid is based on Locality-Preserving Hashing (Indyk et al, 1997), in which adjacent points in a multi-dimensional domain are mapped to nearly-adjacent points in a one-dimensional range.

For example, in a distributed phone book application, one could use a two-dimensional keyword domain, where one dimension is the entries' last name, and the other dimension is the entries' first name. Figure 3 shows how two dimensions can be mapped on a one-dimensional range using a Space Filling Curve (SFC). The SFC passes each combination of the two identifiers exactly once. If the user wants to search for all entries with a last name starting with 'ST' and a first name starting with 'F', then the user simply needs to query the parts of the SFC that lie on the intersection of these two prefixes in the two-dimensional space.

However, as with Skip Graphs, it turns out that the distribution of names in a phone book results in combinations that are very common, while other combinations are very rare. Again, the peers being responsible for common combinations become hot spots in terms of data storage and traffic load. This could be avoided with Squid by introducing many dimensions in order to distribute the entries among many different peers. But introducing many dimensions results in a tangled-up SFC. As a result, many short fragments of the curve need to be processed for each keyword that is not specified in a query. We evaluated Squid and found that this results in a very high number of peers to be queried in order to find an entry.

Mercury

Like Squid, the Mercury approach (Bharambe et al, 2004) supports multi-dimensional keywords. Each dimension is handled within a separate hub, which is a ring-shaped formation of peers. An example of a Mercury hub is illustrated in Figure 4.

The ID range within a hub is ordered linearly, which results in the same load balancing problems as with the other approaches. However, Mercury suggests that peers are moved around dynamically to balance the load. Although this might be a reasonable approach in other scenarios, this raises difficulties in the distributed phone book scenario. First, there are a few very popular last names. A peer being responsible for one of these popular last names cannot be relieved by moving around other peers, and it will stay a hot spot in terms of data load. Second, if peers may choose their position in the overlay deliberately, this raises certain security issues, because an attacker who wants to make a person unreachable can position its peer in a way that it becomes responsible for routing queries to the victim's entry.

Fusion Dictionary

Fusion Dictionaries (Liu et al, 2004) are not a distributed search index, but a load balancing technique that can be combined with search indexes. The idea is to maintain a blacklist of names that are very common, and to cache blacklist entries in large parts of the DHT. If a user queries a last name that is in the blacklist, the query is inter-

Figure 4. Mercury hub

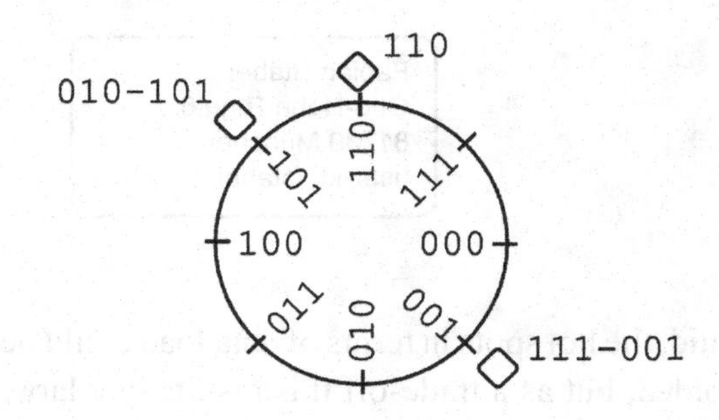

rupted and the user is asked to specify the query more precisely, e.g. by including the first name in the query.

That way, peers being responsible for frequent names are relieved. As the last names are Zipf-distributed, there are only a few names to be included in the blacklist in order to achieve significant load balancing.

However, in spite of the load balancing achieved with fusion dictionaries, the approaches introduced above still do not fulfill the scalability and performance requirements of large scale communication platforms. In this article, we present the EPHT, which is a search index that does not result in overloaded peers. That way it is unnecessary to introduce additional load balancing techniques.

Summary

The brief survey of related work showed that there are several difficulties with previous range query solutions when applied in the distributed phone book scenario. A more detailed overview of search methods in peer-to-peer systems can be found in (Risson et al, 2006). An analysis of arbitrary search in structured peer-to-peer systems was published in (Hautakorpi et al, 2010).

Approaches supporting real multi-dimensional keywords like Mercury and Squid have the problem of hot spots with very popular last names. Additionally, approaches relying on linear keywords instead of real hashing suffer from overloaded peers being responsible for popular prefixes. In

Figure 5. Generating a 32 char identifier for an entry

Fabian Stäber Otto-Hahn-Ring 6 81739 München user-id: fstaber	→	STBE RFAB IANM NCHE NYQU ZNNN DRNX ODRJ

Squid, the hot spots in terms of data load could be avoided, but as a trade-off this results in a large number of peers to be queried to find an entry.

In the following section, we introduce the Extended Prefix Hash Tree as a way of enabling efficient range queries, while preserving the advantages of the Random Oracle Model for hashing, which results in a balanced distribution of the entries among the peers in the DHT.

EXTENDED PREFIX HASH TREE ALGORITHM

Each entry in the distributed phone book is associated with an identifier. The identifier is a fixed-length string, consisting of the capital characters [A-Z]. Identifiers are built by concatenating keywords from the entry. In the example in Figure 5, we used the keywords last name, first name, and city.

The order of the keywords determines the relevance of these keywords for range queries. If identifiers are built as in Figure 5, it is possible to search for the last name without knowing the city, but it is not possible to search for the city without knowing the last name. This corresponds to the hierarchical structure of printed phone books, where entries are ordered by city, last name, first name, etc. In order to allow alternative keyword orders, the application must maintain several trees in parallel.

Special characters like whitespaces or the German ä, ö, ü, ß are omitted. That way, both German names "Müller" and "Möller" map into the same string "MLLER". It is up to the application layer to filter out the right results when a user searched for "Müller".

The identifier length must be sufficient to ensure that a unique identifier can be built for each entry with high probability. In our evaluation, the identifiers were 32 characters long. Identifiers that are longer than that are truncated; identifiers that are shorter are padded with random characters.

Growing the Tree

The structure of an EPHT is shown in Figure 6. There are two parameters that determine the shape of the tree:

1. n is the number of children per node. Each edge is labeled with a character set, like [S-Z]. The partitioning of the alphabet into character sets is fixed and globally known, and cannot be changed dynamically during runtime. n is the number of character sets, which can be any number between 2 and 26. In the section on evaluation, we show that the best performance is achieved with n=26.
2. m is the maximum load of the root node, i.e. the maximum number of entries that can be stored on the root node. If the root node exceeds its maximum load, it splits up into n child nodes and distributes all entries among the children. The maximum load of each child equals the maximum load of the parent node plus one. The reason for incrementing the maximum load is to prevent recursive splits, if all entries happen to be stored on the same child. If a child node's prefix length (see below for the definition

Figure 6. Example of an extended prefix hash tree with n=3 and m=2

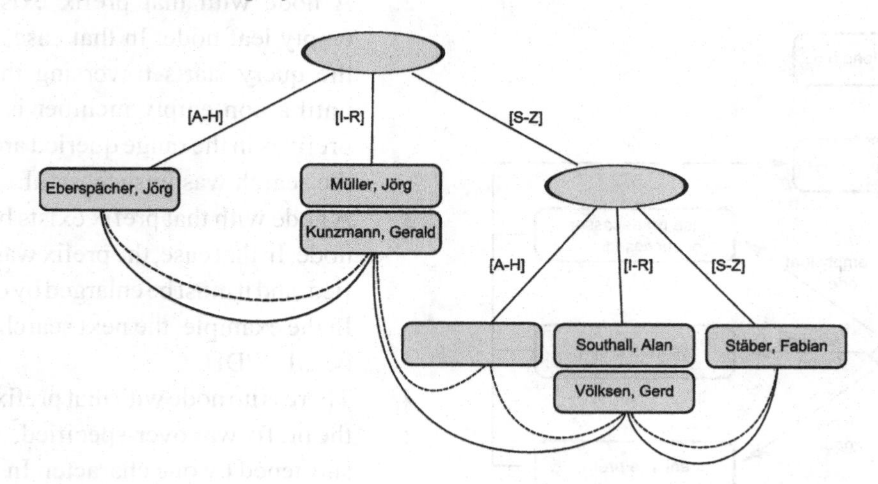

of prefix) equals the identifier length for the entries, then that node cannot split any further, and its maximum load becomes infinite. In the section on evaluation we show that m=100 is a good value.

Each node of the tree is stored as a resource in a DHT, using a hash function operating in the Random Oracle Model. The keyword to be hashed is the prefix of that node in the EPHT, i.e. the sequence of character sets on the path from the root node to the node to be stored. For example, the keyword of the leaf node holding the entry 'Gerd Völksen' in Figure 6 would be '[S-Z][I-R]'. New entries are stored on the leaf node that has the closest matching prefix for the identifier of that entry.

Once a node is split, it becomes an inner node. Inner nodes are kept in the system to indicate the existence of child nodes, but they do not store any data. In particular, inner nodes do not need to store links to their children.

Maintaining the Linked Lists

In addition to the tree structure itself, two doubly linked lists are maintained: one for traversing the non-empty leaf nodes, and the other connecting all leaf nodes, including the empty ones. Each element in a list stores the prefix of its predecessor and successor. The linked list is updated upon the following events:

1. A leaf node splits up into child nodes. In that case, the old leaf node must leave the linked lists, the non-empty new child nodes must join the linked list for non-empty nodes, and all new leaf nodes must join the linked list connecting all leaf nodes. The new nodes learn about their initial successors and predecessors from their parent node.
2. An entry is added to a previously empty leaf node. In that case, that node must join the list for non-empty leaf nodes. The node finds its predecessor and successor using the list connecting all leaf nodes.

Performing Range Queries

Usually, tree algorithms imply that nodes are searched starting at the root node and traversing down the tree to a leaf node. This would mean that the peer holding the root node becomes a bottleneck and single point of failure in a distributed tree structure. EPHTs allow lookups to address

Figure 7. Addressing nodes

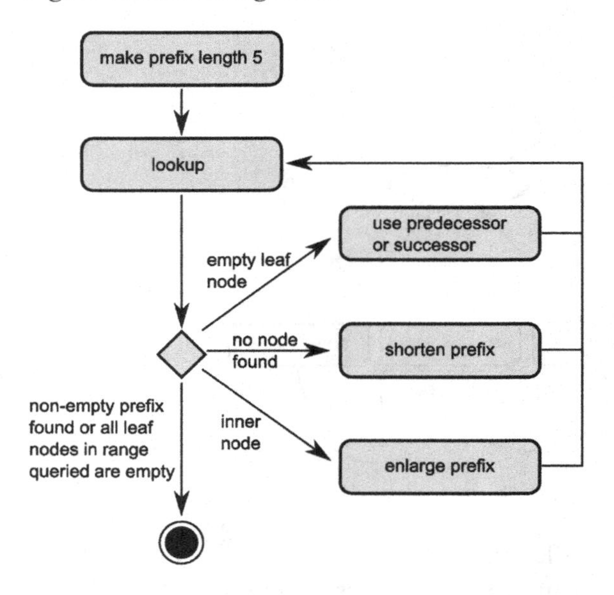

arbitrary nodes directly, using the prefix of the node as the keyword in the DHT.

Range queries are implemented as follows: First, the issuer of a query finds a random, non-empty leaf node lying somewhere in the queried range. Second, the issuer traverses the linked list of non-empty leaf nodes to the left and to the right, subsequently querying the predecessors and successors, until all matching entries are retrieved.

Figure 7 shows how an initial non-empty leaf node is found that can be used as a starting point for traversing the linked list. We exemplify this using a search for all people with the last name 'Olpp'.

Make Prefix Length 5. The first step is to pad the search string with random characters, and to take the first five characters as an initial prefix to start with. In the example, the initial prefix would be OLPPD. In the section on evaluation we will show why 5 is a good initial prefix length.

Lookup. When this prefix is looked up in the DHT, there are four possible results:

1. A node with that prefix exists and is a non-empty leaf node. In that case, the initial node for traversing the linked list is found.
2. A node with that prefix exists and is an empty leaf node. In that case, the issuer of the query starts traversing the linked list until a non-empty member is found. If all prefixes in the range queried are empty, then the search was unsuccessful.
3. A node with that prefix exists but is an inner node. In that case, the prefix was underspecified, and it must be enlarged by one character. In the example, the next search string might be OLPPDH.
4. There is no node with that prefix. This means the prefix was over-specified, and it must be shortened by one character. In the example, the shortened prefix would be OLPP.

In order to decrease latency, the search can be initialized with several different random paddings in parallel. That way, the linked list can be traversed starting from different positions at the same time.

Removing Entries

Entries do not need to be deleted explicitly. Each entry is associated with a lease time. If it is not renewed within that time, it is deleted. That way, users who are no longer part of the system will be removed after some time.

In EPHTs, once a node has split and become an inner node, this node stays an inner node for ever, even if all entries in its sub-tree have timed-out. That means that the EPHT can only grow, but never shrink. This property is in accordance with our use case, as shrinking the tree would only make sense if the service provider operating the distributed phone book application would permanently loose a significant number of customers, or if the distribution of the name's prefixes changes significantly. Both scenarios happen very slowly, and it is feasible to roll out a software update in that case that will built a new tree from scratch. The persistence of inner nodes enables us to implement extensive caching.

Caching

As the EPHT never shrinks, inner nodes are immutable. They will never be deleted or altered. That means that inner nodes can be cached infinitely in the DHT. Whenever a peer learns about the existence of an inner node, it can cache that information and respond when that prefix is queried the next time. Without caching, prefixes that are accessed very frequently would cause a lot of network traffic for the peer being responsible for that prefix. Using caching, this network traffic can be balanced in the DHT.

COMPARISON WITH THE ORIGINAL PREFIX HASH TREES

The Extended Prefix Hash Tree algorithm presented here derives from the Prefix Hash Tree (PHT) algorithm proposed in (Rambhadran et al, 2004). However, the original PHT could not have been used to implement a distributed phone book without the changes presented in this article. The novelty of our work is twofold:

1. The original PHT is a binary tree enabling Bit-wise processing of keywords. Its design does not support caching, and it handles multiple keywords using a Squid-like approach. This does not match the requirements found in the distributed phone book scenario. Therefore, we extended the PHT in several respects, as described below.
2. The EPHT algorithm has several configuration parameters, like the number of child nodes, and the maximum load of a node. We evaluated the Extended PHT with real-world phone book data, and showed how to gain the best performance.

In the rest of this section, we will show the major differences between the EPHT and the PHT algorithm.

- The original PHT is a binary tree. As shown in the evaluation, binary trees do not scale well in a distributed phone book scenario. Therefore, the EPHT is an n-ary tree, and we recommend to use n=26, i.e. the size of the applied alphabet.
- In the original PHT, if the number of entries in a subtree falls below a certain threshold, that subtree collapses into a single leaf node. The EPHT can only grow, but never shrink, which enables us to introduce extensive caching of inner nodes.
- Empty nodes are not handled specially in the PHT algorithm. In the Extended PHT, we introduced an additional linked list skipping the empty nodes to improve performance. This is because we observed that a significant number of prefixes do never appear in user's names, which results in empty leaf nodes for these prefixes.
- The original PHT proposes to handle multiple keywords using Locality-Preserving Hashing, as in Squid. In our application, we simply concatenate the keywords according to their priority, and pad the result with random data.

EVALUATION

In this section, we present the simulation results. The evaluation data is taken from a German phone book CDROM from 1997, because newer electronic phone books restrict data export due to privacy regulations. We used the entries for the city of Munich, which has 620,853 entries. As each peer is supposed to provide only its own entry, the number of peers is equal to the number of entries.

Data Load

The number of entries per peer is one of our key performance indicators, as well-balanced data are the prerequisite for good balancing of the network

Figure 8. Entries per Peer, using n=26 (left), and n=5 (right)

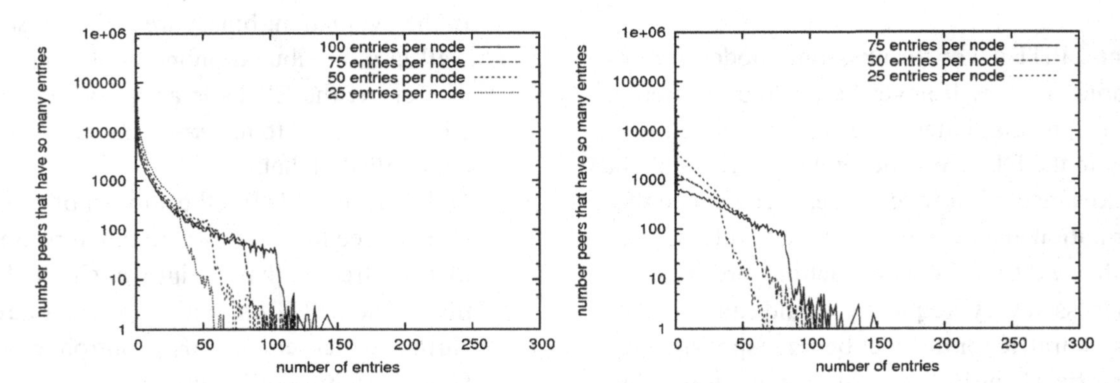

load. Figure 8 shows the number of entries per peer for m in 25, 50, 75, and 100, without replication.

Note that the y-scale showing the number of peers is logarithmic. Nearly all of the 620,853 peers store less than 3 entries. No peer stores more than 150 entries. Assuming an average size of an entry of 128 Bytes, a peer holding 150 entries would store less then 19 kBytes. This is feasible even on embedded devices with a built-in peer-to-peer stack, and it is easily possible to replicate 19 kBytes through current Internet connections.

Prefix Length

In the description of the algorithm above, we said that the initial prefix length to start with when searching in an EPHT is 5 in our dataset. As shown in Figure 9, this is the average prefix length for n in 5, 13, and 26. Only binary EPHTs with n=2 result in a significantly larger average prefix length. If the average prefix length changes over time, e.g. if the number of users or the distribution of names is other than expected, then the initial prefix length needs to be adapted in the search operation.

Network Traffic

The network traffic is evaluated in terms of the number of lookup operations in the DHT that is needed to process a range query[2]. As an example, we queried the prefixes SCHN* which results in

Figure 9. Prefix Length for m=25 (left), and m=100 (right)

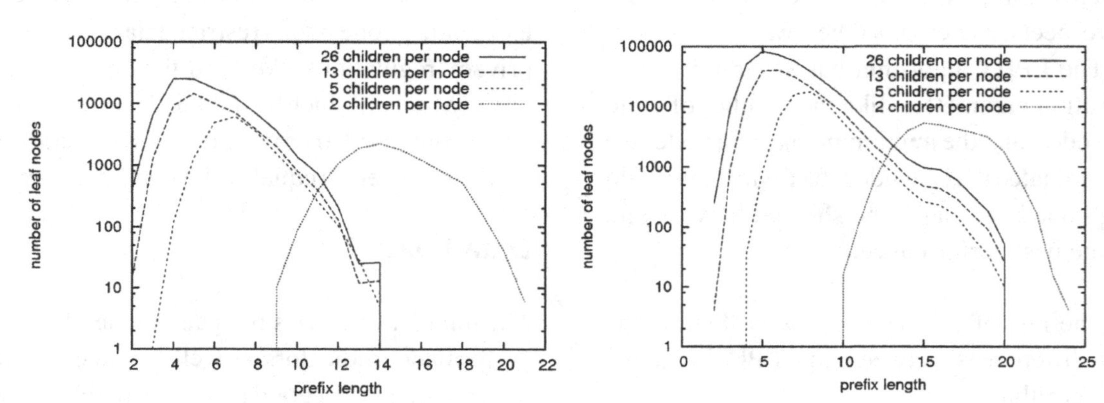

Table 1. Lookup operations

	SCHN*			OLPP*		
	N=5	*n=13*	*n=26*	*n=5*	*n=13*	*n=26*
M=50	*1068*	*958*	*495*	*99*	*3*	*4*
M=100	*590*	*670*	*279*	*39*	*3*	*4*

4683 entries, and OLPP* yielding only a single entry. Of course querying SCHN* is an artificial example, as real-world applications would probably abort that query after a certain number of results is retrieved, and ask the user to formulate the query more specifically. Table 1 shows the number of lookup operations. We did not use any caching.

An increasing maximum load of the root node m results in less nodes to be looked up. With regards to the number of children n we found that more children per node result in a lower number of lookup operations. For example, if the user searches for OLPP* in a tree with n=5, the application searches all entries matching the prefix [K-O][K-O] [P-T] [P-T]. People with a last name starting with Lost would match the same prefix as Olpp. Altogether, the number of matching entries in our phone book is 1801, which explains the overhead of 99 lookups.

These results suggest that the number of children per node *n* should be as large as possible to reduce the number of lookup operations.

Empty Nodes

The percentage of empty nodes is shown in Table 2.

As expected, the number of empty nodes raises with the number of children per node.

Table 2. Empty nodes

	n=5	*n=13*	*n=26*
m=50	*6% of 34,821*	*37% of 94,297*	*60% of 193,801*
m=100	*2% of 18,353*	*29% of 48,757*	*53% of 98,501*

However, even with *n*=26 we got only 60% empty nodes, which is still justifiable in the face of the great reduction of traffic overhead for *n*=26.

Churn

In peer-to-peer terminology, the continuous arrival and disappearance of peers is called churn. The stability of DHTs in the face of churn and the probability of data loss was addressed many times before (Stutzbach 2006, Kunzmann 2009), and we refer the reader to these works for experimental and analytical results on the topic.

The tree nodes of the EPHT are stored as resources on a DHT. DHTs use replication techniques and stabilization protocols to keep the probability of data loss very low, even in typical file-sharing scenarios where the participating peers arrive and disappear very frequently.

The reliability of the EPHT depends on the reliability of the underlying DHT. If the node resources are available on the DHT layer, then the EPHT remains stable. Assuming that VoIP telephones have much longer average online times than file sharing peers, we expect the DHT to be very stable in the distributed phone book scenario.

However, in order to handle the unlikely event of data loss, we propose that the peers look up their own entry on a periodical basis, and re-publish the entry in case it disappeared.

CONCLUSION

In this article, we presented the Extended Prefix Hash Tree as an infrastructure supporting range queries on top of Distributed Hash Tables. The design of the algorithm is driven by the requirements found in a distributed user directory for a commercial VoIP communication platform developed by Siemens. We evaluated the algorithm and showed how to choose the parameters in order to achieve the best performance.

While this article is focused on a specific use case, the methodology and results can be transferred to other scenarios. The algorithm presented here fits specifically in situations where keywords are Zipf-distributed. In the phone book scenario, some last names are very common while other last names are very rare. The EPHT adapts perfectly to this kind of distribution.

The concatenation of keywords provides a simple but powerful approach to handle multiple keywords that are ordered in a hierarchical way.

FUTURE RESEARCH DIRECTIONS

The Extended Prefix Hash Tree algorithm presented in this paper enables the implementation of a distributed user directory for a peer-to-peer-based telephony application. However, apart from user directories, there are more applications that might benefit from a distributed search index.

The evaluation in this article is based on the specific requirements that we derived from a commercial communication platform. When EPHTs are to be applied in other applications, it is a non-trivial task to tell the implications of the algorithm on the specific architecture.

Future research should address this issue and allow for the definition of generic, re-usable components that can be applied on top of peer-to-peer networks. These components are the building blocks fulfilling the application-specific requirements on the distributed infrastructures. A first proposal for the definitions of these components can be found in (Stäber, 2009).

Also, while DHT-based structured overlay networks have many advantages, their string-based approach for registration and lookup carries intrinsic limitations as regards the expressiveness of search. While the extension with range queries and wildcard search seems appropriate for a pure phone book lokup, even a straightforward business directory will require more semantically elaborate queries (e.g., SQL-based or ontology-based queries). One option to achieve this is to combine structured distributed hash tables with super-peer architectures, preserving the robustness and scalability of the overlay while enhancing it with declarative semantic search capability. In (Gerdes et al., 2009), we propose a declarative decentralized query processor and evaluate it in the energy domain. (Stiefel and Müller, 2010) propose the use of an ontology-based query language on top of a DHT architecture for semantic search of digital product models. These approaches will need to be validated and further developed in future work.

REFERENCES

Aspnes, J., & Shah, G. (2003). Skip graphs. In *SODA '03: Proceedings of the Fourteenth Annual ACM SIAM Symposium on Discrete Algorithms*, (pp. 384–393). Philadelphia, PA, USA.

Barsanti, L., & Sodan, A. (2007). Adaptive job scheduling via predictive job resource allocation. In *Proceedings of Job Scheduling Strategies for Parallel Processing* (pp. 115-140).

Bellare, M., & Rogaway, P. (1993). *Random oracles are practical: A paradigm for designing efficient protocols*. In CCS '93 1st ACM Conference on Computer and Communications Security, (pp. 62–73). New York, NY: ACM Press.

Bharambe, A. R., Agrawal, S., & Seshan, S. (2004). *Mercury: Supporting scalable multi-attribute range queries.* In SIGCOMM Symposium on Communications Architectures and Protocols, (pp 353–366). Portland, OR, USA.

Buyya, R., Giddy, J., & Abramson, D. (2000). *An evaluation of economy-based resource trading and scheduling on computational power Grids for parameter sweep applications.* Paper presented at the Second Workshop on Active Middleware Services (AMS2000), Pittsburgh, USA.

Gerdes, C., Eger, K., & Müller, J. P. (2009). *Data-centric peer-to-peer communication in power grids.* Electronic Communications of the EASST 17: Kommunikation in Verteilten Systemen 2009, 2009. *Proceedings of KiVS Global Sensor Networks Workshop* (GSN09).

Hautakorpi, J., & Schultz, G. (2010). *A feasibility study of an arbitrary search in structured peer-to-peer networks.* In ICCCN'10: *Proceedings of the 19th International Conference on Computer Communications and Networks*. Zurich.

Indyk, P., Motwani, R., Raghavan, P., & Vempala, S. (1997). Locality-preserving hashing in multidimensional spaces. In STOC '97: *Proc. of the Twenty-Ninth Annual ACM Symposium on Theory of Computing*, (pp. 618–625). New York, NY: ACM Press.

Liu, L., & Lee, K.-W. (2004). Supporting efficient keyword-based file search in peer-to-peer file sharing systems. In GLOBECOM'04: *Proc. of the IEEE Global Telecommunications Conference*.

Ramabhadran, S., Ratnasamy, S., Hellerstein, J. M., & Shenker, S. (2004). *Prefix hash tree – an indexing data structure over distributed hash tables*. In PODC'04: 23rd Annual ACM Symposium on Principles of Distributed Computing.

Risson, J., & Moors, T. (2006). Survey and research towards robust peer-to-peer networks: Search methods. *Computer Networks*, *50*(17), 3485–3521. doi:10.1016/j.comnet.2006.02.001

Schmidt, C., & Parashar, M. (2004). Enabling flexible queries with guarantees in P2P systems. *IEEE Internet Computing*, *8*(3), 19–26. doi:10.1109/MIC.2004.1297269

Stäber, F. (2009). *Service layer components for decentralized applications*. Doctoral Dissertation at the Clausthal University of Technology

Stiefel, P. D., & Müller, J. P. (2010). A model-based software architecture to support decentral product development processes. In: *Exploring the grand challenges for next generation e-business. Proceedings of the 8th Workshop on eBusiness* (Web 2009). *Volume 52 of Lecture Notes in Business Information Processing*. Springer-Verlag, 2010. To appear.

Stoica, I., Morris, R., Karger, D., Kaashoek, M. F., & Balakrishnan, H. (2001). Chord: A scalable peer-to-peer lookup service for internet applications. In SIGCOMM'01: *Proc. of the 2001 Conference on Applications, Technologies, Architectures, and Protocols for Computer Communications*, (pp. 149–160). San Diego, CA: ACM Press.

Stutzbach, D., & Rejaie, R. (2006). Understanding churn in peer-to-peer networks. In IMC'06: *Proc. of the 6th ACM SIGCOMM on Internet Measurement*, (pp. 189–202). New York, NY: ACM Press.

ADDITIONAL READING

Aberer, K., & Hauswirth, M. (2004). *Peer-to-Peer Systems, Practical Handbook of Internet Computing*. Baton Rouge: Chapman Hall & CRC Press.

Baset, S. A., & Schulzrinne, H. (2004). *An analysis of the Skype Peer-to-Peer Internet telephony protocol. Tech. report*. New York, USA: Columbia University.

Binzenhöfer, A., Staehle, D., & Henjes, R. (2005): *On the stability of Chord-based P2P systems.* In GLOBECOM '05: Proc. of the IEEE Global Telecommunications Conference.

Binzenhöfer, A., & Tran-Gia, P. (2004): *Delay analysis of a Chord-based peer-to-peer file-sharing system*. In ATNAC '04: Proc. of the Australian Telecommunication Networks and Applications Conference.

Biondi, P. and Desclaux F. (2006): *Silver needle in the Skype*. Black Hat Europe 2006.

Dabek, F., Zhao, B., Druschel, P., & Kuiatowicz, J. (2003): *Towards a common API for structured peer-to-peer overlays*. In IPTPS'03: Peer-t-Peer Systems II, Second International Workshop, volume 2734 of Lecture Notes in Computer Science, pages 33—34, Berlin, Heidelberg: Germany, Springer

Eberspaecher, J., & Schollmeier, R. (2005): *Peer-to-Peer systems and applications*. chapter First and Second Generation of Peer-to-Peer Systems, pages 35—56, Springer.

Eyers, T., & Schulzrinne, H. (2000): *Predicting Internet telephony call setup delay*. In IPTel 2000: Proc. of the 1st IP-Telephony Workshop.

Friese, T., Freisleben, B., Rusitschka, S., & Southall, A. (2002): *A framework for resource management in peer-to-peer networks*. Revised Papers from the International Conference NetObjectDays on Objects, Components, Architectures, Services, and Applications for a Networked World, Lecture Notes In Computer Science, volume 2591, Springer, 2002, pages 4—21.

Friese, T., Müller, J. P., & Freisleben, B. (2005). *Self-Healing Execution of Business Processes Based on a Peer-to-Peer Service Architecture*. In: Proc. 18th Int. Conference on Architecture of Computing Systems [Springer.]. *Lecture Notes in Computer Science*, *3432*, 108–123. doi:10.1007/978-3-540-31967-2_8

Ganesan, P., Yang, B., & Garcia-Molina, H. (2004): *One torus to rule them all: multi-dimensional queries in P2P systems*. In WebDB '04: Proc. of the 7th International Workshop on the Web and Databases, ACM Press, pages 19—24.

Garcés-Erice, L., Felber, P. A., Biersack, E. W., Urvoy-Keller, G., & Ross, K. W. (2004): *Data indexing in peer-to-peer DHT networks*. In ICDCS '04: Proc. of the 24th International Conference on Distributed Computing Systems, IEEE Computer Society, pages 200—208.

Guha, S., Daswani, N., & Jain, R. (2006): *An experimental study of the Skype peer-to-peer VoIP system*. In IPTPS '06: Proc. of the 5th International Workshop on Peer-to-Peer Systems.

Gummadi, K., Gummadi, R., Gribble, S., Ratnasamy, S., Shenker, S., & Stoica, I. (2003). The impact of DHT routing geometry on resilience and proximity. In *SIGCOMM '03: Proc. of the conference on Applications, technologies, architectures, and protocols for computer communications* (pp. 381–394). ACM Press. doi:10.1145/863955.863998

Kellerer, W., Kunzmann, G., Schollmeier, R., & Zoels, S. (2006). Structured peer-to-peer systems for telecommunications and mobile environments. *AEÜ. International Journal of Electronics and Communications*, *60*(1), 25–29. doi:10.1016/j.aeue.2005.10.005

Kunzmann, G. (2009): *Performance Analysis and Optimized Operation of Structured Overlay Networks*. Doctoral thesis, Technische Universitaet Muenchen.

Kunzmann, G. and Binzenhoefer A. and Stäber, F. (2008): *Structured overlaynetworks as an enabler for future internet services*. it - Information Technology volume 50, no. 6, pages 376—382.

Lennox, J., & Schulzrinne, H. (2000). *Feature interaction in Internet telephony. Feature Interactions in Telecommunications and Software Systems VI* (pp. 38–50). IOS Press.

Leslie M. and Davies J. and Huffman. (2006): *Replication strategies for reliable decentralized storage*. In ARES'06: Proc. of the First International Conference on Availability, Reliability and Security. pages 740—747.

Li, J., Stribling, J., Morris, R., Kaashoek, M. F., & Gil, T. M. (2005): *A performance vs. cost framework for evaluating DHT design tradeoffs under churn*, In INFOCOM '05: Proc. of the 24th Joint Conference of the IEEE Computer and Communications Societies, pages 225—236.

Liu, L., & Lee, K.-W. (2004): *Keyword fusion to support efficient keyword-based search in peer-to-peer file sharing.* In CCGRID'04: Proc. of the 2004 IEEE International Symposium on Cluster Computing and the Grid, IEEE Computer Society, pages 269—276.

Lua, E. K., Crowcroft, J., Pias, M., Sharma, R., & Lim, S. (2005). A survey and comparison of peer-to-peer overlay network schemes. *IEEE Communications Surveys and Tutorials*, *7*(2), 72–93. doi:10.1109/COMST.2005.1610546

Maymounkov, P., & Mazières, D. (2006): *Kademlia: A peer-to-peer information system based on the xor metric.* In IPTPS'02: Proc of the 1st International Workshop on Peer-to-Peer Systems, pages 53—65, London: UK, Springer

Milojicic, D. S., Kalogeraki, V., Lukose, R., Nagaraja, K., Pruyne, J., Richard, B., et al. (2002): *Peer-to-peer computing.* Technical Report HPL-2002-57, HP Labs, Palo Alto, CA, USA.

Oram, A. (Ed.). (2001). *Peer-to-Peer, Harnessing the Power of Disruptive Technologies*. Sebastopol, CA, USA: O'Reilly.

Ratnasamz, S., Francis, P., Handley, M., Karp, R., & Schenker, S. (2001): A scalable content-addressable network. In SIGCOMM'01: Proc. of the 2001 Conference on Applications, Technologies, Architectures, and Protocols for Computer Communications, pages 161—172, New York, NY: USA, ACM Press.

Reynolds, P., & Vahdat, A. (2003): Efficient peer-to-peer keyword searching, Proceedings of International Middleware Conference, Lecture Notes in Computer Science, vol. 2672, Springer, pages 21—40.

Rhea, S., Geels, D., Roscoe, T., & Kubiatowicz, J. (2003): *Handling churn in a DHT*. Tech. Report UCB/CSD-03-1299, EECS Department, University of California, Berkeley.

Rowstron, A., & Druschel, P. (2001): *Pastry: Scalable, decentralized object location and routing for large-scale peer-to-peer systems.* In Middleware'01: Proc. of the IFIP/ACM International Conference on Distributed Systems, volume 2218 of Lecture Notes in Computer Science, pages 329—350.

Rusitschka, S., & Southall, A. (2003). *The resource management framework: A system for managing metadata in decentralized networks using peer-to-peer technology.* In Agents and Peer-to-Peer Computing. In *Lecture Notes in Computer Science* (*Vol. 2530*, pp. 144–149). Springer.

Sarma, A., Bettstetter, C., Dixit, S., Kunzmann, G., Schollmeier, R., & Nielsen, J. (2006). *Self-organization in communication networks* (pp. 423–451). Wiley.

Seedorf, J. (2006). Security challenges for Peer-to-Peer SIP. *IEEE Network*, *20*, 38–45. doi:10.1109/MNET.2006.1705882

Shu, Y., Ooi, B. C., Tan, K.-L., & Zhou, A. (2005): Supporting multi-dimensional range queries in peer-to-peer systems, In P2P '05: Proc. of the 5th IEEE International Conference on Peer-to-Peer Computing, IEEE Computer Society, pages 173—180.

Sit, E., & Morris, R. (2002): *Security considerations for peer-to-peer distributed hash tables*. In IPTPS '02: Proc. of the 1st International Workshop on Peer-to-Peer Systems.

Spleiss, C., & Kunzmann, G. (2007). *Decentralized supplementary services for Voice-over-IP telephony*. Proceedings of EUNICE 2007. *Lecture Notes in Computer Science*, *4606*(Jul), 62–69. doi:10.1007/978-3-540-73530-4_8

Steinmetz, R., & Wehrle, K. (2005): *What is peer-to-peer about?* In volume 3485 of Lecture Notes in Computer Science, pages 9—16. Berlin, Heidelberg: Germany, Springer.

Stutzbach, D., & Rejaie, R. (2006): *Understanding churn in peer-to-peer networks*. In IMC'06: Proc. of the 6th ACM SIGCOMM on Internet Measurement, pages 189—202, New York, NY, USA: ACM Press.

Tanenbaum, A. (2003). *Computer Networks*. Upper Saddle River, NJ, USA: Prentice Hall International.

Tutschku, K., & Tran-Gia, P. (2005). *Peer-to-peer-systems and applications. chapter Traffic Characteristics and Performance Evaluation of Peer-to-Peer Systems* (pp. 383–397). Springer.

Zhao, B., Huang, L., Stribling, J., Rhea, S., Joseph, A., & Kubiatovicz, J. (2004). Tapestry: A resilient global-scale overlay for service deployment. *IEEE Journal on Selected Areas in Communications*, *22*(1), 41–53. doi:10.1109/JSAC.2003.818784

Zoels, S., Schubert, S., & Kellerer, W. (2006): *Hybrid dht design for mobile environments*. In AP2PC'06: Proc of 5th International Workshop on Agents and Peer-to-Peer Computing, LNCS, Springer.

Zuo, C., Li, R., Shen, H., & Lu, Z. (2009): High Coverage Search in Multi-Tree Based P2P Overlay Network. In ICCCN'09: Proc of the 18th International Conference on Computer Communications and Networks. San Francisco, CA: USA.

KEY TERMS AND DEFINITIONS

Decentralization: Attempt to avoid central services, thus preventing single points of failure.

Directory: A service organizing users

Distributed Hash Table (DHT): Type of decentralized infrastructure providing a hash-table like addressing scheme

Extended Prefix Hash Tree (EPHT): Modified PHT to be used when implementing distributed user directories.

Infrastructure: Underlying algorithm in a distributed application, allowing the nodes to address each other.

IP-Telephony: Telephony over IP networks, mostly using the Voice over IP protocol

Peer-to-Peer: A paradigm in distributed systems, where all nodes may act as both, client and server.

Prefix Hash Tree (PHT): Search algorithm based on DHTs, as proposed by Ramabhadran et al (2004)

ENDNOTES

1. Dr. G. Kunzmann is now working for DOCOMO Communications Laboratories Europe GmbH, Munich, Germany
2. Each lookup operation requires $\log_b n$ messages in the DHT protocol, where n is the number of peers, and b is a parameter depending on the design of the DHT's routing table.

Compilation of References

Aarts, R. (Ed.). (2004). *Liberty ID-WSF interaction service specification*. Liberty Alliance document. Retrieved from http://www.project-liberty.org/

Aberer, K., Mauroux, P. C., Datta, A., Despotovic, Z., Hauswirth, M., Punceva, M., & Schmidt, R. (2003). P-Grid: A self-organizing structured P2P system. *SIGMOD, 32*. ACM.

Abramson, D., & Kommineni, J. (2004). A Flexible IO Scheme for Grid Workflows. In *Proc. of the 18th International Parallel and Distributed Processing Symposium*. Krakow, Poland.

Abramson, D., Foster, I., Giddy, J., Lewis, A., Sosic, R., Sutherst, R., & White, N. (1997). Nimrod Computational Workbench: A Case Study in Desktop Metacomputing. In *Australian Computer Science Conference (ACSC 97)*. Macquarie University, Sydney.

Abramson, D., Sosic, R., Giddy, J., & Hall, B. (1995). Nimrod: A Tool for Performing Parameterised Simulations using Distributed Workstations. In *Proc. of the 4th IEEE Symposium on High Performance Distributed Computing*. Virginia. IEEE Press.

Acunetix. *Google Hacking*. from http://www.acunetix.com/websitesecurity/google-hacking.htm.

Adjie-Winoto, W., Schwartz, E., Balakrishnan, H., & Lilley, J. (1999). The Design and Implementation of an Intentional Naming System. *Proc. 17th ACM SOSP*, Kiawah Island, SC, Dec.

Agrawal, R., & Srikant, R. (1995). *Mining sequential patterns*. Paper presented at the 17th International Conference on Data Engineering.

Agrawal, R., & Srikant, R. (1996). *Mining sequential patterns: Generalizations and performance improvements*. Paper presented at the 5th International Conference on Extending Database Technology: Advances in Database Technology.

Ahronovitz, M., et al. (2010). *Cloud computing use cases*. A white paper produced by the Cloud Computing Use Case Discussion Group. Retrieved from http://groups.google.com/ group/cloud-computing-use-cases

Akyildiz, I. F., Xudong Wang, B., & Weilin Wang, B. (2005). Wireless mesh networks: a survey. *Computer Networks and ISDN Systems*, *47*(4), 445–487.

Aldrich, J. (1997). R. A. Fisher and the making of maximum likelihood 1912-1922. *Statistical Science*, *12*(3), 162–176. doi:10.1214/ss/1030037906

Alfieri, R., Cecchini, R., Ciaschini, V., dell'Agnello, L., Frohner, A., Gianoli, A., et al. Spataro, F. (2004). Voms, an authorization system for virtual organizations. *European Across Grids Conference, LNCS 2970*, (pp. 33-40). Springer, 2004.

Ali, S., Siegel, H. J., Hensgen, D., & Freund, R. F. (1999). Dynamic mapping of a class of independent tasks onto heterogeneous computing systems. *Journal of Parallel and Distributed Computing*, *59*(2), 107–131. doi:10.1006/jpdc.1999.1581

Allen, G., Davis, K., Goodale, T., Hutanu, A., Kaiser, H., & Kielmann, T. (2005). The Grid Application Toolkit: Towards generic and easy application programming interfaces for the Grid. *Proceedings of the IEEE*, *93*(3), 534–550. doi:10.1109/JPROC.2004.842755

Alonso, J., Hernández, V., & Moltó, G. (2006). GMarte: Grid middleware to abstract remote task execution. *Concurrency and Computation*, *18*(15), 2021–2036. doi:10.1002/cpe.1052

Altintas, I., Berkley, C., Jaeger, E., Jones, M., Ludascher, B., & Mock, S. (2004). Kepler: An extensible system for design and execution of scientific workflows. *Proceedings of the 16th International Conference on Scientific and Statistical Database Management* (SSDBM), Santorini Island, Greece. Retrieved from http://kepler-project.org

Altschul, S. F., Madden, T. L., & Schaffer, A. A. (1997). Gapped BLAST and PSI-BLAST: a new generation of protein database search programs. *Nucleic Acids Research*, *25*(17), 3389–3402. doi:10.1093/nar/25.17.3389

Amazon Elastic Compute Cloud EC2. (2007). Retrieved from www.amazon.com/ec2

Amoretti, M., Zanichelli, F., & Conte, G. (2005). *SP2A: A service-oriented framework for P2P-based Grids*. 3rd International Workshop on Middleware for Grid Computing, (pp. 1-6).

Anderson, D. P., Cobb, J., Korpela, E., Lebofsky, M., & Werthimer, D. (2002). Seti@home: an experiment in public-resource computing. *Communications of the ACM*, *45*(11), 56–61. doi:10.1145/581571.581573

Anderson, D. P. (2004). BOINC: A system for public-resource computing and storage. In *Proceedings of Fifth IEEE/ACM International Workshop on Grid Computing*, (pp. 4-10).

Androutsellis-Theotokis, S., & Spinellis, D. (2004). A survey of peer-to-peer content distribution technologies. *ACM Computing Surveys*, *36*(4), 335–371. doi:10.1145/1041680.1041681

Apostolopoulos, G., Peris, V., & Debanjan Saha, D. (1999). Transport Layer Security: How Much Does it Really Cost. *Proceedings of the IEEE INFOCOM*. New York.

Arends, R., Austein, R., Larson, M., Massey, D., & Rose, S. (2005). *DNS security introduction and requirements. RFC 4033. Internet Engineering Task Force*. IETF.

Arnold, K. (2000). *The Jini specification* (2nd ed.). Addison-Wesley.

Asokan, N., & Ginzboorg, P. (2000, November). Key Agreement in Ad Hoc Networks. *Computer Communications*, *23*(17), 1627–1637. doi:10.1016/S0140-3664(00)00249-8

Aspnes, J., & Shah, G. (2003). Skip graphs. In *SODA '03: Proceedings of the Fourteenth Annual ACM SIAM Symposium on Discrete Algorithms*, (pp. 384–393). Philadelphia, PA, USA.

Atkinson, M., DeRoure, D., Dunlop, A., Fox, G., Henderson, P., & Hey, T. (2005). Web Service Grids: An evolutionary approach. *Concurrency and Computation*, *17*(2-4), 377–389. doi:10.1002/cpe.936

Ayres, J., Flannick, J., Gehrke, J., & Yiu, T. (2002). *Sequential pattern mining using a bitmap representation.* Paper presented at the 8th International Conference on Knowledge Discovery and Data Mining Edmonton, Alberta, Canada.

Badia, R. M., Labarta, J. S., Sirvent, R. L., Perez, J. M., Cela, J. M., & Grima, R. (2003). Programming Grid applications with GRID Superscalar. *Journal of Grid Computing*, *1*, 151–170. doi:10.1023/B:GRID.0000024072.93701.f3

Baduel, L., Baude, F., Caromel, D., Contes, A., Huet, F., Morel, M., & Quilici, R. (2006). Grid computing: Software environments and tools. In *Programming, Composing, Deploying on the Grid*, (pp. 205-229)., Berlin, Heidelberg, and New York: Springer

Baker, M., Buyya, R., & Laforenza, D. (2002). *Grids and Grid technologies for wide area distributed computing. SP&E*. John Wiley and Sons, Ltd.

Baker, S. (2007, December 13). Google and the wisdom of clouds. *Business Week*. Retrieved from www.business-week.com/magazine/ content/07_52/ b4064048925836.htm

Bao, L., & Garcia-Luna-Aceves, J. J. (2003). Distributed Dynamic Channel Access Scheduling for Ad Hoc Networks. *Journal of Parallel and Distributed Computing*, *63*(1), 3–14. doi:10.1016/S0743-7315(02)00039-4

Bao, L., & Garcia-Luna-Aceves, J. J. (2001). *A New Approach to Channel Access Scheduling for Ad Hoc Networks*. Proceedings of the 7th annual international conference on Mobile computing and networking (pp. 210 – 221).

Barabasi, A.-L., Freeh, V. W., Jeong, H., & Brockman, J. B. (2001). Parasitic Computing. *Nature*, 412.

Barsanti, L., & Sodan, A. (2007). Adaptive job scheduling via predictive job resource allocation. *Proceedings of the 12th Conference on Job Scheduling Strategies for Parallel Processing*, (pp. 115-140).

Bartosz Baliś, M., & Wegiel, M. (2008). LGF: A flexible framework for exposing legacy codes as services. *Future Generation Computer Systems*, *24*(7), 711–719. doi:10.1016/j.future.2007.12.001

Basin, D., & Doser, J. (2002). SecureUML: A UML-based modeling language for model-driven security. Paper presented at the 5th International Conference on the Unified Modeling Language. *Lecture Notes in Computer Science 2460*.

Basin, D., Doser, J., & Lodderstedt, T. (2003). *Model driven security for process-oriented systems*. Paper presented at the ACM Symposium on Access Control Models and Technologies, Como, Italy.

Batchu, R., Dandass, Y. S., Skjellum, A., & Beddhu, M. (2004). MPI/FT: A Model-Based Approach to Low-Overhead Fault Tolerant Message-Passing Middleware. *Cluster Computing*, *7*(4), 303–315. doi:10.1023/B:CLUS.0000039491.64560.8a

Baur, T., Breu, R., Kalman, T., Lindinger, T., Milbert, A., Poghosyan, G., … Rombert, M. (2009). An interoperable Grid Information System for integrated resource monitoring based on virtual organizations. *Journal of Grid Computing*, *7*(3). Springer.

Bazinet, A., Myers, D., Fuetsch, J., & Cummings, M. (2007). Grid Services Base Library: A high-level, procedural application programming interface for writing Globus-based Grid services. *Future Generation Computer Systems*, *23*(3), 517–522. doi:10.1016/j.future.2006.07.009

BEinGRID. (2008). *Business experiments in grids*. Retrieved from www.beingrid.com

Bell, W. H., Cameron, D. G., Carvajal-Schiaffino, R., Millar, A. P., Stockinger, K., & Zini, F. (2003). *Evaluation of an economy-based file replication strategy for a data Grid*. In International Workshop on Agent based Cluster and Grid Computing at CCGrid 2003. Tokyo, Japan: IEEE Computer Society Press.

Bellare, M., & Rogaway, P. (1993). *Random oracles are practical: A paradigm for designing efficient protocols*. In CCS '93 1st ACM Conference on Computer and Communications Security, (pp. 62–73). New York, NY: ACM Press.

Bellavista, P., & Corradi, A. (2006). *The handbook of mobile middleware*. Auerbach Publications. doi:10.1201/9781420013153

Beltrame, F., Maggi, P., Melato, M., Molinari, E., Sisto, R., & Torterolo, L. (2006). SRB data Grid and compute Grid integration via the EnginFrame Grid portal. *Proceedings of the 1st SRB Workshop*, 2-3 February 2006, San Diego, USA. Retrieved from www.sdsc.edu/srb/Workshop / SRB-handout-v2.pdf

Berendt, B., Günther, O., & Spiekermann, S. (2005). Privacy in e-commerce. *Communications of the ACM*, *48*(4). ACM Press.

Bhanwar, S., & Bawa, S. (2008). *Securing a Grid*. Paper presented at the World Academy of Science, Engineering and Technology.

Bharambe, A. R., Agrawal, S., & Seshan, S. (2004). *Mercury: Supporting scalable multi-attribute range queries*. In SIGCOMM Symposium on Communications Architectures and Protocols, (pp 353–366). Portland, OR, USA.

BIRN. (2008). *Biomedical Informatics Research Network*. Retrieved from www.nbirn.net/index.shtm

Blancquer, I., Hernández, V., Segrelles, D., & Torres, E. (2009). Enhancing privacy and authorization control scalability in the Grid through ontologies. *IEEE Transactions on Information Technology in Biomedicine*, *13*(1), 16–24. doi:10.1109/TITB.2008.2003369

BOINC - Berkeley Open Infrastructure for Network Computing. (2008). http://boinc.berkeley.edu/ (1.5.2008)

Bolosky, W. J., Douceur, J. R., Ely, D., & Theimer, M. (2000). Feasibility of a serverless distributed file system deployed on an existing set of desktop PCs. *ACM SIGMETRICS Performance Evaluation Review*, *28*(1), 34–43. doi:10.1145/345063.339345

Bolze, R., Cappello, F., Caron, E., Dayd'e, M., Desprez, F., & Jeannot, E. (2006). Grid'5000: A large scale and highly reconfigurable experimental grid testbed. *International Journal of High Performance Computing Applications*, *20*(4), 481–494. doi:10.1177/1094342006070078

Boursas, L., & Hommel, W. (2006). Policy-based service provisioning and dynamic trust management in identity federations. In [). IEEE Computer Society.]. *Proceedings of the IEEE International Conference on Communications, ICC*, 2006.

Bouteiller, A., Hérault, T., Krawezik, G., Lemarinier, P., & Cappello, F. (2006). MPICH-V Project: A Multiprotocol Automatic Fault-Tolerant MPI. *International Journal of High Performance Computing Applications*, *20*(3), 319–333. doi:10.1177/1094342006067469

Bradford, P. G., Grizzell, B. M., Jay, G. T., & Jenkins, J. T. (2007). Cap. 4. Pragmatic security for constrained wireless networks. In Xaio, Y. (Ed.), *Security in distributed, Grid, mobile, and pervasive computing* (p. 440). Tuscaloosa, USA: The University of Alabama.

Bramhall, P., & Mont, M. (2005). Privacy management technology improves governance. In *Proceedings of the 12th Annual Workshop of the HP OpenView University Association.*

Bresciani, P., Giorgini, P., Giunchiglia, F., Mylopoulos, J., & Perin, A. (2004). TROPOS: An agent-oriented software development methodology. *Journal of Autonomous Agents and Multi-Agent Systems*, *8*(3), 203–236. doi:10.1023/B:AGNT.0000018806.20944.ef

Brevik, J., Nurmi, D., & Wolski, R. (2004). Automatic methods for predicting machine availability in desktop grid and peer-to-peer systems. In *Proceedings of the 2004 IEEE International Symposium on Cluster Computing and the Grid (CCGRID04)*, (pp. 190–199).

Bruno, R., Conti, M., & Gregori, E. (2005). Mesh Networks: Commodity Multihop Ad Hoc Networks. *IEEE Communications Magazine*, *43*(3), 123–131. doi:10.1109/MCOM.2005.1404606

Buyya, R., Abramson, D., Giddy, J., & Stockinger, H. (2002). Economic models for resource management and scheduling in Grid computing. *Concurrency and Computation*, *14*(13-15), 1507–1542. doi:10.1002/cpe.690

Buyya, R., Abramson, D., & Giddy, J. (2000). Nimrod/G: An architecture for a resource management and scheduling system in a global computational grid. *Proceedings of the 4th International Conference on High Performance Computing in the Asia-Pacific Region.* Retrieved from www.csse.monash.edu.au /~davida/nimrod/ nimrodg.htm

Buyya, R., Giddy, J., & Abramson, D. (2000). *An evaluation of economy-based resource trading and scheduling on computational power grids for parameter sweep applications.* Paper presented at the Second Workshop on Active Middleware Services (AMS2000), Pittsburgh, USA.

Cai, M., & Frank, M. (2004). *RDFPeers: A Scalable Distributed RDF Repository based on A Structured Peer-to-Peer Network.* Paper presented at the Proceedings of the 13th International World Wide Web Conference, New York.

Cai, M., Frank, M., Chen, J., & Szekely, P. (2003). *MAAN: A Multi-attribute Addressable Network for Grid Information Services.* Paper presented at the Proceedings of the 4th International Workshop on Grid Computing.

Canal, P., Constanta, P., Green, C., & Mack, J. (2007). *GRATIA, a resource accounting system for OSG.* CHEP'07, Victoria, British Columbia, Canada. Sep 2007. Enabling Grids for E-Science. from http://www.eu-egee.org/.

Cappello, F., Djilali, S., Fedak, G., Herault, T., Magniette, F., & N'eri, V. (2005). Computing on large-scale distributed systems: Xtrem web architecture, programming models, security, tests and convergence with grid. *Future Generation Computer Systems*, *21*(3), 417–437. doi:10.1016/j.future.2004.04.011

Carriero, N., & Gelernter, D. (1989). How to Write Parallel Programs: A Guide to the Perplexed. *ACM Computing Surveys*, *21*(3), 323–357. doi:10.1145/72551.72553

Carsten, E., Volker, H., & Ramin, Y. (2002). *Economic scheduling in Grid computing.* Paper presented at the 8th International Workshop on Job Scheduling Strategies for Parallel Processing.

Carsten, E., Volker, H., Uwe, S., Ramin, Y., & Achim, S. (2002). *On advantages of Grid Computing for parallel job scheduling.* Paper presented at the 2nd IEEE/ACM International Symposium on Cluster Computing and the Grid.

Casanova, H., Obertelli, G., Berman, F., & Wolski, R. (2000). The AppLeS parameter sweep template: User-level middleware for the Grid. *Proceedings of Supercomputing, 00*, 75–76.

Casanova, H. (2002). Distributed computing research issues in Grid computing. *ACM SIGACT News, 33*(3), 50–70. doi:10.1145/582475.582486

Casanova, H., Legrand, A., Zagorodnov, D., & Berman, F. (2000). Heuristics for scheduling parameter sweep applications in grid environments. *The Ninth IEEE Heterogeneous Computing Workshop (HCW)*,(pp. 349–363).

Castro, J., Kolp, M., & Mylopoulos, J. (2001). *A requirements-driven development methodology*. Paper presented at the 13th Int. Conf. on Advanced Information Systems Engineering, CAiSE'01.

Castro, M., Druschel, P., Kermarrec, A.-M., Nandi, A., Rowstron, A., & Singh, A. (2003). *Splitstream: High-bandwidth Content Distribution in a Cooperative Environment*. Paper presented at the Proceedings of the International Workshop on Peer-to-Peer Systems (IPTPS 2003).

CCI. (2010). *Amazon cluster compute instances*. Retrieved from http://aws.amazon.com/ hpc-applications/

CDO². (2008). *CDOSheet for pricing and risk analysis.* Retrieved from www.cdo2.com

Chadwick, D. W. (2007). *Use of WebDAV for certificate publishing and revocation. Internet Engineering Task.* IETF.

Chan, P., & Abramson, D. (2001). NetFiles: A Novel Approach to Parallel Programming of Master/Worker Applications. In *Proc. of the 5th International Conference and Exhibition on High-Performance Computing in the Asia-Pacific Region (HPCAsia 2001)*, Queensland, Australia.

Chan, P., & Abramson, D. (2007). π-spaces: Support for Decoupled Communication in Wide-Area Parallel Applications. In *Proc. of the Sixth International Conference on Grid and Cooperative Computing*, (pp. 3–10). Urumchi, Xinjiang, China: IEEE.

Chan, P., & Abramson, D. (2008). Netfiles: An Enhanced Stream-based Communication Mechanism. In J. Labarta, K. Joe, & T. Sato (Eds.), *High-Performance Computing, Revised Selected Papers. Sixth International Symposium, ISHPC 2005 and First International Workshop on Advanced Low Power Systems, ALPS 2006, 4759 of Lecture Notes in Computer Science*, (pp. 254–261). Springer-Verlag.

Chanchio, K., & Sun, X.-H. (2004). Communication State Transfer for Mobility of Concurrent Heterogeneous Computing. *IEEE Transactions on Computers, 53*(10), 1260–1273. doi:10.1109/TC.2004.73

Chang, E., & Garcia-Molina, H. (1999). *Medic: A memory and disk cache for multimedia clients.* Paper presented at the IEEE International Conference on Multimedia Computing and Systems, Florence, Italy.

Charleston, M. A., & Perkins, L. (2006). Traversing the tangle: Algorithms and applications for cophylogenetic studies. *Journal of Biomedical Informatics, 39*, 62–71. doi:10.1016/j.jbi.2005.08.006

Chaubal, C. (2003). *Sun Grid engine enterprise edition—software configuration guidelines and use cases.* Sun Blueprints. www.sun.com/blueprints /0703/817-3179.pdf

Chen, J., & Lu, B. (2008). *Load balancing oriented economic Grid resource scheduling*. IEEE Pacific-Asia Workshop on Computational Intelligence and Industrial Application, (pp. 813-817).

Chervenak, A., Deelman, E., Livny, M., Su, M.-H., Schuler, R., Bharathi, S., et al. (September 2007). Data placement for scientific applications in distributed environments. *Proceedings of the 8th IEEE/ACM International Conference on Grid Computing (Grid2007)*.

Chetty, M., & Buyya, R. (2002). Weaving Computational Grids: How Analogous Are They with Electrical Grids? [CiSE]. *Computing in Science & Engineering, 4*(4), 61–71. doi:10.1109/MCISE.2002.1014981

Chien, A. A., Calder, B., Elbert, S., & Bhatia, K. (2003). Entropia: Architecture and performance of an enterprise desktop grid system. *Journal of Parallel and Distributed Computing, 63*(5), 597–610. doi:10.1016/S0743-7315(03)00006-6

Chinnici, R., Gudgin, M., Moreau, J.-J., & Weerawarana, S. (2005). *Web services description language (WSDL) version 2.0 part 1: Core language.* Retrieved from http://www.w3.org/TR/wsdl20

Chlamtac, I., & Farago, A. (1994). Making Transmission Schedules Immune to Topology Changes in Multi-hop Packet Radio Networks. *IEEE/ACM Transactions on Networking, 2*(1), 23 – 29.

Chlamtac, I., & Kutten, S. (1985). *A Spatial-Reuse TDMA/FDMA for Mobile Multi-hop Radio Networks*. Proceedings of IEEE INFOCOM (pp. 389-394).

Chlamtac, I., Farago, A., Myers, A., Syrotiuk, V., & Zaruba, G. (2000). *A performance comparison of hybrid and conventional mac protocols for wireless networks*. Proceedings of VTC 2000 (pp. 201–205).

Choi, J., Noh, S. H., Min, S. L., & Cho, Y. (2000). *Towards application/file-level characterization of block references: A case for fine-grained buffer management.* Paper presented at the ACM SIGMETRICS International Conference on Measurement and Modeling of Computer Systems Santa Clara, California, United States.

Chor, B., & Tuller, T. (2005). Maximum likelihood of evolutionary trees: hardness and approximation. *Bioinformatics (Oxford, England), 21*(1), 97–106. doi:10.1093/bioinformatics/bti1027

Christensen, E., Curbera, F., Meredith, G., & Weerawarana, S. (2001). *Web services description language (WSDL) 1.1.* Retrieved from http://www.w3.org/TR/2001/NOTE-wsdl-20010315

Chu, R., Xiao, N., Zhuang, Y., Liu, Y., & Lu, X. (2006). *A distributed paging RAM Grid system for wide-area memory sharing.* Paper presented at the 20th International Parallel and Distributed Processing Symposium, Rhodes Island, Greece.

Ciaschini, V. (2004). *A VOMS attribute certificate profile for authorization.* Retrieved from http://grid-auth.infn.it/docs/AC-RFC.pdf

Clement, L., Hately, A., von Riegen, C., & Rogers, T. (2004). *UDDI version 3.0.2.* Retrieved from http://uddi.org/pubs/uddi v3.htm

Coca, R. (2011). Security enhancements of GridFTP:Description and Measurements. *Technical Report UVA-SNE-2011-01*, University of Amsterdam.

Condor Project. (n.d.). Retrieved from http://www.cs.wisc.edu /condor/

Conner, W. S., Kruys, J., Kim, K. J., & Zuniga, J. C. (2006). *IEEE 802.11s Tutorial*. Overview of the Amendment for Wireless Local Area Mesh Networking. Intel Corp, Cisco Systems, TMicroelectronics, InterDigital Comm Corp.

Crespo, A., & Garcia-Molina, H. (2003). *Semantic Overlay Networks for P2P Systems*. Technical report, Stanford University.

Cunsolo, V. D., Distefano, S., Puliafito, A., & Scarpa, M. L. (2010). GS3: A Grid storage system with security features. *Journal of Grid Computing, 8*(3). Springer.

Curbera, F., Duftler, M., Khalaf, R., Nagy, W., Mukhi, N., & Weerawarana, S. (2002). Unraveling the Web Services Web: An introduction to SOAP, WSDL, and UDDI. *IEEE Internet Computing, 6*(2), 86–93. doi:10.1109/4236.991449

Dahlin, M. D., Wang, R. Y., Anderson, T. E., & Patterson, D. A. (1994). *Cooperative caching: Using remote client memory to improve file system performance.* Paper presented at the 1st Symposium on Operating Systems Design and Implementation, Monterey, California.

Dai, Y. S., Xie, M., & Poh, K. L. (2002). Reliability analysis of Grid computing systems. *Proceedings of the 2002 Pacific Rim International Symposium on Dependable Computing (PRDC'02), IEEE* (pp. 97-104).

Dail, H., Sievert, O., Berman, F., & Casanova, H. YarKhan, A., Vadhiyar, S., et al. (2004). Scheduling in the Grid application development software project. *In Grid resource management: State of the art and future trends* (pp. 73-98).

Dandamudi, S. (1995). *Performance impact of scheduling discipline on adaptive load sharing in homogeneous distributed system*. 15th IEEE International Conference on Distributed Computing Systems, (pp. 484-492).

Dasarathy, B. (1991). *Nearest neighbor (NN) norms: Nn pattern classification techniques*. IEEE Computer Society Press Tutorial.

Dcache. (n.d.). *Dcache storage system*. Retrieved from http://www.dcache.org/

de Vienne, D. M., Giraud, T., & Martin, O. C. (2007). A congruence index for testing topological similarity between trees. *Bioinformatics (Oxford, England)*, *23*(23), 3119–3124. doi:10.1093/bioinformatics/btm500

DECI. (2010). *DEISA extreme computing initiative*. Retrieved from www.deisa.eu/science/deci

Deering, S., & Hinden, R. (1998). *Internet protocol, version 6 (IPv6) specification. Internet Engineering Task Force*. IETF.

Deerwester, S. C., Dumais, S. T., Landauer, T. K., Furnas, G. W., & Harshman, R. A. (1990). Indexing by Latent Semantic Analysis. *Journal of the American Society for Information Science American Society for Information Science*, *41*(6), 391–407. doi:10.1002/(SICI)1097-4571(199009)41:6<391::AID-ASI1>3.0.CO;2-9

DeFanti, T., Foster, I., Papka, M. E., Stevens, R., & Kuhfuss, T. (1996). Overview of the I-WAY: Wide Area Visual Supercomputing. *International Journal of Super-computing Applications*, *10*, 123–131. doi:10.1177/109434209601000201

DEISA project. (2008). http://www.deisa.org/ (1.5.2008)

DEISA. (2010). *Distributed European infrastructure for supercomputing applications*. Retrieved from www.deisa.eu

Delaittre, T., Kiss, T., Goyeneche, A., Terstyanszky, G., Winter, S., & Kacsuk, P. (2005). GEMLCA: Running legacy code applications as Grid services. *Journal of Grid Computing*, *3*(1-2), 75–90. doi:10.1007/s10723-005-9002-8

Demchenko, Y., de Laat, C., Koeroo, O., & Groep, D. (2008). Re-thinking Grid security architecture. In *Proceedings of Fourth International Conference on eScience*. IEEE Computer Society.

DESHL. (2008). *DEISA services for heterogeneous management layer*. Retrieved from http://forge.nesc.ac.uk/projects /deisa-jra7/

Deshpande, U., Wang, B., Haque, S., Hines, M., & Gopalan, K. (2010). *MemX: Virtualization of cluster-wide memory*. Paper presented at the International Conference on Parallel Processing.

Desprez, F., & Vernois, A. (2007). *Simultaneous scheduling of replication and computation for data-intensive applications on the Grid*. Kluwer Academic Publishers.

Dey, A., & Abowd, G. (2000). *Towards a Better Understanding of Context and Context-Awareness*. Paper presented at the Proceedings of the Workshop on the What, Who, Where, When and How of Context-awareness at CHI 2000. Freenet. http://freenet.sourceforge.net.

D-Grid. (2008). Retrieved from www.d-grid.de/ index.php?id=1&L=1

Dierks, T. (2007). *The Transport Layer Security (TLS) Protocol Version 1.2 Network Resonance, Inc.* Available at http://www.ietf.org /internet-drafts/draft-ietf-tls-rfc4346-bis-07.txt.

Dongarra, J., Foster, I., Fox, G., Gropp, W., Kennedy, K., Torczon, L., & White, A. (2003). *Sourcebook of parallel computing*. Morgan Kaufmann Publishers.

Droms, R. (1997). Dynamic Host Configuration Protocol. *IETF RFC 2131*.

Dror, G. F., Larry, R., Uwe, S., Kenneth, C. S., & Parkson, W. (1997). *Theory and practice in parallel job scheduling*. Paper presented at the Job Scheduling Strategies for Parallel Processing Conference.

Duan, Z., & Gu, Z. (2008). *Dynamic load balancing in Web cache cluster*. 7th International Conference on Grid and Cooperative Computing, (pp. 147-150).

E.C. (1995). Directive 95/46/EC. *European commission data protection regulations overview page*. Retrieved from http://ec.europa.eu/justice_home/fsj/privacy/

Eager, D. L., Lazowska, E. D., & Zahorjan, J. (1986). A comparison of receiver initiated and sender initiated adaptive load sharing. *Performance Evaluation*, *6*(1), 53–68. doi:10.1016/0166-5316(86)90008-8

Eager, D. L., Lazowska, E. D., & Zahorjan, J. (1988). *The limited performance benefits of migrating active processes for load sharing*. 1988 ACM SIGMETRICS Conference on Measurement and Modeling of Computer Systems, (pp. 63-72).

Edwards, W. (2000). *Core Jini* (2nd ed.). Prentice-Hall.

EnginFrame. (2008). *EnginFrame Grid and cloud portal*. Retrieved from www.nice-italy.com

England, D., & Weissman, J. B. (2005). Costs and benefits of load sharing in the computational Grid. In *Proceedings of the Conference on Job Scheduling Strategies for Parallel Processing* (pp. 160-175).

Enterprise Grid Alliance Security Working Group. (2005). *Enterprise Grid security requirements,* version 1.0.

Ephremides, A., & Mowafi, O. A. (1982). Analysis of hybrid access schemes for buffered users probabilistic time division. *IEEE Transactions on Software Engineering, SE-8*, 52–61. doi:10.1109/TSE.1982.234774

Erberich, S., Silverstein, J. C., Chervenak, A., Schuler, R., Nelson, M. D., & Kesselman, C. (2007). Globus medicus - federation of dicom medical imaging devices into healthcare grids. *Studies in Health Technology and Informatics, 126*, 269–278.

Ernemann, C., Hamscher, V., Streit, A., & Yahyapour, R. (2002a). Enhanced algorithms for multi-site scheduling. In *Grid Computing* (pp. 219–231). GRID.

Ernemann, C., Hamscher, V., Streit, A., & Yahyapour, R. (2002b). On effects of machine configurations on parallel job scheduling in computational Grids. *Proceedings of International Conference on Architecture of Computing Systems, ARCS*, (pp. 169-179).

Ernemann, C., Hamscher, V., & Yahyapour, R. (2002). Benefits of global Grid computing for job scheduling. *Proceedings of the Fifth IEEE/ACM International Workshop on Grid Computing (GRID'04)* (pp. 374-379).

Exa. (2008). *PowerFLOW on demand.* Retrieved from http://www.exa.com/ pdf/IBM_Exa_OnDemand_Screen.pdf

Fagg, G. E., & Dongarra, J. (2004). Building and Using a Fault-Tolerant MPI Implementation. *International Journal of High Performance Computing Applications, 18*(3), 353–361. doi:10.1177/1094342004046052

Fahringer, T., & Jugravu, A. (2005). JavaSymphony: A new programming paradigm to control and synchronize locality, parallelism and load balancing for parallel and distributed computing. *Concurrency and Computation, 17*(7-8), 1005–1025. doi:10.1002/cpe.840

Fall, K., & Vradhan, K. (1998). *NS Notes and Documentation*". http://www-mash.cs.berkeley.edu/ns/nsDoc.ps.gz.

Feeley, M. J., Morgan, W. E., Pighin, F. H., Karlin, A. R., Levy, H. M., & Thekkath, C. A. (1995). *Implementing global memory management in a workstation cluster.* Paper presented at the Symposium on Operating Systems Principles, Copper Mountain Resort, Colorado.

Feitelson, D., & Rudolph, L. (1995). Parallel job scheduling: Issues and approaches. In *Proceedings of International Conference on Job Scheduling Strategies for Parallel Processing* (pp. 1-18).

Fernández-Medina, E., Jurjens, J., Trujillo, J., & Jajodia, S. (2009). Special issue: Model-driven development for secure Information Systems. *Information and Software Technology, 51*(5), 809–814. doi:10.1016/j.infsof.2008.05.010

Fernández-Medina, E., & Piattini, M. (2005). Designing secure databases. *Information and Software Technology, 47*(7), 463–477. doi:10.1016/j.infsof.2004.09.013

Field, L. (2008). *Generic Information Provider*. EGEE Middleware Support Group. from http://twiki.cern.ch/twiki/bin/view/EGEE/GIP.

Firesmith, D. G. (2003). Security use cases. *Journal of Object Technology*, 53-64.

Fischer-Huebner, S. (2001). *IT-security and privacy: Design and use of privacy-enhancing security mechanisms*. New York, NY: Springer-Verlag.

Fitzek, F., Angelini, D., Mazzini, G., & Zorzi, M. (2003). Design and Performance of an Enhanced IEEE 802.11 MAC Protocol for Multihop Coverage Extension. *IEEE Wireless Communications, 10*(6), 30–39. doi:10.1109/MWC.2003.1265850

Flechais, I., Sasse, M. A., & Hailes, S. M. V. (2003). *Bringing security home: A process for developing secure and usable systems*. Paper presented at the New Security Paradigms Workshop (NSPW'03), Ascona, Switzerland.

Flouris, M. D., & Markatos, E. P. (1999). The network RamDisk: Using remote memory on heterogeneous NOWs. *Cluster Computing, 2*(4), 281–293. doi:10.1023/A:1019051330479

Foster, I., & Kesselman, C. (1998). *The Grid – Blueprint for a New Computing Infrastructure*. Morgan Kaufmann.

Foster, I., Kesselman, C., & Tuecke, S. (2001). The Anatomy of the Grid: Enabling Scalable Virtual Organizations. *The International Journal of Supercomputer Applications, 15*(3), 200–222. doi:10.1177/109434200101500302

Foster, I., & Kesselman, C. (Eds.). (1999). *The Grid: Blueprint for a new computing infrastructure*. Morgan Kaufmann Publishers.

Foster, I., & Kesselman, C. (Eds.). (2004). *The Grid 2: Blueprint for a new computing infrastructure*. Morgan Kaufmann Publishers.

Foster, I. (2000). Internet computing and the emerging grid. *Nature*. Retrieved from www.nature.com/nature/webmatters/grid /grid.html

Foster, I. (2002). *What is the Grid? A Three Point Checklist*. Argonne National Laboratory, http://www- fp.mcs.anl.gov/~foster/Articles/WhatIsTheGrid.pdf.

Foster, I. (2005). Globus Toolkit version 4: Software for service-oriented systems. *In Network and Parallel Computing - IFIP International Conference, Beijing, China, 3779*, 2-13. Springer.

Foster, I., & Kesselman, C. (1999). Computational Grids. In *The Grid: Blueprint for a New Computing Infrastructure*, (pp. 15–51).

Foster, I., Kesselman, C., Tsudik, G., & Tuecke, S. (1998). *Security Architecture for Computational Grids*. ACM Conference on Computers and Security, (pp. 83-91).

Foster, I., Kesselman, C., Tsudik, G., & Tuecke, S. (1998). A security architecture for computational grids. *Proc. 5th ACM Conf. on Computer and Communication Security*, (pp. 83-92).

Foster, I., Kesselman, C., & Tuecke, S. (2001) The Anatomy of the Grid: Enabling Scalable Virtual Organizations. *International J. Supercomputer Applications, 15*(3).

Foster. (2002). What is the Grid? A three point checklist. *GRIDtoday, 1*(6).

Fox, G., Williams, R., & Messina, P. (1994). *Parallel computing works!* Morgan Kaufmann Publishers.

Freeman, E., Hupfer, S., & Arnold, K. (1999). *JavaSpaces Principles, Patterns, and Practice*. Addison-Wesley.

Freier, A. O., Karlton, P., & Kocher, P. C. (1996). *Internet Draft: The SSL Protocol Version 3.0*. The Internet Engineering Task Force (IETF), Available at http://wp.netscape.com/eng/ssl3/draft302.txt,last accessed in November 2007.

Freund, R. F., Gherrity, R. M., Ambrosius, S., Campbell, M., Halderman, D., Hensgen, E., & Keith, T. Kidd, M. Kussow, Lima, J. D., Mirabile, F. L., Moore, L., Rust, B., & Siegel, H. J. (1998). Scheduling resources in multi-user, heterogeneous, computing environments with SMARTNET. *7th IEEE Heterogeneous Computing Workshop*, (pp. 184–199).

Frey, J., Mori, T., Nick, J., Smith, C., Snelling, D., Srinivasan, L., & Unger, J. (2005). *The open Grid services architecture*, version 1.0. www.ggf.org/ggf_areas _architecture.htm

Fujimoto, N., & Hagihara, K. (2003). Near-optimal dynamic task scheduling of independent coarse-grained tasks onto a computational grid. *32nd Annual International Conference on Parallel Processing (ICPP-03)*, (pp. 391–398).

GAIA. (2010). *European space agency mission*. Gaia overview. Retrieved from http://www.esa.int/esaSC/120377_index_0_m.html

Gamma, E., Helm, R., Johnson, R., & Vlissides, J. (1995). *Design Patterns*. Reading, MA: Addison-Wesley.

Ganeriwal, S., Kumar, R., & Srivastava, M. (2003). *Timing-sync protocol for sensor networks*. Proceedings of the First ACM Conference on Embedded Networked Sensor Systems (SenSys).

Gannon, D., Krishnan, S., Fang, L., Kandaswamy, G., Simmhan, Y., & Slominski, A. (2005). On building parallel and Grid applications: Component technology and distributed services. *Cluster Computing*, *8*(4), 271–277. doi:10.1007/s10586-005-4094-2

Gao, Y., Rong, H., & Huang, J. Z. (2005). Adaptive grid job scheduling with genetic algorithms. *Future Generation Computer Systems*, *21*, 151–161. doi:10.1016/j.future.2004.09.033

Gartner. (2007). *Gartner says worldwide PDA shipments top 17.7 Million in 2006*. Gartner Press Release. Retrieved from http://www.gartner.com/it/page.jsp?id=500898

Gartner. (2009). *Gartner says worldwide mobile phone sales declined 8.6 per cent and smartphones grew 12.7 per cent in first quarter of 2009*. Gartner Press Release. Retrieved from http://www.gartner.com/it/page.jsp?id=985912

GAT. (2005). *Grid application toolkit*. Retrieved from www.gridlab.org/ WorkPackages/wp-1/

Gentzsch, W. (2009). Porting applications to grids and clouds. *International Journal of Grid and High Performance Computing*, *1*(1), 55–77. doi:10.4018/jghpc.2009010105

Gentzsch, W. (2004). Grid computing adoption in research and industry. In Abbas, A. (Ed.), *Grid computing: A practical guide to technology and applications* (pp. 309–340). Charles River Media Publishers.

Gentzsch, W. (2004). Enterprise resource management: Applications in research and industry. In Foster, I., & Kesselman, C. (Eds.), *The Grid 2: Blueprint for a new computing infrastructure* (pp. 157–166). Morgan Kaufmann Publishers.

Gentzsch, W. (2001). *Grid Computing: A New Technology for the Advanced Web*. White Paper, Sun Microsystems, Inc., Palo Alto, CA.

Gentzsch, W. (2007a). *Grid initiatives: Lessons learned and recommendations*. RENCI Report. Retrieved from www.renci.org/publications /reports.php

Gentzsch, W. (2008). *Top 10 rules for building a sustainable Grid*. Grid Thought Leadership Series. Retrieved from www.ogf.org/TLS/?id=1

Gentzsch, W. (2009). *HPC in the cloud: Grids or clouds for HPC?* Retrieved from http://www.hpcinthecloud.com /features/ Grids-or-Clouds-for-HPC-67796917.html

Gentzsch, W. (Ed.). (2007b). *A sustainable Grid infrastructure for Europe*. Executive Summary of the e-IRG Open Workshop on e-Infrastructures, Heidelberg, Germany. Retrieved from www.e-irg.org/meetings /2007-DE/ workshop.html

Gentzsch, W., Girou, D., Kennedy, A., Lederer, H., Reetz, J., Riedel, M., … Wolfrat, J. (2011). DEISA – Distributed European infrastructure for supercomputing applications. *Journal on Grid Computing*. Springer.

Gentzsch, W., Kennedy, A., Lederer, H., Pringle, G., Reetz, J., Riedel, M., et al. Wolfrat, J. (2010). DEISA: E-science in a collaborative, secure, interoperable and user-friendly environment. *Proceedings of the e-Challenges Conference e-2010*, Warsaw.

GEONGrid. (2008). Retrieved from www.geongrid.org

Georg, G., Ray, I., Anastasakis, K., Bordbar, B., Toahchoodee, M., & Houmb, S. H. (2009). An aspect-oriented methodology for designing secure applications. *Information and Software Technology*, *51*(5), 846–864. doi:10.1016/j.infsof.2008.05.004

Gerdes, C., Eger, K., & Müller, J. P. (2009). *Data-centric peer-to-peer communication in power grids*. Electronic Communications of the EASST 17: Kommunikation in Verteilten Systemen 2009, 2009. *Proceedings of KiVS Global Sensor Networks Workshop* (GSN09).

Giorgini, P., Mouratidis, H., & Zannone, N. (2007). Modelling security and trust with secure tropos. In Giorgini, H. M. P. (Ed.), *Integrating security and software engineering: Advances and future visions* (pp. 160–189). Hershey, PA: Idea Group Publishing.

Glite. (n.d.). *Glite middleware*. Retrieved from http://glite.web.cern.ch/glite

Global Grid Forum. (2003). *Usage Record – XML Format*. Globus Toolkit. from http://globus.org.

Globus Toolkit. (n.d.). Retrieved from http://www.globus.org/toolkit

GLOBUS. (2008). *Overview of the Grid security infrastructure.* Retrieved from http://www.globus.org/security/overview.html

Globus. (n.d.). *Globus alliance toolkit homepage*. Retrieved from http://www.globus.org/toolkit/

Glue Working Group. (2007). *GLUE Schema Specification version 1.3 Draft 3*. Gridsite. from http://www.gridsite.org/.

Gniady, C., Butt, A. R., & Hu, Y. C. (2004). *Program-counter-based pattern classification in buffer caching.* Paper presented at the 6th Symposium on Operating Systems Design and Implementation, San Francisco, CA.

Gnutella. http://gnutella.wego.com.

Goel, A. (1985). Software reliability models: Assumptions, limitations, and applicability. *IEEE Transactions on Software Engineering, 11*(12), 1411–1423. doi:10.1109/TSE.1985.232177

Goldman, K. J., Swaminathan, B., McCartney, T. P., Anderson, M. D., & Sethuraman, R. (1995). The Programmers' Playground: I/O Abstraction for User-Configurable Distributed Applications. *IEEE Transactions on Software Engineering, 21*(9), 735–746. doi:10.1109/32.464547

Goloboff, P. (1999). Analyzing Large Data Sets in Reasonable Times: Solutions for Composite Optima. *Cladistics, 15*(4), 415–428. doi:10.1111/j.1096-0031.1999.tb00278.x

Gong, L., Oaks, S., & Traversat, B. (2002). *JXTA in a nutshell a desktop quick reference*. Sebastopol, CA: O'Reilly & Associates.

Goodale, T., Jha, S., Kaiser, H., Kielmann, T., Kleijer, P., & Merzky, A. … Smith, Ch. (2008). *A simple API for Grid applications* (SAGA). Grid Forum Document GFD.90. Open Grid Forum. Retrieved from www.ogf.org/documents /GFD.90.pdf

Google Groups. (2010). *Cloud computing*. Retrieved from http://groups.google.ca/ group/cloud-computing

Google. (2008). *Google app engine*. Retrieved from http://code.google.com/appengine/

Gotthelf, P., Zunino, A., Mateos, C., & Campo, M. (2008). GMAC: An overlay multicast network for mobile agent platforms. *Journal of Parallel and Distributed Computing, 68*(8), 1081–1096. doi:10.1016/j.jpdc.2008.04.002

Gottschling, M., Stamatakis, A., & Nindl, I. (2007). Multiple Evolutionary Mechanisms Drive Papillomavirus Diversification. *Molecular Biology and Evolution, 24*(5), 1242–1258. doi:10.1093/molbev/msm039

Graham, D. (2006). *Introduction to the CLASP process*. Retrieved from https://buildsecurityin.us-cert.gov/daisy/bsi/articles/best-practices/requirements/548.html

Gray, J. (1990). A census of tandem system availability between 1985 and 1990. *IEEE Transactions on Reliability, 39*(4), 409–418. doi:10.1109/24.58719

Grid Computing. (2008). *Info centre.* Retrieved from www.gridcomputing.com

Grid Engine. (2001). *Open source project.* Retrieved from http://sourceforge.net/ projects/gridscheduler/

GridGain Systems. (2008). *GridGain.* Retrieved October 16, 2008, from http://www.gridgain.com.

GridSphere. (2008). Retrieved from www.gridsphere.org/gridsphere/gridsphere

GridWay. (2008). *Metascheduling technologies for the Grid.* Retrieved from www.gridway.org/

Griffioen, J., & Appleton, R. (1994). *Reducing file system latency using a predictive approach.* Paper presented at the USENIX Summer Conference.

Gronkvist, J. (2004). *A distributed scheduling for mobile ad hoc networks a novel approach.* Proceedings of the 15th IEEE International Symposium on Personal, Indoor and Mobile Radio Communications (pp. 964–968).

Gropp, W., & Lusk, E. (2004). Fault Tolerance in Message Passing Interface Programs. *International Journal of High Performance Computing Applications, 18*(3), 363–372. doi:10.1177/1094342004046045

Gu, T., Tan, E., Pung, H. K., & Zhang, D. (2005). *A Peer-to-Peer Architecture for Context Lookup.* Paper presented at the Proceedings of the International Conference on Mobile and Ubiquitous Systems: Networking and Services (MobiQuitous 2005), San Diego, California.

Guan, T., Zaluska, E., & Roure, D. D. (2005). *A Grid service infrastructure for mobile devices*. Paper presented at the First International Conference on Semantics, Knowledge, and Grid (SKG 2005), Beijing, China.

Guha, S., Daswani, N., & Jain, R. (2006). *An experimental study of the Skype peer-to-peer VoIP system*. In The 5th International Workshop on Peer-to-Peer Systems. Retrieved from http://saikat.guha.cc/pub /iptps06-skype.pdf

Guillen-Scholten, J., & Arbab, F. (2005). Coordinated Anonymous Peer-to-Peer Connections with MoCha. In N. Guelfi, G. Reggio, & A. Romanovsky, (Eds.), *Scientific Engineering of Distributed Java Applications, Revised Selected Papers. 4th International Workshop, FIDJI 2004, 3409* of *Lecture Notes in Computer Science*, (pp. 68–77). Springer-Verlag.

Gulbrandsen, A., Vixie, P., & Esibov, L. (2000). *A DNS RR for specifying the location of services (DNS SRV). RFC 2782. Internet Engineering Task Force*. IETF.

Gustafson, J. (1987). Reevaluating Amdahl's law. *Communications of the ACM, 31*, 532–533. doi:10.1145/42411.42415

Guttman, E. (1999). Service location protocol: Automatic discovery of IP network services. *IEEE Internet Computing, 3*(4), 71–80. doi:10.1109/4236.780963

Guttman, E., Perkins, C., & Kempf, J. (1999). *Service templates and schemes. Internet Engineering Task Force*. IETF.

Guttman, E., Perkins, C., Veizades, J., & Day, M. (1999). *Service location protocol, version 2. Internet Engineering Task Force*. IETF.

Hamscher, V., Schwiegelshohn, U., Streit, A., & Yahyapour, R. (2000). Evaluation of job-scheduling strategies for Grid computing. In *Grid Computing* (pp. 191–202). GRID.

Han, J., & Liu, Y. (2006). *Rumor Riding: Anonymizing Unstructured Peer-to-Peer Systems*. Paper presented at the Proceedings of IEEE ICNP, Santa Barbara, CA. Jena 2 - A Semantic Web Framework. http://www.hpl.hp.com/semweb/jena2.htm.

Hansen, H., Bachmann, L., & Bakke, T. A. (2003). Mitochondrial DNA variation of *Gyrodactylus* spp. *Monogenea, Gyrodactylidae* populations infecting Atlantic salmon, grayling, and rainbow trout in Norway and Sweden. *International Journal for Parasitology, 33*(13), 1471–1478. doi:10.1016/S0020-7519(03)00200-5

Harchol-Balter, M., & Downey, A. B. (1997). Exploiting process lifetime distributions for dynamic load balancing. *ACM Transactions on Computer Systems, 15*(3), 253–285. doi:10.1145/263326.263344

Hautakorpi, J., & Schultz, G. (2010). *A feasibility study of an arbitrary search in structured peer-to-peer networks*. In ICCCN'10: *Proceedings of the 19th International Conference on Computer Communications and Networks*. Zurich.

Heinicke, M. P., Duellman, W. E., & Hedges, S. B. (2007). From the Cover: Major Caribbean and Central American frog faunas originated by ancient oceanic dispersal. *Proceedings of the National Academy of Sci*

Henderson, T., Kotz, D., & Abyzov, I. (2004). *The changing usage of a mature campus-wide wireless network*. Proceedings of the Tenth Annual International Conference on Mobile Computing and Networking (MobiCom) (pp. 187–201).

Herveg, J. (2006). *The ban on processing medical data in European law: Consent and alternative solutions to legitimate processing of medical data in healthgrid. Proc. Healthgrid* (*Vol. 120*, pp. 107–116). Amsterdam, The Netherlands: IOS Press.

Herveg, J., Crazzolara, F., Middleton, S. E., Marvin, D. J., & Poullet, Y. (2004). *GEMSS: Privacy and security for a medical Grid*. Paper presented at the HealthGRID 2004, Clermont-Ferrand, France.

Hines, M., Lewandowski, M., Wang, J., & Gopalan, K. (2006). *Anemone: Transparently harnessing cluster-wide memory*. Paper presented at the International Symposium on Performance Evaluation of Computer and Telecommunication Systems, Calgary, Alberta, Canada.

Hines, M., Wang, J., & Gopalan, K. (2006). *Distributed Anemone: Transparent low-latency access to remote memory in commodity clusters*. Paper presented at the International Conference on High-Performance Computing, Bangalore, India.

Hoare, C. (1985). *Communicating Sequential Processes*. Prentice Hall.

Hommel, W. (2005a). Using XACML for privacy control in SAML-based identity federations. In *Proceedings of the 9th Conference on Communications and Multimedia Security (CMS 2005)*. Springer.

Hommel, W. (2005b). An architecture for privacy-aware inter-domain identity management. In *Proceedings of the 16th IFIP/IEEE Distributed Systems: Operations and Management (DSOM 2005)*. Springer.

Housley, R., Polk, W., Ford, W., & Solo, D. (2002). *Certificate and certificate revocation list (CRL) profile*. RFC 3280. Internet Engineering Task Force (IETF).Jenronimo, M., & Weast, J. (2003). *UPnP design by example: A software developer's guide to universal plug and play. Intel Press., ISBN-13*, 978–0971786110.

Hsu, M., & Silberschatz, A. (1991). Unilateral Commit: A New Paradigm for Reliable Distributed Transaction Processing. In *Proc. of the 7th International Conference on Data Engineering*, (pp. 286–293). IEEE Computer Society.

Hu, J., & Klefstad, R. (2006). *Decentralized load balancing on unstructured Peer-2-Peer computing Grids*. 5th IEEE International Symposium on Network Computing and Applications, (pp. 247-250).

Hua, K. A., Jiang, N., Peng, R., & Tantaoui, M. A. (2004). PSP: A Persistent Streaming Protocol for Transactional Communications. In *ICCCAS 2004: Proc. of the 2004 International Conference on Communications, Circuits and Systems, 1*, 529–533. IEEE Computer Society.

Huang, K.-C. (2006). *Performance evaluation of adaptive processor allocation policies for moldable parallel batch jobs*. Paper presented at the Third Workshop on Grid Technologies and Applications.

Huang, K.-C., & Chang, H.-Y. (2006). *An integrated processor allocation and job scheduling approach to workload management on computing Grid.* Paper presented at the 2006 International Conference on Parallel and Distributed Processing Techniques and Applications (PDPTA'06), Las Vegas, USA.

Huang, P. J., Yu, Y. F., Chen, Q. J., Huang, T. L., Lai, K. C., & Li, K. C. (2010). A self-adaptive load balancing strategy for P2P grids. In C. H. Hsu, et al. (Eds.), *ICA3PP 2010, part II, LNCS 6082*, (pp. 348-357). Heidelberg/Berlin, Germany: Springer-Verlag.

Huck, P., Butler, M., Gupta, A., & Feng, M. (2002). A Self-Configuring and Self-Administering Name System with Dynamic Address Assignment. *ACM Transactions on Internet Technology*, *2*(1), 14–46. doi:10.1145/503334.503336

Huda, M. T., Schmidt, W. H., & Peake, I. D. (2005). An agent oriented proactive fault-tolerant framework for Grid computing. *Proceedings of the First International Conference on e-Science and Grid Computing (e-Science '05), IEEE* (pp. 304-311).

Hughes, J., & Maler, E. (2005). *OASIS security assertion markup language (SAML), V2.0 technical overview*. OASIS Security Services Technical Committee Document.

Humphrey, M., Thompson, M. R., & Jackson, K. R. (2005). *Security for Grids*. Lawrence Berkeley National Laboratory. (Paper LBNL-54853).

IBM Websphere MQ. (2008). *The IBM Websphere MQ Family.* [online]. URL: http://www.ibm.com/software/websphere. (March, 2008).

IEEE 802.11 Working Group (1997). *Wireless LAN Medium Access Control (MAC) and Physical Layer (PHY) Specifications*.

IEEE Std 802.11e. (2004). *Medium Access Control (MAC) Enhancements for Quality of Service (QoS)*. IEEE Draft for Wireless Medium Access Control (MAC) and Physical Layer (PHY) Specifications, / Draft 11.0.

Imamura, T., Tsujita, Y., Koide, H., & Takemiya, H. (2000). An Architecture of Stampi: MPI Library on a Cluster of Parallel Computers. In Dongarra, J., Kacsuk, P., & Podhorszki, N. (Eds.), *Recent Advances in Parallel Virtual Machine and Message Passing Interface* (pp. 200–207). Springer. doi:10.1007/3-540-45255-9_29

Inca: User Level Grid Monitoring. from http://inca.sdsc.edu/drupal/.

Indyk, P., Motwani, R., Raghavan, P., & Vempala, S. (1997). Locality-preserving hashing in multidimensional spaces. In STOC '97: *Proc. of the Twenty-Ninth Annual ACM Symposium on Theory of Computing*, (pp. 618–625). New York, NY: ACM Press.

Iosup, A., Jan, M., Sonmez, O., & Epema, D. (2007). On the dynamic resource availability in grids. In *Proceedings of 8th IEEE/ACM International Conference on Grid Computing*, (pp. 26-33).

Iyengar, M. S., & Singhalc, M. (2006). Effect of network latency on load sharing in distributed systems. *Journal of Parallel and Distributed Computing*, *66*(6), 839–853. doi:10.1016/j.jpdc.2005.09.005

Iyer, R. K., & Rossetti, D. J. (1985). Effect of system workload on operating system reliability: A study on IBM 3081. *IEEE Transactions on Software Engineering*, *11*(12), 1438–1448. doi:10.1109/TSE.1985.232180

Jacob, B., Ferreira, L., Bieberstein, N., Gilzean, C., Girard, J.-Y., Strachowski, R., & Yu, S. (2003). *Enabling applications for Grid computing with Globus*. IBM Redbook. Retrieved from www.redbooks.ibm.com /abstracts/ sg246936.html?Open

Jacobson, I., Booch, G., & Rumbaugh, J. (1999). *The unified software development process*. Addison-Wesley Professional.

Jain, N., Das, S. R., & Nasipuri, A. (2000). *A Multichannel CSMA MAC Protocol with Receiver-Based Channel Selection for Multihop Wireless Networks*. Proceedings of the 10th IEEE International Conference on Computer Communications and Networks (pp. 432-439).

Jameel, H., Kalim, U., Sajjad, A., Lee, S., & Jeon, T. (2005). *Mobile-to-Grid middleware: Bridging the gap between mobile and Grid environments*. Paper presented at the European Grid Conference EGC 2005, Amsterdam, The Netherlands.

Jana, D., Chaudhuri, A., & Bhaumik, N. B. (2009). Privacy and anonymity protection in computational Grid services. *International Journal of Computer Science and Applications*, *6*(1), 98–107.

Jha, S., Kaiser, H., El Khamra, Y., & Weidner, O. (2007). *Design and implementation of network performance aware applications using SAGA and Cactus*. 3rd IEEE Conference on eScience and Grid Computing, Bangalore, India, 10-13 Dec, (pp. 143-150).

Jiang, S., Petrini, F., Ding, X., & Zhang, X. (2006). *A locality-aware cooperative cache management protocol to improve network file system performance*. Paper presented at the 26th IEEE International Conference on Distributed Computing Systems, Lisbon, Portugal.

Job Description Language Attributes. (n.d.). Retrieved from http://auger.jlab.org/jdl /PPDG_JDL.htm

John, T., Uwe, S., Joel, L. W., & Philip, S. Y. (1994). *Scheduling parallel tasks to minimize average response time*. Paper presented at the fifth annual ACM-SIAM Symposium on Discrete algorithms.

Johnson, R. (2005). J2EE development frameworks. *Computer*, *38*(1), 107–110. doi:10.1109/MC.2005.22

Johnson, B. K., & Ram, D. J. (2001). DP: A Paradigm for Anonymous Remote Computation and Communication for Cluster Computing. *IEEE Transactions on Parallel and Distributed Systems*, *12*(10), 1052–1065. doi:10.1109/71.963417

Jones, M. (2003). *Grid Security - An overview of methods used to create a secure grid.* Retrieved from http://www.cse.buffalo.edu/faculty/miller/Courses/Grid-Seminar/Security.pdf.

JPPF. (2008). *Java Parallel Processing Framework.* Retrieved October 16, 2008, from http://www.jppf.org.

JSDL. (n.d.). *Job submission description language (jsdl) specification, v. 1.0.* Retrieved from http://www.gridforum.org/documents/GFD.56.pdf

Jürjens, J. (2005). *Secure systems development with UML*. Springer.

Jurjens, J. (2001). *Towards development of secure systems using UMLsec*. Paper presented at the Fundamental Approaches to Software Engineering (FASE/ETAPS).

Jurjens, J. (2002). *UMLsec: Extending UML for secure systems development*. Paper presented at the 5th International Conference on the Unified Modeling Language (UML), Dresden, Germany.

Jürjens, J., Schreck, J., & Bartmann, P. (2008). *Model-based security analysis for mobile communications*. Paper presented at the International Conference on Software Engineering, Leipzig, Germany.

Kahn, G. (1974). The Semantics of Simple Language for Parallel Programming. In *Proc. of the 1974 IFIP Congress*, (pp. 471–475).

Kaler, C., & Nadalin, A. (Eds.). (2003). *Web services federation language (WS-Federation)*. Web Services Specifications Document.

Kalra, D., Singleton, P., Ingram, D., Milan, J., MacKay, J., Detmer, D., & Rector, A. (2005). Security and confidentiality approach for the clinical e-science framework (clef). *Methods of Information in Medicine*, *44*(2), 193–197.

Kalyanakrishnam, M., Kalbarczyk, Z., & Iyer, R. (1999). Failure data analysis of a LAN of Windows NT based computers. In *Proceedings of the 18th IEEE Symposium on Reliable Distributed Systems (SRDS99)*, (pp. 178-187).

Karmarkar, A., Hadley, M., Mendolsohn, N., Lafon, Y., Gudgin, M., Moreau, J. J., & Nielsen, H. (2007). *SOAP version 1.2 part 1: Messaging framework* (2nd ed.). Retrieved from http://www.w3.org/TR/2007/REC-soap12-part1-20070427/

Karonis, N. T., Toonen, B., & Foster, I. (2002). MPICH-G2: A Grid-enabled implementation of the Message Passing Interface. *Journal of Parallel and Distributed Computing*, *63*, 551–563. doi:10.1016/S0743-7315(03)00002-9

Kirchler, W., Schiffers, M., & Kranzlmüller, D. (2009). Harmonizing the management of virtual organizations despite heterogeneous Grid middleware – assessment of two different approaches. In *Proceedings of the Cracow Grid Workshop*.

Kleinberg, J. (2000). *The Small-World Phenomenon: an Algorithm Perspective*. Paper presented at the Proceedings of the 32nd ACM Symposium on Theory of Computing. LimeWire. http://www.limewire.com/english/content/home.shtml.

Kolonay, R., & Sobolewski, M. (2004). *Grid interactive service-oriented programming environment*. Paper presented at the Concurrent Engineering: The Worldwide Engineering Grid, Tsinghua, China.

Koubias, S. A., & Haralabidis, H. C. (1996). Mition: A mac-layer hybrid protocol for multi-channel real-time lans. *Proceedings of the Third IEEE International Conference on Electronics, Circuits, and Systems* (pp. 327 – 330).

Krishna Rana, Y., Hua Liu, B., Nyandoro, A., & Jha, S. (2006). Bandwidth Aware Slot Allocation in Hybrid MAC. *Proceedings of 31st IEEE Conference on Local Computer Networks* (pp. 89 – 96).

Kruchten, P. (2000). *The rational unified process: An introduction* (2nd ed.). Addison-Wesley.

Kuhn, D. R., Coyne, E. J., & Weil, T. R. (2010). Adding attributes to role-based access control. *IEEE Security*, June 2010.

Kumar, A., & Qureshi, S. R. (2008, March 29). *Integration of mobile computing with Grid computing: A middleware architecture*. Paper presented at the 2nd National Conference on Challenges & Opportunities in Information Technology (COIT-2008), Mandi Gobindgarh, India.

Kwok-Yan, L., Xi-Bin, Z., Siu-Leung, C., Gu, M., & Jia-Guang, S. (2004). Enhancing Grid security infrastructure to support mobile computing nodes. *Lecture Notes in Computer Science*, *2908*, 42–54. doi:10.1007/978-3-540-24591-9_4

Kyasanur, P., Jungmin, C., Chereddi, S., & Vaidya, N. H. (2006). Multichannel mesh networks: challenges and protocols. *IEEE Wireless Communication*, *13*(2), 30–36. doi:10.1109/MWC.2006.1632478

Laure, E., Stockinger, H., & Stockinger, K. (2005). Performance engineering in data Grids. *Concurrency and Computation*, *17*(2-4), 171–191. doi:10.1002/cpe.923

Laure, E., Fisher, S., & Frohner, A. (2006). Programming the Grid with gLite. *Computational Methods in Science and Technology*, *12*(1), 33–45.

Lederer, H. (2008). DEISA2: Supporting and developing a European high-performance computing ecosystem. *Journal of Physics*, *125*. doi:10.1088/ 1742-6596/125/1/011003.

Lee, I., Tang, D., Iyer, R., & Hsueh, M.-C. (1993). Measurement-based evaluation of operating system fault tolerance. *IEEE Transactions on Reliability*, *42*(2), 238–249. doi:10.1109/24.229493

Lee, C. (2003). Grid programming models: Current tools, issues and directions. In Berman, G. F., & Hey, T. (Eds.), *Grid computing* (pp. 555–578). USA: Wiley Press. doi:10.1002/0470867167.ch21

Legendre, P., Desdevises, Y., & Bazin, E. (2002). A Statistical Test for Host-Parasite Coevolution. *Systematic Biology, 51*(2), 217–234. doi:10.1080/10635150252899734

Legendre, P., & Anderson, M. J. (1998). DistPCOA program description, source code, executables, and documentation: http://www.bio.umontreal.ca/Casgrain/en/labo/distpcoa.html

Legrand, I. (2007). *MonALISA: An Agent Based, Dynamic Service System to Monitor, Control and Optimize Distributed Systems*. CHEP'07, Victoria, British Columbia, Canada. Sep 2007. MonALISA Repository for Alice. from http://pcalimonitor.cern.ch/map.jsp.

Lei, S., Yuyan, S., & Lin, W. (2007). *Effect of scheduling discipline on CPU-MEM load sharing system*. 6th International Conference on Grid and Cooperative Computing, (pp. 242-249).

LHC – Large Hadron Collider Project. (2008). http://lhc.web.cern.ch/lhc/

LHC. (n.d.). *LHC computing grid project*. Retrieved from http://lcg.web.cern.ch/LCG

Li, J., & Vuong, S. (2006). *Grid resource discovery based on semantic P2P communities*. 2006 ACM Symposium on Applied Computing, (pp. 754-758).

Li, J., Haas, Z. J., Sheng, M., & Chen, Y. (2003). *Performance Evaluation of Modified IEEE 802.11 MAC for Multi-Channel Multi-Hop Ad Hoc Network*. Proceedings of the 17th International Conference on Advanced Information Networking and Applications. (pp. 312–317).

Li, M., Lee, W. C., Sivasubramaniam, A., & Lee, D. L. (2004). *A Small World Overlay Network for Semantic Based Search in P2P*. Paper presented at the Proceedings of the Second Workshop on Semantics in Peer-to-Peer and Grid Computing, in conjunction with the World Wide Web Conference.

Li, M., Lee, W.-C., & Sivasubramaniam, A. (2006). *DPTree: a Balanced Tree Based Indexing Framework for Peer-to-Peer Systems*. Paper presented at the Proceedings of IEEE ICNP, Santa Barbara, CA.

Li, S., & Tahvildari, L. (2006). JComp: A reuse-driven componentization framework for Java applications. *In 14th IEEE International Conference on Program Comprehension (ICPC'06),* (pp. 264-267). IEEE Computer Society.

Li, Y., & Mascagni, M. (2003). Improving performance via computational replication on a large-scale computational Grid. *Third IEEE International Symposium on Cluster Computing and the Grid (CCGrid'03), Tokyo, Japan* (pp. 442-448).

Li, Y., Yang, Y., & Zhu, R. (2009). *A hybrid load balancing strategy of sequential tasks for computational Grids*. IEEE International Conference on Networking and Digital Society, (pp. 112-117).

Liang, G. (2008). *Adaptive load balancing algorithm over heterogeneous workstations*. 7th International Conference on Grid and Cooperative Computing, (pp. 169-174).

LinuxForum. (n.d.). *Linux filesystem hierarchy,* 1.10. Retrieved from http://www.linuxforum.com /linux-filesystem/proc.html

Litke, A., Skoutas, D., & Varvarigou, T. (2004). *Mobile Grid computing: Changes and challenges of resource management in a mobile Grid environment*. Paper presented at the 5th International Conference on Practical Aspects of Knowledge Management (PAKM 2004).

Litzkow, M., Livny, M., & Mutka, M. (1988). Condor - a hunter of idle workstations. In *Proceedings of the 8th International Conference of Distributed Computing Systems*, (pp. 104–111).

Liu, B., Liu, Z., & Towsley, D. (2003). On the Capacity of Hybrid Wireless Networks. *Proc. of IEEE Infocom*.

Liu, L., & Lee, K.-W. (2004). Supporting efficient keyword-based file search in peer-to-peer file sharing systems. In GLOBECOM'04: *Proc. of the IEEE Global Telecommunications Conference*.

Liu, L., Wu, Z., Ma, Z., & Cai, Y. (2008). *A dynamic fault tolerant algorithm based on active replication*. Seventh International Conference on Grid and Cooperative Computing, China (pp. 557-562).

Liu, Y., Liu, X., Xiao, L., Ni, L. M., & Zhang, X. (2004). *Location-aware Topology Matching in P2P Systems*. Paper presented at the Proceedings of IEEE INFOCOM, Hong Kong, China.

Long, D., Muir, A., & Golding, R. (1995). A longitudinal survey of internet host reliability. In *Proceedings of the 14th Symposium on Reliable Distributed System (SRDS95)*, (pp. 2-9).

Lőrincz, L. C., Kozsik, T., Ulbert, A., & Horváth, Z. (2005). A method for job scheduling in Grid based on job execution status. *Multiagent and Grid Systems - An International Journal 4 (MAGS) 1*(2), 197-208.

Luther, A., Buyya, R., Ranjan, R., & Venugopal, S. (2005). Peer-to-peer Grid computing and a. NET-based Alchemi framework. In M. Guo (Ed.), High performance computing: Paradigm and infrastructure. Wiley Press, USA. Retrieved from www.alchemi.net

Luther, A., Buyya, R., Ranjan, R., & Venugopal, S. (2005). *Alchemi: A. netbased enterprise grid computing system*. In International Conference on Internet Computing, (pp. 269-278).

Magee, J., Kramer, J., & Sloman, M. (1989). Constructing Distributed Systems in Conic. *IEEE Transactions on Software Engineering, 15*(6), 663–675. doi:10.1109/32.24720

Maheshwari, R., Gupta, H., & Samir, R. (2006). *Multichannel MAC Protocols for Wireless Networks*. Proceedings of the 3rd IEEE Communication Society on Sensor and Ad Hoc Communications Networks (pp. 393-401).

Malin, B. (2002). *Compromising privacy with trail re-identification: The Reidit algorithms.* (CMU Technical Report, CMU-CALD-02-108), Pittsburgh.

Manion, F. J., Robbins, R. J., Weems, W. A., & Crowley, R. S. (2009). Security and privacy requirements for a multi-institutional cancer research data grid: An interview-based study. *BMC Medical Information and Decision Making, 9*(31).

Mao, W., Martin, A., Jin, H., & Zhang, H. (2009). Innovations for Grid security from trusted computing – protocol solutions to sharing of security resource. *LNCS 5087*. Springer. Mont, M., Pearson, S., & Bramhall, P. (2003). *Towards accountable management of identity and privacy: Sticky policies and enforceable tracing services*. (Report No. HPL-2003-49). Bristol, UK: HP Laboratories.

Markatos, E. P., & Dramitinos, G. (1996). *Implementation of a reliable remote memory pager.* Paper presented at the USENIX Annual Technical Conference, San Diego, CA.

Mateos, C., Zunino, A., & Campo, M. (2005). Integrating intelligent mobile agents with Web Services. *International Journal of Web Services Research, 2*(2), 85–103. doi:10.4018/jwsr.2005040105

Mateos, C., Zunino, A., & Campo, M. (2008a). A survey on approaches to gridification. *Software, Practice & Experience, 38*(5), 523–556. doi:10.1002/spe.847

Mateos, C., Zunino, A., & Campo, M. (2008b). JGRIM: An approach for easy gridification of applications. *Future Generation Computer Systems, 24*(2), 99–118. doi:10.1016/j.future.2007.04.011

Mateos, C. (2008). *An approach to ease the gridification of conventional applications. Doctoral dissertation.* Universidad del Centro de la Provincia de Buenos Aires, Argentina. Retrieved October 16, 2008, from http://www.exa.unicen.edu.ar/~cmateos/files/phdthesis.pdf.

MATLAB. (2010). *Amazon Web Services for high-performance cloud computing – MATLAB. Solving Ax=b.* Retrieved from http://aws.typepad.com/aws /2010/09/high-performance-cloud-computing-nasa-matlab.html

Maymounkov, P., & Mazieres, D. (2002). *Kademlia: A peer-to-peer Information System based on the XOR metric*.

McClatchey, R., Anjum, A., Stockinger, H., Ali, A., Willers, I., & Thomas, M. (2007, March). Data intensive and network aware (DIANA) Grid scheduling. *Journal of Grid Computing, 5*(1), 43–64. doi:10.1007/s10723-006-9059-z

McGinnis, L., Wallom, D., & Gentzsch, W. (Eds.). (2007). *2nd International Workshop on Campus and Community Grids*. Retrieved from http://forge.gridforum.org/ sf/go/doc14617?nav=1

McGough, S., Lee, W., & Das, S. (2008). A standards based approach to enabling legacy applications on the Grid. *Future Generation Computer Systems, 24*(7), 731–743. doi:10.1016/j.future.2008.02.004

Medina, A., Lakhina, A., Matta, I., & Byers, J. (2001). *BRITE: An approach to universal topology generation.* Paper presented at the International Workshop on Modeling, Analysis and Simulation of Computer and Telecommunications Systems, Cincinnati, Ohio.

Meier-Kolthoff, J. P., Auch, A. F., Huson, D. H., & Göker, M. (2007). COPYCAT: Co-phylogenetic Analysis tool. *Bioinformatics (Oxford, England)*, *23*(7), 898–900. doi:10.1093/bioinformatics/btm027

Meinilä, M., Kuusela, J., Zietara, M. S., & Lumme, J. (2004). Initial steps of speciation by geographic isolation and host switch in salmonid pathogen *Gyrodactylus salaris (Monogenea: Gyrodactylidae)*. *International Journal for Parasitology*, *34*(4), 515–526. doi:10.1016/j.ijpara.2003.12.002

Merkle, D., & Middendorf, M. (2005). Reconstruction of the cophylogenetic history of related phylogenetic trees with divergence timing information. *Theory in Biosciences*, *123*(4), 277–299. doi:10.1016/j.thbio.2005.01.003

Mickens, J. W., & Noble, B. D. (2006). Exploiting availability prediction in distributed systems. In *Proceedings of the 3rd Conference on Networked Systems Design & Implementation (NSDI06)*, (pp. 6-19).

Microsoft. (2008). *Microsoft Message Queueing.* [online]. URL: http://www.microsoft.com/windowsserver2003/technologies/msmq/default.mspx (March, 2008).

Microsoft. (n.d.). *Peer name resolution protocol.* Retrieved from http://technet.microsoft.com/en-us/library/bb726971.aspx

Milojičić, D. S., Douglis, F., Paindaveine, Y., Wheeler, R., & Zhou, S. (2000). Process migration. *ACM Computing Surveys*, *32*(3), 241–299. doi:10.1145/367701.367728

Mockapetris, P. (1987). *Domain names - implementation and specification. RFC 1035. Internet Engineering Task Force*. IETF.

Mohsin, M., & Prakash, R. (2002). IP Address Assignment in a Mobile Ad Hoc Network. *IEEE Military Communications Conference (MILCOM 2002)*, *2*(10), 856-861.

Mont, M. (2004). *Dealing with privacy obligations in enterprises*. (Report No. HPL-2004-109). Bristol, UK: HP Laboratories.

Montagnat, J., Frohner, A., Jouvenot, D., Pera, C., Kunszt, P., & Koblitz, B. (2007). A secure grid medical data manager interfaced to the glite middleware. *Journal of Grid Computing*, *6*(1).

Morselli, R., Bhattacharjee, B., Srinivasan, A., & Marsh, M. A. (2005). *Efficient Lookup on Unstructured Topologies*. Paper presented at the Proceedings of ACM PODC, Las Vegas, NV, USA.

Moses, T. (Ed.). (2005). *OASIS eXtensible access control markup language 2.0, core specification.* OASIS XACML Technical Committee Standard.

Mouratidis, H. (2004). *A security oriented approach in the development of multiagent systems: Applied to the management of the health and social are needs of older people in England.* University of Sheffield.

Mouratidis, H., & Giorgini, P. (2006). *Integrating security and software engineering: Advances and future vision.* Hershey, PA: IGI Global.

Mujumdar, M., Bheevgade, M., Malik, L., & Patrikar, R. (2008). *High performance computational Grids - fault tolerance at system level.* International Conference on Emerging Trends in Engineering and Technology (ICETET) (pp. 379-383).

Mutka, M. W., & Livny, M. (1988). Profiling workstations' available capacity for remote execution. In *Proceedings of the 12th IFIP WG 7.3 International Symposium on Computer Performance Modelling, Measurement and Evaluation*, (pp. 529–544).

Myers, M., & Schaad, J. (2007). *Certificate management over CMS (CMC) transport protocols. Internet Engineering Task Force*. IETF.

Myers, A. D. (2002). Hybrid MAC Protocols For Mobile Ad Hoc Networks. *PhD thesis, Computer Science, University of Texas at Dallas.*

MyGrid. (2008). Retrieved from www.mygrid.org.uk

Nabrizyski, J., Schopf, J. M., & Weglarz, J. (2003). Grid resource management: State of the art and future trends. In Nabrizyski, J., Schopf, J. M., & Weglarz, J. (Eds.), *International series in operations research and management*. Kluwer Academic Publishers Group.

Nadeem, F., Prodan, R., & Fahringer, T. (2008). Characterizing, modeling and predicting dynamic resource availability in a large scale multi-purpose grid. In *Proceedings of the 2008 8th IEEE International Symposium on Cluster Computing and the Grid (CCGRID08)*, (pp. 348-357).

Naedele, M. (2003). Standards for XML and Web Services Security. *Computer, 36*(4), 96–98. doi:10.1109/MC.2003.1193234

Nagaratnam, N., Janson, P., J. Dayka, Nadalin, A., Siebenlist, F., Welch, V., et al. (2003). *The security architecture for open Grid services.*

Nagel, W. E., Kröner, D. B., & Resch, M. M. (2007). *High Performance Computing in Science and Engineering 07*. Berlin, Heidelberg, New York: Springer.

NCSA. (2008). *MyProxy credential management service.* Retrieved from http://grid.ncsa.uiuc.edu/myproxy/ca/

NEESGrid. (2008). Retrieved from www.nees.org/

Nejdl, W., Wolpers, M., Siberski, W., Schmitz, C., Schlosser, M., Brunkhorst, I., & Lser, A. (2003). *Super-peer-based Routing and Clustering Strategies for RDF-based Peer-to-Peer Networks*. Paper presented at the Proceedings of the 12th World Wide Web Conference.

Nesargi, S., & Prakash, R. (2002). MANETconf: Configuration of Hosts in a Mobile Ad Hoc Network. *Proceedings of INFOCOM'02,* (pp. 1059-1068.L).

Neuroth, H., Kerzel, M., & Gentzsch, W. (Eds.). (2007). *German Grid initiative D-Grid*. Universitätsverlag Göttingen Publishers. Retrieved from www.d-grid.de/ index.php?id=4&L=1

Newhall, T., Amato, D., & Pshenichkin, A. (2008). *Reliable adaptable network RAM.* Paper presented at the International Conference on Cluster Computing.

Niederberger, R., & Alessandrini, V. (2004). DEISA: Motivations, strategies, technologies. In *Proceedings of the International Supercomputer Conference 2004*.

NIST. (2007). *Special publication 800-88: Guidelines for media sanitization by the national institute of standards and technology*. Retrieved from http://csrc.nist.gov/publications/nistpubs/#sp800-88

NPB. (2010). *NAS parallel benchmark.* Retrieved from http://www.nas.nasa.gov/Resources /Software/npb.html

Nurmi, D., Brevik, J., & Wolski, R. (2005). Modeling machine availability in enterprise and wide-area distributed computing environments. In *Proceedings of the 11th International Euro-par Conference*, (pp. 432-441).

OCCI. (2010). *Open Cloud Computing Interface working group at OGF*. Retrieved 2010 from http://forge.ogf.org/sf/ projects/occi-wg

OGF. (2008). *Open Grid forum*. Retrieved from www.ogf.org

Olabarriaga, S. D., Nederveen, A. J., Snel, J. G., & Belleman, R. G. (2006). *Towards a virtual laboratory for FMRI data management and analysis. Proc. HealthGrid 2006* (*Vol. 120*, pp. 43–54). Amsterdam, The Netherlands: IOS Press.

Oleszkiewicz, J., Xiao, L., & Liu, Y. (2004). *Parallel network RAM: Effectively utilizing global cluster memory for large data-intensive parallel programs.* Paper presented at the International Conference on Parallel Processing, Montreal, Quebec, Canada.

Oliveira, L., Sales, L., Loureiro, E., Almeida, H., & Perkusuch, A. (2006). Filling the gap between mobile and service-oriented computing: issues for evolving mobile computing towards wired infrastructures and vice versa. *International Journal of Web and Grid Services*, *2*(4), 355–378. doi:10.1504/IJWGS.2006.011710

Ong, S. H. (2003). *Grid Computing: Business Policy and Implications*. Master's Thesis, MIT, Cambridge, MA.

Open Grid Forum. (2006). *The open Grid services architecture*, version 1.5 o.

Open Group. (2009). *TOGAF™ version 9 - the open group architecture framework*. Retrieved from http://www.opengroup.org/architecture/togaf9-doc/arch/

Open Science Grid Consortium. from http://www.opensciencegrid.org/.

Open, C. A. (2008a). *LibPKI: The easy PKI library*. Retrieved from http://www.openca.org/projects/libpki/

Open, C. A. (2008b). *OpenCA-NG: The next generation CA*. Retrieved from http://www.openca.org/projects/ng/

Open, C. A. Labs. (2008c). *OpenCA's PKI resource discovery package*. Retrieved from http://www.openca.org/projects/prqpd/

Oppenheimer, D., Ganapathi, A., & Patterson, D. A. (2003). Why do internet services fail, and what can be done about it? In *Proceedings of USENIX Symposium on Internet Technologies and Systems (USITS 03)*, (p. 1).

OptorSim. (n.d.). *Simulating data access optimization algorithms.* Retrieved from http://edg-wp2.web.cern.ch/edg-wp2/optimization/ optorsim.html

OSG Grid Operations Center. from http://www.grid.iu.edu/.

OSG Resource and Service Validation Project. from http://rsv.grid.iu.edu/documentation/.

Ousterhout, J., Agrawal, P., Erickson, D., Kozyrakis, C., Leverich, J., & Mazieres, D. (2010). The case for RAMClouds: Scalable high-performance storage entirely in DRAM. *ACM SIGOPS Operating Systems Review*, *43*(4), 92–105. doi:10.1145/1713254.1713276

Padmanabhan, A. (2007). *OSG Information Services – A Discussion.* Presentation at OSG Site Administrators Meeting, Dec 2007.

Pakin, S., & Johnson, G. (2007). *Performance analysis of a user-level memory server.* Paper presented at the International Conference on Cluster Computing.

Pala, M. (2008). *PKI resource discovery protocol (PRQP). Internet Engineering Task Force.* IETF.

Pala, M., & Smith, S. W. (2007). AutoPKI: A PKI resources discovery system. *Public Key Infrastructure: EuroPKI 2007.* [Springer-Verlag.]. *LNCS*, *4582*, 154–169.

Pala, M. (2010). *A proposal for collaborative Internet-scale trust infrastructures deployment: The public key system.* 9th Symposium on Identity and Trust on the Internet (IDTrust 2010). Gaithersburg, MD: NIST.

Pala, M., & Smith, S. W. (2008). PEACHES and peers. *5th European PKI Workshop: Theory and Practice. LNCS 5057*, (pp. 223-238). Springer-Verlag.

Pan, Y., Lu, W., Zhang, Y., & Chiu, K. (2007). *A static load-balancing scheme for parallel XML parsing on multicore CPUs.* 7th IEEE International Symposium on Cluster Computing and the Grid, (pp. 351-362).

Papadimitratos, P., & Haas, Z. J. (2005). Secure Routing for Mobile Ad Hoc Networks. *Advances in Wired and Wireless Communication, IEEE/Sarnoff Symposium*, (pp. 168-171).

Parallel Workloads Archive. (n.d.). Retrieved from http://www.cs.huji.ac.il/labs/ parallel/workload/

Paranhos, D., Cirne, W., & Brasileiro, F. (2003). Trading cycles for information: Using replication to schedule bag-of-tasks applications on computational grids. *International Conference on Parallel and Distributed Computing (Euro-Par). Lecture Notes in Computer Science*, *2790*, 169–180.

Patel, J. K., Kapadia, C. H., & Owen, D. B. (1976). *Handbook of statistical distributions*. Marcel Dekker, Inc.

Patterson, D. A. (2004). Latency lags bandwith. *Communications of the ACM*, *47*(10), 71–75. doi:10.1145/1022594.1022596

Paventhan, A., Takeda, K., Cox, S., & Nicole, D. (2007). MyCoG.NET: A multi-language CoG toolkit. *Concurrency and Computation*, *19*(14), 1885–1900. doi:10.1002/cpe.1133

Paxson, V., & Floyd, S. (1997). Why we don't know how to simulate the Internet. In *Proceedings of the 29th Conference on Winter Simulation*, (pp. 1037–1044).

Pei, J., Han, J., Mortazavi-Asl, B., & Pinto, H. (2001). *PrefixSpan: Mining sequential patterns efficiently by prefix-projected pattern growth.* Paper presented at the 17th International Conference on Data Engineering.

Perez, J.M., Bellens, P., Badia, R.M., & Labarta, J. (2007). CellSs: Programming the Cell/ B.E. made easier. *IBM Journal of R&D, 51*(5).

Perkins, C. E. (2002). Mobile IP. *Communications Magazine, IEEE, 40*(5), 66–82. doi:10.1109/MCOM.2002.1006976

Perkins, C., & Royer, E. (1999). Ad Hoc On-Demand Distance Vector Routing. *In 2nd IEEE Workshop on Selected Areas in Communication, 2,* 90–100.H.

Pettersson, J. S., Fischer-Hübner, S., Danielsson, N., Nilsson, J., Bergmann, M., Clauss, S., et al. Krasemann, H. (2005). Making PRIME usable. In *Proceedings of the Symposium on Usable Privacy and Security (SOUPS)*. ACM Press.

Pfitzmann, B. (2002). Privacy in browser-based attribute exchange. In *Proceedings of the ACM Workshop on Privacy in Electronic Society (WPES 2002)*. ACM Press.

P-GRADE portal. (n.d.). Retrieved from http://www.lpds.sztaki.hu /pgrade/

P-GRADE. (2003). *Parallel Grid run-time and application development environment.* Retrieved from www.lpds.sztaki.hu /pgrade/

Phan, T., Huang, L., Ruiz, N., & Bagrodia, R. (2005). Integrating mobile wireless devices into the computational Grid. In Ilyas, M., & Mahgoub, I. (Eds.), *Mobile computing handbook*. Auerbach Publications.

Phinjaroenphan, P., Bevinakoppa, S., & Zeephongsekul, P. (2005). A method for estimating the execution time of a parallel task on a Grid node. *Lecture Notes in Computer Science, 3470*, 226–236. doi:10.1007/11508380_24

Pitzmann, A., & Köhntopp, M. (2001). *Anonymity, unobservability, and pseudonymity — a proposal for terminology. Designing Privacy Enhancing Technologies* (pp. 1–9). LNCS.

Plank, J., & Elwasif, W. (1998). *Experimental assessment of workstation failures and their impact on checkpointing systems.* Twenty-Eighth Annual International Symposium on Fault-Tolerant Computing, (pp. 48-57).

Popp, G., Jürjens, J., Wimmel, G., & Breu, R. (2003). *Security-critical system development with extended use cases.* Paper presented at the Tenth Asia-Pacific Software Engineering Conference (APSEC'03).

Portal, C. H. R. O. N. O. S. (2004). Retrieved from http://portal.chronos.org/ gridsphere/gridsphere

Powers, C., & Schunter, M. (2003). *Enterprise privacy authorization language*. W3C member submission. Retrieved from http://www.w3.org/Submission /2003/SUBM-EPAL-20031110/

Prabhakar, S., Ribbens, C., & Bora, P. (2002). *Multifaceted web services: An approach to secure and scalable grid scheduling.* Proceedings of Euroweb, Oxford, UK.

PRACE. (2008). *Partnership for advanced computing in Europe*. Retrieved from www.prace-project.eu/

PRAGMA-Grid. (2008). http://www.pragma-grid.net/ (1.5.2008)

Proactive. (2005). *Proactive manual*, rev.ed. 2.2. Proactive, INRIA. Retrieved from http://www-sop.inria.fr / oasis/Proactive/

Purtilo, J. M. (1994). The POLYLITH Software Bus. *ACM Transactions on Programming Languages and Systems, 16*(1), 151–174. doi:10.1145/174625.174629

Qiang, Q., Jacob, L., Radhakrishna Pillai, R., & Prabhakaran, B. (2002). MAC Protocol Enhancements for QoS Guarantee and Fairness over the IEEE 802.11 Wireless LAN. *Proceeding of the Conference on Computer Communication Network (ICCNC).*

Rajendran, V., Obraczka, K., & Garcia-Luna-Aceves, J. J. (2003). Energy-Efficient, Collision-Free Medium Access Control for Wireless Sensor Networks. *Proceedings of the First ACM Conference on Embedded Networked Sensor Systems (SenSys).*

Ramabhadran, S., Ratnasamy, S., Hellerstein, J. M., & Shenker, S. (2004). *Prefix hash tree – an indexing data structure over distributed hash tables.* In PODC'04: 23rd Annual ACM Symposium on Principles of Distributed Computing.

Ranganathan, K., & Foster, I. (2003). Computation scheduling and data replication algorithms for data Grids. In Nabrzysk, J., Schopf, J., Weglarz, J., Nabrzysk, J., Schopf, J., & Weglarz, J. (Eds.), *Grid resource management: State of the art and future trends* (pp. 359–373). Kluwer Academic Publishers Group.

Ratnasamy, S., Francis, P., Handley, M., Karp, R., & Shenker, S. (2001). *A Scalable Content Addressable Network.* Paper presented at the Proceedings of ACM SIGCOMM.

Raza, Z., & Vidyarthi, D. P. (2009). GA based scheduling model for computational Grid to minimize turnaround time. *International Journal of Grid and High Performance Computing, 1*(4), 70–90. doi:10.4018/jghpc.2009070806

Raza, Z., & Vidyarthi, D. P. (2008). *Maximizing reliability with task scheduling in a computational Grid.* Second International Conference on Information Systems Technology and Management(ICISTM), Dubai, UAE.

RDF. http://www.w3.org/RDF. World Wide Web Consortium: Resource Description Framework. RDFStore. http://rdfstore.sourceforge.net.

RDQL. http://www.w3.org/Submission/2004/SUBM-RDQL-20040109/.

Reddy, M. V., Srinivas, A. V., Gopinath, T., & Janakiram, D. (2006). Vishwa: A Reconfigurable P2P Middleware for Grid Computations. In *Proc. of the 2006 International Conference on Parallel Processing (ICPP 2006)*. IEEE Press.

Ren, X., & Eigenmann, R. (2006). Empirical studies on the behavior of resource availability in fine-grained cycle sharing systems. In *Proceedings of 2006 International Conference on Parallel Processing*, (pp. 3-11).

Resch, M., Rantzau, D., & Stoy, R. (1999). Metacomputing Experience in a Transatlantic Wide Area Application Test bed. *Future Generation Computer Systems*, *5*(15), 807–816. doi:10.1016/S0167-739X(99)00028-X

Rhee, I., Warrier, A., Aia, M., & Min, J. (2005). ZMAC: a Hybrid MAC for Wireless Sensor Networks. *Proceedings of the 3rd international conference on Embedded networked sensor systems* (pp. 90-101).

Richmond, M., & Hitchens, M. (1997). A new process migration algorithm. *ACM SIGOPS Operating Systems Review*, *31*(1), 31–42. doi:10.1145/254784.254790

Ricklefs, R. E., Fallon, S. M., & Birmingham, E. (2004). Evolutionary relationships, cospeciation, and host switching in avian malaria parasites. *Systematic Biology*, *53*(1), 111–119. doi:10.1080/10635150490264987

Risson, J., & Moors, T. (2006). Survey and research towards robust peer-to-peer networks: Search methods. *Computer Networks*, *50*(17), 3485–3521. doi:10.1016/j.comnet.2006.02.001

Ronquist, F., & Huelsenbeck, J. (2003). MrBayes 3: Bayesian phylogenetic inference under mixed models. *Bioinformatics (Oxford, England)*, *19*(12), 1572–1574. doi:10.1093/bioinformatics/btg180

Rood, B., & Lewis, M. (2007). Multi-state grid resource availability characterization. In *Proceedings of 8th IEEE/ACM International Conference on Grid Computing*, (pp. 42-49).

Rosado, D. G., Fernández-Medina, E., & López, J. (2009b). Obtaining security requirements for a mobile Grid system. *International Journal of Grid and High Performance Computing*, *1*(3), 1–17. doi:10.4018/jghpc.2009070101

Rosado, D. G., Fernández-Medina, E., & López, J. (2011a). Towards an UML extension of reusable secure use cases for mobile Grid systems. *IEICE Transactions on Information and Systems*, *94-D*(2), 243–254.

Rosado, D. G., Fernández-Medina, E., & López, J. (2011b). Security services architecture for secure mobile Grid systems. *Journal of Systems Architecture. Special Issue on Security and Dependability Assurance of Software Architectures*, *57*(3), 240–258.

Rosado, D. G., Fernández-Medina, E., López, J., & Piattini, M. (2010a). Analysis of secure mobile Grid systems: A systematic approach. *Information and Software Technology*, *52*, 517–536. doi:10.1016/j.infsof.2010.01.002

Rosado, D. G., Fernández-Medina, E., López, J., & Piattini, M. (2010b). Developing a secure mobile Grid system through a UML extension. *Journal of Universal Computer Science*, *16*(17), 2333–2352.

Rosado, D. G., Fernández-Medina, E., López, J., & Piattini, M. (2011). (in press). Systematic design of secure mobile Grid systems. *Journal of Network and Computer Applications*. doi:10.1016/j.jnca.2011.01.001

Rosado, D. G., Fernández-Medina, E., & López, J. (2009a). *Applying a UML extension to build use cases diagrams in a secure mobile Grid application.* Paper presented at the 5th International Workshop on Foundations and Practices of UML, in conjunction with the 28th International Conference on Conceptual Modelling, ER 2009, Gramado, Brasil.

Rosado, D. G., Fernández-Medina, E., & López, J. (2009c). *Reusable security use cases for mobile Grid environments.* Paper presented at the Workshop on Software Engineering for Secure Systems, in conjunction with the 31st International Conference on Software Engineering, Vancouver, Canada.

Rosado, D. G., Fernández-Medina, E., López, J., & Piattini, M. (2008). *PSecGCM: Process for the development of secure Grid computing based systems with mobile devices.* Paper presented at the International Conference on Availability, Reliability and Security (ARES 2008), Barcelona, Spain.

Rowstron, A. (1998). WCL: A Co-ordination Language for Geographically Distributed Agents. *World Wide Web (Bussum)*, *1*(3), 167–179. doi:10.1023/A:1019263731139

Rowstron, A., & Druschel, P. (2001). Pastry: Scalable. Distributed Object Location and Routing for Large-scale Peer-to-Peer Systems. *Lecture Notes in Computer Science*, *2218*, 161–172.

Saara Väärtö, S. (Ed.). (2008). *Advancing science in Europe*. DEISA – Distributed European Infrastructure for Supercomputing Applications. EU FP6 Project. Retrieved from www.deisa.eu/press/ DEISA-AdvancingScience InEurope.pdf

Sabin, G., Lang, M., & Sadayappan, P. (2007). Moldable parallel job scheduling using job efficiency: An iterative approach. In *Proceedings of the Conference on Job Scheduling Strategies for Parallel Processing* (pp. 94-114).

SAGA. (2006). *SAGA implementation homepage*. Retrieved from http://fortytwo.cct.lsu.edu:8000/SAGA

Sajjad, A., Jameel, H., Kalim, U., Han, S. M., Lee, Y.-K., & Lee, S. (2005). *AutoMAGI - an autonomic middleware for enabling mobile access to Grid infrastructure.* Paper presented at the Joint International Conference on Autonomic and Autonomous Systems and International Conference on Networking and Services - (icas-icns'05).

Salzberg, S. L., Kingsford, C., & Cattoli, G. (2007). Genome analysis linking recent European and African influenza (H5N1) viruses. *Emerging Infectious Diseases*, *13*(5), 713–718.

Sarkar, P., & Hartman, J. (1996). *Efficient cooperative caching using hints.* Paper presented at the Symposium on Operating Systems Design and Implementation, Seattle, WA.

Saroiu, S., Gummadi, P., & Gribble, S. (2002). *A Measurement Study of Peer-to-Peer File Sharing Systems*. Paper presented at the Proceedings of Multimedia Computing and Networking.

Sathya, S. S., Kuppuswami, S., & Ragupathi, R. (2006). *Replication strategies for data Grids.* International Conference on Advanced Computing and Communications ADCOM, India (pp. 123-128).

Scarfone, K., & Mell, P. (2009) An analysis of CVSS version 2 vulnerability scoring. *Proceedings of the 3rd. Int'l Symposium on Empirical Software Engineering and Measurement (ESEM'09),* (pp. 516-525).

Schiffers, M., Ziegler, W., Haase, M., Gietz, P., Groeper, R., Pfeiffenberger, H., et al. Grimm, C. (2007). Trust issues in Shibboleth-enabled federated Grid authentication and authorization infrastructures supporting multiple Grid middleware. In *Proceedings of IEEE eScience 2007 and International Grid Interoperability Workshop 2007 (IGIIW 2007)*. IEEE Computer Society.

Schmidt, C., & Parashar, M. (2004). Enabling flexible queries with guarantees in P2P systems. *IEEE Internet Computing*, *8*(3), 19–26. doi:10.1109/MIC.2004.1297269

Schroeder, B., & Gibson, G. A. (2006). A large-scale study of failures in high-performance computing systems. In *Proceedings of the International Conference on Dependable Systems and Networks (DSN06)*, (pp. 249-258).

Seymour, K., Nakada, H., Matsuoka, S., Dongarra, J., Lee, C., & Casanova, H. (2002). Overview of GridRPC: A remote procedure call API for Grid computing. *Proceedings of the Third International Workshop on Grid Computing* [Baltimore, MD: Springer.]. *Lecture Notes in Computer Science*, *2536*, 274–278. doi:10.1007/3-540-36133-2_25

Sgaravatto, M. (2005). *CEMon Service Guide*. from https://edms.cern.ch/document/585040.

SGI. (2010). *Cyclone: HPC cloud results on demand.* Retrieved from http://www.sgi.com/products/hpc_cloud/ cyclone /index.htm

Shah, R., Veeravalli, B., & Misra, M. (2007). On the design of adaptive and decentralized load balancing algorithms with load estimation for computational Grid Environments. *IEEE Transactions on Parallel and Distributed Systems*, *18*(12), 1675–1686. doi:10.1109/ TPDS.2007.1115

Shan, J., Chen, G., He, J., & Chen, X. (2002). *Grid society: A system view of Grid and P2P environment.* International Workshop on Grid and Cooperative Computing, (pp. 19-28).

Sharmin, M., Ahmed, S., & Ahamed, S. I. (2006). *An Adaptive Lightweight Trust Reliant Secure Resource Discovery for Pervasive Computing Environments*. Proceedings of the Fourth Annual IEEE International Conference on Pervasive Computing and Communications, March, 258-263.

Shi, X., Yang, Z., Peir, J.-K., Peng, L., Chen, Y.-K., Lee, V., et al. (2006). *Coterminous locality and coterminous group data prefetching on chip-multiprocessors.* Paper presented at the 20th International Parallel and Distributed Processing Symposium, Rhodes Island, Greece.

Shudo, K., Tanaka, Y., & Sekiguchi, S. (2005). *P3: P2P-based middleware enabling transfer and aggregation of computational resources*. IEEE International Symposium on Cluster Computing and the Grid, (pp. 259- 266).

SIMDAT. (2008). *Grids for industrial product development*. Retrieved from www.scai.fraunhofer.de / about_simdat.html

Smarr, L., & Catlett, C. E. (1992). Metacomputing. *Communications of the ACM*, *35*(6), 44–52. doi:10.1145/129888.129890

So, J., & Vaidya, N. (2004). Multi-Channel MAC for Ad Hoc Networks: Handling Multi-Channel Hidden Terminals Using A Single Transceiver. *Proceedings of the 5th ACM international symposium on Mobile ad hoc networking and computing* (pp. 222 – 233).

Soh, H., Shazia Haque, S., Liao, W., & Buyya, R. (2006). Grid programming models and environments. In Dai, Y.-S. (Eds.), *Advanced parallel and distributed computing* (pp. 141–173). Nova Science Publishers.

Song, S., Kwok, Y. K., & Hwang, K. (2005). *Trusted Job Scheduling in Open Computational Grids: Security-Driven Heuristics and A Fast Genetic Algorithms*. Proceedings of International Symposium Parallel and Distributed Processing, Denver, Colorado.

Sonmez, O., Mohamed, H., & Epema, D. (2010). On the benefit of processor coallocation in multicluster Grid systems. *IEEE Transactions on Parallel and Distributed Systems*, (June): 778–789. doi:10.1109/TPDS.2009.121

Sotomayor, B., & Childers, L. (2006). *Globus toolkit 4 - programming Java services*. Morgan Kaufmann Publishers.

Spantzel, A., Squicciarini, A., & Bertino, E. (2005). *Integrating federated digital identity management and trust negotiation*. (Report No. 2005-46). Purdue University.

Srividya, S., Vijay, S., Rajkumar, K., Praveen, H., & Sadayappan, P. (2002). *Effective selection of partition sizes for moldable scheduling of parallel jobs*. Paper presented at the 9th International Conference on High Performance Computing.

Stäber, F. (2009). *Service layer components for decentralized applications*. Doctoral Dissertation at the Clausthal University of Technology

Stamatakis, A. (2006). RAxML-VI-HPC: maximum likelihood-based phylogenetic analyses with thousands of taxa and mixed models. *Bioinformatics (Oxford, England)*, *22*(21), 2688–2690. doi:10.1093/bioinformatics/btl446

Stamatakis, A., Auch, A. F., Meier-Kolthoff, J., & Göker, M. (2007). AxPcoords & parallel AxParafit: statistical co-phylogenetic analyses on thousands of taxa. *BMC Bioinformatics*, *8*, 405. doi:10.1186/1471-2105-8-405

Stamatakis, A., Hoover, P., & Rougemont, J. (2008). (in press). A Rapid Bootstrapping Algorithm for the RAxML Web Servers. *Systematic Biology*. doi:10.1080/10635150802429642

Steel, C., Nagappan, R., & Lai, R. (2005). Chapter 8-the alchemy of security design methodology, patterns, and reality checks. In *Core security patterns: Best practices and strategies for J2EE™, Web services, and identity management* (pp. 10-88). Prentice Hall PTR/Sun Micros.

Sterck, H. D., Markel, R. S., & Knight, R. (2005). A Lightweight, Scalable Grid Computing Framework for Parallel Bioinformatics Applications. In *HPCS'05: Proc. of the 19th International Symposium on High Performance Computing Systems and Applications*. IEEE Press.

Sterck, H. D., Markel, R. S., Pohl, T., & Rüede, U. (2003). A Lightweight Java Taskspaces Framework for Scientific Computing on Computational Grids. In *SAC2003: Proc. of the ACM Symposium on Applied Computing*, (pp. 1024–1030). New York, NY, USA: ACM Press.

Stevens, W. R. (1998). *Unix Network Programming: Networking APIs: Sockets and XTI, 1* (2nd ed.). Prentice-Hall PTR.

Stiefel, P. D., & Müller, J. P. (2010). A model-based software architecture to support decentral product development processes. In: *Exploring the grand challenges for next generation e-business. Proceedings of the 8th Workshop on eBusiness* (Web 2009). *Volume 52 of Lecture Notes in Business Information Processing*. Springer-Verlag, 2010. To appear.

Stockinger, H., Pagni, M., Cerutti, L., & Falquet, L. (2006). Grid Approach to Embarrassingly Parallel CPU-Intensive Bioinformatics Problems. *2nd IEEE International Conference on e-Science and Grid Computing (e-Science 2006)*, IEEE Computer Society Press, Amsterdam, The Netherlands.

Stoica, I., Morris, R., Karger, D., Kaashoek, M. F., & Balakrishnan, H. (2001). Chord: A scalable peer-to-peer lookup service for internet applications. In SIGCOMM'01: *Proc. of the 2001 Conference on Applications, Technologies, Architectures, and Protocols for Computer Communications*, (pp. 149–160). San Diego, CA: ACM Press.

Streit, A., Bergmann, S., Breu, R., Daivandy, J., Demuth, B., & Giesler, A. … Lippert, T. (2009). UNICORE 6, a European Grid technology. In W. Gentzsch, L. Grandinetti, & G. Joubert (Eds.), High-speed and large scale scientific computing, (pp. 157-176). IOS Press.

Stutzbach, D., & Rejaie, R. (2006). Understanding churn in peer-to-peer networks. In IMC'06: *Proc. of the 6th ACM SIGCOMM on Internet Measurement*, (pp. 189–202). New York, NY: ACM Press.

Subrata, R., Zomaya, A. Y., & Landfeldt, B. (2008). Game-theoretic approach for load balancing in computational Grids. *IEEE Transactions on Parallel and Distributed Systems*, *19*(1), 66–76. doi:10.1109/TPDS.2007.70710

Sudha, S., Savitha, K., & Sadayappan, P. (2003). *A robust scheduling strategy for moldable scheduling of parallel jobs*.

Sun. (2010). *Sun Network.com, SunGrid, and Sun utility computing, now under Oracle*. Retrieved from www.sun.com/service/sungrid/

SURA Southeastern Universities Research Association. (2007). *The Grid technology cookbook. Programming concepts and challenges*. Retrieved from www.sura.org/cookbook/gtcb/

Sweeney, L. (2002). K-anonymity: A model for protecting privacy. *International Journal of Uncertainty. Fuzziness and Knowledge-Based Systems*, *10*(5), 557–570. doi:10.1142/S0218488502001648

SWITCH. (2008). *SWITCH pki, an X.509 public key infrastructure for the Swiss higher education system*. Retrieved from http://www.switch.ch/pki/

Syslog-ng Logging System. from http://www.balabit.com/network-security/syslog-ng/.

Taiwan UniGrid. (n.d.). Retrieved October 13, 2009, from http://www.unigrid.org.tw/index.html.

Talukder, A., & Yavagal, R. (2006). Security issues in mobile computing. In *Mobile computing*. McGraw-Hill Professional.

Tanenbaum, A. S. (2007). *Modern operating systems* (3rd ed.). Prentice Hall.

Tanenbaum, A. S., & Steen, M. V. (2007). *Distributed Systems: Principles and Paradigms*. Pearson Prentice Hall, 2 edition.

Tang, Z., Birdwell, J. D., & Chiasson, J. (2008). Resource-constrained load balancing controller for a parallel database. *IEEE Transactions on Control Systems Technology*, *16*(4), 834–840. doi:10.1109/TCST.2007.916305

Tang, C. Q., Xu, Z. C., & Dwarkadas, S. (2003). *Peer-to-Peer Information Retrieval Using Self-Organizing Semantic Overlay Networks*. Paper presented at the Proceedings of ACM SIGCOMM 2003, Karlsruhe, Germany.

Tarricone, L., & Esposito, A. (2005). *Grid computing for electromagnetics*. Artech house Inc.

Taura, K., Kaneda, K., Endo, T., & Yonezawa, A. (2003). Phoenix: A Parallel Programming Model for Accommodating Dynamically Joining/Leaving Resources. In *PPoPP '03: Proc. of the Ninth ACM SIGPLAN Symposium on Principles and Practice of Parallel Programming*, (pp. 216–229), New York, NY, USA: ACM Press.

TAVERNA. (2008). *The Taverna workbench* 1.7. Retrieved from http://taverna.sourceforge.net/

Teragrid. from http://www.teragrid.org/.

Thain, D., Tannenbaum, T., & Livny, M. (2005). Distributed computing in practice: The Condor experience. *Concurrency and Computation*, *17*(2-4), 323–356. doi:10.1002/cpe.938

Thain, D., Tannenbaum, T., & Livny, M. (2003). Condor and the grid. In Berman, F., Fox, G., & Hey, A. (Eds.), *Grid computing: Making the global infrastructure a reality* (pp. 299–335). New York, NY, USA: John Wiley & Sons Inc.

The DataGrid Project. (n.d.). Retrieved from http://eu-datagrid.web.cern.ch /eu-datagrid/

Thomas, P. L., & Menzies, J. G. (1997). Cereal smuts in Manitoba and Saskatchewan, 1989-95. *Canadian Journal of Plant Pathology*, *19*(2), 161–165. doi:10.1080/07060669709500546

Thompson, J. D., Higgins, D. G., & Gibson, T. J. (1994). CLUSTAL W: improving the sensitivity of progressive multiple sequence alignment through sequence weighting, position-specific gap penalties and weight matrix choice. *Nucleic Acids Research*, *22*(22), 4673–4680. doi:10.1093/nar/22.22.4673

Tian, J., & Dai, Y. (2007). *Understanding the dynamic of peer-to-peer systems*. In Sixth International Workshop on Peer-to-Peer Systems (IPTPS2007).

Tiang, H. (2003). *Grid Computing as an Integrating Force in Virtual Enterprises*. Master's Thesis, MIT, Cambridge, MA.

Tierney, B. L., Gunter, D., & Schopf, J. M. (2007). *The CEDPS Troubleshooting Architecture and Deployment on the Open Science Grid*. J. Phys.: Conf. Ser. 78 012075, SciDAC 2007. Virtual Data Toolkit (VDT). from http://www.cs.wisc.edu/vdt/.

Tonellotto, N., Yahyapour, R., & Wieder P. H.(2006). A Proposal for a Generic Grid Scheduling Architecture. *Core GRID TR-0025.*

TOP500 List. (2008). http://www.top500.org/ (1.5.2008).

Travostino, F., Daspit, P., Gommans, L., Jog, C., de Laat, C. T. A. M., & Mambretti, J. (2006). *S*eamless live migration of virtual machines over the man/wan. *Future Generation Computer Systems*, *22*(8), 901–907. doi:10.1016/j.future.2006.03.007

TRIANA. (2003). *The Triana project*. Retrieved from www.trianacode.org/

Trujillo, J., Soler, E., Fernández-Medina, E., & Piattini, M. (2009). An engineering process for developing secure data warehouses. *Information and Software Technology*, *51*(6), 1033–1051. doi:10.1016/j.infsof.2008.12.003

Tschantz, M. C., & Krishnamurthi, S. (2006). Towards reasonability properties for access-control policy languages. In *Proceedings of SACMAT 2006*. ACM Press.

Tschumperlé, D., & Deriche, R. (2003). Vector-valued image regularization with PDE's: A common framework for different applications. *In IEEE Conference on Computer Vision and Pattern Recognition (CVPR '03), Madison, WI, USA, 1*, 651-656. IEEE Computer Society.

Tzamaloukas, A., & Garcia-Luna-Aceves, J. J. (2001). A Receiver-Initiated Collision-Avoidance Protocol for Multi-Channel Networks. *Proceedings of the 20th IEEE INFOCOM* (pp. 189-198).

U.S. Congress (1996). *Health insurance portability and accountability act,* 1996.

U.S. Safe Harbor Framework. (n.d.). Retrieved from http://www.export.gov/safeharbor/

UNICORE. (2008). *Uniform interface to computing resources*. Retrieved from www.unicore.eu/

UPnP forum. (2008). *Universal plug and play specifications.* Retrieved from http://www.upnp.org/resources/

Vaidya, N. H. (2002). *Weak Duplicate Address Detection in Mobile Ad Hoc Networks.* MIBIHOC2002, June.

van Heiningen, W., MacDonald, S., & Brecht, T. (2008). Babylon: middleware for distributed, parallel, and mobile Java applications. *Concurrency and Computation*, *20*(10), 1195–1224. doi:10.1002/cpe.1264

van Nieuwpoort, R. V., Maassen, J., Wrzesinska, G., Hofman, R. F. H., Jacobs, C. J. H., Kielmann, T., & Bal, H. E. (2005). Ibis: a Flexible and Efficient Java-based Grid Programming Environment. *Concurrency and Computation*, *17*(7–8), 1079–1107. doi:10.1002/cpe.860

Van 't Noordende, G. J., Brazier, F. M. T., & Tanenbaum, A. S. (2004). *Security in a mobile agent system.* 1st IEEE Symp. on Multi-Agent Security and Survivability, Philadelphia.

Van 't Noordende, G., Balogh, A., Hofman, R., Brazier, F. M. T., & Tanenbaum, A. S. (2007). *A secure jailing system for confining untrusted applications*. 2nd Int'l Conf. on Security and Cryptography (SECRYPT), (pp. 414-423). Barcelona, Spain.

Vanderwiel, S. P., & Lilja, D. J. (2000). Data prefetch mechanisms. *ACM Computing Surveys*, *32*(2), 174–199. doi:10.1145/358923.358939

Venugopal, S., Buyya, R., & Winton, L. (2004). A Grid service broker for scheduling distributed data-oriented applications on global grids. *Proceedings of the 2nd workshop on Middleware for Grid computing*, (pp. 75–80). Toronto, Canada. Retrieved from www.Gridbus.org/broker

Verbeke, J., Nadgir, N., Ruetsch, G., & Sharapov, I. (2002). Framework for peer-to-peer distributed computing in a heterogeneous, decentralized environment. In *Proceedings of Third International Workshop on Grid Computing*, (pp. 1-12).

Vidyarthi, D. P., Sarker, B. K., Tripathi, A. K., & Yang, L. T. (2009). *Scheduling in distributed computing systems*. Springer. doi:10.1007/978-0-387-74483-4

Vincze, G., Novák, Z., Pap, Z., & Vida, R. (2008). *RESERV: A distributed, load balanced Information System for Grid applications*. 8th IEEE International Symposium on Cluster Computing and the Grid, (pp. 596-601).

Vishwanath, V., Burns, R., Leigh, J., & Seablom, M. (2008). Accelerating tropical cyclone analysis using LambdaRAM, a distributed data cache over wide-area ultra-fast networks. *Future Generation Computer Systems*, *25*(2), 184–191. doi:10.1016/j.future.2008.07.005

Vivas, J. L., López, J., & Montenegro, J. A. (2007). Grid security architecture: Requirements, fundamentals, standards, and models. In Xiao, Y. (Ed.), *Security in distributed, Grid, mobile, and pervasive computing* (p. 440). Tuscaloosa, USA.

Voelker, G. M., Anderson, E. J., Kimbrel, T., Feeley, M. J., Chase, J. S., Karlin, A. R., et al. (1998). *Implementing co-operative prefetching and caching in a globally-managed memory system.* Paper presented at the Joint International Conference on Measurement and Modeling of Computer Systems, Madison, Wisconsin, United States.

von Laszewski, G., Gawor, J., Lane, P., Rehn, N., & Russell, M. (2003). Features of the Java Commodity Grid Kit. *Concurrency and Computation*, *14*(13-15), 1045–1055. doi:10.1002/cpe.674

Walfredo, C., & Francine, B. (2002). Using moldability to improve the performance of supercomputer jobs. *Journal of Parallel and Distributed Computing*, *62*(10), 1571–1601.

Walfredo, C., & Francine, B. (2000). *Adaptive selection of partition size for supercomputer requests.* Paper presented at the Workshop on Job Scheduling Strategies for Parallel Processing.

Walker, E. (2008). *Benchmarking Amazon EC2 for high-performance scientific computing*. Retrieved from http://www.usenix.org/ publications/login/ 2008-10/openpdfs/walker.pdf

Walls, C., & Breidenbach, R. (2005). *Spring in action*. Greenwich, Connecticut, USA: Manning Publications Co.

Walton, J., & Whicker, L. (1996) Virtual Enterprise: Myth and Reality. *Journal of Control*, (pp. 22-25).

Wang, H., Takizawa, H., & Kobayashi, H. (2007). A dependable peer-to-peer computing platform. *Future Generation Computer Systems*, *23*(8), 939–955. doi:10.1016/j.future.2007.03.004

Wang, N., Liu, X., He, J., Han, J., Zhang, L., & Xu, Z. (2007). *Collaborative memory pool in cluster system*. Paper presented at the International Conference on Parallel Processing.

Wason, T. (Ed.). (2004). *Liberty identity federation framework ID-FF architecture overview*. Liberty Alliance Specification. Retrieved from http://www.project-liberty.org/

Weiler, S., & Ihren, J. (2006). *Minimally covering NSEC records and DNSSEC online signing. Internet Engineering Task Force*. IETF.

Welch, V., Barton, T., Keahey, K., & Siebenlist, F. (2005). Attributes, anonymity, and access: Shibboleth and Globus integration to facilitate Grid collaboration. In *Proceedings of the Internet2 PKI R&D Workshop*.

Welch, V., Siebenlist, F., Foster, I., Bresnahan, J., Czajkowski, K., Gawor, J., et al. (2003). *Security for Grid services*. Paper presented at the 12th IEEE International Symposium on High Performance Distributed Computing (HPDC-12 '03).

Weniger, K., & Zitterbart, M. (2004). Mobile ad hoc networks – current approaches and future directions. *Network, IEEE, 18*(4), 6–11. doi:10.1109/MNET.2004.1316754

WGBO. (1994). *Dutch ministry of health, welfare and sport – WGBO*. Retrieved from http://www.hulpgids.nl/wetten/wgbo.htm

Witten, I. H., & Frank, E. (2005). *Data mining: Practical machine learning tools and techniques* (2nd ed.). Morgan Kaufmann.

WMO. (1998). *Dutch ministry of health, welfare and sport - WMO.* Retrieved from http://www.healthlaw.nl/wmo.html.

World Community Grid. (2008). http://www.worldcommunitygrid.org/ (1.5.2008).

Wrzesinska, G., van Nieuwport, R., Maassen, J., Kielmann, T., & Bal, H. (2006). Fault-tolerant scheduling of fine-grained tasks in Grid environments. *International Journal of High Performance Computing Applications, 20*(1), 103–114. doi:10.1177/1094342006062528

Wu, Y. J., Lin, S. J., Lai, K. C., Huang, K. C., & Wu, C. C. (2008). *Distributed dynamic load balancing strategies in P2P Grid systems*. 5th Workshop on Grid Technologies and Applications, (pp. 95-102).

Wyckoff, P., McLaughry, S. W., Lehman, T. J., & Ford, D. A. (1998). T Spaces. *IBM Systems Journal, 37*(3), 454–474. doi:10.1147/sj.373.0454

Xia, Y., Chen, S., & Korgaonkar, V. (2006). *Load balancing with multiple hash functions in peer-to-peer networks*. IEEE 12th International Conference on Parallel and Distributed Systems, (pp. 411-420).

Xie, T., & Qin, X. (2007). Performance Evaluation of a New Scheduling Algorithm for Distributed Systems with Security Heterogeneity. *Journal of Parallel and Distributed Computing, 67*, 1067–1081. doi:10.1016/j.jpdc.2007.06.004

Xu, L. (2005). *Hydra: A platform for survivable and secure data storage systems*. ACM StorageSS.

Xu, J., Kalbarczyk, Z., & Iyer, R. (1999). Networked Windows NT system field failure data analysis. In *Proceedings of 1999 Pacific Rim International Symposium on Dependable Computing*, (pp. 178-185).

Xu, Z., & Bhuyan, L. (2006). *Effective load balancing in P2P systems*. 6th IEEE International Symposium on Cluster Computing and the Grid, (pp. 81-88).

Yang, C.-L., Lebeck, A. R., Tseng, H.-W., & Lee, C.-H. (2004). Tolerating memory latency through push prefetching for pointer-intensive applications. *ACM Transactions on Architecture and Code Optimization, 1*(4), 445–475. doi:10.1145/1044823.1044827

Yang, C. T., Li, C. T., Chiang, W. C., & Shih, P. C. (2005). *Design and implementation of TIGER Grid: An integrated metropolitan-scale Grid environment*. 6th International Conference on Parallel and Distributed Computing Applications and Technologies, (pp. 518-520).

Yanmin, Z., Jinsong, H., Yunhao, L., & Ni, L. M. Chunming, H., & Jinpeng, H. (2005). *TruGrid: A self-sustaining trustworthy Grid.* Paper presented at the 25th IEEE International Conference on Distributed Computing Systems Workshops, 2005.

Ying, Z., Ananda, A. L., & Jacob, L. (2003). A QoS Enabled MAC Protocol for Multi-Hop Ad Hoc Wireless Networks. *Proceeding of IEEE International Conference on Performance, Computing, and Communications* (IPCCC).

Zhang, X., Freschl, J., & Schopf, J. (2007). Scalability analysis of three monitoring and information systems: MDS2, R-GMA, and Hawkeye. *Journal of Parallel and Distributed Computing, 67*(8), 883–902. doi:10.1016/j.jpdc.2007.03.006

Zhang, J., & Honeyman, P. (2008). *Performance and availability tradeoffs in replicated file systems.* Eighth IEEE International Symposium on Cluster Computing and the Grid, Lyon, France (pp. 771-776).

Zhang, J., Zhou, G., Huang, C., Son, S. H., & Stankovic, J. A. (2007). TMMAC: An Energy Efficient Multi- Channel MAC Protocol for Ad Hoc Networks. *Proceedings of IEEE International Conference on Communications* (pp. 24-28).

Zhang, Y., & Dao, S. (1995). A 'Persistent Connection' Model for Mobile and Distributed Systems. In *ICCCN '95: Proc. of the 4th International Conference on Computer Communications*, (pp. 300–307). IEEE Computer Society.

Zhang, Y., Li, D., Chu, R., Xiao, N., & Lu, X. (2007). *PIBUS: A network memory-based peer-to-peer IO buffering service*. Paper presented at the 6th International IFIP-TC6 Conference on Ad Hoc and Sensor Networks, Wireless Networks, Next Generation Internet.

Zhao, B. Y., Huang, L., Stribling, J., Rhea, S. C., Joseph, A. D., & Kubiatowicz, J. D. (2004). Tapestry: A Resilient Global-scale Overlay for Service Deployment. *IEEE Journal on Selected Areas in Communications*, *22*(1), 41–53. doi:10.1109/JSAC.2003.818784

Zhong, X., & Xu, C.-Z. (2004). A Reliable Connection Migration Mechanism for Synchronous Transient Communication in Mobile Codes. In *Proc. of the 2004 International Conference on Parallel Processing*. IEEE Press.

Zhou, G., Huang, C., Yan, T., He, T., Stankovic, J. A., & Abdelzaher, T. (2006). MMSN: Multi-Frequency Media Access Control for Wireless Sensor Networks. *Proceedings of the 25th IEEE INFOCOM* (pp. 1-13).

Zhu, F., Mutka, M., & Mi, L. (2003). *Splendor: A secure, private, and location-aware service discovery protocol supporting mobile services* (pp. 235–242). Pervasive Computing and Communications.

Zwickl, D. (2006). Genetic algorithm approaches for the phylogenetic analysis of large biological sequence datasets under the maximum likelihood criterion. *PhD Thesis*, The University of Texas at Austin.

About the Contributors

Emmanuel Udoh is a Professor of Computer Science at the Indiana Institute of Technology, USA. He received his PHD degree in Information Technology and Master of Business Administration (MBA) degree from Capella University, USA. Moreover, he is also a PhD holder in Geology from the University of Erlangen, Germany.

* * *

David Abramson has been involved in computer architecture and high performance computing research since 1979. Previous to joining Monash University in 1997, he has held appointments at Griffith University, CSIRO, and RMIT. At CSIRO he was the program leader of the Division of Information Technology High Performance Computing Program, and was also an Adjunct Associate Professor at RMIT in Melbourne. He served as a program manager and chief investigator in the Co-operative Research Centre for Intelligent Decisions Systems and the Co-operative Research Centre for Enterprise Distributed Systems. Abramson is currently an ARC professorial fellow; Professor of Computer Science in the faculty of Information Technology at Monash University, Australia, and associate director of the Monash e-Research Centre. Abramson has served on committees for many conferences and workshops, and has published over 150 papers and technical documents. He has given seminars and received awards around Australia and internationally and has received over $3.6 million in research funding. He also has a keen interest in R&D commercialization and consults for Axceleon Inc, who produce an industry strength version of Nimrod, and Guardsoft, a company focused on commercializing the Guard relative debugger. Abramson's current interests are in high performance computer systems design and software engineering tools for programming parallel, distributed supercomputers.

Ashish Agarwal is currently a doctoral student at the Tepper School of Business, Carnegie Mellon University. He holds a Bachelor's degree in Materials Engineering from the Indian Institute of Technology, Bombay, India and a Master's in Engineering from the Massachusetts Institute of Technology. His research interests include distributed computing, wireless networks, and economics of Information Systems.

Alexander Auch works as freelance software developer and consultant for industry as well as academia. He has received a Master's degree in bioinformatics from the University of Tübingen in 2005, and is currently working on his doctoral thesis.

Marcelo Campo received a PhD degree in Computer Science from UFRGS, Porto Alegre, Brazil. He is a full Associate Professor at the Computer Science Department and Head of the ISISTAN. He is also a research fellow of the CONICET. He has over 70 papers published in conferences and journals about software engineering topics.

Philip Chan is a doctoral candidate under the faculty of Information Technology at Monash University. He is currently a project officer at the Research Support Services of the Information Technology Services Division, where he develops various solutions to assist scientific end-users of the Monash Campus Grid facility. He is on research leave-of-absence from the Software Technology Department, College of Computer Studies, De La Salle University, Manila, Philippines where he has served as an assistant professor since 1995. His research interests include parallel/distributed programming techniques, distributed shared memory, and concurrency theory. In 1995, he was a visiting fellow at the United Nations University/International Institute for Software Technology (UNU/IIST) where he developed duration calculus formal specifications for real-time schedulers. Chan is a member of ACM and EATCS.

Shreyas Cholia is a software engineer at the National Energy Research Scientific Computing Center (NERSC) of Lawrence Berkeley National Laboratory. He is primarily responsible for managing the grid and science gateway infrastructure at NERSC, in conjunction with the Open Science Grid. Cholia has been involved with various grid projects since 2002, including the Open Science Grid, HPSS-GridFTP integration project, the Grid-File-Yanker, SGE Gratia development, the NERSC Online CA, the OSG MPI integration effort, and the NERSC Web Toolkit. He has a Bachelor's degree from Rice University in Computer Science and Cognitive Sciences. Prior to his work at NERSC, Shreyas was a developer and consultant for IBM with the HPSS project.

Rui Chu was born in 1979. He received the B.S. degree in Computer Science from the National University of Defense Technology in 2001, and the M.S. degree and PhD degree in Computer Science from the National Laboratory for Parallel and Distributed Processing, in 2003 and 2008, respectively, and was employed as a one-year visiting scholar of the Hong Kong University of Science and Technology in 2007-2008. He achieved the Outstanding Winner in Mathematical Contest in Modeling held by COMAP in 2000, and was involved in several projects including network computing environment architecture, Grid software data resource management, Grid file system technology, et cetera. He is currently an Assistant Professor in National Laboratory for Parallel and Distributed Processing, where he mainly worked on the Internet-based virtual computing environment project. His research interests include parallel and distributed computing, cloud computing, real-time operating system for sensor network, and so on. He is a member of the ACM and China Computer Federation.

Yeh-Ching Chung received a B.S. degree in Information Engineering from Chung Yuan Christian University in 1983, and the M.S. and Ph.D. degrees in Computer and Information Science from Syracuse University in 1988 and 1992, respectively. He joined the Department of Information Engineering at Feng Chia University as an Associate Professor in 1992 and became a full professor in 1999. From 1998 to 2001, he was the chairman of the department. In 2002, he joined the Department of Computer Science at National Tsing Hua University as a Full Professor. His research interests include parallel and distributed processing, cluster systems, Grid computing, multi-core tool chain design, and multi-core embedded systems. He is a member of the IEEE Computer Society and ACM.

Messaoud Doudou is a PhD student at the University of Science and Technology Houari Boumediène (USTHB). He obtained a Master degree and an Engineer degree in Computer Science from the same university. He is also a research member at the Center of Research on Scientific and Technical Information (CERIST) in Algiers. His research interest includes mesh networks, security, sensor networks, and QoS.

Eduardo Fernández-Medina (Eduardo.fdezmedina@uclm.es) holds a PhD and an MSc in Computer Science from the University of Sevilla. He is associate Professor at the Escuela Superior de Informática of the University of Castilla-La Mancha at Ciudad Real (Spain), his research activity being in the field of security in databases, datawarehouses, Web services and Information Systems, and also in security metrics. Fernández-Medina is co-editor of several books and chapter books on these subjects, and has several dozens of papers in national and international conferences (DEXA, CAISE, UML, ER, etc.). Author of several manuscripts in national and international journals (Information Software Technology, Computers And Security, Information Systems Security, etc.), he leads the GSyA research group of the Information Systems and Technologies Department at the University of Castilla-La Mancha, in Ciudad Real, Spain. He belongs to various professional and research associations (ATI, AEC, ISO, IFIP WG11.3 etc.).

Edgar Gabriel is an Assistant Professor in the Department of Computer Science at the University of Houston, Texas, USA. He got his PhD and Dipl.-Ing. in Mechanical Engineering from the University of Stuttgart. His research interests are message passing systems, high performance computing, parallel computing on distributed memory machines, and Grid computing.

Wolfgang Gentzsch is dissemination advisor for the DEISA Distributed European Initiative for Supercomputing Applications. From 2008 to 2010, Dr. Gentzsch was a member of the Board of Directors of the Open Grid Forum standards organization. Before, he was the Chairman of the German D-Grid Initiative; managing director of MCNC Grid and Data Center Services in Durham; Adjunct Professor of Computer Science at Duke University; and visiting scientist at RENCI Renaissance Computing Institute at UNC Chapel Hill, North Carolina. During this time, he was also a member of the US President's Council of Advisors for Science and Technology, PCAST. In 2000, he joined Sun Microsystems in Menlo Park, CA, as the senior director of Grid Computing. Before, he was the President, CEO, and CTO of start-up companies Genias and Gridware, and a Professor of Mathematics and Computer Science at the University of Applied Sciences in Regensburg, Germany. Dr. Gentzsch studied mathematics and physics at the Technical Universities in Aachen and Darmstadt, Germany.

Markus Göker received his Diploma in Biology in July 1999 from the University of Heidelberg. In December 2003 he received his PhD for research on "Molecular and light microscopical investigations into the phylogeny of the obligate biotrophic Peronosporales" from the University of Tübingen. Since then he has been working as a postdoctoral researcher at the Institute of Organismic Botany/Mycology in Tübingen. His research interests include evolution, taxonomy and co-phylogenetic analyses of plant-parasitic fungi with a focus on downy mildews and smut fungi, and phylogenetic inference with alignment-free approaches, and from sequences with intra-individual variability. He has been particularly interested in compiling very large host-parasite datasets to conduct co-phylogenetic tests.

Tao Gu is currently an Assistant Professor at University of Southern Denmark. He received his PhD degree in Computer Science from National University of Singapore (NUS). His research interests involve ubiquitous and pervasive computing, wireless sensor networks and peer-to-peer computing. He published more than 30 journal and conference papers in these areas in the past five years. He frequently served as a technical committee member in many international conferences in ubiquitous/pervasive computing such as PERCOM and Mobiquitous.

Amar Gupta is Thomas R. Brown endowed Chair of Management and Technology; Professor of Entrepreneurship, Management Information Systems, management of organizations, and Computer Science; all at the University of Arizona. In addition, he is visiting Professor at MIT for part of the year. Earlier, he was with the MIT Sloan School of Management (1979-2004); for half of this 25-year period, he served as the founding co-director of the productivity from Information Technology (PROFIT) initiative. He has published over 100 papers, and serves as a founding associate editor of ACM Transactions on Internet Technology. His most recent book is an edited one and is entitled: Outsourcing and Offshoring of Professional Services: Business Optimization in a Global Economy. He holds a Bachelor's degree in Electrical Engineering from the Indian Institute of Technology, Kanpur, India; a Master's in Management from the Massachusetts Institute of Technology; and a Doctorate in Computer Science from the Indian Institute of Technology, Delhi, India. He is a senior member of IEEE and a member of ACM.

Wolfgang Hommel has a PhD in Computer Science from Ludwig Maximilians University, Munich, and heads the network services planning group at the Leibniz Supercomputing Centre (LRZ) in Germany. His current research focuses on IT security and privacy management in large distributed systems, including identity federations and Grids. Emphasis is put on a holistic perspective, i.e. the problems and solutions are analyzed from the design phase through software engineering, deployment in heterogeneous infrastructures, and during the operation and change phases according to IT service management process frameworks, such as ISO/IEC 20000-1. Being both a regional computing centre for higher education institutions with more than 100,000 users and a national supercomputing centre, the LRZ offers a plethora of real world scenarios and large projects to apply and refine the research results in practice.

Zoltán Horváth received his MSc in Mathematics, Physics and Computer Science in 1986 at Eötvös Loránd University (Budapest, Hungary). He received his PhD (title: "A Relational Model of Parallel Programs") in 1996 and completed his habilitation process (title: "Verification of Distributed Functional Programs") in 2004 at the same university. He is head of Department of Programming Languages and Compilers since 2003, and full Professor since 2008. Between 2007 and 2010, he was vice-dean for scientific affairs and international relations of Faculty of Informatics. Since 2010 he is vice-rector for international relations. He is the leader of the Budapest Associate Node of EIT ICT Labs.

Kuo-Chan Huang received his B.S. and Ph.D. degrees in Computer Science and Information Engineering from National Chiao-Tung University, Taiwan, in 1993 and 1998, respectively. He is currently an Assistant Professor in Computer and Information Science Department at National Taichung University, Taiwan. He is a member of ACM and IEEE Computer Society. His research areas include parallel processing, cluster and Grid computing, and workflow computing.

Po-Jung Huang received the BS and MS degree in Computer and Information Science from the National Taichung University, Taiwan. His research interests include P2P computing, Grid computing, load balancing, and cloud computing.

Hiroaki Kobayashi is currently a Director and Professor of Cyberscience Center and a Professor of Graduate School of Information Sciences, Tohoku University. His research interests include high-performance computer architectures and their applications. He received the B.E. Degree in Communication Engineering, and the M.E. and D.E. Degrees in Information Engineering from Tohoku University in 1983, 1985, and 1988 respectively. He is a senior member of IEEE CS, and a member of ACM, IEICE and IPSJ.

Matthijs R. Koot is a PhD-student in the System and Network Engineering group at the University of Amsterdam. His research interests include privacy and anonymity, and the application of privacy-enhancing technologies to Grid environments. His teaching interests include intrusion detection and honeypots.

Tamás Kozsik received his PhD (summa cum laude) in Computer Science in 2006 at Eötvös Loránd University (Budapest, Hungary), where he works as Associate Professor and vice-dean for scientific affairs and international relations of Faculty of Informatics. Since 1992 he has been teaching programming languages, as well as distributed and concurrent programming. His research fields are program analysis and verification, refactoring, type systems and distributed systems. His PhD thesis investigated the integration of logic-based and type system based verification of functional programs.

Gerald Kunzmann studied Electrical and Information Engineering at the Technische Universität München (TUM) in Munich, Germany with special focuses on information and communication technology. In his diploma thesis at the Institute of Communication Networks, he evaluated a distributed network monitoring service realized at the edges of the network. Following, he researched novel, distributed, and self-organizing routing protocols and network architectures for the Next Generation Internet. Then, he worked in several industrial projects; amongst others, he evaluated and enhanced a serverless VoIP communication architecture developed by Siemens AG. Within the scope of the BMBF funded G-Lab project, he developed a scalable, hierarchical Internet mapping architecture based on distributed hash tables. In his dissertation at the TUM, he evaluated the performance of structured overlay networks and developed novel concepts and algorithms for their optimized operation. Dr. Kunzmann is now working as researcher for NTT DOCOMO Communication Laboratories Europe GmbH in Munich, where he is evolving mobile services and networks for the realization of a society where everyone can live a safe, secure, and comfortable life, filled with richness, beyond borders, and across generations.

Cees de Laat is Professor and leader of the System and Network Engineering research group at the University of Amsterdam. Research in his group includes optical/switched networking for Internet transport of massive amounts of data in TeraScale eScience applications, Semantic web to describe networks and associated resources, distributed cross organization Authorization architectures and systems security, and privacy of information in distributed environments. He serves in the Open Grid Forum as IETF Liaison and is acting co-chair of the Grid High Performance Networking Research Group (GHPN-RG) and is chair of GridForum.nl and boardmember of ISOC.nl. He is co-founder and organizer of several of the past meetings of the Global Lambda Integrated Facility (GLIF) and founding member of CineGrid.org. http://www.science.uva.nl/~delaat

Kuan-Chou Lai received his MS degree in Computer Science and Information Engineering from the National Cheng Kung University in 1991, and the PhD degree in Computer Science and Information Engineering from the National Chiao Tung University in 1996. Currently, he is an Associate Professor in the Department of Computer and Information Science at the National Taichung University. His research interests include parallel processing, system architecture, P2P, cluster computing, Grid computing, and cloud computing. He is a member of the IEEE and the IEEE Computer Society.

Javier Lopez (jlm@lcc.uma.es) received his M.S. and PhD. degrees in Computer Science in 1992 and 2000, respectively, from the University of Malaga, where he currently is a Full Professor. His research activities are mainly focused on network security and critical information infrastructures, and he leads national and international research projects in those areas. He is also Co-Editor in Chief of Springer's International Journal of Information Security (IJIS), a member of the editorial boards of international journals, and the Spanish representative on the IFIP Technical Committee 11 on security and protection in Information Systems.

László Csaba Lőrincz received his MSc in Computer Science at Eötvös Loránd University (Budapest, Hungary). The title of his thesis was "Optimization of Data Access on Clusters and Data Grids -- Strategy, aspect, extension of JDL." László started his PhD studies in 2004 under the supervision of Zoltán Horváth. His research areas are distributed systems, data Grids, and multiagent systems. He has also been working at different IT companies since 2000.

Xicheng Lu was born in 1946. He received the B.S degree in Computer Science from Haerbin Military Engineering College in 1970, and was employed as a visiting scholar of the University of Massachusetts in 1982-1984. He is currently the Fellow of the Chinese Academy of Engineering, the Chief Director of National Laboratory for Parallel and Distributed Processing, and the Professor and Ph.D. supervisor of the National University of Defense Technology. His current research interests include computer architecture, computer networks, and parallel and distributed technology. He has published more than 100 papers in referred journals, conferences, and books including IEEE Transactions on Software Engineering, IEEE Transactions on Neural Networks, IEEE Transactions on Knowledge and Data Engineering, et cetera, and has served as the Program Chair or Program Committee Member of international conferences such as ICDCS, GCC, APPT, and ICCNMC. He is a senior member of China Computer Federation, and member of ACM and IEEE.

Cristian Mateos received a PhD degree in Computer Science from UNICEN, Tandil, Argentina, in 2008. He is a full Teacher Assistant at the computer science department of UNICEN and a research fellow of the CONICET. His recent thesis was on solutions to ease Grid application development and tuning through dependency injection and policies.

Jan Meier-Kolthoff is employed as a software developer in bioinformatics at a medium-sized biotech company in Bavaria, Germany. In March 2007 Jan received a Master's degree in Bioinformatics from Eberhard Karls Universität Tübingen. Despite his job-related occupation he is still highly interested and involved in scientific challenges.

Jörg P. Müller holds a Chair for Business Information Technology at Clausthal University of Technology; currently, he is Head of the Department of Informatics at CUT. Previously, Prof. Dr. Müller was Principal Researcher at Siemens AG, John Wiley & Sons, Zuno Ltd., Mitsubishi Electric, and the German Artificial Intelligence Research Center (DFKI). He holds a Ph.D. from Saarbrücken University and an M.Sc. in Computer Science from Kaiserslautern University. Within the last fifteen years, he has published over 160 papers on intelligent agents and multi-agent systems, business Information Systems, and distributed computing. His current research interests include methods, technologies and applications for multiagent systems, decentral (P2P) coordination and resource management, enterprise interoperability, and model-driven business process automation.

Yoshitomo Murata is currently a post-doctoral fellow in Cyberscience Center, Tohoku University. His research interests include distributed computing systems and their applications. He received the B.E. Degree in Mechanical Engineering, and the M.S. and Ph.D. Degrees in Information Sciences from Tohoku University in 2003, 2005 and 2008, respectively. He is a member of the IEEE CS.

Guido J. van 't Noordende is a researcher at the University of Amsterdam, The Netherlands. He is finishing his Ph.D. in Computer Science at the Vrije Universiteit Amsterdam while working in the System and Network Engineering group at the University of Amsterdam. His research interests include security and privacy in distributed systems, in particular for (bio)medical applications, and security of electronic medical record systems.

Silvia D. Olabarriaga is Assistant Professor of the Academic Medical Center of the University of Amsterdam. She leads the e-Bioscience group of the Bioinformatics Laboratory, which researches the design, development, deployment and evaluation of advanced infrastructures to enable and enhance biomedical research. Her main research interest lies on improving the usability of production e-infrastructures for the benefit of the biomedical researcher as end-user. Research topics include scientific workflows, data management, resource selection, monitoring, and fault tolerance. She also actively participates in various initiatives that promote the adoption of e-infrastructures for biomedical research, including the HealthGrid Association and the European Life Science Grid Community.

Massimiliano Pala joined Dartmouth College as a Post-Doctoral Research Fellow with the Computer Science department and the Institute for Security, Technology, and Society in 2007. He received his Ph.D. from the Poilitecnico di Torino in Computer Engineering in March 2007. In addition to his activity as Security and PKI consultant, he is actively involved in standardization bodies like IETF. In 1998, he started the OpenCA project and he still leads its development and management as Director of the OpenCA Labs. Today, his work is focused on usable security with special regards to Internet trust infrastructures.

Mario Piattini has an MSc and a PhD in Computer Science from the Politechnical University of Madrid. He is a Certified Information System Auditor from the ISACA (Information System Audit and Control Association). Full Professor at the Escuela Superior de Informática of the Castilla-La Mancha University (Spain), and author of several books and papers on databases, software engineering, and Information Systems, Piattini leads the ALARCOS research group of the Information Systems and Technologies Department at the University of Castilla-La Mancha, in Ciudad Real, Spain. His research interests are: advanced database design, database quality, software metrics, object- oriented metrics and software maintenance. His e-mail address is Mario.Piattini@uclm.es

R. Jefferson Porter has worked for many years as an experimental research Physicist in the field of relativistic heavy ion physics. During his career, he has participated in several scientific computing infrastructure projects that focused on challenges in data-intensive distributed computing. He is currently on staff with NERSC at LBNL as a member of the validation and integration team for the Open Science Grid.

Hung Keng Pung is an Associate Professor in the Department of Computer Science, National University of Singapore. He heads the Networks Systems and Services Laboratory as well as holding a joint appointment as a principal scientist at the Institute of Infocomm Research in Singapore. His areas of research are context-aware systems, service-oriented computing, quality of service management, protocols design, and networking.

Zahid Raza is currently an Assistant Professor in the School of Computer and Systems Sciences, Jawaharlal Nehru University, India. He has a Master degree in Electronics, Master's degree in Computer Science, and Ph.D. in Computer Science. Prior to joining Jawaharlal Nehru University, he served as a Lecturer in Banasthali Vidyapith University, Rajasthan, India. His research interest is in the area of Grid computing and has proposed a few models for job scheduling in a computational Grid. He is a member of IEEE.

Scott Rea joined DigiCert Inc. as the Sr. PKI architect in September 2009; he previously performed a similar role for Dartmouth College from May 2004, and still advises them in that capacity. Rea is also responsible for the implementation and operation of the Research & Education Bridge Certificate Authority (REBCA) which facilitates trust between disparate PKIs in the education and research fields. He received his MS from Queensland University of Technology in Information Security in March 1999, and has since been working in his field of discipline as a PKI/Security consultant/expert. Rea is a founding member and current Chair of the Americas Grid Policy Management Authority (TAGPMA) which is a member constituent of the International Grid Trust Federation (IGTF), and for which Rea is also currently serving as Chair. Rea is active in multiple initiatives for authorization and authentication services, including REBCA, the US Federal PKI, and the Four Bridges Forum (4BF).

Michael M. Resch is a Full Professor of High Performance Computing at the Department of Energy Technology, Process Engineering, and Biological Engineering of the University of Stuttgart in Germany. He received his Dipl.-Ing. (MSc) in Technical Mathematics from the Technical University of Graz, Austria, and his PhD in Engineering from the University of Stuttgart, Germany. His research interests include supercomputing architectures and software, Grid computing, and simulation in research and industrial development.

Imed Romdhani is a Lecturer in Networking in the School of Computing at Napier University in Edinburgh, UK since June 2005. He received his PhD degree from the University of Technology of Compiegne, France in May 2005. While working toward his PhD, he was a Research Engineer with Motorola Labs Paris for four years. He obtained a Master's degree in networking from Louis Pasteur University of Strasbourg (ULP), France in 2001 and an Engineering degree in Computer Science from the National School of Computer Sciences (ENSI), Tunis, Tunisia in 1998. His research interest includes IP multicast, mobile IP, moving network (NEMO), mesh networks, IP security, and QoS.

David G. Rosado (david.grosado@uclm.es) has an MSc and PhD. in Computer Science from the University of Málaga (Spain) and from the University of Castilla-La Mancha (Spain), respectively. His research activities are focused on security for Information Systems and mobile Grid computing. He has published several papers in national and international conferences on these subjects. He is a member of the GSyA research group of the Information Systems and Technologies Department at the University of Castilla-La Mancha, in Ciudad Real, Spain.

Po-Chi Shih received the B.S. and M.S. degrees in Computer Science and Information Engineering from Tunghai University in 2003 and 2005, respectively. He is now studying Ph.D. degree at Computer Science in National Tsing Hua University.

Sean Smith is currently in the Department of Computer Science at Dartmouth College; previously, he worked as a scientist for IBM and for Los Alamos National Lab. His research interests focus on human and hardware aspects of security.

Fabian Stäber was a research student at Siemens Corporate Technology. He received his Ph.D. from the Clausthal University of Technology, and his M.Sc. in Computer Science from the University of Erlangen-Nürnberg. His focus is on transferring research results from the field of peer-to-peer-based infrastructures to industrial applications. In his Ph.D. thesis, he presented a service layer for decentralized applications. That service layer allows for the definition of re-usable components that can be used as building blocks when fulfilling the requirements of emerging industrial applications. The resulting architecture and methodology was applied to different application scenarios, including IP telephony, distributed power generation, and business collaboration. Before his Ph.D., Dr. Fabian Stäber was a graduate-trainee at the Siemens Industry sector. He did his Diploma thesis at Siemens Transportation Systems, evaluating a new communication infrastructure for rail vehicles. Dr. Fabian Stäber recently became a consultant at MGM technology partners GmbH in Munich, Germany.

Alexandros Stamatakis received his Diploma in Computer Science in March 2001 from the Technical University of Munich. In October 2004, he received his PhD for research on "Distributed and Parallel Algorithms and Systems for Inference of Huge Phylogenetic Trees based on the Maximum Likelihood Method" also from the Technical University of Munich. From January 2005 to June 2006, he worked as postdoctoral researcher at the Institute of Computer Science in Heraklion, Greece. In July 2006 he joined Bernard Moret's group at the Swiss Federal Institute of Technology at Lausanne as a PostDoc. In January 2008 he moved back to Munich to set up a junior research group funded under the auspices of the Emmy-Noether program by the German Science Foundation (DFG), at the bioinformatics department of the Ludwig-Maximilians University of Munich. His main research interest are: technical and algorithmic solutions for inference of huge phylogenetic trees, applications of high performance computing techniques in bioinformatics, and challenging phylogenetic analyses of real-world datasets in collaboration with biologists.

Heinz Stockinger has been working in Grid projects in Europe (CERN, etc.) and in the USA (Stanford Linear A Accelerator Center) in various technical, scientific, and management functions. Heinz is affiliated with the Swiss Institute of Bioinformatics where he works on diverse Grid subjects. He has been appointed "Privatdozent" at the University of Vienna - leading the Research Lab for Computational Technologies and Applications in 2005. Currently, he is also a lecturer at the Swiss Federal Institute of Technology in Lausanne (EPFL). Heinz holds a PhD degree in Computer Science and Business Administration from the University of Vienna, Austria.

Hiroyuki Takizawa is currently an Associate Professor in Graduate School of Information Sciences, Tohoku University. His research interests include high-performance computing systems and their applications. He received the B.E. Degree in Mechanical Engineering, and the M.S. and Ph.D. Degrees in Information Sciences from Tohoku University in 1995, 1997, and 1999, respectively. He is a member of the IEEE CS, the IEICE, and the IPSJ.

Djamel Tandjaoui is a Researcher at the Center of Research on Scientific and Technical Information (CERIST) in Algiers, Algeria since 1999. He received his PhD degree from the University of Science and Technology Houari Boumediène (USTHB), Algiers in 2005. He obtained a Master's degree and an Engineer degree in Computer Science from the same university. Currently, he is member of Basic Software Laboratory at CERIST. His research interest includes mobile networks, mesh networks, sensor networks, ad hoc networks, QoS and security.

Attila Ulbert received his MSc in Computer Science (with distinction) in 1999, and his PhD in Informatics (summa cum laude) in 2004, both at Eötvös Loránd University (Budapest, Hungary). The title of his PhD thesis was "Pluggable Semantic Elements and Semantic Extensions in Distributed Object Systems." Attila Ulbert has taken part in several research projects on Grid systems since 2002. His main research areas are: distributed systems, data Grids, and telecommunication. Currently he is working for Ericsson as Software Architect.

Deo Prakash Vidyarthi, received Master degree in Computer Application from MMM Engineering College Gorakhpur and PhD in Computer Science from Banaras Hindu University, Varanasi. He was associated with the Department of Computer Science of Banaras Hindu University, Varanasi for more than 12 years. He joined JNU in 2004 and currently works as Associate Professor in the School of Computer & Systems Sciences, Jawaharlal Nehru University, New Delhi. Dr. Vidyarthi has published around 50 research papers in various peer reviewed International Journals and Transactions (including IEEE, Elsevier, Springer, World Scientific, IGI, Inderscience, etc.) and around 25 research papers in the proceedings of peer-reviewed international conferences in India and abroad. He has authored a book (Research Monograph) entitled "Scheduling in Distributed Computing Systems: Design, Analysis and Models" published by Springer, USA released in December, 2008. He has contributed chapters in many edited books. He is in the editorial board of two International Journals and in the reviewer's panel of many International Journals. Dr. Vidyarthi is member of IEEE, International Society of Research in Science and Technology (ISRST), USA and senior member of the International Association of Computer Science and Information Technology (IACSIT), Singapore. His research interests include parallel and distributed system, Grid computing, and mobile computing.

Hong Wang is currently with Nomura International (Hong Kong) Ltd, as a Senior Associate in Fixed Income Analytics Group. Prior to Nomura, he was a Ph.D. student in the Graduate School of Information Sciences of Tohoku University. During his Ph.D. study, Hong did research in the field of workflow management, fault tolerance, and performance optimization for Peer-to-Peer computing platforms. He received his B.S. (2003) in Computer Science from Peking University in China, and the M.S. (2006) and Ph.D. (2009) in Information Sciences from Tohoku University in Japan.

Nong Xiao was born in 1969. He received the B.S. degree and the PhD degree in computer science from the National University of Defense Technology in 1990 and 1996 respectively. He is currently a Professor and Ph.D. supervisor of the National University of Defense Technology. His research interests include Grid computing, ubiquitous computing, P2P computing, parallel computer architecture, distributed computing, and wireless sensor network. He is the director of the key projects "Distributed Data Management in Large-scale network storage environment," and "Dynamic Scalable Architecture for Data-intensive Computing in Network Environment" funded by National Natural Science Foundation of China, and senior member in projects such as China National Grid, National High Performance Computing Environment in China, Spatial Information Grid, et cetera. He was employed as the Special Expert on high performance computing and kernel software by the Ministry of Science and Technology of China, and served as Program Co-Chair of the 5th International Conference on Grid and Cooperative Computing.

You-Fu Yu received the BBA degree in Computer Science and Information Management from the Providence University, and the MS degree in Computer and Information Science from the National Taichung University, Taiwan. He is currently a project assistant in the National Taichung University. His research interests include P2P computing, Grid computing, resource discovery and cloud computing.

Daqing Zhang is a Professor at Institute TELECOM & Management SudParis, France. He obtained his PhD from University of Rome "La Sapienza" and University of L'Aquila, Italy in 1996. His research interests include pervasive healthcare, service-oriented computing, context aware systems, et cetera. He has published more than 90 papers in referred journals, conferences, and books. Zhang was the program chair of First International Conference of Smart Home and Health Telematics (ICOST2003) in Paris, France. He served as the General Co-Chair of ICOST2004 (Singapore), ICOST2005 (Canada), SH 2008 (China) and WISH 2008 (Australia), respectively. He also served in the technical committee for conferences such as UbiComp, Pervasive, PerCom, et cetera. Zhang was a frequent invited speaker in various international events such as pHealth, Net@Home, OSGi World Congress, e/home, SH, et cetera.

Alejandro Zunino received a PhD degree in Computer Science from UNICEN in 2003. He is a full Assistant Professor at the Computer Science Department of UNICEN and a research fellow of the CONICET. He has published over 30 papers in journals and conferences.

Index

A

B

C

D

E

F

G

O

P

Q

R

S

T

U

V

W